Frontispiece by Ruth Weisberg

Foundations of Measurement

VOLUME II
Geometrical, Threshold, and Probabilistic Representations

Patrick Suppes
Stanford University

David H. Krantz
Columbia University

R. Duncan Luce
University of California, Irvine

Amos Tversky

DOVER PUBLICATIONS, INC.
Mineola, New York

Bibliographical Note

This Dover edition, first published in 2007, is an unabridged and slightly corrected republication of the work originally published by Academic Press, Inc., San Diego and London, in 1989.

International Standard Book Number

ISBN-13: 978-0-486-45315-6
ISBN-10: 0-486-45315-4

Manufactured in the United States by LSC Communications
4500056989
www.doverpublications.com

... geometry was first discovered among the Egyptians and originated in the remeasuring of their lands. This was necessary for them because the Nile overflows and obliterates the boundary lines between their properties. It is not surprising that the discovery of this and the other sciences had its origin in necessity, since everything in the world of generation proceeds from imperfection to perfection. Thus they would naturally pass from sense-perception to calculation and from calculation to reason.

—Proclus, *Commentary on Euclid* (5th century A.D.)

Table of Contents

13. Axiomatic Geometry and Applications

14. Proximity Measurement

15. Color and Force Measurement

16. Representations with Thresholds

17. Representation of Choice Probabilities

Preface

Sometime toward the end of 1969 or the beginning of 1970 we came to realize that the material covered by this treatise was of such extent that the originally projected single volume would require two volumes. Because most of the work in Volume I was by then complete, we concentrated on finishing it, and it was published in 1971. There (pages 34–35) we listed the chapter titles of the projected Volume II, which we expected to complete by 1975. Fifteen years later, when we finally brought the work to a close, our plan had changed in several respects. First, the total body of material far exceeded a reasonably sized second volume, so, at the urging of the publisher, the manuscript expanded into two volumes, for a total of three. Second, the chapter on statistical methods was not written, largely because the development of statistical models for fundamental measurement turned out to be very difficult. Third, we decided against a summary chapter and scattered its contents throughout the two volumes. Fourth, Volume III came to include two chapters (19 and 20) based on research that was carried out in the late 1970s and throughout the 1980s. Volume II discusses or references many results that have been obtained since 1971 by the large number of persons working in the general area of these volumes.

As in Volume I, we attempt to address two audiences with different levels of interest and mathematical facility. We hope that a reader interested in the main ideas and results can understand them without having to read proofs, which are usually placed in separate sections. The proofs themselves are given somewhat more fully than would be appropriate for a purely mathematical audience because we are interested in reaching those in scientific disciplines who have a desire to apply the mathematical results of measurement theories. In fact, several chapters of Volume II contain extensive analyses of relevant experimental data.

Volume II follows the same convention used in Volume I of numbering definitions, theorems, lemmas, and examples consecutively within a chapter. Volume III deviates from this with respect to lemmas, which are numbered consecutively only within the theorem that they serve. The reason for this departure is the large number of lemmas associated with some theorems.

Volumes II and III can be read in either order, although both should probably be preceded by reading at least Chapters 3 (Extensive Measurement), 4 (Difference Measurement), and 6 (Additive Conjoint Measurement) of Volume I. Within Volume II, Chapter 12 (Geometrical Representations) should probably precede Chapters 13 (Axiomatic Geometry), 14 (Proximity Measurement), and 15 (Color and Force Measurement). Within Volume III, much of Chapter 21 (Axiomatization) can be read in isolation of the other chapters. Chapter 22 (Invariance and Meaningfulness) depends only on the definability section of Chapter 21 as well as on Chapters 10 (Dimensional Analysis and Numerical Laws), 19 (Nonadditive Representations), and 20 (Scale Types). Chapter 19 should precede Chapter 20.

Acknowledgments

A number of people have read and commented critically on parts of Volume II. We are deeply indebted to them for their comments and criticisms; however, they are in no way responsible for any errors of fact, reasoning, or judgment found in this work. They are: Mathew Alpern, Maya Bar-Hillel, Thom Bezembinder, Ruth Corbin, Robyn M. Dawes, Zoltan Domotor, Jean-Claude Falmagne, Peter C. Fishburn, Scott Ferguson, Rob Goldblatt, Macduff Hughes, Leo Hurvich, Tarow Indow, Dorothea Jameson, Yves LeGrand, Anthony A. J. Marley, Brent Mundy, Harriet Pollatsek, Raymond A. Ravaglia, Fred S. Roberts, Halsey Royden, Shmuel Sattath, Brian Sherman, and L. W. Szczerba.

In addition, each of us has taught aspects of this material in classes and seminars over the years, and students have provided useful feedback ranging from detecting minor and not so minor errors to providing clear evidence of our expository failures. Although we do not name them individually, their contribution has been substantial.

Others who contributed materially are the many secretaries who typed and retyped portions of the manuscript. To them we are indebted for accuracy and patience. They and we found that the process became a good deal less taxing in the last eight years with the advent of good word processing.

With chapters primarily authored by different people and over a long span, the entire issue of assembling and culling a reference list is daunting. We were fortunate to have the able assistance of Marguerite Shaw in this endeavor, and we thank her.

Financial support has come from many sources over the years, not the last being our home universities. We list in general terms the support for each of the authors.

Luce: Numerous National Science Foundation grants to The University of California, Irvine, and Harvard University (the most recent being IRI-1969-72); The Institute for Advanced Study, Princeton, New Jersey, 1969–1972; AT&T Bell Laboratories, Murray Hill, New Jersey, 1984–85; and The Center for Advanced Study in the Behavioral Sciences, Stanford, California, 1987–88 (funds from NSF grant BNS-8700864 and the Alfred P. Sloan Foundation).

Tversky: The Office of Naval Research under contracts N00014-79-C-0077 and N00014-84-K-0615 to Stanford University.

Chapter 11 Overview

This introductory chapter describes the topics covered in this volume and relates them to Volume I. The theme of Volume I was measurement in one dimension. We considered structures having a natural concatenation operation, representable by a sum or by a weighted average, and structures in which the combined effect of several factors on some unidimensional, ordered attribute is representable by an additive or a polynomial combination rule.

Chapters 12 through 15 of this volume discuss multidimensional geometrical representations of a variety of kinds. The last two chapters, 16 and 17, return to the general theme of unidimensional representations, but with the added feature of a threshold or error component. We turn first to the overview of the chapters, grouped as indicated.

11.1 GEOMETRY UNIT

Our first unit, Chapters 12 through 15, deals with geometrical representations, many of which are vectorial. Historically, the earliest example of numerical representation was the invention of analytic geometry, which provides coordinate-vector representations for qualitative geometrical structures formulated in terms of such primitives as points, lines, comparative distances, and angles. The unidimensional representations of Volume I

1

were, in fact, a later development, in which foundational work was extended from geometry to other domains, first in the physical sciences and then in the social sciences. Later still, new approaches to the foundations of the classical geometries were created to accommodate applications in physical science (distance in special relativity), perception (color measurement), and social science (multidimensional scaling). In this unit, we cover both the older and the newer aspects of geometrical measurement.

Of course, in one sense, and at a much earlier time, Euclid's *Elements* was concerned with the unidimensional representation of irrational numbers as geometrical quantities. That original Greek development of geometry was highly axiomatic in character, but it inverted our concept of representation. It emphasized the representation of numbers by geometrical entities rather than the other way around. This had to be so since, at the time, there was no theory of irrational numbers and hence no way to represent, say, the side and diagonal of a square within the same numerical system. Only after the invention of irrationals was it possible for Descartes to propose a full representation of geometrical entities by numbers. Moreover, this proposal was not fully justified until the last part of the nineteenth century, when the basis for constructing real numbers from rational ones was rigorously formulated by Dedekind, Weierstrass, Cantor, and others.

There are, in fact, two quite distinct developments to be considered: analytic geometry, which formulates the spaces of numerical geometrical structures that potentially may serve to represent qualitative geometrical structures, and synthetic geometry, which develops the axiomatic theories of those qualitative structures. The pattern is the same as in measurement theory: a representation theorem shows how to embed a qualitative structure isomorphically into some family of numerical structures, and the corresponding uniqueness theorem describes the different ways that the embedding is possible. Chapter 12 describes the candidate representations, called analytic geometries, and Chapter 13 discusses a number of the classical axiomatizations and the corresponding representation and uniqueness theorems. In the last part of Chapter 13 we present two interesting extensions of these results: the axiomatization of space-time in special relativity and the axiomatization of visually perceived spatial relations (subjective three-dimensional space).

Throughout Chapters 13, 14, and 15, the character of the theorems is similar to those we encountered in Volume I. One begins with a qualitative structure that is thought to capture some empirical domain. In each chapter, the primitives are somewhat different, reflecting different substantive concerns, but the representations are all in terms of classical, numerically formulated geometries such as affine, Euclidean, and projective. In most cases, the representation involves vectors in Re^n, even though the dimensionality is not always evident in the primitives. Of course, such dimensionality is captured in various ways by the axioms satisfied by the primitives. Chapter 12 provides a detailed formulation of these potential representing vectorial structures.

The presentations in Chapters 12 and 13 differ in two respects from Volume I. First, we did not include in Volume I a chapter on the ordered real numbers, corresponding to Chapter 12, because we assumed familiarity with the elementary properties of numbers. Second, most of that volume was devoted to axiomatic structures, their representation and uniqueness theorems, and the proofs of those theorems. We are able to report the analogous geometrical results in a single chapter by not including proofs. We compress the treatment because the area is extremely well developed, with many thousands of papers and numerous excellent books devoted to the theorems and their proofs, and because the proofs would require an extensive development of synthetic geometry.

11.1.1 Geometrical Representations (Chapter 12)

Chapter 12 is devoted to an exposition of many standard mathematical structures. There are no theories of measurement in this chapter; we discuss only the representing structures, not the empirical structures nor how they are mapped into their representations.

Recall that in Cartesian or analytic geometry a point in the plane is represented by a pair of numbers (x_1, x_2) and distance is calculated using the Pythagorean formula $[(x_1 - y_1)^2 + (x_2 - y_2)^2]^{1/2}$. In classical or axiomatic Euclidean geometry, however, points are abstract entities represented by letters $P, Q, \ldots$, and distances are represented by expressions of the form $\overline{PQ}$. Numbers form no part of the primitive concepts or axioms. The bridge between these two subjects is provided by the Pythagorean Theorem, $\overline{PQ}^2 + \overline{QR}^2 = \overline{PR}^2$, where PQ and QR are the legs and PR is the hypotenuse of a right triangle. In this volume, we treat the Pythagorean Theorem as the basis for a formal representation, in which "abstract" point P is mapped into numerical pair (x_1, x_2). In other words, we treat analytic geometry as a numerical representation or model whereas the axioms of classical geometry give a qualitative characterization of the structures that can be so modeled.

Chapter 12, then, contains a brief exposition of various sorts of analytic geometry: the Euclidean, which has just been mentioned, and others that provide important representations for various empirical relational structures, including the analytic versions of projective, affine, and non-Euclidean metric geometries. It also discusses other structures that do not involve coordinates. Of course, we cannot cover all analytic geometry; excellent books on the subject exist, e.g., Busemann and Kelly (1953) and Postnikov (1982). Our exposition consists mainly of definitions, with some discussion of principal properties and some allusions to how and where the structure being defined is used as a representation. Collecting this material in one chapter allows us to avoid breaking up the exposition in Chapters 13–15 with definitions of the representing structures. Another advantage, we hope, is that the reader will perceive the relationships and the distinctions among various geometrical representations and will obtain some feel for the range of available geometrical representations.

The last part of Chapter 12 is concerned with metric geometry developed in terms of a distance function rather than in terms of coordinates. The obvious empirical structure to be represented by a metric is a qualitative congruence or proximity structure, i.e., a set A and a quaternary relation on A with the relation interpreted either as qualitative equidistance (congruence) or as ordering of qualitative distance (proximity).

Metric spaces possessing nothing more than a distance function are too general to be of much interest as numerical representations. Consequently, the concern in Chapter 12 is to develop more specialized cases. For example, a subsection (12.3.1) is devoted to metrics that possess geodesics. The prototypical example is the geometry of a two-dimensional curved surface, embedded in three-dimensional Euclidean space. One gets a metric geometry on such a surface by using a distance function that is equal to the length of the shortest path on the surface between two points.

The critical feature of surfaces that are embeddable in three-dimensional space is the fact that in a sufficiently small region the surface can be approximated very well by a tangent plane in three-dimensional Euclidean space. When other smoothness properties are assumed, together with such approximations, the metric is Riemannian and has a number of rich features that we sketch in Section 12.3.3. We do not provide a thorough introduction to Riemannian geometry but present only the elementary parts, especially those that we use in Chapter 13 in the study of visual space, such as the theory of Gaussian curvature.

Finally, in Section 12.3.4 we survey some other metrics that are used in the proximity models studied in Chapter 14.

11.1.2 Axiomatic Synthetic Geometry (Chapter 13)

A basic test of the adequacy of a qualitative axiom system for a particular branch of geometry, such as affine, projective, or Euclidean geometry, is that it should lead to the expected numerical or vectorial representation with the expected uniqueness properties. In the nineteenth century such representations were especially important in settling contro-versial questions about the consistency of non-Euclidean geometry.

As we have already remarked, the problem of proving a representation theorem in geometry is, in principle, the same as for the one-dimensional systems of measurement considered in Volume I. The problem of unique-ness is also conceptually the same, but it assumes a more intricate form because the groups of transformations under which the properties of a representation are invariant are considerably more complex.

Chapter 13 begins with a section on the affine notion of order, which is the familiar ternary relation of betweenness for points on the line. The projective notion of order is the less familiar quaternary relation of separa-tion. We allot more space to the theory of separation because it is less familiar and less accessible in standard books on geometry (see Figures 1 and 2 of Chapter 13). In fact, the purely linear theory of separation, which gives rise to uniqueness up to projective monotonicity, is ordinarily not studied at all. Yet it is a natural extension of earlier results on order.

A section on the projective plane and another on projective spaces follow. Projective planes are first axiomatized in terms of the set of points, the set of lines, and the binary relation of incidence. A second axiomatiza-tion is given that includes the projective concept of order mentioned previously. Some space is devoted to informal discussion of finite geome-tries and to the extension of representation theorems to structures that are a generalization of the field of real numbers. Perhaps the most interesting feature of these developments is the interplay between algebraic and geometric properties. The classic example is Hilbert's result that a projec-tive plane satisfies Pappus' proposition (see Figure 5 of Chapter 13) if and only if the multiplication operation of its representing division ring is commutative—a division ring is a straightforward generalization of an algebraic field. We also discuss more recent examples relating geometrical properties of the projective plane to the associativity of the underlying representing operations.

Section 13.5 extends the axiomatic approach to projective geometry to the case of three-dimensional ordered projective space. The primitives remain the same except that now, in addition to the set of points and lines, the set of planes is introduced. The two primitive relations remain the binary relation of incidence and the quaternary relation of separation.

The next section, 13.6, develops the classical theory of affine and absolute spaces. An affine space is a space that can be characterized just in terms of order in the sense of betweenness or in terms of the concept of lines being parallel. Absolute spaces are a generalization of Euclidean spaces. Most of the propositions proved in Book 1 of Euclid, for example, hold in absolute geometry as well as in Euclidean geometry. Absolute spaces are richer than affine spaces in that they possess a relation of congruence. The axiom that does not hold in absolute spaces is Euclid's famous fifth postulate. One form of the axiom is that, given a point and a line on which the point does not lie, there exists through the point at most one line in the plane that does not meet the given line. This axiom separates hyperbolic and Euclidean geometry, both of which satisfy all the other axioms of absolute geometry. We develop the elementary theory of hyperbolic geometry in some detail because of its extensive application in the study of visual perception.

Section 13.7 is devoted to elliptic spaces. We give a qualitative axiomatization of spherical or double elliptic spaces, but because of their greater complexity we do not give a complete axiomatization of single elliptic spaces, although their properties are discussed, because again they are a topic of some interest in visual perception. (Single elliptic space is derived from spherical space by identifying as a single point each pair of antipodal points: the North Pole has as its antipodal point the South Pole, etc.)

The remainder of Chapter 13 is devoted to applications of axiomatic geometry. The first of these sections, 13.8, is devoted to axiomatization and representation of classical space-time. Here and in the next section on restricted relativistic space-time, the basic concept of a four-dimensional affine space plays a central role. In the relativistic case there is a large axiomatic literature, and we attempt to give a somewhat detailed sense of the variety of axiomatic approaches. The axioms that we present are meant to be easy to comprehend and physically motivated. Axioms that begin with simpler primitives and that have a very complex formulation are discussed but not described in any technical detail.

In the case of classical space-time, beyond the affine concept of order introduced as usual by the ternary relation of betweenness, the two additional primitive concepts are, first, the set of inertial lines and, second, the relation of congruence. The relation of congruence does not hold between arbitrary pairs of points but only between pairs of simultaneous points. The fundamental concept of simultaneity in classical physics is not introduced as a primitive relation but is easily definable in terms of the concept of an inertial line.

Qualitative axioms for restricted relativistic space-time assume the same general affine structure as well as the set of inertial lines but add as a primitive the set of optical lines; of course, the relation of congruence is restricted in a different way from the classical case. However, in both instances a fundamental axiom is that certain subsets have exactly the structure of three-dimensional Euclidean space. In the classical case these are sets of points, each of which is simultaneous with a given point. In the relativistic case, a standard construction with optical parallelograms is used to identify in similar fashion the sets of points generated by a given point and an inertial line. Again, these sets are postulated to have the structure of a Euclidean three-dimensional space. There is also in Section 13.9 a discussion, from the standpoint of uniqueness of representation, of the classical Lorentz transformations of special relativity.

In Section 13.10, the application is to human visual perception; geometrical relational structures are abstracted from subjective judgments of collinearity, parallelism, etc., rather than from physical measurements. The literature on visual perception is large, and we attempt to survey only the most directly relevant results. In this long section, we hope to give a sense of the interplay between theory and experiment in a tradition that goes back to the last part of the nineteenth century. The main emphasis is on the Luneburg theory of binocular vision because it has been well developed mathematically and extensively tested experimentally. In addition, we review some of the more recent studies that move away from the Luneburg theory to the consideration, for example, of qualitative axioms violated in a way that rules out any of the standard elementary classical spaces treated earlier, and also some recent studies concerned with the use of concepts from projective geometry.

11.1.3 Proximity Measurement (Chapter 14)

In Chapter 13, incidence of points and lines, collinearity of three points, and congruence of pairs of points are the most important geometrical relations. But one of the most common empirical relations, especially in social sciences, is ordering of pairs. Such a quaternary relation we call a *proximity* relation. An ordering of pairs includes congruence (as its symmetric part) but does not directly impose any linearity structure. Hence, the axiomatizations in Chapter 13 are unsuited for scientific applications that seek geometrical representations for proximity relations.

Chapter 4 dealt with the special case of one-dimensional representations for proximity (the various difference structures). In Chapter 14, we construct more general geometrical representations for proximities. Neither incidence nor collinearity of points is used as a primitive relation. Since a straight line is the shortest distance between two points (in affine metrics), it is plausible that straight lines can be defined in terms of shorter and longer distances, that is, an ordering of pairs. This is accomplished in the early sections of the chapter. The representing structure is a metric space in which any two points are joined by a straight segment along which distance is additive.

The middle sections of Chapter 14 combine the ordering of pairs with a factorial structure for elements that enter the pairs. The representations have coordinate axes that correspond to this factorial structure. This material relates closely to additive conjoint measurement, and that part of the chapter can be thought of as a simultaneous generalization of Chapters 4 and 6. Specifically, we axiomatize the representation $\Sigma F_i(|\varphi_i(a_i) - \varphi_i(b_i)|)$ that is monotonically related to the proximity $\delta(a_1 \cdots a_n, b_1 \cdots b_n)$ of a and b. We also treat several more general representations, including ones not necessarily additive across factors or not subtractive within a factor, and we consider some special representations, including the combination of factorial structure with the segmental additivity condition imposed on representations in the first part of the chapter. We include among these special cases an axiomatization of the familiar r^{th} root of the sum of rth powers of differences.

Much work has been devoted to developing computer algorithms that seek approximate representations of proximity relations for finite sets of points in low-dimensional metric spaces, most often Euclidean. This is called multidimensional scaling, and is widely used in the social sciences. Chapter 14 does not treat such algorithms; rather we focus on the qualitative conditions that must be satisfied by proximity relations for various representations. In Section 14.6 we illustrate the use of ordinal methods to test the validity of basic geometrical assumptions that underlie multidimensional scaling.

The last sections of this chapter discuss set-theoretical representations of proximity in which the dissimilarity of objects is expressed as a linear combination of the measures of their common and their distinctive features. Finally, the use of tree structures to represent distance is discussed and illustrated.

In applying conjoint measurement to obtain an additive set-theoretical representation of proximity, it was necessary to solve a problem that had been left open in Section 6.5.3 of Volume I, namely, the extension of conjoint measurement to nonfactorial structures. We replaced the standard factorial assumption by a more general axiom of factorial connectedness (Definition 14.18). Since this generalization is relevant to many applications of conjoint measurement, we describe it briefly here, using the notation of Chapter 6; the description can be omitted without loss of continuity.

Factorial connectedness states that any two elements $(a, p) \succsim (b, q)$, with $a, b \in A$ and $p, q \in P$, are connected by a finite sequence of elements (c_i, r_i), $i = 0, \ldots, n$, such that $(c_0, r_0) = (a, p)$, $(c_n, r_n) = (b, q)$, $(c_{i-1}, r_{i-1}) \succsim (c_i, r_i)$, $i = 1, \ldots, n$, and any two adjacent elements in the sequence belong to a common Cartesian product. This assumption permits a straightforward application of additive conjoint measurement to domains whose image under the representation is a convex set (see Figure 6 on p. 275 of Volume I).

11.1.4 Color and Force Measurement (Chapter 15)

Force configurations and lights with mixed wavelengths are both describable using infinite-dimensional vectors: force elements (measured extensively) acting in every possible direction, or radiant energy density (again, measured extensively) at each possible wavelength. In both cases, there is a three-dimensional representation obtained by factoring equivalences out of the infinite-dimensional description: equivalences of resultant force, or equivalences for human color matching. Chapter 15 formulates as axioms the empirical laws that permit these three-dimensional representations; they are virtually the same for resultant force and human colorimetry. The most important law is additive closure: adding a light to both sides of a human color match produces another match, adding two force configurations that produce equilibria produces another equilibrium.

Color equivalences are perceptual, not physical; no purely optical phenomenon reveals them. Some forms of defective human color vision lead to representations that are two- or even one-dimensional. The low number of dimensions reflects something important about color vision. That this low dimensionality is not a necessary feature of all perception is demonstrated by perception of pitch and timbre. The infinite-dimensional space of sounds seems not to be reduced by the perceiver to a low-dimensional perceptual space of hearing.

Equilibrium is the physical state of force configurations that corresponds to human matching of colors. Except for the fact that colors correspond only to a convex cone of vectors whereas forces correspond to a full vector space, the qualitative structure and representation theorems are the same in the two cases. Similar qualitative structures are found for subspaces—components of force and of color—although they do differ in that force is assumed to be isotropic and color is not.

11.2 THRESHOLD AND ERROR UNIT

Our second unit, composed of Chapters 16 and 17, deals with the problem of error from the perspective of the foundations of measurement. The problem arises because the standard statistical methods are not readily applicable when the data are inherently ordinal but are imperfect for one reason or another. The difficulty stems from the fact that at the outset all possible monotonic—or really, nearly all monotonic—transformations are under consideration, and the question is which best approximates a particular type of representation. If after some monotonic transformation the error is, say, additive, it will fail to be additive in any transformation that is nonlinearly related; and so the usual type of statistical model cannot plausibly be assumed to describe the untransformed data. Moreover, there is no good reason to suppose that a transformation that produces a simple (e.g., additive) algebraic measurement representation will also produce a simple (e.g., additive) error representation. This point is explored in Section 22.6.5.

11.2.1 Representations with Thresholds (Chapter 16)

Chapter 16 axiomatically addresses one source of error, namely, a failure of transitivity of indifference in the ordering relation. The kind of inconsistency envisaged is this: a succession of indifferences that may arise, for example, because of limits of human or instrumental discrimination can lead to end points that are by no means indifferent. One can experience this in many ways, not all of which are failures of discrimination in any usual sense. Suppose one is looking at real estate. An agent can sometimes slowly undermine one's resolve on price by successively showing you houses increasing in price by "small" increments of, let us say, $5000. You may not be especially sensitive to such a price inflation until you realize that the price of the latest "bargain" is some $30,000 higher than the first one you saw, a difference no longer within your "threshold of indifference." (Such problems are discussed in Section 17.2.4 in terms of a lexicographic semiorder.)

A main focus of Chapter 16 is to model such nontransitive indifference in an algebraic fashion. The resulting representation is much like the classical one except that indifference is no longer represented by identical numerical values (which necessarily implies that it is an equivalence relation) but by values that lie within an interval—a threshold. In spirit, these models are very much like those of Volume I, but more complex.

The theory is quite fully developed for the ordinal case. It assumes that the ordering relation decomposes into a strict ordering that is transitive and an indifference relation that is symmetric but not (in general) transitive. The strict ordering and indifference relation are such, however, that a weak order exists that is compatible with them in the following sense. First, two entities that are strictly ordered in the given order are also strictly ordered in the compatible weak order. Second, the given indifference relation gives rise to nonequivalence in the weak order only if some indirect way can be found to order the elements in terms of the given order. For example, if $a \sim b$, $b \sim c$, and $a \succ c$ in the given order, we conclude indirectly that a is greater than b. The ordinary theory of representations of weak orders gives rise to a numerical function that describes the indirect weak order adequately, and in terms of that and ~ it is possible to define upper and lower numerical thresholds in such a way that the three functions recode all of the ordering information. For historical reasons having to do with Fechner's definition of a sensory scale, there is special interest in the conditions that yield a representation with constant thresholds. The analogous threshold theory for extensive and conjoint cases is less fully developed. The problem is that it is relatively difficult to make both the underlying measure and the thresholds combine in a simple fashion. This is closely related to a classical problem in sensory thresholds, namely, whether or not the threshold is proportional to the scale value (Weber's law). If it is, then additivity of the measure yields additivity of the threshold. In any other case, if the one is additive the other cannot be. The difficulty is analogous to that of meshing random error with a measurement representation.

Random-variable rather than algebraic representations are developed in the final section (16.8) of the chapter. The axioms for extensive measurement in Section 16.8.2 use the idea of qualitative raw moments and their ordering as a basis from which to derive the existence of a unique probability distribution for an object's extensive attribute. Qualitative characterizations in terms of moments are given for the uniform, Bernoulli, and beta distributions.

11.2.2 Representations of Choice Probabilities (Chapter 17)

Chapter 17 approaches the error problem by introducing an explicit probabilistic structure. Instead of dealing with an ordering over all alternatives, one supposes that for each finite subset B of alternatives there is a probability $P(a; B)$ of alternative a being selected from B as exhibiting to the greatest degree the attribute in question. Sometimes only the binary probabilities are studied. Then it is customary to write $P(a, b)$ for $P(a; \{a, b\})$. The algebraic models of Chapter 16 can be viewed as approximations to the probability model obtained by using a cutoff λ to distinguish strict inequality from indifference:

$$a \sim b \quad \text{iff} \quad |P(a, b) - \tfrac{1}{2}| \leqslant \lambda \quad \text{and}$$
$$a \succ b \quad \text{iff} \quad P(a, b) > \lambda + \tfrac{1}{2}.$$

A variety of probabilistic choice models are investigated in the chapter. The most general (binary) model is the analogue of a weak order, which assumes that if $P(a, b) \geqslant 1/2$ and $P(b, c) \geqslant 1/2$, then $P(a, c) \geqslant 1/2$. This property, called weak stochastic transitivity, represents one generalization of algebraic transitivity. A stronger condition, called strong stochastic transitivity, requires the conclusion $P(a, c) \geqslant \max[P(a, b), P(b, c)]$, given the same hypotheses. This condition is necessary for the existence of a real-valued function φ on the set of alternatives and a function F of two real variables such that for every pair of alternatives a and b,

$$P(a, b) = F[\varphi(a), \varphi(b)].$$

Certain special cases of this form are related to Fechner's threshold definition of a sensory scale in the case where φ and a function f of one variable can be chosen so that

$$P(a, b) = f[\varphi(a) - \varphi(b)].$$

Several results about such representations, in particular, conditions on P that lead to them, are examined in the chapter.

A third type of ordinal model supposes that associated to each alternative a is a random variable U_a that represents the momentary value or magnitude of that alternative. Choices are based on making a single observation for each of the random variables in the set of alternatives under consideration and selecting the alternative with the largest observed value. If the joint distribution of the random variables is known, then the probabilities can be calculated, which is done in Chapter 17 for some examples. The conditions under which such a representation exists and their relation to the previous models are also discussed.

Unlike the usual statistical approach to measurement, the major thrust of Chapters 16 and 17 is to treat error variability as an inherent part of the measurement process. This perspective is somewhat different from the one that assumes that the algebraic models are basically correct and that the observations are rendered imperfect by extraneous factors. The latter approach definitely invites efforts to refine the process of measurement: to better control any extraneous factors so as to improve the approximation to the true model. Such refining of measurement is typical of much physics and has, in some cases, reduced measurement errors by many orders of magnitude.

The methods developed in this section suggest a different approach for improving measurement, an approach that is based on algebraic (Chapter 16) or probabilistic (Chapter 17) refining techniques rather than on the development of more accurate instruments. It appears that both approaches to the error problem can prove useful in many contexts.

Chapter 12   Geometrical Representations

12.1 INTRODUCTION

We shall describe a variety of structures that are considered as representations in Chapters 13, 14, and 15. Collecting them here permits economy in explanation and avoids later interruption; it also permits a systematic review and comparison of various sorts of analytic geometry, which some readers will find useful. Though we define many geometries, we develop little by way of geometrical theorems. Instead, we cite standard textbooks, where thorough treatments of various analytic geometries can be found. (For a broad introduction to modern geometry, we recommend Dubrovin, Fomenko, and Novikov, 1984, 1985.)

Section 12.2 gives a fairly comprehensive overview of linear (i.e., vector) representations. These include geometries with linear but no metric structure, such as projective and affine geometry, and ones with compatible linear and metric structures, such as Hilbert geometry and Minkowski geometry. We include a subsection on groups of permissible transformations.

Section 12.3 covers metrics without accompanying linear structure. G-spaces form a broad class for which interesting geometrical theory exists; Riemannian spaces are the best developed subtype. A final subsection touches briefly on non-G-space metrics.

13

12.2   VECTOR REPRESENTATIONS

We introduce this section by discussing the most familiar geometrical structure: n-dimensional Euclidean geometry with orthogonal coordinates. This consists of the pair $\langle \mathrm{Re}^n, \delta \rangle$ where δ is the Euclidean distance function: for x, y in Re^n,

$$\delta(x, y) = \left[\sum_{i=1}^{n} (x_i - y_i)^2 \right]^{1/2}.$$

Euclidean geometry is not concerned with distance alone; it deals with many other geometrical constructs such as lines and planes, circles and polygons, angles and parallels. The geometrical objects most easily defined in terms of distance are the n-spheres. The sphere with center x and radius α simply consists of all points y such that $\delta(x, y) = \alpha$. It turns out, indeed, that all the other kinds of geometrical objects can be defined in terms of distances, but the definitions of the linear objects such as lines and planes are quite complicated. It is simpler to use the algebraic structure of Re^n to define some kinds of geometrical objects. This structure consists of two operations, coordinate-wise vector addition, denoted $+$, and coordinate-wise multiplication of vectors by elements of Re, denoted $\cdot$. The context generally makes clear whether $+$ denotes addition in Re or in Re^n.

Lines, planes, etc., can be defined using the algebraic apparatus alone, without the distance function. Thus the line through x with direction y consists of all elements of form $x + ty$ for t in Re (we usually suppress the dot in writing formulas involving multiplication by scalars). The plane through x generated by directions y, z consists of all elements of form $x + ty + uz$, for t, u in Re, etc.

The study of those objects that involve only the algebraic structure on Re^n, and not the distance function δ, is called *n-dimensional affine geometry*. Affine geometry has no general concept of congruence (since that involves equality of distances) or perpendicularity (since that involves the idea of shortest distance from a point to a line), but it does have parallelism. For example, two lines through different points x, x', but both in direction y, are parallel.

We can consider an even weaker form of geometry, in which the operation $\cdot$ is suppressed, by regarding as equivalent any two nonzero elements of Re^n that are related by scalar multiplication, i.e., that have coordinates proportional. An equivalence class $[x] = \{ tx \mid t \neq 0 \}$ can be thought of as a pure direction in Re^n (but unoriented since $[x] = [-x]$). This is called $(n - 1)$-*dimensional projective geometry*. (The reduction in dimension corresponds to the loss of one degree of freedom in Re^n, e.g., any nonzero x can be normalized so that one of its coordinates is unity since multiplication by a scalar is irrelevant.) As we shall see, projective geometry still allows discussion of certain linear objects, but the concept of parallelism is lost since the expression $x + ty$ makes no sense; one no longer has independent concepts of point and direction.

Affine and projective geometry have importance for us in two ways. First, the numerical structures that we have just discussed provide representations for certain empirical relational structures in which empirical relations of linearity are given, but without any relation that corresponds to distance. Second, one can enrich the affine or projective structure with Euclidean structure or with various non-Euclidean distance functions. This leads to whole classes of possible representations: the various *affine metrics* and *projective metrics*. Affine metrics of particular importance are those of the Minkowski geometries (Section 12.2.6). Projective metrics include the classical hyperbolic and elliptic non-Euclidean geometries.

It should also be mentioned that the pure theory of projective geometry is one of the greatest jewels of human intellect; our presentation necessarily obscures some of its beauty by emphasizing the representation aspect and failing to develop the full geometry. For example, in this chapter we do not even touch upon the theorems of Desargues and Pappus; these appear only in the following chapter as qualitative conditions necessary for a projective representation.

The reader may have noticed that we made no use of coordinates in the above discussion subsequent to the definition of the distance function δ. Everything was expressed in vector notation, e.g., $x + ty$ rather than $(x_1 + ty_1, \ldots, x_n + ty_n)$. In fact, it is more geometrical to work in an abstract vector space over Re, avoiding the use of coordinates altogether. A *vector space over Re* is a triple $\langle V, +, \cdot \rangle$, where $+$ is a binary operation with all the properties of coordinatewise addition (i.e., $\langle V, + \rangle$ is a commutative group) and $\cdot$ is a function from $Re \times V$ to V, with the abstract properties of scalar multiplication. Euclidean distance can also be discussed in an abstract vector space without reference to coordinates. The concept of finite dimension can be treated abstractly using the theory of linear dependence—the dimension is the maximum number of linearly independent vectors (see Definition 2); and any n-dimensional vector space over Re is isomorphic to $\langle Re^n, +, \cdot \rangle$. Thus we can break our representation theorems for projective, affine, or Euclidean geometry into two parts: a representation into an abstract vector space and a representation of the latter onto $\langle Re^n, +, \cdot \rangle$.

We organize this section as follows. First we give the definition of an abstract n-dimensional vector space (2.1) and state without proof the representation and uniqueness theorem using n-dimensional coordinates. This broadens the class of "numerical" relational structures that may be used. In the next three sections we define analytic projective, affine, and Euclidean geometries using vector spaces, and we discuss the automorphism group of each of these structures. (An automorphism of a structure is an isomorphism of the structure onto itself.) In Section 2.5 we compare these automorphism groups, relating them to problems of invariance and meaningfulness as they arise in these geometries; in essence, we restate in modern terms some of the ideas in Klein's famous Erlangen address of 1872 (see Klein, 1893). The final subsections describe some of the interesting alternatives to the Euclidean distance function, i.e., alternative affine and projective metrics, including the important Minkowski metrics.

This section does not contain all our examples of analytic geometry with coordinates in Re^n, only those in which the geometric structures are defined from, or are highly compatible with, the vector structure. Essentially, we cover those representations whose uniqueness is given by a subgroup of the projective group. Other analytic geometries are introduced in Section 12.3, including the important Riemannian geometries.

12.2.1 Vector Spaces

We shall define the notion of a *finite-dimensional vector space over a base field*. In the familiar examples, the base field is the field of real numbers, but the vector-space theory is exactly the same for any field F, culminating in a representation using F^n with coordinatewise numerical operations exactly like those of Re^n. The choice of field is quite significant in geometry, however, because it corresponds closely to aspects of the empirical relational structure. The most important distinction is between ordered fields, such as the reals and the rationals, and unordered fields, such as the complex numbers or various finite fields. Once we are representing an unordered empirical structure, there is no need to use an ordered field in our numerical structure; this becomes necessary only if there is a primitive or definable ordering along lines in the geometry. Despite this introduction, we shall place little emphasis on fields other than the real numbers; but we do want the reader to be aware that there is a large geometrical literature using other fields. The pure vector-space theory, which we develop in this section, is the same for any field.

A vector space over F is a triple $\langle V, +, \cdot \rangle$, where V is a nonempty set, $+$ is a binary operation on V (vector addition), and $\cdot$ is a function from $F \times V$ into V (multiplication of a vector by a scalar to yield another vector), satisfying the same algebraic properties as $\langle \text{Re}^n, +, \cdot \rangle$. These defining properties are set forth in the next definition.

DEFINITION 1. *Let F be a field. The triple* $\langle V, +, \cdot \rangle$ *is a* vector space *over F iff* + *is a closed binary operation on V and* $\cdot$ *is a function from* $F \times V$ *into V such that for some element* $0 \in V$, *all* $x, y, z \in V$, *and all* $t, u \in F$, *the following axioms hold:*

1.   (i) $(x + y) + z = x + (y + z)$
     (ii) $x + y = y + x$
    (iii) $0 + x = x$
    (iv) *There exists* $-x \in V$ *such that* $x + (-x) = 0$
2.   (i) $t \cdot (x + y) = t \cdot x + t \cdot y$
     (ii) $(t + u) \cdot x = t \cdot x + u \cdot x$
    (iii) $(tu) \cdot x = t \cdot (u \cdot x)$
    (iv) $1 \cdot x = x$

Axiom 1 states that $\langle V, + \rangle$ is a commutative group with identity element 0. Axiom 2(i) states that multiplication by t is a mapping of $\langle V, + \rangle$ into itself; 2(iv) states that multiplication by the number 1 is the identity mapping. The other parts of Axiom 2 show how addition and multiplication operations in the field F correspond to the natural addition and function composition operations for mappings of $\langle V, + \rangle$ into itself. The reader can easily verify that all these properties are valid for $\langle \text{Re}^n, +, \cdot \rangle$ over Re where $+, \cdot$ are defined componentwise.

To give the reader a perspective on models other than $\langle F^n, +, \cdot \rangle$ that satisfy this definition, we present three examples.

EXAMPLE 1. $F = \text{Re}$, V is the set of $l \times m$ matrices with entries in Re. The operation $+$ for two matrices is defined by adding componentwise: $(a_{ij}) + (b_{ij}) = (a_{ij} + b_{ij})$, and multiplication by a scalar t means multiplying each of the lm entries by t. Of course, we can regard this as equivalent to Re^{lm} and just arrange the vectors in l rows of length m rather than one row of length lm.

EXAMPLE 2. $F = \text{Re}$, V is the set of real-valued functions on Re^3 that have continuous second partial derivatives and that satisfy Laplace's equation,

$$\sum_{i=1}^{3} \frac{\partial^2 f}{\partial x_i^2} = 0.$$

Define $+$ and $\cdot$ in the usual way for functions, e.g., $h = f + g$ iff for all x, $h(x) = f(x) + g(x)$. The point of this example is that properties such as existence of and continuity of derivatives and satisfaction of a particular linear differential equation are closed under addition and scalar multiplication; so + and · map into V. Obviously, there are numerous examples of this kind, some more and some less complex than this one.

EXAMPLE 3. F = Re or Ra, V is the set of complex numbers with usual addition, and · is ordinary multiplication of a complex number by a real or a rational number. In this case, we could also let V itself be the base field. In short, any field is a vector space over any subfield; one merely restricts multiplication so that one factor is always in the subfield.

Next we turn to the definition of linear independence and finite dimensionality.

DEFINITION 2. *Let* $\langle V, +, \cdot \rangle$ *be a vector space over F. The vectors* $x^{(1)}, \ldots, x^{(m)} \in V$ *are* linearly independent *iff for any* $t_1, \ldots, t_m \in F$, *if* $t_1 x^{(1)} + \cdots + t_m x^{(m)} = 0$, *then all* $t_i = 0$.

(i) *V has* dimension $\geq n$ *over F iff the following holds: If* $x^{(1)}, \ldots, x^{(m)}$ *are linearly independent with* $m < n$, *then there exist* $x^{(m+1)}, \ldots, x^{(n)} \in V$ *such that the full set* $x^{(1)}, \ldots, x^{(n)}$ *is linearly independent.*

(ii) *V has* dimension $\leq n$ *over F iff there is no set of* $n + 1$ *linearly independent vectors.*

(iii) *V has* dimension = n *over F iff V has dimension* $\geq n$ *and V has dimension* $\leq n$, *i.e., (i) and (ii) hold with the same n. In that case, a* basis *is any set of n linearly independent vectors.*

One might imagine that a vector space could contain some set of n or more linearly independent vectors, but also some other set of $n - 1$ or fewer linearly independent vectors that is maximal, that is, that cannot be supplemented by another vector without violating linear independence. This is excluded by the following fundamental result.

LEMMA 1. *Let* $\langle V, +, \cdot \rangle$ *be a vector space over F with* $V \neq \{0\}$. *Either V has dimension* $\geq n$ *over F for every* $n \in I^+$ *(V is then* infinite-dimensional *over F), or there exists* $n \in I^+$ *such that V has dimension* = n *over F.*

In the infinite-dimensional case, we can construct an infinite set of vectors such that every finite subset is linearly independent. Otherwise, there exists n such that every set of linearly independent vectors can be enlarged to size exactly n, but no larger.

Example 1 is a vector space with dimension lm and, as we pointed out, is isomorphic in an obvious way to Re^{lm}. Example 2 has infinite dimension over Re: for every n, we can find functions $f_1, \ldots, f_n$ of three variables that satisfy Laplace's equation and do not satisfy a linear equation $\sum_{i=1}^n t_i f_i \equiv 0$. If we considered only functions of one variable that are twice differentiable, we would again have an infinite-dimensional space over Re. But functions of one variable that satisfy a particular second-order differential equation such as $d^2f/dx^2 = 0$ constitute a two-dimensional vector space. In particular, $f_1(x) \equiv 1$ and $f_2(x) \equiv x$ are a basis when $d^2f/dx^2 = 0$ since any linear function can be written as $f = t_1 f_1 + t_2 f_2$. If the equation is $d^2f/dx^2 = -x$, then $\sin x$ and $\cos x$ form a basis. Finally, in Example 3 the complex numbers have dimension 2 over Re (1 and $i = \sqrt{-1}$ form a basis) but are infinite dimensional over Ra.

A proof of Lemma 1 is to be found in any standard book on finite-dimensional vector spaces. Lemma 1 is moderately deep; the other results about finite dimensionality follow easily from it and include the following representation theorem (any V of dimension n is isomorphic to F^n) and uniqueness theorem (in which the automorphisms of F^n are characterized by $n \times n$ matrices with determinant $\neq 0$). The theorem is stated in such a way as to include the fact that F^n itself has dimension n over F.

THEOREM 1. *A vector space* $\langle V, +, \cdot \rangle$ *has dimension n over F iff there exists an isomorphism* $\varphi = (\varphi_1, \ldots, \varphi_n)$ *of* $\langle V, +, \cdot \rangle$ *onto* $\langle F^n, +, \cdot \rangle$.

If φ *is an isomorphism of* $\langle V, +, \cdot \rangle$ *onto* $\langle F^n, +, \cdot \rangle$, *then* φ' *is another such iff there exists an* $n \times n$ *matrix* (α_{ij}) *with entries in F and determinant* $\neq 0$, *such that*

$$\varphi'_i = \sum_{j=1}^{n} \alpha_{ij}\varphi_i, \qquad i = 1, \ldots, n.$$

We shall not prove this well-known theorem, but it may help some readers to have a general outline of the proof.

First, we should comment on the definition of isomorphism and automorphism in vector spaces. If V, V' are vector spaces over the *same* field F, a mapping $\varphi: V \to V'$ is a *homomorphism* if and only if it satisfies

$$\varphi(x + y) = \varphi(x) + \varphi(y)$$
$$\varphi(tx) = t\varphi(x).$$

Moreover, φ is an *endomorphism* if it maps V into itself, an *isomorphism* if it is one-to-one and onto V', and an *automorphism* if it is an isomorphism of V onto itself. For some purposes it would be better to introduce a slightly broader concept, in which V, V' are vector spaces over distinct fields F, F'; a homomorphism would then be a pair (φ, σ) where φ maps V into V', σ is a homomorphism of F into F', and the second condition above is replaced by $\varphi(tx) = \sigma(t)\varphi(x)$. An automorphism would involve a mapping of V onto itself and an automorphism of the field F. However, the most important field we consider, Re, has no automorphisms except the identity. Therefore we keep the narrower definition, having in mind mainly isomorphisms and automorphisms of vector spaces over Re.

OUTLINE OF THE PROOF OF THEOREM 1. If $x^{(1)}, \ldots, x^{(n)}$ form a basis for V, then it is easily shown that any x in V has a unique representation $x = \sum_{i=1}^{n} t_i x^{(i)}$; the mapping φ: $x \to (t_1, \ldots, t_n)$ is an isomorphism onto F^n. Conversely, let $y^{(i)}$ in F^n be the ith coordinate vector, i.e., $y^{(i)}$ has 1 in the ith place and 0 elsewhere. It is easily shown that $y^{(1)}, \ldots, y^{(n)}$ form a basis for F^n; and if φ maps V isomorphically onto F^n, then the vectors $\varphi^{-1}(y^{(i)})$ form a basis for V. This completes the representation part of the theorem.

If φ, φ' are isomorphisms of V onto F^n, then $\varphi'\varphi^{-1}$ and $\varphi(\varphi')^{-1}$ are automorphisms of F^n. It is easy to show, using the coordinate-vector basis just described, that every endomorphism of F^n can be represented by an $n \times n$ matrix; so $\varphi_i' = \sum_{i=1}^{n} \alpha_{ij}\varphi_j$ and likewise $\varphi_j = \sum_{j=1}^{n} \beta_{jk}\varphi_k'$. It now remains only to show that automorphisms are characterized among the endomorphisms by nonzero determinant. If $\varphi'\varphi^{-1}$ and $\varphi(\varphi')^{-1}$ are inverse automorphisms, then the matrix product $(\alpha)(\beta)$ is the identity matrix and has determinant 1. Since the determinant is multiplicative, (α) and (β) each have determinant $\neq 0$. Conversely, if (α) has nonzero determinant, it yields an automorphism of F^n because (α) maps the unit vector $y^{(k)}$ onto the kth column vector $\alpha^{(k)} = (\alpha_{1k}, \ldots, \alpha_{nk})$, and if the columns of (α) are not linearly independent then elementary theorems on determinants show that the determinant is 0. $\Diamond$

The preceding representation and uniqueness theorem has two main uses in this volume. In the remainder of this section and in the next chapter, it allows us to move freely back and forth between geometrical representations in an abstract finite-dimensional vector space and more numerical representations in F^n. Second, in Chapter 15, we obtain a vector representation for color and for force, and this theorem allows us to move between the more natural coordinate-free representation and the more commonly encountered coordinate representation.

It should be mentioned also that there are analogues to the preceding theory using base structures with weaker algebraic properties than fields. First, and most important for our purpose, is the theory for division rings, which are like fields except that multiplication is not commutative. One must distinguish then between F^n as a left vector space with multiplication $tx = (tx_1, \ldots, tx_n)$ and as a right vector space with multiplication $xt = (x_1 t, \ldots, x_n t)$; for each of these, there is a set of results on linear independence, finite dimensionality, and representation parallel to those above.

See Jacobson (1953, Chap. 1) for details. Second, in Chapter 13 we state a representation theorem for a projective plane that satisfies Desargues' property but not Pappus' property, using a vector space over a division ring. Third, more general coordinate structures and their applications in geometry are also briefly described in Chapter 13. From the standpoint of measurement, all this provides a deeper insight into those empirical properties of geometrical structures that give rise to a particular algebraic representation. It is worth pursuing in detail, though we do not do so here.

12.2.2 Analytic Affine Geometry

It is only a small step from an n-dimensional vector space V to n-dimensional affine geometry. We define geometrical objects—*points, lines, planes*, etc.—as certain subsets of V. We define geometrical relations of *incidence* in terms of set inclusions of the defining subsets of V; and if the underlying field F is ordered, we obtain an ordered affine geometry by defining *betweenness* on lines in terms of the ordering of F. The automorphisms of V map points onto points, lines onto lines, etc., and preserve incidence and betweenness, so they are also automorphisms of the geometrical structure. Translations of V (i.e., transformations that add a constant vector to each vector) are also geometrical automorphisms. Conversely, as we shall see in Theorem 2, every automorphism of the geometrical structure (*affinity*) is induced by an automorphism of V plus a translation; thus the affinities can be represented (relative to a basis) by $n \times (n + 1)$ matrices with the first n columns linearly independent. The definitions are all exactly what one would expect on the basis of the usual analytic representations of points, lines, and planes in Re^3, and the reader who thinks in terms of Re^3 will not be misled by the following details.

A point is any subset consisting of a single element $x^{(0)}$ of V; a line is any subset consisting of all elements of form $x^{(0)} + tx^{(1)}$, where $x^{(0)}$ and $x^{(1)}$ are fixed, with $x^{(1)} \neq 0$, and t varies through F; a plane consists of all elements $x^{(0)} + t_1 x^{(1)} + t_2 x^{(2)}$, where $x^{(1)}$, $x^{(2)}$ are linearly independent, and t_1, t_2 vary through F; etc. The general term covering all these objects is *k-dimensional linear variety*, where $k = 0, 1, \ldots, n$. So the concept of a linear variety is just a natural generalization of the concepts of point, line, and plane for dimensions greater than two. More formally, we have the following definition.

DEFINITION 3. *Let $\langle V, +, \cdot \rangle$ be an n-dimensional vector space over a field F.*

(i) *Let $x^{(0)}$ be any element of V and if $1 \leq k \leq n$, let $x^{(1)}, \ldots, x^{(k)}$ be any linearly independent elements of V. The k-dimensional (linear) variety*

$A[x^{(0)}, x^{(1)}, \ldots, x^{(k)}]$ is defined by

$$A[x^{(0)}, x^{(1)}, \ldots, x^{(k)}] = \begin{cases} \{x^{(0)}\} & \text{if } k = 0, \\ \left\langle x^{(0)} + \sum_{i=1}^{k} t_i x^{(i)} \,\middle|\, t_1, \ldots, t_k \in F \right\rangle & \text{if } k \geqslant 1. \end{cases}$$

The set of all k-dimensional linear varieties is denoted $\mathscr{A}_k$ (in particular $\mathscr{A}_n = \{V\}$). The set of all linear varieties $\bigcup_{k=0}^{n} \mathscr{A}_k$ is denoted $\mathscr{V}$.

(ii)   Two linear varieties are incident iff they have unequal dimension and either is included in the other; we write $A \; I \; B$ if $A \in \mathscr{A}_k$, $B \in \mathscr{A}_l$, $k \neq l$, and either $A \subset B$ or $B \subset A$. (Obviously, incidence is defined symmetrically; if $k < l$, we can only have $A \subset B$.)

(iii)   An n-dimensional analytic affine geometry over F consists of a structure $\langle V, +, \cdot, \mathscr{A}_0, \ldots, \mathscr{A}_n, I \rangle$, where $\langle V, +, \cdot \rangle$ is an n-dimensional vector space over F and $\mathscr{A}_0, \ldots, \mathscr{A}_n, I$ are defined by (i) and (ii) above. (I is a binary relation on the set $\mathscr{V} = \bigcup_{k=0}^{n} \mathscr{A}_k$.)

(iv)   An affinity is a mapping ξ of $\mathscr{V}$ into itself such that for every A, $B \in \mathscr{V}$,

$$\begin{array}{lll} A \in \mathscr{A}_k & \text{iff} & \xi(A) \in \mathscr{A}_k \\ A \; I \; B & \text{iff} & \xi(A) \; I \; \xi(B). \end{array}$$

(Note that ξ is not a mapping of V into V but is rather a mapping of $\mathscr{V}$ into $\mathscr{V}$, geometrical objects to other such.)

The simplest example is, of course, the affine plane $\langle \text{Re}^2, +, \cdot, \mathscr{A}_0, \mathscr{A}_1, \mathscr{A}_2, I \rangle$, where a point in $\mathscr{A}_0$ consists of just one pair (x_1, x_2), a line in $\mathscr{A}_1$ is any set of the form

$$\left\{ (x_1, x_2) \,\middle|\, \frac{(x_2 - x_2^{(0)})}{(x_1 - x_1^{(0)})} = m = \text{slope} \right\},$$

and a point is incident with a line iff it satisfies the defining equation of the line. An affinity maps each point of $\mathscr{A}_0$ into another point and every straight line onto another, and if a point is incident with a line, the image point and image line are also incident. Every linear map of Re^2 onto itself induces an affinity; the converse is true but not at all obvious.

An analytic affine geometry is used as a representation for an empirical relational structure in which there are arbitrary sets $P_0, P_1, \ldots, P_n$, interpreted as points, lines, etc., with an abstract incidence relation I. See Section 13.6.1.

Important geometrical concepts that can be defined in affine geometry are the *spans* and *intersections* of linear varieties and the concept of *parallel*. We are familiar with the idea that two distinct points determine (or span) a unique line, three noncollinear points span a unique plane, etc., that two lines intersect in a point or not at all, and that two planes in the same three-dimensional space intersect in a line or are parallel. All these facts are easily developed in the general case. For example, the set-theoretic intersection of linear varieties is either empty or a linear variety; the *span* of varieties $A, B, \ldots$ is the intersection of all varieties that contain all of $A, B, \ldots$. For example, the span of any $k + 1$ linearly independent points is a k-dimensional variety; the span of two lines is two-dimensional iff they are parallel or intersect in a unique point.

For our purposes, the most important of the deep theorems in analytic affine geometry is the characterization of affinities in terms of automorphisms of the underlying vector space plus translations.

THEOREM 2. *Let* $\langle V, +, \cdot, \mathscr{A}_0, \ldots, \mathscr{A}_n, I \rangle$ *be an n-dimensional analytic geometry with* $n \geqslant 2$. *A mapping* ξ *on* $\mathscr{V}$ *is an affinity iff there exists an automorphism* α *of* $\langle V, +, \cdot \rangle$ *and a vector* $y^{(0)}$ *such that* $\xi(A) = \{\alpha(x) + y^{(0)} | x \in A\}$ *for all* $A \in \mathscr{V}$.

To be correct, either we must add to this theorem the hypothesis that the underlying field is Re or some other field with no nontrivial automorphisms, or else we must use the broader concept of vector-space automorphism, which includes automorphisms of the base field as well. For our purposes, it is quite natural to take the field to be Re.

This theorem is crucial for uniqueness results in Chapter 13 since if we have two different representations φ, φ' of an abstract or empirical affine geometry into an analytic affine geometry, then $\varphi' \varphi^{-1}$ is an affinity, and thus we obtain the uniqueness result $\varphi' = \alpha\varphi + y^{(0)}$ where α is a vector-space automorphism.

Another valuable application of this theorem is to show quickly which geometrical concepts are meaningful in affine geometry and which are not. Parallelism is meaningful because it is preserved under transformations of the form

$$y_i = \xi_i(x) = \sum_j \alpha_{ij} x_j + y_i^{(0)};$$

perpendicularity is not preserved nor are orderings of distance or area. (Distances of collinear or parallel segments can be compared in affine geometry; see Section 12.2.6.) We characterize this concept of meaningfulness systematically in Section 2.5.

The proof of Theorem 2 follows readily from the corresponding theorem for projective geometry (Theorem 3 below); see, e.g., Busemann and Kelly (1953, p. 84) for the reduction of the affine to the projective theorem (they treat $n = 2$, but the general case is similar).

If the field F is an ordered field, then it can be used to define a betweenness relation on each line. Suppose that $z^{(i)} = x^{(0)} + t_i x^{(1)}$, $i = 1, 2, 3$. Then we define $z^{(1)}|z^{(2)}|z^{(3)}$ iff either $t_1 > t_2 > t_3$ or $t_3 > t_2 > t_1$. This definition does not depend on $x^{(0)}$ and $x^{(1)}$ as such, only on the line; for if $\{x^{(0)} + tx^{(1)}\}$ and $\{y^{(0)} + uy^{(1)}\}$ define the same line, then there are u_0, u_1 in F, with $u_1 \neq 0$, such that $x^{(0)} = y^{(0)} + u_0 y^{(1)}$ and $x^{(1)} = u_1 y^{(1)}$; hence $x^{(0)} + tx^{(1)} = y^{(0)} + (u_0 + u_1 t)y^{(1)}$. The mapping $t \to u_0 + u_1 t$ either preserves or inverts order in F, depending on the sign of u_1; and in either case betweenness is preserved.

DEFINITION 4. *An n-dimensional analytic ordered affine geometry is a structure $\langle V, +, \cdot, \mathscr{A}_0, \ldots, \mathscr{A}_n, I, B \rangle$ such that Definition 3(iii) is satisfied, the base field F is ordered, and the ternary relation $B = \{(z^{(1)}, z^{(2)}, z^{(3)}) \mid z^{(1)}|z^{(2)}|z^{(3)}\}$ is defined as above.*

Such a structure can be used as a representation for an empirical relational structure with abstract sets $P_0, \ldots, P_n$, an incidence relation I, and a ternary relation B on P_0 (all satisfying suitable axioms, of course). An affinity of an ordered affine geometry is required to preserve the ternary relation B as well as the other requirements of Definition 3(iv). However, Theorem 2 remains unchanged since vector-space automorphisms and translations do preserve betweenness.

12.2.3 Analytic Projective Geometry

To construct a projective geometry from a vector space V, we regard as equivalent any two nonzero vectors that are proportional. This concept of equivalence is central. A *point* is a single equivalence class, i.e., a set $\{tx \mid t \in F, t \neq 0\}$. (In affine terms, a projective point is a line through the origin.) A projective *line* is a set of such equivalence classes or points, namely, all those that can be obtained from two distinct points by linear operations. Because we lose one degree of freedom by ignoring the distinction between proportional elements of V, we obtain an n-dimensional analytic projective geometry from an $(n + 1)$-dimensional vector space.

If this seems like a strange idea, the thought behind it is not hard to explain. Consider ordinary three-dimensional space, and choose an arbitrary point 0 as an origin from which to project and an arbitrary plane P that does not contain 0 as our "projective" plane. (See Figure 1.) Each

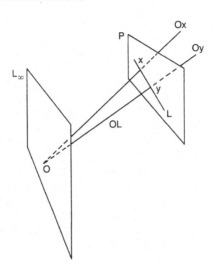

FIGURE 1. 0 is the origin in 3-space. $0x$ and $0y$ are projective points corresponding to affine points x, y in the plane P; $0L$ is the projective line determined by $0x$ and $0y$ corresponding to the affine line L in P. From the standpoint of P, L_∞ is the projective line at infinity, the plane through 0, parallel to P.

point x in P determines a unique line $0x$ through 0 and x; we consider x as the projection of line $0x$ into the plane. Denote by L_∞ the plane through 0 parallel to P. Any line through 0 that lies in L_∞ does not intersect P and therefore has no projection into P; but any other line through 0 does intersect P. Thus we have a one-to-one correspondence between points of P and lines through 0, except for lines in L_∞.

Next, consider 0 as the origin of coordinates. Then any line through 0 is an equivalence class of 3-tuples with respect to proportionality; i.e., if x and y are nonzero vectors on the same line through 0, $y = tx$ for some nonzero scalar t. Thus, there is a one-to-one correspondence between points x in P and equivalence classes of 3-tuples, except for 3-tuples that lie in the plane L_∞. The equivalence class of 3-tuples corresponding to a point x in P constitutes the *homogeneous coordinates* for x.

Finally, consider lines in P. A line L in P, together with 0, determines a *plane* through 0 that intersects P in L. We can thus identify every plane through 0, except L_∞, with a line L in P; L is the projection of the plane $0L$. In short, we have taken an ordinary affine plane P and set up a correspondence of its points to homogeneous coordinates or lines through 0, and its lines to planes through 0. Thus, it is not unreasonable to think of a two-dimensional space in terms of proportionality equivalence classes from three-dimensional space, by projecting through 0.

The advantage of this interpretation for a plane P is that it is easy to add on the concept of "points at infinity." In the affine plane P, lines may be parallel. It is tempting to remove this exception and say that any two lines in P do intersect, either in an ordinary point or a point at infinity. If we interpret the lines L and L' in P as planes $0L$ and $0L'$, these two planes do intersect: they contain at least 0 in common and thus contain a line through 0. If L and L' intersect in a point x in P, then the intersection of planes $0L$ and $0L'$ is the line $0x$, which is the correspondent to point x. But if L and L' are parallel in P, the line of intersection of $0L$ and $0L'$ must be parallel to P, i.e., it must be a line through 0 lying in plane L_∞. Therefore we interpret the plane L_∞ as the "line at infinity" of P and the lines in L_∞ through 0 as "points at infinity." We have now completed the one-to-one correspondence—*every* line through 0 is interpreted as a point of the extended plane P, and every plane through 0 is interpreted as a line—and simultaneously we have removed the exceptional nature of parallels. Every two lines in the extended, or projective, plane intersect in a unique point. If we now drop the plane P and consider only the geometry of lines and planes through 0, the plane L_∞ and lines in it are no different from any others; the exceptional nature of the points at infinity arises only when we try to project onto some particular P.

There are many other points of view from which projective geometry can be considered, but the above should suffice as an introduction. We turn now to the definition. The reader should note the close parallelism with Definition 3.

DEFINITION 5. *Let $\langle V, +, \cdot \rangle$ be an $(n + 1)$-dimensional vector space over a field F.*

(i) *Let $x^{(0)}, \ldots, x^{(k)}$ be any $k + 1$ linearly independent elements of V, where $0 \leqslant k \leqslant n$. The k-dimensional projective variety $L[x^{(0)}, \ldots, x^{(k)}]$ is just the $(k + 1)$-dimensional linear subspace $\{\Sigma_{i=0}^{k} t_i x^{(i)} | t_0, \ldots, t_k \in F\}$. The set of all k-dimensional projective varieties is denoted $\mathscr{L}_k$, and $\bigcup_{k=0}^{n} \mathscr{L}_k$ is denoted $\mathscr{P}$.*

(ii) *Projective varieties L, M are incident, denoted $L I M$, iff $L \in \mathscr{L}_k$, $M \in \mathscr{L}_l$, $k \neq l$, and either $L \subset M$ or $M \subset L$.*

(iii) *An n-dimensional analytic projective geometry over F consists of a structure $\langle V, +, \cdot, \mathscr{L}_0, \ldots, \mathscr{L}_n, I \rangle$ where $\langle V, +, \cdot \rangle$ is an $(n + 1)$-dimensional vector space over F and $\mathscr{L}_0, \ldots, \mathscr{L}_n, I$ are defined by (i) and (ii) above.*

(iv)   *A projectivity is a mapping ξ of $\mathcal{P}$ into itself such that for every L,
$M \in \mathcal{P}$,*

$$L \in \mathcal{L}_k \qquad \text{iff} \qquad \xi(L) \in \mathcal{L}_k \,,$$
$$LIM \qquad \text{iff} \qquad \xi(L)\,I\,\xi(M).$$

In comparison with Definition 3, we have added a dimension by changing n to $n + 1$ and then have taken it away by considering only linear varieties through the origin, not allowing parallel translations through $x^{(0)}$ as in Definition 3. Thus, a one-dimensional subspace $\{t_0 x^{(0)} | t_0 \in F\}$ becomes an element of $\mathcal{L}_0$, or a point; two points, generated respectively by $x^{(0)}$ and $x^{(1)}$, yield a line $L = \{t_0 x^{(0)} + t_1 x^{(1)}\}$ in $\mathcal{L}_1$; etc.

To obtain the usual representation of the projective plane over Re, we take $\langle \mathrm{Re}^3, +, \cdot, \mathcal{L}_0, \mathcal{L}_1, \mathcal{L}_2, I \rangle$. A point in $\mathcal{L}_0$ has homogeneous coordinates $\{(tx_1, tx_2, tx_3) | t \neq 0\}$, where $(x_1, x_2, x_3) \neq (0, 0, 0)$. A line in $\mathcal{L}_1$ is a set

$$\{(t_0 x_1^{(0)} + t_1 x_1^{(1)},\, t_0 x_2^{(0)} + t_1 x_2^{(1)},\, t_0 x_3^{(0)} + t_1 x_3^{(1)}) \mid (t_0, t_1) \neq (0, 0)\},$$

where $x^{(0)}$ and $x^{(1)}$ are linearly independent 3-tuples.

If $x^{(0)}$ and $x^{(1)}$ both satisfy the equation

$$\nu_1 x_1 + \nu_2 x_2 + \nu_3 x_3 = 0,$$

where ν_1, ν_2, and ν_3 are numbers, then so does any linear combination $t_0 x^{(0)} + t_1 x^{(1)}$; thus, if x_0 and x_1 are also linearly independent, then the line generated by them is characterized precisely by the above equation: the well-known equation of a plane in Re^3 through the origin.

Notice that any proportional triple $(t\nu_1, t\nu_2, t\nu_3)$ generates the same plane in Re^3 (or line in the projective plane) as does (ν_1, ν_2, ν_3) provided $t \neq 0$. Thus we could also introduce a correspondence between a projective line and an equivalence class of homogeneous coordinates. If we now simply reverse the roles of the 3-tuples x and ν, interpreting the former as lines and the latter as points, we obtain a mapping of $\mathcal{P}$ into itself that sends $\mathcal{L}_0$ into $\mathcal{L}_1$ and $\mathcal{L}_1$ into $\mathcal{L}_0$ and preserves incidence relations between $\mathcal{L}_0$ and $\mathcal{L}_1$. Such a mapping is called a *correlation*; it is like a projectivity, but reverses the roles of points and lines, which, in the projective plane, are entirely dual. Because the incidence relations are symmetric (two distinct points determine exactly one line and two distinct lines determine exactly one point), the representations of points and lines can be interchanged, and every true assertion in plane projective geometry remains true if it is expressed symmetrically in terms of incidence and the terms *point* and *line* are interchanged.

In n-dimensional projective geometry, duality is also valid, with $\mathscr{L}_0$ exchanged for $\mathscr{L}_{n-1}$, $\mathscr{L}_1$ for $\mathscr{L}_{n-2}$, and, more generally, $\mathscr{L}_i$ for $\mathscr{L}_j$, where $i + j = n - 1$. Thus, in projective 3-space (Re^4, etc.), points and planes are interchangeable—any three linearly independent points are incident with a unique plane and vice versa—whereas lines are self-dual.

An analytic projective geometry is used as a representation for the same general sort of structure as an analytic affine geometry: an empirical relational structure with sets $P_0, \ldots, P_n$, interpreted as points, lines, etc., with an abstract incidence relation I. The choice of which representation to use, affine or projective, depends, of course, on the properties of the relation I. If P_i, P_j are dual relative to I, for $i + j = n - 1$, then the projective representation is the only one possible; if duality fails, then only the affine representation is possible.

As in the affine case, the key to understanding uniqueness and meaningfulness lies in the characterization of incidence-preserving transformations of the geometrical objects. We have the following fundamental theorem.

THEOREM 3. *Let* $\langle V, +, \cdot, \mathscr{L}_0, \ldots, \mathscr{L}_n, I \rangle$ *be an n-dimensional analytic projective geometry with* $n \geqslant 2$. *A mapping* ξ *on* $\mathscr{P}$ *is a projectivity iff there exists an automorphism* α *of* $\langle V, +, \cdot \rangle$ *such that* $\xi(L) = \{\alpha(x) | x \in L\}$ *for all* $L \in \mathscr{P}$.

A proof for the projective plane over Re is found in Busemann and Kelly (1953, Chap. 1); Schreier and Sperner (1961, Chap. 6) treat the general case. As in the case of Theorem 2, we must either restrict this theorem to vector spaces over Re or other fields with no nontrivial automorphisms, or else we must use the broader concept of vector-space automorphism, including automorphisms of F.

To understand this theorem better, let us consider the case of the plane $n = 2$, and compare it with Theorem 2 for the affine plane. Let α be an automorphism of Re^3, i.e., a mapping of the form $y_i = \sum_{j=1}^{3}\alpha_{ij}x_j$, $i = 1, 2, 3$, where (α_{ij}) has nonzero determinant. Any such mapping sends each line and plane through the origin into another such and thus induces a projectivity of the projective plane $\langle \text{Re}^3, +, \cdot, \mathscr{L}_0, \mathscr{L}_1, \mathscr{L}_2, I \rangle$. The theorem states that every projectivity is obtained in this way. Clearly, two such automorphisms induce the same projectivity if and only if they are strictly proportional, i.e., $\alpha'_{ij} = \gamma\alpha_{ij}$, where γ is a nonzero element of Re. Thus the projectivities of the plane can be considered as an 8-parameter group since some a_{ij}, for example, can always be taken equal to 1 by an appropriate choice of γ.

To obtain an affine plane, we proceed as in the introduction to this section, projecting from the origin onto a particular plane in Re^3 that does not contain the origin. Take this plane to be $x_3 = 1$. Thus every projective point $\{(tx_1, tx_2, tx_3)\}$ such that $x_3 \neq 0$ projects onto the affine point $(x_1/x_3, x_2/x_3, 1)$; the points with $x_3 = 0$ constitute L_∞ and have no affine counterparts. Let $x_1' = x_1/x_3$ and $x_2' = x_2/x_3$ be the affine coordinates corresponding to a projective point not in L_∞. A projectivity (α_{ij}) sends a projective point x that is not in L_∞ into another projective point whose affine coordinates are $(y_1/y_3, y_2/y_3)$, or

$$\left(\frac{\alpha_{11}x_1 + \alpha_{12}x_2 + \alpha_{13}x_3}{\alpha_{31}x_1 + \alpha_{32}x_2 + \alpha_{33}x_3}, \frac{\alpha_{21}x_1 + \alpha_{22}x_2 + \alpha_{23}x_3}{\alpha_{31}x_1 + \alpha_{32}x_2 + \alpha_{33}x_3} \right)$$

$$= \left(\frac{\alpha_{11}x_1' + \alpha_{12}x_1' + \alpha_{13}}{\alpha_{31}x_1' + \alpha_{32}x_2' + \alpha_{33}}, \frac{\alpha_{21}x_1' + \alpha_{22}x_2' + \alpha_{23}}{\alpha_{31}x_1' + \alpha_{32}x_2' + \alpha_{33}} \right).$$

These expressions are defined only if the denominator $y_3 = \alpha_{31}x_1' + \alpha_{32}x_2' + \alpha_{33}$ is nonzero. Thus a general projectivity transforms the affine plane by mapping (x_1', x_2') into ratios of linear nonhomogeneous functions of x_1', x_2' with a common denominator, provided that the denominator is nonzero. The affine points such that $\alpha_{31}x_1' + \alpha_{32}x_2' + \alpha_{33} = 0$ are mapped outside the affine plane into projective points with $y_3 = 0$, i.e., L_∞. Any two lines in the affine plane that intersect on this critical line are mapped into parallels in the affine plane; so general projectivities do not preserve parallelism.

To preserve parallelism and map the plane $x_3 = 1$ onto itself, the projectivity must have a denominator in the preceding expression that never vanishes, i.e., $\alpha_{31} = \alpha_{32} = 0$ and $\alpha_{33} \neq 0$; we can then normalize so that $\alpha_{33} = 1$ and obtain the 6-parameter affine subgroup, consisting of 3×3 matrices of form

$$\begin{pmatrix} \alpha_{11} & \alpha_{12} & \alpha_{13} \\ \alpha_{21} & \alpha_{22} & \alpha_{23} \\ 0 & 0 & 1 \end{pmatrix}$$

such that the 2×2 determinant $\alpha_{11}\alpha_{22} - \alpha_{21}\alpha_{12} \neq 0$. This is, of course, precisely the result of Theorem 2: affinities are obtained from an automorphism of Re^2 plus a translation by the vector $y^{(0)} = (\alpha_{13}, \alpha_{23})$.

The generalizations to $n > 2$ should now be obvious. The projectivities can be considered to be a $(n + 1)^2 - 1$ parameter group of equivalence classes of automorphisms of Re^{n+1} (the -1 comes from the proportionality factor). The affinities are those automorphism classes that map a particular

n-dimensional subspace (say, $x_{n+1} = 0$) onto itself; they constitute an $n(n + 1)$-parameter subgroup.

We turn finally to consideration of the additional geometrical relation that can be defined for an analytic projective geometry over an ordered field. We shall consider the problem of ordering the points on a projective line. To lessen subscripts, we represent the line by $\{tx + uy | t, u \in F\}$. Each point on it is identified by a ratio u/t; this ratio takes all values in F and also the value ∞ when $t = 0$.

The ordering or betweenness of the ratios u/t is not invariant under different representations of the same line, i.e., different choices of x, y. What does remain invariant is a quaternary function from the line to F, namely, the *cross ratio*. If $r_i = u_i/t_i$, $i = 1, 2, 3, 4$, and all r_i are distinct and in F (finite), then the cross ratio is defined by

$$\frac{(r_1 - r_3)/(r_1 - r_4)}{(r_2 - r_3)/(r_2 - r_4)}.$$

A more general definition, which allows one of the r_i to be ∞, or $t_i = 0$, is given by

$$\frac{(u_1 t_3 - u_3 t_1)/(u_1 t_4 - u_4 t_1)}{(u_2 t_3 - u_3 t_2)/(u_2 t_4 - u_4 t_2)}.$$

(If the points are all distinct, then none of the expressions in parentheses, which are all 2×2 determinants, can vanish.) It can be shown that the cross ratio of four collinear projective points is invariant under every projectivity that maps their line into itself, hence, under any change in representation for the line. If the cross ratio is positive, we can think of the first two points as falling into the same "region of slopes" relative to the last two: i.e., if r_1 is greater than both r_3 and r_4, then r_2 is either greater than both or less than both; if r_1 is between r_3 and r_4, so is r_2. If the cross ratio is negative, the first two points fall into opposite regions, or *separate* the last two. We define a quaternary relation S of *separation*: An ordered quadruple of four distinct collinear projective points is in relation S iff the cross ratio is *negative*, i.e., < 0 in the ordering of F. Qualitative axioms for the relation of separation are given in Section 13.2.2; see also Figures 1 and 2 of Chapter 13.

DEFINITION 6. *An n-dimensional analytic ordered projective geometry is a structure* $\langle V, +, \cdot, \mathscr{L}_0, \ldots, \mathscr{L}_n, I, S \rangle$ *such that Definition 5(iii) is satisfied, the base field F is ordered, and the quaternary relation S is defined as above.*

For representations using such an analytic geometry, see Section 13.4. The projectivities of an ordered projective geometry are required to preserve S, but this does not change Theorem 3 since automorphisms of F^{n+1} do preserve the sign of the cross ratio of four distinct collinear points when the field F is an ordered field.

12.2.4 Analytic Euclidean Geometry

In projective geometry we introduce geometrical objects of dimensions 0 through n by using the algebraic structure of an $(n + 1)$-dimensional vector space, restricting ourselves to linear homogeneous subspaces. To enrich this geometry with a concept of parallelism, we delete a particular n-dimensional subspace (L_∞), or equivalently we start with an n-dimensional vector space and consider not only the homogeneous subspaces but parallel translations of them. A logical next step after parallelism would be a concept of perpendicularity. We cannot define this solely in terms of the operations $t, \cdot$, however, because perpendicularity is not preserved by general vector-space automorphisms.

In the vector space F^n it is customary to define perpendicularity $x \perp y$ by the equation $\sum_{i=1}^{n} x_i y_i = 0$. This defines a binary relation with the additional convention that each pair of distinct coordinate vectors is perpendicular, i.e., the coordinates in F^n are orthogonal coordinates. A more general procedure is to characterize the essential properties of the function $\sum_{i=1}^{n} x_i y_i$, which leads to a satisfactory theory of perpendicularity. The mapping $(x, y) \rightarrow \sum_{i=1}^{n} x_i y_i$ has domain $F^n \times F^n$ and range F; it is *bilinear* (linear in each variable x or y); it is *symmetric*; and if F is an ordered field, it is *positive definite* (its value at (x, x) is > 0, except for $x = 0$). These abstract properties are sufficient. In fact, suppose V is any vector space over Re. A function from $V \times V$ into Re that is bilinear, symmetric, and positive definite is called a *Euclidean scalar product*. It is not hard to show that such a function can always be represented by $\sum_{i=1}^{n} \varphi_i(x) \varphi_i(y)$ using a suitable isomorphism φ of F of V onto Re^n.

A Euclidean scalar product determines a binary relation $\perp$ on V, and it is easily shown that two different Euclidean scalar products determine the same binary relation if and only if they are proportional. Thus, the perpendicularity relation determines the scalar product up to multiplication by a constant.

From a scalar product we can define the usual Euclidean distance function: $\delta(x, y) = [\sum_{i=1}^{n} (x_i - y_i)^2]^{1/2}$ is just the square root of the value of the scalar product at the pair $(x - y, x - y)$. So a binary relation of perpendicularity, with the properties expected from Euclidean geometry, determines the Euclidean distances uniquely except for a unit of measurement.

It seems as though perpendicularity is essentially a metric rather than an algebraic property. A concept of perpendicularity with many familiar properties can be introduced into metric geometries other than the Euclidean (Busemann, 1955, Chap. 3); but if we try to introduce it algebraically by means of a scalar product in a vector space, then we already have essentially determined the Euclidean metric.

We now introduce the formal definitions. It is convenient to use the notation $[x, y]$ for the value of a scalar product and $[\ ,\ ]$ for the function itself.

DEFINITION 7.   Let $\langle V, +, \cdot \rangle$ be a vector space over Re.

(i)   A function $[\ ,\ ]$ from $V \times V$ into Re is a Euclidean scalar product iff it satisfies the following three properties for x, y, z in V and t in Re:

1.   (Bilinearity) $[x + y, z] = [x, z] + [y, z]$

$$[x, y + z] = [x, y] + [x, z]$$

$$[tx, y] = t[x, y] = [x, ty].$$

2.   (Symmetry) $[x, y] = [y, x]$.
3.   (Positive definiteness)   If $x \neq 0$, then $[x, x] > 0$.

(ii)   Let $[\ ,\ ]$ be a Euclidean scalar product. Define x perpendicular to y ($x \perp y$) iff $[x, y] = 0$.

(iii)   Let $[\ ,\ ]$ be a Euclidean scalar product. The Euclidean norm $\|x\|$ is the nonnegative-valued function defined by

$$\|x\| = [x, x]^{1/2}.$$

The Euclidean metric $\delta(x, y)$ is the nonnegative function defined by

$$\delta(x, y) = \|x - y\| = [x - y, x - y]^{1/2}.$$

We define a quaternary relation of congruence on V, denoted $xy \approx x'y'$, by the equation $\delta(x, y) = \delta(x', y')$.

We have already described the most familiar example of a Euclidean scalar product, the function $\sum_{i=1}^{n} x_i y_i$ on $\mathrm{Re}^n \times \mathrm{Re}^n$. To prove that every Euclidean scalar product can be represented in this way via an isomorphism φ of V into Re^n such that $[x, y] = \sum_{i=1}^{n} \varphi_i(x) \varphi_i(y)$, one constructs an orthonormal basis in V; the details are described in any standard book on linear algebra.

The concept of norm is important because many non-Euclidean metric geometries are defined by using a function on a vector space that has the same formal properties as ‖ ‖ but does not come from a scalar product. An example is the Minkowski p-norm, defined in Re^n by $[\sum|x_i|^p]^{1/p}$, where $1 \le p < \infty$.

We now turn to the definitions of classes of transformations that preserve the various geometrical concepts we have introduced.

DEFINITION 8. *Let* $\langle V, +, \cdot \rangle$ *be a vector space over* Re, *let* $\mathscr{A}_0, \ldots, \mathscr{A}_n$, I, B *be the ordered affine relations* (*Definitions* 3, 4), *let* [,] *be a Euclidean scalar product on* V, *and let* ‖ ‖, δ, $\perp$, $\approx$ *be as defined by* [,] (*Definition* 7).

(i) *An affinity* ξ *is a* similarity *iff either of the following two equivalent conditions holds*:
   (a) $\xi(x) - \xi(0) \perp \xi(y) - \xi(0)$ *iff* $x \perp y$.
   (b) $\xi(x)\xi(y) \approx \xi(x')\xi(y')$ *iff* $xy \approx x'y'$.

(ii) *A mapping* ξ *of* V *into itself is a* motion *iff it preserves distance*, i.e., $\xi(x)\xi(y) \approx xy$.

(iii) *A mapping* ξ *of* V *into itself is an* orthogonality *iff it preserves* [,], i.e., $[\xi(x), \xi(y)] = [x, y]$.

Although we have not explicitly required that a motion or an orthogonality preserve the affine structure, this property follows from the definition. In fact, similarities, motions, and orthogonalities are successively narrower subgroups of the affinities. The following theorem specifies these groups completely.

THEOREM 4. *Assume the structures of Definition* 8.

(i) *Every motion is a similarity, and every similarity has the form* $\alpha\xi$, *where* $\alpha \in \mathrm{Re}^+$ *and* ξ *is a motion.*

(ii) *Every orthogonality is a motion, and every motion has the form* $\xi + z$, *where* $z \in V$ *and* ξ *is an orthogonality.*

(iii) *Let* φ *be an isomorphism of* V *onto* Re^n *such that* $[x, y] = \sum_{i=1}^{n}\varphi_i(x)\varphi_i(y)$. *Then* ξ *is an orthogonality of* V *iff there is an* $n \times n$ *matrix* $\alpha = (\alpha_{ij})$, *satisfying*

$$\sum_{k=1}^{n} \alpha_{ik}\alpha_{jk} = \begin{cases} 1, & i = j \\ 0, & i \ne j \end{cases}$$

such that $\varphi_i\xi(x) = \sum_{j=1}^{n}\alpha_{ij}\varphi_j(x)$, i.e., $\xi = \varphi^{-1}\alpha\varphi$.

Thus orthogonalities are specified by orthogonal matrices, i.e., matrices that satisfy the $(n + 1)n/2$ constraints given above and, hence, form an

$(n - 1)n/2$-parameter subgroup of the affinities. A motion is obtained from an orthogonality by a translation, and a similarity is obtained from an orthogonality by a translation and multiplication by a scalar. In the next subsection we illustrate this theorem further by exhibiting a nested sequence of five groups of matrices: the projectivities, affinities, similarities, motions, and orthogonalities of the plane.

From the standpoint of measurement it would be most natural to define an *analytic Euclidean geometry* as consisting of an analytic ordered affine geometry over Re with the addition of either a congruence relation $\approx$ or a perpendicularity relation $\perp$, defined from a Euclidean scalar product. Such a structure would be the natural representation for an abstract empirical relational structure in which the primitive concepts are linear varieties, incidence, betweenness, and either congruence or perpendicularity. The natural uniqueness theorem would involve the group of similarities, which preserves $\approx$ or $\perp$; distances themselves need not be preserved since the only empirically meaningful relations would be *comparisons* among distances, which are invariant under multiplication by a scalar. This is, in fact, the Euclidean measurement theory we shall present in Section 13.6.2. Traditionally, however, Euclidean analytic geometry is associated with the group of motions rather than with similarities: *size* is considered as a meaningful geometrical property. This arises in connection with another important concept of metric geometry, namely, that of *mobility*, from which the name *motions* is taken.

The reader may recall from secondary-school geometry that congruence theorems for triangles are often proved by arguments involving "superposition" of one triangle on another. To explicate this, we really need to work with *two copies* of the Euclidean space, say V and V', together with a congruence relation that is considered as an equivalence relation on $(V \times V) \cup (V' \times V')$. That is, we have assertions of form $xy \approx zw$ in $V \times V$, or $x'y' \approx z'w'$ in $V' \times V'$; but we also can assert $xy \approx x'y'$, where x, y are in V and x', y' in V'. Empirically, we conceptualize V as a space of movable points (i.e., as equipped with mappings into itself) and V' as a space of fixed objects. Thus, to move a box across a room, we think of the box as a subset of V, and we think of all the fixed objects in the room as subsets of V'. We also consider a fixed one-to-one correspondence between V and V', i.e., a mapping g of V onto V' such that $xy \approx g(x)g(y)$. This correspondence *superimposes* the two spaces. Thus, if we say we move a box $B \subset V$ to a position B' on a tabletop in V', we mean that we perform a mapping ξ of B into V such that the composition with the superposition mapping sends B onto B', i.e., $g\xi(B) = B'$.

We can now see that a *motion* is a mapping of V onto itself that preserves the extended congruence $\approx$ on $(V \times V) \cup (V' \times V')$, i.e., $xy \approx$

$x'y'$ iff $\xi(x)\xi(y) \approx x'y'$. Euclidean geometry is unusual but not unique among metric geometries in that it has the property of *free mobility*. If B_1 and B_2 are any two congruent subsets of V (i.e., there is a mapping ξ of B_1 onto B_2 such that $xy \approx \xi(x)\xi(y)$), then B_1 and B_2 can be superimposed by a motion, i.e., ξ can be extended to a motion on all of V, and the superposition is given by $g(B_2) = g\xi(B_1)$. To appreciate this property, one must recognize its rarity. The metric $\delta(x, y) = |x_1 - y_1| + |x_2 - y_2|$ in Re^2 does not have it: the only motions[1] have the form $(x_1, x_2) \to (\pm x_1, \pm x_2)$ or $(x_1, x_2) \to (\pm x_2, \pm x_1)$; thus the sets $B_1 = \{(0,0),(1,0)\}$ and $B_2 = \{(0,0),(1/2,1/2)\}$, which are obviously congruent, cannot be superimposed by a motion. We shall discuss this topic further in connection with the Helmholtz–Lie problem on the nature of space (Section 3.2).

To summarize, mappings that preserve the Euclidean relations of perpendicularity or congruence in a single space are similarities. Mappings that preserve congruence relations between sets in one (mobile) space and another (fixed) space are motions. We leave it to the reader to give empirical interpretations for the two special subgroups of mappings that generate orthogonalities and translations.

12.2.5 Meaningfulness in Analytic Geometry

There are three reasons why issues of meaningfulness and invariance under permissible transformations merit special discussion in a geometrical setting. First, these issues arose in nineteenth-century geometry and have exerted a prolonged and extraordinary influence on the development of geometry and of mathematical physics. Second, the groups of permissible transformations that occur in geometry are much "larger" than those that are typically of interest in unidimensional measurement and, consequently, the invariant relations are rather different from those found in the unidimensional cases. Third, and perhaps most important, it is generally the case that the groups of geometrical transformations are much more obvious than the relations to be represented or preserved, and it is consequently common to *define the geometry in terms of the group*, to determine subsequently the relations that are invariant for that group, and only later to axiomatize, designating certain relations as primitive and others as definable in terms of the primitive relations. The most dramatic example of this last point is found in algebraic topology, in which the group of interest is intuitively appealing (transformations that "stretch but do not tear"), but the invariant

[1] A motion would again be required to preserve extended congruence with two copies of the space, and thus must satisfy $\delta(x, y) = \delta[(\xi(x), \xi(y)]$ in all of Re^2.

relations are not obvious and indeed have not been fully characterized
except for some special cases.

We start the discussion with the second of the points just raised: to wit,
we review the groups of permissible transformations introduced in the three
preceding subsections. (They are permissible because they preserve the
relevant geometrical concepts.) For concreteness we restrict attention to
$n = 2$, i.e., plane geometry. The five groups involved are simply ordered by
inclusion:

$$\text{projective} \supset \text{affine} \supset \text{similarity} \supset \text{motion} \supset \text{orthogonal}$$

(This list could easily be extended and turned into a partial rather than a
simple order.)

A two-dimensional projective geometry over Re can be represented by
homogeneous coordinates $\{(tx_1, tx_2, tx_3)\}$ in Re^3, and the representation is
unique up to automorphisms of Re^3. We can represent any permissible
transformation of the homogeneous coordinates by a matrix of form

$$\begin{pmatrix} t\alpha_{11} & t\alpha_{12} & t\alpha_{13} \\ t\alpha_{21} & t\alpha_{22} & t\alpha_{23} \\ t\alpha_{31} & t\alpha_{32} & t\alpha_{33} \end{pmatrix} \tag{1}$$

with determinant $\neq 0$. This really designates an equivalence class in which
t is any number $\neq 0$; thus one of the nonzero α_{ij} can be normalized to 1,
and (1) is an 8-parameter group.

We can represent two-dimensional affine geometry by projecting all
homogeneous coordinates with $x_3 \neq 0$ into the plane $x_3 = 1$, i.e., by
coordinates $(x_1, x_2, 1)$. Any permissible transformation can then be written
as a matrix,

$$\begin{pmatrix} \alpha_{11} & \alpha_{12} & \alpha_{13} \\ \alpha_{21} & \alpha_{22} & \alpha_{23} \\ 0 & 0 & 1 \end{pmatrix}, \tag{2}$$

with $\alpha_{11}\alpha_{22} - \alpha_{12}\alpha_{21} \neq 0$. We interpret this as

$$\begin{aligned} x_1' &= \alpha_{11}x_1 + \alpha_{12}x_2 + \alpha_{13} \\ x_2' &= \alpha_{21}x_1 + \alpha_{22}x_2 + \alpha_{23} \\ 1 &= \qquad\qquad\qquad\quad 1, \end{aligned}$$

thus $(\alpha_{13}, \alpha_{23})$ can be considered as a translation in the plane Re^2 whereas
the upper left 2×2 matrix is an automorphism of Re^2. We have a
6-parameter subgroup of (1).

A similarity transformation can be represented by the special form

$$\begin{pmatrix} \alpha\cos\theta & \pm\alpha\sin\theta & \alpha_{13} \\ \alpha\sin\theta & \mp\alpha\cos\theta & \alpha_{23} \\ 0 & 0 & 1 \end{pmatrix} \tag{3}$$

with $\alpha \neq 0$. If the sign is taken as $-$ in row 1, and $+$ in row 2, then this is interpretable as a rotation through angle θ, change of scale by factor α, and translation by $(\alpha_{13}, \alpha_{23})$; with the other choice of signs, we add a reflection in the x-axis (after the rotation but before the translation). Since the choice of including the reflection or not is just binary, we have a 4-parameter subgroup of (1) or (2).

In a motion, $\alpha = 1$, so we have a 3-parameter subgroup given by

$$\begin{pmatrix} \cos\theta & \pm\sin\theta & \alpha_{13} \\ \sin\theta & \mp\cos\theta & \alpha_{23} \\ 0 & 0 & 1 \end{pmatrix}. \tag{4}$$

Finally, an orthogonality omits the translation, and we are reduced to a 1-parameter subgroup

$$\begin{pmatrix} \cos\theta & \pm\sin\theta & 0 \\ \sin\theta & \mp\cos\theta & 0 \\ 0 & 0 & 1 \end{pmatrix}. \tag{5}$$

Working backwards through these groups, we can verify that (5) preserves the scalar product $x_1 y_1 + x_2 y_2$:

$$\begin{aligned} x_1' y_1' + x_2' y_2' &= (x_1\cos\theta \pm x_2\sin\theta)(y_1\cos\theta \pm y_2\sin\theta) \\ &\quad + (x_1\sin\theta \mp x_2\cos\theta)(y_1\sin\theta \mp y_2\cos\theta) \\ &= x_1 y_1 + x_2 y_2. \end{aligned}$$

Clearly (4) does not preserve scalar products of x, y but does preserve scalar products of *differences* of vectors since the translation drops out and (4) reduces to (5) for differences. Hence, (4) preserves the Euclidean metric, which is just the square root of the scalar product of $x - y$ with itself.

Transformation (3) multiplies Euclidean distance by $|\alpha|$, so it preserves ordering of distances and, in particular, the quaternary congruence relation. More generally, (3) multiplies the scalar product of differences between vectors by $|\alpha|$ and thus preserves perpendicularity of difference vectors.

Affinities (Equation 2) preserve equality of the slope of the line joining two points, i.e., relations of the form

$$\frac{y_2 - x_2}{y_1 - x_1} = \frac{z_2 - w_2}{z_1 - w_1},$$

since

$$\frac{y_2' - x_2'}{y_1' - x_1'} = \frac{(\alpha_{21}y_1 + \alpha_{22}y_2 + \alpha_{23}) - (\alpha_{21}x_1 + \alpha_{22}x_2 + \alpha_{23})}{(\alpha_{11}y_1 + \alpha_{12}y_2 + \alpha_{13}) - (\alpha_{11}x_1 + \alpha_{12}x_2 + \alpha_{13})}$$
$$= \frac{\alpha_{21} + \alpha_{22}(y_2 - x_2)/(y_1 - x_1)}{\alpha_{11} + \alpha_{12}(y_2 - x_2)/(y_1 - x_1)}.$$

Thus parallelism is a relation invariant under affinities.

Finally, projectivities preserve linear dependence relations: there exist t, u, v such that $tx_i + uy_i + vz_i = 0$, $i = 1, 2, 3$, if and only if the same holds for x_i', y_i', z_i', $i = 1, 2, 3$, where the primed coordinates are calculated from the unprimed ones by multiplying by matrix (1).

Obviously, a subgroup preserves any relation that the larger group preserves; so, for example, similarities preserve parallelism and linear dependence as well as congruence and the perpendicularity of difference vectors.

Consider the difference between the following two relations T_1 and T_2:

$$T_1 = \{(x, y, z) \mid \delta(x, y) = 3\delta(x, z)\}$$
$$= \{(x, y, z) \mid (x_1 - y_1)^2 + (x_2 - y_2)^2 = 9[(x_1 - z_1)^2 + (x_2 - z_2)^2]\}$$
$$T_2 = \{(x, y, z) \mid (x_1 - y_1)^3 + (x_2 - y_2)^3 = 9[(x_1 - z_1)^3 + (x_2 - z_2)^3]\}.$$

Relation T_1 is invariant under similarities whereas T_2 is not. Note, however, that if these relations were considered to be 6-ary relations on Re, they both would be invariant under interval scale (one-dimensional affine) transformations, $t' = \alpha t + \beta$. One can write these latter transformations as transformations of the plane by using the matrix form

$$\begin{pmatrix} \alpha & 0 & \beta \\ 0 & \alpha & \beta \\ 0 & 0 & 1 \end{pmatrix}. \tag{6}$$

This makes it clear that (6) is a subgroup of (3), obtained by setting $\cos \theta = 1$, the binary sign $= +$, and $\alpha_{13} = \alpha_{23} = \beta$. The choice of sign and the equality of α_{13} and α_{23} do not matter; what does matter is that x_1' depends only on x_1, etc., i.e., the angle of rotation is 0. The fact that the group of geometrical transformations (3) includes rotations is what substantially reduces the invariant relations compared with the unidimensional measurement situation.

The preceding discussion achieves one goal, namely, the review of some of the most important groups of permissible transformations in geometry and an indication of the sharp restriction of the invariant numerical relations when the groups in question include rotations. We now return to the question of meaningfulness and ask, in particular, what does it mean to say that relation T_2 is meaningless in Euclidean geometry?

First, consider the analytic Euclidean structure as a representation for an empirical structure $\mathscr{A}$. The objects in $\mathscr{A}$ could consist of two sets, denoted P_0 (points) and P_1 (lines) with a binary incidence relation $I \subset (P_0 \times P_1) \cup (P_1 \times P_0)$, a ternary betweenness relation B on P_0, and a binary perpendicularity relation $\perp$ on P_1. With suitable axioms we can construct a representation φ mapping P_0 onto Re^2 and P_1 onto the set of affine lines of Re^2, with I and B going into the corresponding affine relations in Re^2 and with $\perp$ being represented by perpendicularity of the Re^2 lines relative to the usual scalar product in Re^2. Such a representation of the structure $\mathscr{A} = \langle P_0, P_1, I, B, \perp \rangle$ would be unique up to similarities [Equation (3)]. If we now tried to define a relation S_2 on P_0 by means of the above relation T_2, i.e., $(a, b, c) \in S_2$ iff $(\varphi(a), \varphi(b), \varphi(c)) \in T_2$, S_2 would be ill-defined, because T_2 is not invariant under similarities. There would be a triple a, b, c and two equally good representations φ, φ' such that φ maps the triple into T_2 and φ' maps it outside T_2.

To put this another way, consider the problem of defining a relation S_1 on P_0 corresponding to relation T_1 above. First, we construct a homomorphism φ. This construction would begin by designating a point a_0 in P_0 to be mapped into the origin, $\varphi(a_0) = (0, 0)$, and another, distinct point a_1 to be mapped to $(0, 1)$. The rest of the construction would proceed, and φ would be uniquely defined in terms of P_0, P_1, I, B, $\perp$, a_0, and a_1. Then T_1, hence $S_1 = S(\varphi, T_1)$, would be defined. Finally, the fact that T_1 is invariant under similarities would entail that the same S_1 would be obtained from any other φ', hence, from any alternative choice of a_0 and a_1. Thus S_1 would be defined in terms of P_0, P_1, I, B, $\perp$ only. Moreover, S_1 would be invariant under any automorphism of $\mathscr{A}$, i.e., any mapping of P_0, P_1 onto themselves that preserves I, B, $\perp$. By contrast, T_2 leads to a definition of $S_2(\varphi, T_2)$ in terms of P_0, P_1, I, B, $\perp$, a_0, a_1, but this relation is not independent of a_0, a_1 since T_2 is not invariant under similarities. Therefore, $S_2(\varphi, T_2)$ cannot be defined in terms of P_0, P_1, I, B, $\perp$ alone; for if it could be, then a_0, a_1 would be eliminable from the definition and, hence, different choices of a_0, a_1 corresponding to different representations would lead to the same relation, which we know is false. Furthermore, $S_2(\varphi, T_2)$ is not invariant under automorphisms of $\mathscr{A}$. For, by way of contradiction, suppose it were. We know that for any distinct b_0, b_1 we can find an automorphism α with $\alpha(a_i) = b_i$, $i = 0, 1$. Using this automorphism we can translate each step of the definition of $S_2(\varphi, T_2)$ in terms of P_0, P_1, I, B, $\perp$, a_0, a_1 into a definition of $S_2(\varphi', T_2)$ in terms of P_0, P_1, I, B, $\perp$, b_0, b_1, and this would necessarily satisfy $(a, b, c) \in S_2(\varphi, T_2)$ iff $[\alpha(a), \alpha(b), \alpha(c)] \in S_2(\varphi', T_2)$. But if S_2 were invariant under α, then we would obtain $S_2(\varphi, T_2) = S_2(\varphi', T_2)$, which we know to be false.

To summarize, we say that T_2 is meaningless in Euclidean geometry because it is not definable in terms of the basic relations that constitute the subject matter of Euclidean geometry: incidence, betweenness, and perpendicularity. It is not preserved under similarities, which are the transformations for which the analytic representations of incidence, betweenness, and perpendicularity are preserved, and it has no correspondence to a relation S_2, on the abstract structure $\mathscr{A}$, which is invariant under automorphisms of $\mathscr{A}$. Relation T_1, by contrast, is meaningful in Euclidean geometry: the contraries of all the above equivalent statements hold. But T_1 is not meaningful in affine geometry. We pass from affine to Euclidean geometry by adding the relation $\perp$ to the abstract structure or by adding its representation by means of a Euclidean scalar product to the analytic structure Re^2. Using $\perp$ or $[\ ,\ ]$, we can define T_1 and the corresponding abstract relation S_1.

Problems of meaningfulness are severe in applications of geometry because whenever a Cartesian coordinate system is used, one tends to assume that all Euclidean relations, and only those, are defined and usable. One loses sight of the empirical relations actually represented by the Cartesian coordinates, which alone determine whether numerical relations are meaningful. A good example of this problem is the use of the projective plane as a chromaticity chart in color vision (see Section 15.4.2). Only empirical linear-dependence relations go into the construction of this chart, and scalar multiples are mapped into the same point; so, in fact, the projective group applies. Yet it is common for people who look at the chart to think about colors in terms of Euclidean distances on the page.

We have emphasized the measurement-theory point of view: we conceive of a geometry as a specific sort of relational structure and of other relations as meaningful if they can be defined in terms of given relations. From another point of view, the transformation group is the fundamental geometrical entity. Thus one is given a set A (perhaps Re^n or some more abstract structure such as an n-dimensional manifold) and a group Γ of transformations—one-to-one mappings of A onto itself. This defines a geometry: all relations on A that are invariant under Γ. The problem then is to character-ize such relations. In particular, one would wish to axiomatize the geometry by finding relations $S_1, \ldots, S_m$ that can be used as primitives. The S_i must be invariant under Γ, and Γ must be the largest group for which $S_1, \ldots, S_m$ are all invariant. Then any other S that is invariant under Γ can be defined in terms of $S_1, \ldots, S_m$.

This view was expressed very clearly by Klein in his Erlangen address of 1872 (see Klein, 1893—we have made minor changes in the quoted English translation). He argued that geometrical or spatial properties can be recog-nized by their invariance under transformations that themselves are given as primitives:

> For geometric properties are, from their very idea, independent of the position occupied in space by the configuration in question, of its absolute magnitude, and finally of the sense in which its parts are arranged. The properties of a configuration remain therefore unchanged by any motions of space, by transformation into similar configurations, by transformation into symmetrical configurations with regard to a plane (reflection), as well as by any combination of these transformations. The totality of all these transformations we designate as the *principal group* of space-transformations; *geometric properties are not changed by the transformations of the principal group*. And, conversely, *geometric properties are characterized by their remaining invariant under the transformations of the principal group*. For, if we regard space for the moment as immovable, etc., as a rigid manifold, then every figure has an individual character; of all the properties possessed by it as an individual, only the properly geometric ones are preserved in the transformations of the principal group. (p. 218)

Klein thus took the *similarity* transformations [Equation (3) for the plane] to be the primitive geometrical concept in Euclidean geometry, where Euclidean geometry in turn is regarded as a theory of our observations of the physical world. Instead of summarizing these physical observations as a relational structure, e.g., involving incidence, betweenness, and perpendicu-larity, he summarized them by a transformation group under which the observable relations are invariant. He quickly went on to formulate the mathematical problem of generalized geometry:

> As a generalization of geometry arises then the following comprehensive problem:
> *Given a manifold and a group of transformations of the same; to investigate the configura-tions belonging to the manifold with regard to such properties as are not altered by the transformations of the group.* (p. 218)

This approach has much to recommend it. From a purely mathematical standpoint, interesting and sometimes very deep problems arise in this way. As mentioned earlier, one particular group (the transformations that "stretch but do not tear," i.e., one-to-one bicontinuous transformations) poses particularly deep problems. Instances do occur when the group under which invariance is to be required is derived first and then other invariants are found later. An example is the Lorentz group, the transformations of four-dimensional affine space (interpreted as space-time) of special relativity, which are studied in some detail in Section 13.9.

From a measurement point of view, it will not do to specify the transformation group as a group acting in Re^n or in some Re^n manifold (a structure with local Re^n coordinates) because this already assumes that the measurement problem has been solved. One of the main goals, rather, is to start with a qualitative geometrical structure and show that it is isomorphic to an analytic geometry. Nonetheless, it is quite possible to do this using the transformation group as the basic qualitative structure. This is one of the principal achievements of the theory of topological groups. We do not pursue these ideas further in this work; we merely wish to relate the uniqueness and meaningfulness results for classical foundations of geometry, which are developed in these chapters, to the other main approaches to this subject.

12.2.6 Minkowski Geometry

In this and the next section, we briefly consider some alternatives to the Euclidean scalar product that can be used to define additional geometrical relations in an analytic affine or projective geometry. This section covers geometries in which the metric is derived from a *norm* that has some of the characteristic properties of the Euclidean norm, $\|x\| = (\Sigma x_i^2)^{1/2}$, but that need not come from a scalar product. As we shall indicate, these normed geometries are, in a sense, the only natural enrichments of a full affine geometry to a metric geometry. The next section deals with geometries that are only partly compatible with the affine structure, as well as with enrichment of the full projective structure to a metric geometry; these cases include the classical non-Euclidean geometries, hyperbolic and elliptic. The material in this section is based on Busemann (1955, Sect. 17).

DEFINITION 9. *Let* $\langle V, +, \cdot \rangle$ *be a vector space over* **Re**.

(i) *A* norm *is a function from* V *to* **Re** *such that the following properties hold*:
1. $\|x\| > 0$ *unless* $x = 0$.
2. $\|tx\| = |t|\,\|x\|$.
3. $\|x + y\| \leqslant \|x\| + \|y\|$.
A norm is strongly convex *iff* $<$ *holds in Axiom 3 except for* x, y *linearly dependent.*

(ii) *A* Minkowski geometry *is a structure consisting of an n-dimensional analytic ordered affine geometry over* Re, *together with a norm on the vector space. In such a geometry, the metric is defined by* $\delta(x, y) = \|x - y\|$.

The most familiar examples of norms are the $\mathscr{L}^p$ norms, given in Re^n by the equation (for $1 \leqslant p < \infty$)

$$\|x\|_p = \left[\sum_{i=1}^{n} |x_i|^p \right]^{1/p},$$

or, in certain vector spaces of integrable functions, by the analogous equation

$$\|f\| = \left[\int |f(u)|^p \, du \right]^{1/p}.$$

For $\mathscr{L}^p$ norms with $1 < p < \infty$, strong convexity holds; this result is the well-known *Minkowski inequality*. (For $p < 1$, the Minkowski inequality reverses, and so Axiom 3 fails.) The space $\langle \text{Re}^n, +, \cdot \rangle$, with its ordered affine structure and an $\mathscr{L}^p$ norm, is the standard example of n-dimensional Minkowski geometry. For $p = 2$, we have the Euclidean norm, whose special feature is that the function $(1/2)[\|x + y\|^2 - \|x\|^2 - \|y\|^2]$ is bilinear and, hence, can be taken as a scalar product; the analogous construction fails to be bilinear for $p \neq 2$. For $p \to \infty$, $\|x\|_p$ approaches $\sup\{|x_i| \,|\, 1 \leqslant i \leqslant n\}$, or $\|f\|_p$ approaches $\sup\{|f(u)|\}$; this also defines a norm, the $\mathscr{L}^\infty$ norm, or *supremum norm*, which is used to define uniform convergence of functions. The supremum norm, like the $\mathscr{L}^1$ norm, is not strongly convex.

Because of Axioms 2 and 3, any norm is a convex function, i.e.,

$$\|tx + (1 - t)y\| \leqslant t\|x\| + (1 - t)\|y\|, \qquad \text{for} \quad 0 < t < 1.$$

Conversely, any real-valued convex function on V is a norm provided that it satisfies Axioms 1 and 2. A norm cannot be strictly convex (inequality always holding for $0 < t < 1$ above) because Axiom 2 implies that $=$ holds for x, y linearly dependent; however, *strong* convexity asserts that this is the only exception to strict inequality. The geometrical importance of strong convexity will be pointed out in Section 12.3.1 on *G*-spaces.

Axioms 2 and 3 of Definition 9 show the compatibility of the norm with the underlying vector structure. This compatibility of the affine and the metric structures emerges more clearly in the following discussion, which also shows how to construct general Minkowski geometries, not limited to the $\mathscr{L}^p$ family.

The norm $\|x\|$ is also the distance $\delta(x, 0)$, and thus it gives a special role to the origin. This is removed by the metric, which by its definition is *translation-invariant*: $\delta(x + z, y + z) = \delta(x, y)$. That is, the translation group $\xi_z(x) = x + z$ is a subgroup of the automorphism group for any Minkowski metric geometry in that it preserves affine structure and distances.

A *sphere* with center $x^{(0)}$ and radius α is defined as $\{x|\delta(x, x^{(0)}) = \alpha\}$. Clearly, any sphere can be obtained from the *unit sphere* $\{x|\|x\| = \delta(x, 0) = 1\}$ by first multiplying by α and then translating by $x^{(0)}$; thus *all spheres are similar*. Minkowski geometries are in this sense homogeneous with respect to location and size transformations. Therefore, a Minkowski geometry is specified completely by designating *any convex and radially symmetric $(n - 1)$-dimensional surface in* Re^n as the sphere with radius 1 about its center of symmetry. If $x^{(0)}$ is the center of symmetry, then $\|x\|$ is defined by the property that $\|x\|^{-1}x + x^{(0)}$ lies on the surface in question. A Minkowski geometry is Euclidean if and only if the surface in question is an ellipsoid; if so, the ellipsoid's principal axes can be taken as the directions of an orthogonal coordinate system.

A Minkowski geometry has the further property that any affine line is isometric to one-dimensional Euclidean geometry. Specifically, consider the mapping

$$x(t) = x^{(0)} + \frac{t}{\|x^{(1)}\|}x^{(1)}$$

of Re onto the affine line with direction $x^{(1)}$ through $x^{(0)}$. This has the property that $\delta[x(t), x(u)] = |t - u|$ and $|t - u|$ is of course the Euclidean distance on Re^1.

This brings us to a point that could have been raised in our earlier section: although the congruence relation $\approx$ of point pairs in Euclidean geometry is not invariant under affinities, a partial congruence relation, limited to collinear quadruples, is invariant, i.e., *distance along a line is definable in affine geometry* over Re. Obviously, distances of parallel segments can also be compared. For a metric geometry to be fully compatible with the affine structure, its congruence relation should be an extension of this partial, affine-invariant congruence relation. The Minkowski geometries have this property. Busemann (1955) showed that the Minkowski geometries are the only ones that are compatible with affine structure in the sense defined above; in fact, he used a slightly weaker compatibility condition as a definition of Minkowski geometry and then went on to show that such geometries are characterized by convex, positive-homogeneous functions (norms).

A Minkowski geometry with a strongly convex norm has a perpendicularity relation for lines but it need not be symmetric. Line M is *perpendicular* to L if it intersects L at x and if for every y in M, x is the point of L closest to y. The reader can easily verify that in the $\mathscr{L}^4$ metric in Re^2, the perpendiculars to a line with slope ν have slope $-|\nu|^{-1/3}$, e.g., any line with slope 8 has as its perpendiculars lines with slopes $-1/2$, but the latter have as their perpendiculars lines with slope $\sqrt[3]{2}$. Blaschke's Theorem (Busemann, 1955, p. 103) states that in dimensions $n > 2$, any strongly convex Minkowski geometry with symmetric perpendicularity is Euclidean.

Finally, we turn to the uniqueness question for Minkowski geometry. Let V_1, V_2 be n-dimensional vector spaces over Re with norms $\| \ \|_1$, $\| \ \|_2$ and metrics δ_1, δ_2. An isomorphism is a mapping ξ of V_1 onto V_2 that preserves the affine relations and the metric, i.e., for the latter, $\delta_2[\xi(x), \xi(y)] = \delta_1(x, y)$. It is easy to show that $\xi' = \xi - \xi(0)$ preserves the norm, i.e., ξ is composed of a vector-space isomorphism that preserves the norm, plus a translation.

By taking fixed bases in V_i and considering the mapping ξ' as a mapping of coordinates relative to the bases, we reduce the uniqueness problem to that of finding all norm-preserving linear transformations of Re^n onto itself, for a given norm $\| \ \|$ on Re^n. The solution to this problem depends on the particular norm, but two general facts can be mentioned. The identity $x \rightarrow x$ and the reflection in the origin $x \rightarrow -x$ are the only such mappings in common to all norms; and *all such norm-preserving mappings are Euclidean orthogonalities*. These facts can be proved in a way that is quite informative, and so we shall sketch the proof.

Let B denote the unit ball $\{x | \|x\| \leqslant 1\}$. By Loewner's Theorem (Busemann, 1955, p. 90) there exists a *unique* ellipsoid E of minimum volume containing B. Therefore the mapping is an orthogonality.

To prove that any particular orthogonality, except $x \rightarrow x$ and $x \rightarrow -x$, is not norm-preserving for some norm, we use the fact that any orthogonality can be expressed as a product of reflections in various hyperplanes through the origin. In the plane, for example, each rotation ξ is the product of two reflections ρ_1, ρ_2 in lines through 0, and if the rotation is not 180° (i.e., not $x \rightarrow -x$), then the two lines of symmetry are not orthogonal. We can then construct a convex curve that has the first line, say L_1, as an axis of symmetry, plus radial symmetry about 0, but for which the second line is not an axis of symmetry. Define $\| \ \|$ to have this curve as unit sphere. Then ρ_1 is norm-preserving and ρ_2 is not; hence ξ is not, otherwise, $\rho_2 = \rho_1^{-1}\xi$ would have to be.

For the particular case of the $\mathscr{L}^p$ norms, it is obvious that we can permute the n principal axes; i.e., if σ is a one-to-one mapping of

$(1, \ldots, n)$ onto itself, then $x_i' = \pm x_{\sigma(i)}$, $i = 1, \ldots, n$ preserves the norm. These are, in fact, the only norm-preserving mappings.

Thus, except for the Euclidean case, uniqueness for a representation by a Minkowski geometry is apt to be quite sharp; for the $\mathscr{L}^p$ geometries, it amounts to permutation and reflection of axes plus changing the origin and perhaps the overall unit of distance (i.e., multiplying by a positive scalar). A Minkowski geometry that admits all rotations is necessarily Euclidean.

12.2.7 General Projective Metrics

In Euclidean geometry we derive the metric from a norm that is derived from a scalar product; in the Minkowski geometries we derive the metric from a more general sort of norm, and in this section we use more general metrics that do not come from norms. We consider two kinds of geometry. *Straight spaces* are those in which the metric is defined in an open subset X of an n-dimensional affine geometry, and the portion of any affine line that intersects X is isometric to a Euclidean line. *Spaces of elliptic type* are those in which the metric is defined on the entire n-dimensional projective geometry, and each projective line is isometric to a Euclidean circle. (Hamel's Theorem asserts the impossibility of hybrid types, in which part but not all of a "hyperplane at ∞" is omitted from a projective geometry so that there could be some projective and some affine metric lines.)

We give just the definitions of these two types, together with a few general remarks, and then provide examples of each: the *Hilbert geometries*, one of which can be defined in the interior of any closed convex surface in affine space, and which include the hyperbolic geometries, defined in the interiors of ellipsoids; and the *classical elliptic geometry*, defined on the entire projective space. The material presented in this section is based on the books by Busemann and Kelly (1953) and Busemann (1955).

DEFINITION 10.

(i) *Let X be a set. A real-valued function δ on $X \times X$ is a* metric *iff the following properties hold*:
   1. $\delta(x, y) > 0$ *unless $x = y$ and $\delta(x, x) = 0$.*
   2. $\delta(x, y) = \delta(y, x)$.
   3. $\delta(x, y) + \delta(y, z) \geqslant \delta(x, z)$.

(ii) *Let δ be a metric on X. A subset L of X is a* metric straight line *iff there exists a function f from Re onto L such that $\delta(f(t), f(u)) = |t - u|$. A subset C of X is a* metric great circle *iff there exists $r > 0$ and a function f from $[0, 2\pi r]$ onto C such that $\delta(f(t), f(u)) = \inf\{|t - u|, 2\pi r - |t - u|\}$.*

(iii) *Let X be an open subset of Re^n and δ be a metric on X. The pair $\langle X, \delta \rangle$ is a* straight projective metric space *iff the following properties hold*:
1. *If x, y, z are not collinear, then $\delta(x, y) + \delta(y, z) > \delta(x, z)$.*
2. *If L is an affine line that intersects X, $L \cap X$ is a metric straight line.*

(iv) *Let P^n be the n-dimensional projective space over Re and δ be a metric on P^n. The pair $\langle P^n, \delta \rangle$ is a* projective metric space of elliptic type *iff the following properties hold*:
1. *If x, y, z are not collinear, then $\delta(x, y) + \delta(y, z) > \delta(x, z)$.*
2. *If C is a projective line, then it is a metric great circle.*

Part (i) of this definition is very well known. The axioms of a general metric are very weak; for example, they can be satisfied trivially on any set X by setting $\delta(x, x) = 0$ and $\delta(x, y) = 1$ for $y \neq x$. Nonetheless, there exist geometric theories in which each of them is weakened. If $>$ is replaced by $\geq$ in Axiom 1, we obtain a *pseudometric*. A single pseudometric is of little interest because $\delta(x, y) = 0$ is an equivalence relation, and one can replace every x by its equivalence class to obtain a metric. However, families of pseudometrics can be of interest. More generally, a pseudometric can be of interest when there is another structure not preserved by the equivalence classes. Some work on *nonsymmetric* metrics, in which Axiom 2 is dropped (and Axiom 3 is strengthened by symmetrizing), has been done but is not used in this volume.

The last axiom is the triangle inequality. One way to think about this is to consider $\delta(x, y)$ as a function of y. It cannot change too rapidly: the difference between $\delta(x, y)$ and $\delta(x, z)$ cannot exceed the distance between y and z. If this is violated, then straight-line segments lose their special nature (in which the difference between $\delta(x, y)$ and $\delta(x, z)$ is as large as possible), and geometry changes its character. Sometimes the *ultrametric* inequality $\delta(x, z) \leq \sup\{\delta(x, y), \delta(y, z)\}$ is used as a replacement to develop a very different sort of theory. Section 12.3 is devoted to metrics that do not necessarily have any affine or projective structure.

To get a better appreciation of metric straight lines and of straight spaces, consider the following example. Let X be the interior of the unit circle in Re^2, and let φ be a mapping of X onto Re^2 obtained by stretching each radius into an infinite ray from the origin; specifically, set

$$\varphi_i(x_1, x_2) = \frac{x_i}{\left[1 - (x_1^2 + x_2^2)\right]^{1/2}}, \qquad i = 1, 2.$$

Define a metric δ on X by setting $\delta(x, y) = \|\varphi(x) - \varphi(y)\|$, the

Euclidean distance between $\varphi(x)$ and $\varphi(y)$. By definition the mapping φ is an isometry, and so in some sense $\langle X, \delta \rangle$ is identical to the Euclidean geometry of the plane. But considered as an open subset of Re^2, $\langle X, \delta \rangle$ does not satisfy (iii) of Definition 10. Each *diameter* of X is a metric straight line since φ simply extends it to a full line through the origin, and φ^{-1} plays the role of the function f required in (ii). But affine straight lines in Re^2 that do not pass through the origin intersect X in chords of the circle; and since φ maps these chords onto curves in Re^2, the chords are not metric straight lines in $\langle X, \delta \rangle$. We can further note that any strongly convex Minkowski geometry does satisfy (iii), with $X = \mathrm{Re}^n$. The $\mathscr{L}^1$ and $\mathscr{L}^\infty$ geometries are not strongly convex, however, and violate property 1 of (iii).

Finally, in (iv), note that the elements of $X = P^n$ are homogeneous coordinates, or lines, through the origin in Re^{n+1}. Thus it is very natural to consider distances between elements of a projective line (i.e., distances between coplanar lines through the origin in Re^{n+1}) as arcs of a circle.

When X is restricted to a proper subset of Re^n, the incidence relation of linear varieties changes its character drastically. It is no longer true, for example, that any two coplanar lines either intersect or are affine parallels; if they intersect outside X, then from the standpoint of the geometry of X, they do not intersect. Euclid's parallel postulate, valid in affine geometry, breaks down: through a point outside a given line, there can be infinitely many lines that do not intersect the given line. And since each of these lines is isometric to a full Euclidean line, one can travel indefinitely large distances (in terms of the metric δ) without reaching an intersection. Clearly the distances along an affine line do not preserve the partial congruence relation defined within affine geometry; so straight projective metric spaces as subsets of Re^n cannot be fully compatible with the affine structure of Re^n. Indeed, their automorphisms need not come from affinities on Re^n, but what can be shown in justification of the term *projective metric* is that *any automorphism can be obtained from a projectivity* of Re^{n+1}. Thus we are guaranteed a reasonably sharp and manageable uniqueness theory for representations in any projective metric space.

EXAMPLE 1.   *Hilbert geometry.*   Let X be the interior of a closed convex surface S in Re^n. For any distinct x, y in S, the affine line through x, y intersects S in two points v, w. Define

$$\delta(x, y) = \frac{K}{2} \left| \log R(x, y, v, w) \right| = \frac{K}{2} \left| \log \frac{\|x - v\| / \|x - w\|}{\|y - v\| / \|y - w\|} \right|,$$

where R is the cross ratio (Section 12.2.3), $\| \ \|$ is the Euclidean norm, and K is a constant that permits identification with the metrics of classical non-Euclidean geometries (examples are given below). Define $\delta(x, x) = 0$. It is easy to see that δ satisfies Axioms 1 and 2 of Definition 10(i) and that on any line iff $x|y|z$, then $\delta(x, y) + \delta(y, z) = \delta(x, z)$. Furthermore, $\delta(x, y) \rightarrow \infty$ as x or y approaches one of the intersection points v or w; hence it is quite easy to construct a mapping f from Re to the segment from v to w such that $\delta(f(t), f(u)) = |t - u|$. The triangle inequality [Axiom 3 of Definition 10(i)] is not obvious, but its proof is not too hard (Busemann and Kelly, 1953, p. 158) and includes the fact that strict inequality holds if x, y, z are not collinear. Hence Definition 10(iii) is satisfied, and $\langle X, \delta \rangle$ is a straight projective metric space.

The automorphisms of $\langle X, \delta \rangle$ are obtained by regarding Re^n as part of n-dimensional projective space and considering the mappings induced by projectivities (not necessarily affinities). Any projectivity that maps the entire set X onto itself induces an automorphism of $\langle X, \delta \rangle$, and every automorphism is obtained thus.

Just as a Minkowski geometry can be defined in Re^n with any (centrally symmetric) closed convex surface as unit sphere, so a Hilbert geometry can be defined in the interior of any closed convex surface; and the automorphisms are analogous also, since the Minkowskian automorphisms are orthogonalities that map the unit sphere onto itself (plus translations), and the Hilbertian automorphisms are projectivities that map the interior of the surface onto itself. In both cases, the automorphism group can be severely limited; and in both, the special case of an ellipsoid yields the richest automorphism group and the classical geometry. In the Minkowskian case, an ellipsoid yields Euclidean geometry; here, the specialization to an ellipsoid gives *hyperbolic* (Lobachevskian) geometry, one of the two classic non-Euclidean geometries.

In the special case of plane hyperbolic geometry, it is convenient to take S as the circle, $x_1^2 + x_2^2 = 1$. The distance $\delta(x, y)$, defined by the logarithm of the cross ratio, takes the simplified form

$$\delta(x, y) = \frac{K}{2} \cosh^{-1} \left[\frac{1 - x_1 y_1 - x_2 y_2}{(1 - x_1^2 - x_2^2)^{1/2}(1 - y_1^2 - y_2^2)^{1/2}} \right].$$

It is useful in plane hyperbolic geometry to define a bilinear function on Re^3: $H(x, y) = x_1 y_1 + x_2 y_2 - x_3 y_3$. This is symmetric but not a Euclidean scalar product since it is not positive definite. If we identify (x_1, x_2) in Re^2 with its nonhomogeneous projective coordinate in the plane

$x_3 = 1$, i.e., as $(x_1, x_2, 1)$, then we get

$$\delta(x, y) = \frac{K}{2} \cosh^{-1}\left[\frac{-H(x, y)}{|H(x, x)|^{1/2}|H(y, y)|^{1/2}}\right].$$

The projectivities that map the circle $x_1^2 + x_2^2 = 1$ onto itself, and hence constitute automorphisms, can be described in terms of H: they are the matrices (α_{ij}) whose rows $\alpha^{(i)} = (\alpha_{i1}, \alpha_{i2}, \alpha_{i3})$ satisfy

$$H(\alpha^{(1)}, \alpha^{(1)}) = H(\alpha^{(2)}, \alpha^{(2)}) = -H(\alpha^{(3)}, \alpha^{(3)}) = 1,$$
$$H(\alpha^{(i)}, \alpha^{(j)}) = 0 \quad \text{if} \quad i \neq j.$$

This is similar to the group of orthogonalities in Re^3 except for the substitution of H for the Euclidean scalar product and the normalization to -1 in the third row.

These transformations can be written as mappings of the inhomogeneous plane circle, $x_1^2 + x_2^2 = 1$, given as follows:

$$x_1' = x_1 \cos\theta \pm x_2 \sin\theta$$
$$x_2' = x_1 \sin\theta \mp x_2 \cos\theta$$
$$x_3' = 1,$$

which are the rotations

$$x_1' = \frac{x_1 \sec\varphi + \tan\varphi}{\pm x_1 \tan\varphi \pm \sec\varphi}$$
$$x_2' = \frac{x_2}{\pm x_1 \tan\varphi \pm \sec\varphi}$$
$$x_3' = 1,$$

which are mappings that send the line $x_2 = 0$ into itself; and all products of these mappings.

EXAMPLE 2.   *Elliptic geometry.*   Let $[x, y]$ and $\|x\|$ be the usual Euclidean scalar product and norm in Re^{n+1}; define δ in P^n, the n-dimensional projective geometry over Re, by

$$\delta(\{tx\}, \{ty\}) = K \cos^{-1}\frac{|[x, y]|}{\|x\| \, \|y\|}.$$

That is, the distance between two projective points, thought of as lines through 0 in Re^{n+1}, is just K times the smaller angle between the lines. This can be shown to be a metric and to satisfy (iv) of Definition 10. The automorphisms are precisely the orthogonalities of Re^{n+1}.

12.3 METRIC REPRESENTATIONS

The main goals in metric geometry are to define various geometrical objects and to develop theory in terms of the distance function or metric. The simplest geometrical objects to define are the spheres: loci of points at a fixed distance from a given point. Linear objects are harder to define and may not exist or may not have all the properties expected of linear objects in vector representations.

Because of the very sparseness of required structure, a metric geometry can be an attractive "numerical" structure for representation theorems. The natural empirical structure to be represented by a metric is a qualitative proximity structure, i.e., a set A with a quaternary relation on A, interpreted as an ordering of pairs with respect to their distance or proximity. Such representations are studied in Chapter 14.

To be of real interest, a metric geometry must possess some sort of regularity. The kind of regularity that has been most studied in the mathematical literature is the existence of an additive segment (i.e., a curve along which distances are additive) joining every pair of points.

The prototype of a metric geometry that lacks vector structure but which possesses additive segments is the geometry of a two-dimensional curved surface embedded in three-dimensional Euclidean space. Major questions about such a surface concern intrinsic concepts of curvature and distance (i.e., the shortest paths on the surface between any two points), for example, great circle arcs on a sphere or an arc of a helix between two points on a cylinder. One can define a metric geometry on such a surface by setting $\delta(x, y)$ equal to the length of the shortest path from x to y; the shortest path itself then constitutes a metric straight-line segment since arc lengths are additive along it. When the surface in question is *smooth* [for example, when it is described by an equation in Re^3 of form $f(x_1, x_2, x_3) = 0$, where f has continuous third-partial derivatives], we have an example of Riemannian geometry.

The critical feature of this prototypical example is the fact that any sufficiently small region can be approximated arbitrarily well by a tangent plane in Euclidean 3-space so that local geometrical properties are Euclidean. Thus, in the region of the surface near any particular point, we can introduce a two-dimensional coordinate system, and if we stay close enough to the point in question, the geometry is nearly that of the Euclidean plane. Abstracting this feature leads to the concept of geometry on a manifold, a structure in which a local coordinate system can be introduced around any point, and the local geometry is approximated by an analytic geometry in those coordinates. Such geometrical manifolds still permit introduction of the global concept of shortest path between two points and thus form a subtype of metrics with additive segments joining pairs of points. The study of such geometrical manifolds is called *differential geometry* and is undoubtedly the dominant topic in the modern research literature on geometry.

The first two subsections of this section develop the above ideas in greater detail. Dubrovin *et al* (1984) is a good reference. Once we leave these topics, we are in more or less unchartered waters; yet there are some other interesting metrics that have been proposed as representations for qualitative proximity structures. We describe this heterogeneous material in the final subsection.

12.3.1 General Metrics with Geodesics

Many aspects of metric geometry are well illustrated by the properties of a cylindrical surface, so we shall start by discussing this example.

Example of cylindrical surface. Specifically, let X be the surface of a circular cylinder in Re^3 with its axis the x_3-axis and radius 1; i.e., $X = \{x | x_1^2 + x_2^2 = 1, x_3 \text{ arbitrary}\}$. In the remainder of this example it will be understood that $x, y, \dots$ denote elements of X.

Figure 2 depicts a vertical strip of width 2π in the plane. This strip could be cut out and wrapped tightly around the cylinder X, the two sides of the

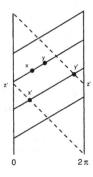

FIGURE 2. Helices on a cylinder. Surface of a right circular cylinder of unit radius, cut open and flattened into the Euclidean plane. The sloping solid and dashed lines show parts of two helices that wind around the cylinder with shallow or steep pitch respectively.

strip just meeting in a vertical line on the cylinder. A horizontal line in the
strip goes onto a (horizontal) circle on X. Each system of diagonal lines
shown in the figure is mapped onto a helix in X. This figure shows that
points x, y can be joined by a shortest curve on the cylinder, namely, the
arc of the helix depicted by the solid line-segment joining x and y in the
figure. Points x' and y', however, have a horizontal distance of more than π
in the figure; the shortest curve joining them on the cylinder is not the arc
of the same helix but is the arc of the dashed-line helix through the point z'
(which is doubly represented at the edges of the strip in the figure).

Clearly, the lengths of shortest curves on the cylinder are the same as
lengths of line segments in a strip on the plane. If we measure distance
$\delta(x, y)$ on X, not along a chord cutting inside the cylinder in Re^3 but by
the length of the shortest curve in X joining x and y, then locally, at least,
the cylinder is geometrically equivalent to the Euclidean plane. The helices
on a cylinder (as well as the circles and vertical lines) are *geodesic* curves,
curves that in any local region are metrically equivalent to the Euclidean
line.

Curves and geodesics. Motivated by this example, we want to consider in
a general metric space $\langle X, \delta \rangle$ the concepts of curve, arc length, and
geodesic curve and to assume that the metric δ on X is the intrinsic metric,
i.e., the arc length of the shortest curve joining the two points. These ideas
are developed rigorously in Chapter 1 of Busemann (1955); we give a brief
version here of the essential definitions.

The definition of curve is tricky, however, because of the possibility of
pathological curves that may intersect or retrace themselves, multiply, and
even fill up a higher-dimensional region. Thus we do not think of a curve as
a subset of X but rather as an ordered tracing of a path on X. That is, we
define a *parametric curve* by means of a definite mapping from a one-
dimensional parameter space—an interval in Re—into the space X. Such a
curve is required to be continuous with respect to the metric on X. We then
define the *length* of a parametric curve; when this is finite, the curve is
rectifiable and can be parametrized by the arc length from one endpoint. A
rectifiable curve, geometrically, is an equivalence class of parametrizations,
all having the same arc-length parametrization. The culmination of all this
is the theorem that assures us that any two points that can be joined by
some rectifiable curve can be joined by a shortest curve. This is valid for
any metric space that satisfies the topological assumption that any closed
and bounded set is compact (explicit definitions of these concepts are given
below). In our cylinder example, the existence of shortest curves was
obvious because they exist in the plane and are carried from the plane onto
the cylinder by "wrapping" a strip of the plane around the cylinder. But the
topological theorem just cited shows that shortest curves exist much more

generally. The existence of shortest curves, in turn, motivates the definition of metrics in which the shortest curves are isometric to Euclidean line segments and of metrics in which, in addition, any additive segment can be extended to a geodesic curve.

DEFINITION 11.

(i) *Let δ be a metric on X (Definition 10). A* parametric curve *in $\langle X, \delta \rangle$ is a function f that maps a closed interval $[\alpha, \beta] \subset$ Re *into X and which is* continuous.[2] *The parametric curve f is said to* join *$f(\alpha)$ and $f(\beta)$, its endpoints.*

(ii) *Let f be a parametric curve in $\langle X, \delta \rangle$ with domain $[\alpha, \beta]$. Its* length $L(f)$, *is the supremum of the sums*

$$\sum_{i=1}^{r} \delta[f(\alpha_{i-1}), f(\alpha_i)]$$

over all finite sequences $\alpha_0, \alpha_1, \ldots, \alpha_r$ such that

$$\alpha = \alpha_0 \leqslant \alpha_1 \leqslant \cdots \leqslant \alpha_r = \beta.$$

If $L(f) < \infty$, then f is rectifiable.

(iii) *Let f be a rectifiable parametric curve in $\langle X, \delta \rangle$, with domain $[\alpha, \beta]$ and length $L(f) = l$. The* arc-length parametrization *of f is the function g from $[0, l]$ into X defined by[3] $g(\gamma) = f(\tau)$ iff $L(f_\tau) = \gamma$, where f_τ is the rectifiable parametric curve obtained by restricting f to the interval $[\alpha, \tau]$.*

(iv) *Two rectifiable parametric curves are* equivalent *iff their arc-length parametrizations coincide. A* rectifiable curve *is an equivalence class of rectifiable parametric curves.*

(v) *A rectifiable curve is a* segment *iff its length is equal to the distance between its endpoints.*

(vi) *A function g from Re into X is a* geodesic *iff its restriction to every sufficiently small interval about any $\alpha \in$ Re is a segment.*

[2] Topological concepts in general metric spaces, such as continuity, may be unfamiliar to some readers, so we provide definitions in footnotes throughout this section. The concepts involved will be seen to be closely analogous to the usual definitions in Re or Re^n.

A function f mapping $[\alpha, \beta]$ into X is *continuous* iff for any $t \in [\alpha, \beta]$ and any $\epsilon > 0$, there exists $\delta > 0$ such that if $u \in [\alpha, \beta]$ and $|t - u| < \delta$, then $\delta(f(t), f(u)) < \epsilon$.

[3] That is, $g(\gamma)$ is the point $x = f(\tau)$ in X such that the curve f, restricted to the subinterval $[\alpha, \tau]$, has length γ.

We have already commented on the need to think of a curve as a definite tracing of a path, i.e., as a parametrized set of points rather than merely as a subset of X. The definition of length generalizes the usual definition for a curve in a higher-dimensional Euclidean space. The curve is approximated by a sum of "chordal" distances $\delta(x^{(i-1)}, x^{(i)})$, where $x^{(i)}$ is the endpoint of the ith chord. Because of the triangle inequality, if the approximation is refined by adding intermediate endpoints, the sum can only remain constant or increase; the supremum over arbitrarily fine approximations is the curve's length. For a *segment* [part (v) of the definition], additivity of successive distances is valid, and so the "coarsest approximation"—just the two endpoints—already gives the curve's length.

The example of a helix on a cylinder shows that a geodesic need not be a segment over arbitrary portions of Re. The helix $g(\theta) = (\cos\theta, \sin\theta, \theta\alpha)$, through $(0, 1, 0)$ with pitch α, is a segment in any interval of length $\leqslant \pi$.

Part (iii) of this definition requires some lemmas: the function g must be shown to be well defined and continuous on the interval $[0, l]$. That is, for each γ between 0 and l, there must be at least one value of τ such that the length of f from α to τ is γ; and $f(\tau)$ must be uniquely determined and vary continuously with γ. This is all fairly straightforward; see Busemann (1955, pp. 19–24).

An important result is the following.

THEOREM 5. *Let δ be a metric on X, and suppose that every closed and bounded subset of X is compact.*[4] *If there exists a rectifiable curve with*

---

[4]An open sphere of a metric space $\langle X, \delta \rangle$ with center c and radius r is the set $S(c, r)$ of points of X such that

$$S(c, r) = \{ x \mid \delta(c, x) < r \}.$$

A subset Y of X is *open* iff each y in Y is the center of an open sphere that is contained in Y, i.e., there is a positive number r_y such that

$$S(y, r_y) \subseteq Y.$$

A subset Y of X is *closed* iff for any x in the complement of Y, $X - Y$, there is a positive number r_x such that

$$S(x, r_x) \subseteq X - Y.$$

Under this definition both X and the empty set are closed. A subset Y of X is *bounded* if $\sup\{\delta(x, y) \mid x, y \in Y\}$ is finite; and Y is *compact* if for any infinite sequence $y_1, y_2, \ldots$ in Y there exists $y \in Y$ such that for every $\epsilon > 0$, $\delta(y_i, y) < \epsilon$ for infinitely many i. The assumption that every closed, bounded subset is compact is called *finite compactness*. It is true in any finite-dimensional Euclidean space and in most other geometrical settings. It is false in infinite-dimensional metric spaces.

endpoints x, y in X, then there exists a curve with minimum length. If every pair of points can be joined by a rectifiable curve, let $\tilde{\delta}(x, y)$ *denote the minimum length of curves joining x and y. Then* $\tilde{\delta}$ *is a metric on X; every rectifiable curve in* $\langle X, \delta \rangle$ *is also a rectifiable curve in* $\langle X, \tilde{\delta} \rangle$, *and conversely; and the arc lengths are equal for* δ *and* $\tilde{\delta}$. *For* $\langle X, \tilde{\delta} \rangle$, *any curve of minimal length is a segment, and the arc-length parametrization of the curve of minimum length from x to y is an isometry from* $[0, \tilde{\delta}(x, y)]$ *into X.*

Thus, if we replace δ by the arc length $\tilde{\delta}$ of the shortest join, then the shortest curve from x to y satisfies

$$\tilde{\delta}[g(\gamma), g(\gamma')] = |\gamma - \gamma'|.$$

Additive segments and G-spaces. For many purposes, then, we can restrict attention to metric spaces in which every pair of points can be joined by an additive segment. The following definition deals with this class of metrics and with the subclass in which segments can be prolonged to geodesic curves.

DEFINITION 12.   *Let* δ *be a metric on X.*

(i) $\langle X, \delta \rangle$ *is M-convex*[5] *iff for any distinct x, z* $\in$ *X, there exists y* $\in$ *X distinct from x and z such that* $\delta(x, z) = \delta(x, y) + \delta(y, z)$.

(ii) $\langle X, \delta \rangle$ *is a metric space with additive segments iff for any x, y* $\in$ *X there is a segment joining them.*

(iii) $\langle X, \delta \rangle$ *is a G-space iff the following conditions hold.*
    1. $\langle X, \delta \rangle$ *is a metric space with additive segments.*
    2.   (*Uniqueness of Prolongation*)   *If g is an isometric mapping of* $[0, \alpha]$ *into X (i.e., the arc-length parametrization of a segment) and* $\beta > \alpha > 0$, *then there is at most one isometry of* $[0, \beta]$ *into X that coincides with g on* $[0, \alpha]$.
    3.   (*Local Prolongability*)   *For each w* $\in$ *X there is a radius* $\rho(w) > 0$ *such that for any x, y whose distances from w are* $< \rho(w)$, *any segment joining x, y can be prolonged.*
    4.   (*Finite Compactness*)   *Any closed, bounded subset of X is compact.*[6]

The axioms in (iii) are essentially the ones given by Busemann (1955, p. 37) except that we formulate them to emphasize segments whereas Busemann's formulation uses only additive point triples, i.e., triples x, y, z

[5] M stands for Karl Menger.
[6] See footnote for Theorem 5, p. 55.

of distinct points satisfying $\delta(x, z) = \delta(x, y) + \delta(y, z)$. Finite compactness (iii, Axiom 4) and M-convexity imply the existence of additive segments; thus Axiom 1 can be weakened to M-convexity in the presence of Axiom 4. Axioms 2 and 3 can be weakened to assert that if $x \neq y$, then for any $\gamma > 0$ there is at most one solution z to the equations

$$\delta(x, z) = \delta(x, y) + \delta(y, z)$$
$$\delta(y, z) = \gamma;$$

and for x, y sufficiently close to any point w, and γ sufficiently small, a solution z exists.

Numerous general theorems about G-spaces are found in Busemann's book, together with detailed development of the geometry of special types of G-spaces. The most important general theorem is the *prolongability of any segment to a unique geodesic*.

In Chapter 14, we establish representation theorems in which the representing structure is a metric space with additive segments or a G-space. Since these theorems do not involve any coordinate systems, the uniqueness results are far less detailed than in the cases for which the representing geometry is an analytic one. Essentially, the uniqueness involves arbitrary isometries of one G-space on another plus changes of the unit distance, $\delta' = \alpha\delta$. A detailed characterization of G-space isometries is not possible at this level of generality; such results can be expected only for special cases. For example, all the projective metric spaces in the previous section are G-spaces. Chapter 6 of Busemann's book studies the special properties of G-spaces with groups of isometries comparable in richness to those of Euclidean space.

The next section considers G-spaces satisfying an important property of mobility, which we use later in the study of visual spaces in Chapter 13.

12.3.2 Elementary Spaces and the Helmholtz–Lie Problem

Riemann's famous lecture of 1854 (published in 1866–1867), "Ueber die Hypothesen, welche der Geometrie zu Grunde liegen," was responded to by Helmholtz (1868) more than a decade later in an equally famous paper, "Ueber die Thatsachen, die der Geometrie zu Grunde liegen." Helmholtz argued that although arbitrary Riemannian spaces are conceivable, actual *physical* space has as an essential feature the free mobility of solid (i.e., rigid) bodies.

Adapting the concept of motion [Definition 8(ii)] to this more general setting, we say that a mapping of φ of A onto itself is a *motion* or an *isometry* if it preserves distance, that is, $\delta[\varphi(x), \varphi(y)] = \delta(x, y)$. We pointed out in Section 2.4 that free mobility of a rigid object within space could be expressed in terms of existence of an isometry between any two congruent subsets of the space. Helmholtz argued that such free mobility could occur only in a very few Riemannian spaces. Helmholtz based his analysis on four axioms, which we describe informally following Freudenthal (1965). The first axiom asserts that space is an n-dimensional manifold with differentiability properties. The second axiom asserts there is a metric with motions as isometric transformations. The third axiom asserts the free mobility of solid bodies, which means that if φ is an isometric mapping of a set B of points onto a set B' (in the same space), then φ can be extended to a motion of the whole space. The fourth axiom requires that the motion should be periodic (and not spiraling). This is often called the *monodrony axiom*.

Helmholtz claimed to have proved that the only spaces satisfying his four axioms are the Euclidean, hyperbolic, and spherical spaces. Sophus Lie (1886) noticed a gap in Helmholtz's proof. Lie strengthened the axioms and solved the problem. Some years later, Weyl (1923) weakened Lie's assumptions. The details of the many subsequent contributions to the problem of weakening the axioms and essentially retaining Helmholtz's solution are to be found in Busemann (1955, Sect. 48) and Freudenthal (1965). We outline here the solution under unnecessarily strong assumptions given by Busemann (1955) for G-spaces, which were defined in the preceding section. The basic aim of the modern work is to eliminate differentiability assumptions, which are extraneous to the problem of characterizing the spaces that have free mobility of solid bodies. The spaces having this property, together with properties characteristic of G-spaces, are often called *elementary* spaces. We note that double elliptic spaces and spherical spaces are the same. Single elliptic spaces treat a pair of antipodal points as a single point.

THEOREM 6. *Let $\mathfrak{A} = \langle A, \delta \rangle$ be a G-space such that for any four points a, a', b, c with $\delta(a, b) = \delta(a', b)$ and $\delta(a, c) = \delta(a', c)$ there is a motion φ of $\mathfrak{A}$ with the property that $\varphi(b) = b$, $\varphi(c) = c$, and $\varphi(a) = a'$. Then the space $\mathfrak{A}$ is elementary, i.e., $\mathfrak{A}$ is Euclidean, hyperbolic, or spherical.*

An intuitively even more appealing result is obtained by Wang (1951) for spaces of dimensions 2 or 3, or indeed for n odd.

THEOREM 7. *Let $\mathfrak{A} = \langle A, \delta \rangle$ be a G-space of dimension 2, 3, or any odd number such that for any four points, a, b, a', b' with $\delta(a, b) = \delta(a', b')$ there is a motion φ of $\mathfrak{A}$ with $\varphi(a) = a'$ and $\varphi(b) = b'$. Then $\mathfrak{A}$ is Euclidean, hyperbolic, spherical, or single elliptic.*

The motions required in this theorem seem to be the natural ones to expect. The property formulated is called *pairwise transitivity* of the group of motions. It is worth noting that the explicit invariance of rigid bodies is not required in the hypothesis of the theorem. The dimensional restriction is due to the fact that for even dimensions greater than two, there are G-spaces with the group of motions being pairwise transitive but the spaces are not elementary in the sense of Theorem 7. (For examples, see Busemann, 1955, Sect. 53.)

Projected motions and visual shape perception. Because we live in a world that (to a high degree of approximation) has freely mobile rigid objects, it is natural to assume that a change that could have been generated by an isometric transformation was in fact so generated. Such an assumption is computationally useful when a three-dimensional transformation is projected onto a two-dimensional receptive surface, as in human or in robot vision. Ullman and Fremlin (see Ullman, 1979) proved, for example, that any three distinct orthographic projections of four noncoplanar points in a metrically rigid configuration are sufficient to determine that configuration. Ullman's general postulate, that any projected transformation that has a unique intepretation as a rigid body moving in space is to be interpreted as such, is useful for computing shape, but is probably too strong for human shape perception (Hochberg, 1988).

12.3.3 Riemannian Metrics

This section presents an introduction to Riemannian geometry. The simplest examples are associated with two-dimensional surfaces in three-dimensional space, for example, geometry on a sphere or on a saddle-shaped surface. On such surfaces, any two points can be connected by a shortest curve, and the distance between the two points is *defined* to be the length of that curve. Since the length of the shortest curve from x to z is the sum of the lengths from x to y and from y to z, where y is any intermediate point on the curve, such a curve is an additive segment. Thus, the geometry is that of a metric with additive segments, and in fact, it can be shown to be a G-space (Definition 12). However, it is a special kind of G-space, one whose geometry, in any sufficiently small region, is approximately Euclidean. This approximation stems from the fact that at any point on the surface, there is a Euclidean plane tangent to the surface. The geometry near that point is well approximated by the geometry of the tangent plane. This is a familiar fact, since we use plane Euclidean geometry very successfully locally on the surface of our approximately spherical planet.

The first part of this section sketches the geometry of two-dimensional surfaces, first developed by Gauss. We include definitions of the first and second fundamental forms and the Gaussian curvature. Later, we indicate how to extend the idea of a local Euclidean approximation to metric spaces more general than two-dimensional surfaces. Such spaces are called Riemannian manifolds.[7]

Parametrized surfaces in Re^3. We represent a surface (or portion of a surface) by a vector equation

$$x = x(u), \tag{1}$$

where $x = (x_1, x_2, x_3)$ and $u = (u_1, u_2)$. The parameter vector u is restricted to some subset of Re^2, and Equation (1) comprises three functions, one for each component x_i.

For example, a sphere of radius R can be represented using latitude u_1 and longitude u_2 as parameters:

$$x_1 = R \cos u_1 \cos u_2$$
$$x_2 = R \cos u_1 \sin u_2$$
$$x_3 = R \sin u_1.$$

The mapping between a point x on the surface and the corresponding two-dimensional parameter vector u is called a *chart* of the surface, the inverse mapping from u to x is called a *parametrization* of the surface. The latitude–longitude chart maps the sphere to the rectangle in Re^2 given by

$$-\frac{\pi}{2} \leqslant u_1 \leqslant +\frac{\pi}{2}$$
$$0 \leqslant u_2 < 2\pi.$$

The surfaces for which the theory can be developed easily are those whose parametrizations are sufficiently smooth. For reasons that will become clear, it is often assumed that the function x at least has continuous third partial derivatives, that is, $\partial^3 x_i/\partial u_j^3$ and $\partial^3 x_i/\partial u_j^2 \, \partial u_k$ exist and are continuous for $i, j, k \in \{1, 2, 3\}$. For the sphere, and indeed for most important examples, the parametrizations and charts actually have partial derivatives of every order.

---

[7]Finsler spaces, which are still more general, use a local Minkowskian approximation. Thus, in Finsler geometry, sufficiently small hyperspheres about a point approach some arbitrary radially symmetric convex shape—see Section 12.2.6—whereas in Riemannian geometry, this limiting shape is an ellipsoid, characteristic of Euclidean geometry.

We now consider a smooth curve on the surface (1) and derive an expression for its length as an integral calculated within the parameter domain (u_1, u_2). This introduces the first fundamental form, or metric tensor.

According to Definition 11, a parametric curve is a continuous function $y = y(t)$ from a closed interval $[\alpha, \beta]$ in Re into the surface X. Where possible, we write such a function in terms of the parameter vector u, i.e.,

$$y = y(t) = x(u_1(t), u_2(t)), \tag{2}$$

where the vector function $u(t)$ maps the interval in Re into the parameter domain, and has continuous derivatives with respect to t. We use the notation $\dot{y}$ for the vector of derivatives dy_i/dt, and likewise $\dot{u}$ is the derivative of $u(t)$. A classical result from elementary calculus expresses the length of the curve between $t = \alpha$ and $t = \beta$ as an integral:

$$\int_\alpha^\beta [\dot{y}, \dot{y}]^{1/2} \, dt = \int_\alpha^\beta (\dot{y}_1^2 + \dot{y}_2^2 + \dot{y}_3^2)^{1/2} \, dt. \tag{3}$$

We use the $[\,,]$ notation for the ordinary Euclidean scalar product in Re^3. Equation (3) of course comes from the basic definition of length of a parametric curve in a metric space (Definition 11), together with the fact that the summands in (ii) of that definition can be expressed in this instance in the form

$$\left(\frac{(\Delta y_1)^2}{(\Delta t)^2} + \frac{(\Delta y_2)^2}{(\Delta t)^2} + \frac{(\Delta y_3)^2}{(\Delta t)^2} \right)^{1/2} \Delta t.$$

The supremum of sums of such terms is approached as Δt goes to 0, and is the integral in Equation (3).

To calculate $\dot{y}$ in terms of $\dot{u}$, we differentiate Equation (2) with respect to t, obtaining the matrix equation:

$$\dot{y} = \frac{\partial x}{\partial u}(u(t))\dot{u}. \tag{4}$$

It may also be useful to write Equation (4) in terms of the coordinates of y and u:

$$\dot{y}_i = \frac{\partial x_i}{\partial u_1}(u(t))\dot{u}_1 + \frac{\partial x_i}{\partial u_2}(u(t))\dot{u}_2, \qquad i = 1, 2, 3.$$

The matrix of partial derivatives, $\partial x/\partial u = (\partial x_i/\partial u_j)$, is called the *Jacobian* of $x(u)$ at u, or, in more modern terminology, is simply the *derivative* of the vector function $x = x(u)$.

Substituting Equation (4) into Equation (3), the integral (3) becomes:

$$\int_\alpha^\beta \left[\frac{\partial x}{\partial u}\dot{u}, \frac{\partial x}{\partial u}\dot{u} \right]^{1/2} dt = \int_\alpha^\beta \left[\frac{\partial x^T}{\partial u}\frac{\partial x}{\partial u}\dot{u}, \dot{u} \right]^{1/2} dt$$

$$= \int_\alpha^\beta [g(u)\dot{u}, \dot{u}]^{1/2}\, dt$$

$$= \int_\alpha^\beta \left(g_{11}(u)\dot{u}_1^2 + 2g_{12}(u)\dot{u}_1\dot{u}_2 + g_{22}(u)\dot{u}_2^2 \right)^{1/2} dt.$$

$$(5)$$

Here, the superscript T denotes the transposed matrix. To get from the expression to the left of the first equality in (5) to the expression on the right, terms are regrouped, so that we change from the scalar product

$$\left[\frac{\partial x}{\partial u}\dot{u}, \frac{\partial x}{\partial u}\dot{u} \right]$$

in Re^3 to the equivalent scalar product

$$\left[\frac{\partial x^T}{\partial u}\frac{\partial x}{\partial u}\dot{u}, \dot{u} \right]$$

in Re^2. The latter parts of Equation (5) rewrite this scalar product in Re^2 using a notation that has become classic: the g_{jk} are the elements of a 2×2 symmetric matrix that varies as a function of u and is defined implicitly by Equation (5). Since the matrix g represents (in these particular coordinates) the crucial *metric tensor*, we give its explicit definition in the following equations (for $j, k \in \{1, 2\}$):

$$g_{jk} = \left[\frac{\partial x}{\partial u_j}, \frac{\partial x}{\partial u_k} \right] = \sum_{i=1}^3 \frac{\partial x_i}{\partial u_j}\frac{\partial x_i}{\partial u_k}.$$

$$(6)$$

Alternatively, in matrix form:

$$g(u) = \frac{\partial x^T}{\partial u}\frac{\partial x}{\partial u}.$$

$$(6')$$

We can think of the metric tensor $g(u)$ as a variable Euclidean scalar product in Re^2, which we also denote as $\langle,\rangle_u$. This is defined for any vectors $v, w \in \text{Re}^2$ by:

$$\langle v, w\rangle_u = [g(u)v, w]. \qquad (7)$$

Thus, the length of the curve can be written as

$$\int_\alpha^\beta [\langle \dot{u}(t), \dot{u}(t)\rangle_{u(t)}]^{1/2} \, dt.$$

The function $\langle,\rangle_u$ does in fact satisfy the properties of a Euclidean scalar product (Definition 7). It expresses the Euclidean geometry of the tangent plane at the point u, using coordinates that are rotated relative to (u_1, u_2) —axes of the ellipse given by $[g(u)v, v] = $ constant.

We have introduced the matrix $g(u)$ and associated scalar product $\langle,\rangle_u$ relative to the particular parametrization $x(u)$ for the surface X. In the latter part of this subsection we indicate how the metric tensor can be introduced without depending on a particular parametrization and how its representations relative to different parametrizations transform one into another.

In the literature on surface theory the scalar product $\langle,\rangle_u$ is called the *first fundamental form*, and its matrix $g(u)$ is often written in the form

$$\begin{bmatrix} E & F \\ F & G \end{bmatrix},$$

where E, F, G are functions of u, that is, $E = g_{11}, F = g_{12}, G = g_{22}$.

When the function $x(u)$ that represents the surface is assumed to have continuous third derivatives, then its derivative, $\partial x/\partial u$, and hence also the metric tensor must have continuous second derivatives. Thus, the metric tensor varies smoothly from point to point, a fact that is important for some subsequent developments.

To continue the example of the sphere, parametrized by latitude and longitude, we calculate the derivative:

$$\frac{\partial x}{\partial u} = \begin{bmatrix} -R \sin u_1 \cos u_2 & -R \cos u_1 \sin u_2 \\ -R \sin u_1 \sin u_2 & R \cos u_1 \cos u_2 \\ R \cos u_1 & 0 \end{bmatrix}.$$

Equation (6) now leads to

$$g_{11} = R^2$$
$$g_{12} = 0$$
$$g_{22} = R^2 \cos^2 u_1 .$$

Summarizing the results so far, we see that lengths of curves, and hence, distances on a surface, can be calculated using the Euclidean scalar product $\langle , \rangle_u$, represented by the 2×2 matrix $g_{jk}(u)$, in the neighborhood of the parameter vector u in Re^2. This is done by integrating the arc-length differential,

$$ds = [\langle \dot{u}(t), \dot{u}(t) \rangle_{u(t)}]^{1/2} dt, \qquad (8)$$

along the curve $u = u(t)$ in the parameter plane or chart of the surface in Re^2. Thus, the metric tensor specifies a local two-dimensional Euclidean approximation to the non-Euclidean geometry on the surface, from which the metric can be recovered.

Two major gaps have been left in this discussion of the metric tensor. First, we have not discussed the transformations induced by changes in parametrization. Second, there may be curves, even geodesic segments on the surface that do not have the representation of Equation (2), that is, cannot be represented as a curve for the particular parameter chart under consideration. For a sphere with latitude and longitude parameters, for instance, a curve that crosses the 0 meridian would leap discontinuously from the chart boundary at 2π back to the boundary at 0. There are also special problems at the North and South poles. What this means is that a full treatment of the sphere requires multiple parametrizations. One might want to arrange enough charts so that the geodesic segment connecting any two points lies entirely in some parameter domain. Alternatively, the calculation of a curve length may be continued off the end of one chart by means of another chart. To close this second gap, then, involves closing the first one; we must be in a position to consider alternative parametrizations, and transformations of the metric tensor from one parametrization to another. This approach is sketched in the latter part of this subsection in a more general context.

Curvature. We complete the portion on surface theory by developing the second fundamental form and the Gaussian curvature. This topic can be approached by either extrinsic or intrinsic methods. Both can be illustrated by considering how we know that a spherical surface is non-Euclidean whereas a cylinder is essentially Euclidean. A sphere lies entirely on one side of any tangent plane, except for the point of tangency, but a cylinder touches any tangent plane along a full line. This is one way to tell for the sphere has positive curvature everywhere while the cylinder has 0 curvature everywhere. Curvature is negative at a saddle point, where the surface lies partly on one side and partly on the other side of a tangent plane. To use this method we have to go outside the surface itself, considering points (such as those in the tangent plane) that do not form part of the surface. It is therefore an *extrinsic* method, one that makes essential use of the embedding of the surface in Re³. A rather different method is to consider the angles of a triangle whose sides are geodesic curves on the surface, for example, arcs of great circles on the sphere. In the plane or cylinder, the angles sum to 180°, but on the sphere the sum is greater, another sign of positive curvature. This method can be used without going outside the surface, because the geodesics (shortest curves) are determined entirely by the metric tensor, and the angles are likewise determined by the metric tensor (using the usual definition of angle from a Euclidean scalar product). This is an *intrinsic* method. Intrinsic methods are intellectually satisfying and also generalize to situations where the geometry is specified only by local Euclidean approximation around each point, without a specific embedding in a higher-dimensional Euclidean space. But the geometric intuitions underlying extrinsic methods are sometimes simpler, as the above example suggests.

We first consider an extrinsic measure of curvature for curves on the surface $x = x(u)$; this leads to the definition of the second fundamental form.

A straight line has second derivative 0; thus, curvature can be thought of as measuring the deviation of the second derivative from 0. For a curve given by Equation (2), the second derivative with respect to t is obtained by differentiating Equation (4):

$$\ddot{y} = \left(\frac{\partial^2 x}{\partial u_1^2} \dot{u}_1^2 + 2 \frac{\partial^2 x}{\partial u_1 \, \partial u_2} \dot{u}_1 \dot{u}_2 + \frac{\partial^2 x}{\partial u_2^2} \dot{u}_2^2 \right)$$
$$+ \left(\frac{\partial x}{\partial u_1} \ddot{u}_1 + \frac{\partial x}{\partial u_2} \ddot{u}_2 \right). \tag{9}$$

The two vectors $\partial x / \partial u_i$ that occur in the latter part of Equation (9) are the respective tangent vectors to the curves through $x(u)$ holding u_i constant, $i = 1, 2$; hence, they lie in the tangent plane at $x(u)$, so they are perpendicular to the surface normal vector. Denoting the unit vector normal to the surface at $x(u)$ by $\eta(u)$, we obtain the *normal curvature* of $y(t)$, i.e., the projection of $\ddot{y}$ on η, as the ordinary (Re³) scalar product of these two vectors. Using Equation (9) for $\ddot{y}$ and noting that the tangent-plane components are orthogonal to η, we derive

$$[\ddot{y}, \eta] = b_{11}\dot{u}_1^2 + 2b_{12}\dot{u}_1\dot{u}_2 + b_{22}\dot{u}_2^2, \tag{10}$$

where the symmetric 2×2 matrix (b_{jk}), a function of u, is given by:

$$b_{11} = L = \left[\frac{\partial^2 x}{\partial u_1^2}, \eta \right]$$

$$b_{12} = M = \left[\frac{\partial^2 x}{\partial u_1 \partial u_2}, \eta \right]$$

$$b_{22} = N = \left[\frac{\partial^2 x}{\partial u_2^2}, \eta \right]. \tag{11}$$

The matrix $b(u)$ is called the *second fundamental form*. The notation L, M, N for the coefficients of this form is traditional, along with E, F, G for the first fundamental form. The scalar product represented by this matrix is not in general positive definite, that is, $[b(u)v, v]$ need not be strictly positive for $v \neq 0$. Note that since $b(u)$ is defined using second derivatives, the assumption of continuous third derivatives for x assures that b has continuous first derivatives.

To illustrate again with the sphere, parametrized by latitude and longitude, the vectors of second derivatives are

$$\frac{\partial^2 x}{\partial u_1^2} = (-R \cos u_1 \cos u_2, -R \cos u_1 \sin u_2, -R \sin u_1)$$

$$\frac{\partial^2 x}{\partial u_1 \partial u_2} = (R \sin u_1 \sin u_2, -R \sin u_1 \cos u_2, 0)$$

$$\frac{\partial^2 x}{\partial u_2^2} = (-R \cos u_1 \cos u_2, -R \cos u_1 \sin u_2, 0).$$

The unit normal for a sphere is in the direction of the radius vector, so it is given by

$$\eta(u) = \frac{x(u)}{R} = (\cos u_1 \cos u_2, \cos u_1 \sin u_2, \sin u_1).$$

These last sets of equations can be substituted into Equation (11) to yield

$$b_{11} = -R$$
$$b_{12} = 0$$
$$b_{22} = -R\cos^2 u_1.$$

We now turn to the Gaussian curvature, which can be calculated extrinsically using the second fundamental form. There are several geometric definitions of Gaussian curvature. An intrinsic definition can be based on the observation that a triangle made up of geodesics has interior angles totalling π radians in the plane but that other totals occur on curved surfaces. Small geodesic triangles lie nearly in a plane, and so have interior angles whose sum differs little from π, but it can be shown that the *ratio* of the deviation from π to the area of a triangle approaches a limiting value as the vertices all approach the same point on the surface. This limiting value is the Gaussian curvature, denoted K.

Extrinsically, if one measures the component of curvature along the normal to the surface for any curve through a point (Equation 10), either this is constant (at a point where the surface is sphere-like) or it attains a maximum and a minimum value in different tangential directions. These extrema are the *principal* curvatures at that point. They can be calculated as the eigenvalues of the 2×2 matrix $g(u)^{-1}b(u)$. The Gaussian curvature can be shown to be equal to the product of these principal curvatures. At a point where the Gaussian curvature is positive, the principal curvatures are both positive, and the surface lies entirely on one side of the tangent plane. At a point with negative curvature, the maximum is positive and the minimum negative; the surface is saddle-shaped near such a point.

From the preceding extrinsic definition, the Gaussian curvature at a point, $K(u)$, can be calculated as the determinant of $g^{-1}b$, that is to say, the ratio of the determinants of the second and first fundamental forms. Thus,

$$K = \frac{b_{11}b_{22} - b_{12}^2}{g_{11}g_{22} - g_{12}^2} = \frac{LN - M^2}{EG - F^2}. \tag{12}$$

For the sphere, the previous calculations of g and b can be inserted into Equation 12 to yield

$$K = \frac{1}{R^2}.$$

Thus the sphere has *constant* positive curvature, which goes to 0 as $R \to \infty$.

The first and second fundamental forms can be used to characterize a smooth surface fully in Re^3. When the parametrizations have continuous third derivatives, the functions g and b have continuous first derivatives, and satisfy two families of differential equations, known as the Gauss and Codazzi equations; conversely, any $g(u)$ and $b(u)$ that satisfy these equations are the first and second forms for a surface, which is determined uniquely up to Euclidean motions.

Thus, representation by a surface can be axiomatized via empirical relations representable by appropriate matrix-valued functions g and b. For example, Silberstein (1946) suggested that g for a surface on which color discrimination is uniform could be obtained from Gaussian error distributions for color matching. Gaussian curvature has also been used in axiomatizing non-Euclidean representations of perceived spatial relations, as we show in detail in Section 13.10.

Riemannian manifolds. The concept of a local Euclidean approximation, specified by means of a Euclidean scalar product that varies smoothly from point to point on a surface, can be generalized readily from 2 to m dimensions. There are three steps required for this generalization. The first is to introduce a concept of *differentiable coordinates* in a set X that may not be a subset of any Re^n. Second, using such coordinates, it becomes possible to introduce an *intrinsic concept of tangent vector*. Thirdly one assumes a *scalar product* at each point of X, which applies to the tangent vectors at that point. The formula giving length of a curve as an integral of a scalar product $\langle , \rangle_u$ can then be generalized. We assume throughout that X is a metric space, in order to use the concept of open set introduced in Section 12.3.1

A *chart* for X is a one-to-one mapping $u = \varphi(x)$ of an open subset of X (see footnote 4) onto an open subset of Re^m. The inverse of a chart is a *parametrization*, denoted $x = \varphi^{-1}(u)$. An *atlas* is a collection of charts whose domains cover all of X. Examples of these concepts in the case of the sphere were given earlier.

An example will show the usefulness of these concepts. Let X be the set of *perceived colors*, with a metric that corresponds to dissimilarity of colors. Perceived color varies with context, and indeed, the range of attainable color percepts varies with context, but for any fixed context and any fixed choice of three independent colors as primaries for additive color mixture, the variable mixtures of the primaries provide a parametrization of an open subset of X. The mapping of a color into its coordinates, the amounts of the three primaries required to match, constitutes a chart for an open subset of X. Since color percepts lack any natural algebraic structure there is no very satisfactory way to consider them directly as a subset of Re^n for any n, but it is very satisfactory to make this set of percepts into a three-dimensional manifold by charts derived from three-primary matching. A fuller discussion of this will be found in Section 15.4.4.

One cannot directly speak of a parametrization as possessing the needed partial derivatives, since one does not necessarily assume an algebraic structure in X with which to define such derivatives. Instead, one requires that each parametrization be differentiable from the perspective of any chart whose domain overlaps the range of the parametrization. That is, if φ and ψ are charts, then $\varphi \circ \psi^{-1}$ (the function composition of a chart with a parametrization) is a mapping from one region of Re^m to another such region, hence, one can assume that its m component functions have the required partial derivatives.

In the example of perceived color, one requires that if two different color-matching setups are used to match some of the same colors (giving rise to charts φ and ψ with overlapping domains) then the amount of each primary in one setup, say φ_i, can be regarded as a function of the amounts ψ_1, ψ_2, ψ_3 of the three primaries used in the other setup; this function will be assumed to be appropriately smooth, e.g., to have continuous k^{th} partial derivatives.

A *differentiable structure of class k* for X consists of a maximal atlas for which all function compositions of charts with parametrizations have continuous k^{th} derivatives. Any atlas can be expanded to its maximal atlas by including all possible charts that overlap the given charts with the appropriate class of differentiability.

The next step is to make use of differentiable structure to define the m-dimensional tangent hyperplane at each point of X. A curve f mapping the interval $[\alpha, \beta]$ into X is said to be differentiable at $x = f(t)$ if the function composition $\varphi \circ f$ is differentiable at t for some chart φ (and therefore, for every chart) whose domain includes x. Two curves through x are tangentially equivalent at x if they are both differentiable at x and have identical derivatives for some (hence, for every) chart. The *tangent space* at x consists of the equivalence classes of curves differentiable at x. We will denote this set by V_x.

For any chart φ whose domain includes x, we denote by $d\varphi_x$ the mapping which sends each equivalence class of a curve f at x into its derivative $d(\varphi \circ f)/dt$, evaluated at $f^{-1}(x)$. This mapping, which sends V_x into Re^m, can be used to define the vector-space operations in V_x (easily shown to be independent of the choice of φ); it is then a vector-space isomorphism between V_x and Re^m. Thus we have an m-dimensional vector space or hyperplane of tangents to X at each point x. The coordinates of an element in V_x depend on the selection of a chart. If the coordinates are

$$(d\psi_1, \ldots, d\psi_m)$$

relative to the chart ψ, then relative to φ they are transformed into

$$d\varphi_i = \sum_{j=1}^{m} \frac{\partial(\varphi \circ \psi^{-1})_i}{\partial u_j}(\psi(x)) \, d\psi_j. \tag{13}$$

This is the classical transformation equation for a vector field (contravariant tensor of degree 1). It can be written more efficiently as a matrix equation,

$$d\varphi = \frac{\partial(\varphi \circ \psi^{-1})}{\partial u} \, d\psi. \tag{13'}$$

Having a tangent vector space V_x at each x, we can now introduce what is called a *covariant tensor field of degree 2*. This is simply an assignment of a Euclidean scalar product $\langle,\rangle_x$ to each tangent space V_x, varying smoothly from point to point, that is, the $m \times m$ matrix representation of the scalar product, relative to some (and hence every) chart has the required partial derivatives with respect to the parameter u. That is, if f and $\bar{f}$ are curves through x, then the scalar product of the tangents to f and $\bar{f}$ at x is given as

$$\langle f, \bar{f} \rangle_x = \left[(g_\varphi) \, d\varphi_x f, \, d\varphi_x \bar{f} \right]$$
$$= \sum_{i=1}^{m} \sum_{j=1}^{m} (g_\varphi)_{ij}(\varphi(x)) \frac{d\varphi_i f}{dt}(f^{-1}(x)) \frac{d\varphi_j \bar{f}}{dt}(\bar{f}^{-1}(x)).$$

Each element of the matrix g_φ is an appropriately smooth function of the chart coordinates or parameters $(\varphi_1(x), \ldots, \varphi_m(x))$. It is not hard to derive, parallel to Equation (13), the classical transformation formula for the matrix representation g when the chart φ replaces chart ψ:

$$g_\varphi = \left(\frac{\partial(\varphi \circ \psi^{-1})}{\partial u} \right)^T g_\psi \frac{\partial(\varphi \circ \psi^{-1})}{\partial u}.$$

This is the standard transformation for a second-degree covariant tensor.

With the introduction of differentiable structure, tangent spaces, and smoothly varying scalar products, we have all the apparatus for intrinsic Riemannian geometry. In the section on surface theory, the scalar product on each tangent plane was induced by the metric in the enclosing space Re^3. For the intrinsic theory of surfaces, this is the only use made of Re^3. In the more general theory of intrinsic Riemannian manifolds, we simply assume a differentiable structure for X and some specified second degree covariant tensor field on X, denoted $\langle , \rangle_x$. Associated with this field of scalar products is a natural metric:

$$D(x, y) = \inf_{\substack{f \text{ differentiable} \\ f(0)=x \\ f(1)=y}} \int_0^1 (\langle f, f \rangle_{f(t)})^{1/2} \, dt.$$

The infimum is over all differentiable curves mapping the interval $[0, 1]$ into X and connecting x to y.

We call the metric space $\langle X, \delta \rangle$ *Riemannian* if there is a differentiable structure and a second degree covariant tensor field on X such that $\delta = D$, the natural Riemannian metric.

There is a vast literature on Riemannian spaces (for an overview, see Klingenberg, 1982). Among the basic topics that we have not touched on are covariant differentiation, singular points, geodesic curves (for which the infimum in the preceding equation is actually achieved for points not too far apart), and the generalization of the concept of intrinsic curvature.

12.3.4 Other Metrics

As its title implies, this section covers a miscellany of metric spaces that fall outside the main lines of development of geometry but that have been proposed or used as representing structures for proximity relations. The first two models, based on the chordal metric and Mulholland's inequality, respectively, can be viewed as modifications or generalizations of the $\mathcal{L}^p$ metric. The remainder of this section is devoted to the discussion of metrics defined on finite graphs.

Chordal metrics. If a set X is embedded in a metric space $\langle Y, \delta \rangle$, then δ is a metric when restricted to $X \times X$. We call this a chordal metric because the distance in X is measured through Y rather than along a shortest path in X. For examples we can take any surface in Euclidean space with distance measured along the chord in the including space instead of along a geodesic segment on the surface. To take a specific application, proximity relations among tones varying in pitch are sometimes represented as a helix (a one-dimensional subset X of a three-dimensional space Y). Tones separated by an octave are close together along the chord that connects a point on one turn of the helix to the corresponding point on the next turn. This application indicates why chordal metrics are sometimes attractive: two objects can be put in close proximity even though the shortest path connecting them within the space X is long. Whenever the ordering of proximities violates the axioms for a metric with additive segments, a chordal metric might at least be considered.

Unfortunately, there seems to be no theory of chordal metrics that could be used to help decide in a qualitative way which metrics might be appropriate as representations or which metrics are equivalent from an ordinal standpoint. For example, the ordering of chordal distances between pairs of points of a sphere is obviously the same as the ordering of corresponding arc lengths on the sphere; but we know of no general way to tell whether two metrics are monotonically equivalent.

Mulholland's inequality. Recall that the $\mathscr{L}^p$ metric is based on the norm $\|x\|_p = [\Sigma|x_i|^p]^{1/p}$ (Section 12.2.6). The function $\delta_p(x, y)$ is defined as the norm of the vector difference, i.e., $\|x - y\|_p$. For this function, the triangle inequality reduces to the Minkowski inequality (valid for $1 \leqslant p < \infty$):

$$[\Sigma|x_i + y_i|^p]^{1/p} \leqslant [\Sigma|x_i|^p]^{1/p} + [\Sigma|y_i|^p]^{1/p}.$$

One way to generalize this is to replace the pth power by a more general function f, replacing the norm by the "distance" from the origin,

$$\delta_f(x, 0) = f^{-1}[\Sigma f(|x_i|)],$$

and defining other "distances" by vectorial translation to the origin, i.e.,

$$\delta_f(x, y) = \delta_f(x - y, 0).$$

The question that needs to be answered first is, when does δ_f satisfy the properties of a metric?

It is not hard to satisfy the positivity requirement by requiring that $f(0) = 0$ and f be strictly increasing on Re^+; symmetry results from the symmetry of the absolute-value function $|x_i| = |-x_i|$. The triangle inequality generalizes the Minkowski inequality

$$f^{-1}[\Sigma f(|x_i + y_i|)] \leqslant f^{-1}[\Sigma f(|x_i|)] + f^{-1}[\Sigma f(|y_i|)]. \tag{14}$$

Unfortunately, necessary and sufficient conditions on f such that this inequality is satisfied are not known and are perhaps very difficult. The best result we know of is a quite general sufficient condition, established by

Mulholland (1947, p. 297):

Inequality (14) *holds for all* $x, y \in \text{Re}^n$, $n = 1, 2, \ldots$ *whenever the following three conditions are satisfied*:

(i) f *is continuous and strictly increasing, and* $f(0) = 0$;

(ii) f *is convex on* $[0, \infty)$;

(iii) $\log f(e^t)$ *is convex for* $t \in (-\infty, \infty)$.

Just as the Minkowski inequality reverses for $0 < p < 1$, so does its generalization, inequality (14), if *concave* replaces *convex* in (ii) and (iii) and if, in addition, f is assumed to be unbounded (so f^{-1} is defined for arbitrary Σf). This was shown also in Mulholland's paper (p. 301).

An example of a family of functions that satisfy the conditions (i)–(iii) is given by

$$f(t) = q(e^{rt} - 1)^p \qquad (q, r > 0; \quad p \geqslant 1).$$

This family includes the $\mathscr{L}^p$ metrics as a limiting case for which $r \to 0$ (let $q = r^{-p}$ and then let $r \to 0$). However, for any $r > 0$, this family fails to satisfy the homogeneity condition of a power function, i.e., the functional equation $f(\alpha t) = g(\alpha)f(t)$; and therefore[8] the function $\delta_f(x, 0)$ fails to satisfy the homogeneity property of a norm [part (i2) of Def. 9, i.e., $\delta_f(\alpha x, 0) = |\alpha|\delta_f(x, 0)$]. Other examples of families satisfying the criteria are found in Mulholland's paper.

One way to regard this material is to observe that the $\mathscr{L}^p$ metric can be generalized either by giving up additivity across coordinates but preserving homogeneity along every line through the origin (this yields a general norm, or the Minkowski geometries of Section 12.2.6), or by giving up homogeneity but keeping the additive focus in a special coordinate system. Mulholland's inequality shows the extent to which the latter is possible. The first kind of generalization, to Minkowski geometries, has been very important mathematically because it is compatible with the affine structure: algebraic lines act like metric straight lines. For the generalization discussed here, however, there is little geometric work. The coordinate axes play a very special role in that they display metric additivity, but there may not be any other directions in which the lines display such additivity.

Nonetheless, there are representations in which the coordinate axes do have a natural role. For example, in the axiom systems in Chapter 14, in

_____
[8]See Exercise 18 and Theorem 14.10.

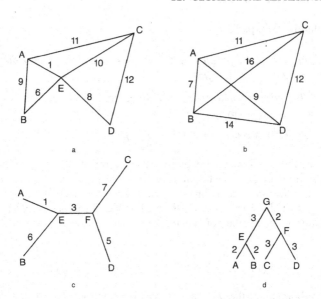

FIGURE 3. (a) Weighted graph with five vertices (connected); (b) complete weighted graph with the same distances among A, B, C, D as in (a); (c) weighted tree with six vertices and the same distances among A, B, C, D as in (a) and (b); (d) an ultrametric tree.

which the basic set of objects has a Cartesian product structure, metrics of the sort discussed here occur as representations.

Weighted graphs. A method of defining distances among a finite set of points is illustrated in Figure 3a: add up the numbers along the path connecting the two points, which minimizes the sum. Thus the distance from A to D is 9, rather than 23, etc.

In graph-theory terminology, we have a finite set X of *vertices* ($A, \ldots, E$ in the figure); a finite set $\mathscr{B}$ of *branches*, each element of $\mathscr{B}$ being an unordered pair of distinct elements of X [thus $\mathscr{B} = \{(A, C), (A, E), (B, E), \ldots\}$ in the figure]; and a positive-valued function W on $\mathscr{B}$, the *weight function* (shown by the numbers on the branches). The triple $\langle X, \mathscr{B}, W \rangle$ is a *weighted* graph. There are generalizations in which W need not be positive-valued, but we shall not cover these. Since W is defined on unordered pairs, it is symmetric as a function of two variables.

An *edge-sequence* is an alternating finite sequence $x_0, b_1, x_1, b_2, \ldots, b_n, x_n$ of vertices x_i and branches b_j such that $b_i = (x_{i-1}, x_i)$, $i = 0, \ldots, n$. For example, $A, (A, E), E, (E, C), C, (C, D), D, (D, E), E$

is an edge-sequence in Figure 3a. The subsequence of branches in an edge-sequence is a *path*. The *length of a path* is the sum $\sum_{i=1}^{n}W(b_i)$ of the weights of its branches; the path just illustrated has length 31. A weighted graph is *connected* if every two vertices are included in some edge-sequence. This is satisfied in Figure 3a. In a connected weighted graph, any two vertices are included in an edge-sequence with minimum path length, and this minimum is the value of δ. Defining $\delta(x, x) = 0$, we obtain a metric on X.

Figure 3b illustrates another connected weighted graph. This one is *complete* (a branch for every pair of distinct vertices). It is related to the graph in (a) in an interesting way: the distances among the four points A, B, C, D are identical in the two graphs. In fact, (b) was formed by assigning $\delta(x, y)$ from (a) as the branch weight $W(x, y)$ in (b), for x, y in $\{A, B, C, D\}$. Since δ satisfies the triangle inequality, each branch in (b) is a minimum path connecting its vertices; so $\delta = W$ for any distinct x, y in (b).

The set $\{A, B, C, D\}$ is a subset of the set X in graph (a), and the metric is just that induced[9] by the metric δ on $X \times X$. The construction going from (a) to (b) is quite general. Given any connected weighted graph $\langle X, \mathscr{B}, W \rangle$ with metric δ and any nonempty subset Y of X, we obtain a metric on Y by restricting δ to pairs in $Y \times Y$; and then, given a finite metric space, we can always realize it with a complete weighted graph.

What this shows, however, is that there can be more than one way to realize a finite metric space $\langle Y, \delta \rangle$ by means of a connected weighted graph. Doing it with a complete graph, as in (b), is trivial in a sense because there is no interesting structure in the graph that is not equally clear in the metric. A realization as a subset of a larger graph, as in (a), can be more interesting because introducing the point E shows that some of the distances in (b) can be broken down into sums. The existence of such a decomposition is a constraint on the metric shown in (b). Perhaps the most interesting realization of (b) as a subset of a larger graph is the *tree realization* shown in Figure 3c. Obviously, this has a lot of structure: all the distances among A, B, C, D are broken down as sums.

More formally, we define an *elementary path* to be a path whose edge-sequence contains all distinct vertices. Note that the minimum path length always occurs for an elementary path. There can be more than one elementary path connecting two vertices; this occurs if and only if there are

---

[9]Note that the induced metric on a subset is quite different from the metric on a *subgraph*. The latter may not even exist: deleting vertices (and the branches containing them) may disconnect a graph. In Figure 3a, the subgraph obtained by deleting E is still connected, but the distances in it are not at all the same as those in Figure 3b.

circuits (i.e., edge-sequences whose vertices are all distinct except for the first and last) that are equal and whose branches are all distinct. For example, in Figure 3a, two elementary paths from A to D can be combined to form a circuit. A *weighted tree* is a connected weighted graph without circuits, i.e., with exactly one elementary path joining any two distinct vertices.

In the tree realization, (c), it is obvious that

$$\delta(A, C) + \delta(B, D) = \delta(A, D) + \delta(B, C).$$

This is because both sums count every branch once except (E, F), which is counted twice; and these equal sums are larger than the third possible combination

$$\delta(A, B) + \delta(C, D),$$

since that sum omits (E, F) entirely. We have thus derived a necessary condition for a metric δ on $\{A, B, C, D\}$ to admit a tree realization. This condition turns out to be of general importance, and we formulate it in a definition.

DEFINITION 13. *A metric space $\langle Y, \delta \rangle$ satisfies the* tree inequality *iff any four points of Y can be relabeled as x, y, u, v such that*

$$\delta(x, y) + \delta(u, v) \leqslant \delta(x, u) + \delta(v, y) = \delta(x, v) + \delta(u, y).$$

We note that, in the presence of the other axioms for a metric, the tree inequality (also called the additive inequality or the four-point condition) implies the triangle inequality since $\delta(x, z) + \delta(y, y) \leqslant \delta(x, y) + \delta(y, z)$. The significance of the tree inequality stems from the fact that it is not only necessary but also sufficient for the representation of distances by a tree. It has been shown by several authors (see, e.g., Buneman, 1971; Dobson, 1974; Hakimi and Yau, 1964; Patrinos and Hakimi, 1972) that a finite metric space $\langle Y, \delta \rangle$ can be represented by a tree if and only if the tree inequality (Definition 13) is satisfied. In this case, there exists a mapping that associates each point in Y with a vertex of a weighted tree so that the distance between the points in Y is given by the path-length distance between the respective vertices of the tree. Moreover, the tree is essentially unique.

Another inequality that leads to a particular type of tree representation is the ultrametric inequality mentioned in Section 12.2.7. We restate it more formally in a manner analogous to the tree inequality.

DEFINITION 14. *A metric space* $\langle Y, \delta \rangle$ *satisfies the* ultrametric inequality *iff any three points of Y can be relabeled as* x, y, z *such that*

$$\delta(x, z) = \delta(y, z) \geqslant \delta(x, y).$$

It is easy to verify that the ultrametric inequality implies the tree inequality. Furthermore, unlike the tree inequality which is based on the ordering of sums of distances, the ultrametric inequality is a purely ordinal property. Moreover, it has been shown by several authors (e.g., Hartigan, 1967; Jardine and Sibson, 1971; S.C. Johnson, 1967) that the ultrametric inequality is both necessary and sufficient for the representation of $\langle Y, \delta \rangle$ by an ultrametric tree. An ultrametric tree is a weighted tree with a distinct vertex, called the root, which is equidistant from all terminal vertices (i.e., leaves) of the tree. In Figure 3d, the root is G, and the leaves are A, B, C, and D. If the ultrametric inequality holds, then there exists a mapping that associates each point in Y with a terminal vertex of an ultrametric tree such that the distances between the points are given by the path-length distances between the respective terminal vertices. The ultrametric tree is of special interest because it induces a hierarchical clustering of Y. A nonempty subset $X \subset Y$ is a cluster iff for all $x, y \in X$ and $z \in Y - X$, $\max \delta(x, y) < \min \delta(x, z)$. It is easy to verify that in an ultrametric tree each external vertex belongs to one (or more) clusters and that the collection of clusters is hierarchical in the sense that any two clusters are either disjoint or else one contains the other.

EXERCISES

1. Consider Example 2 following Definition 1. Prove the infinite dimensionality of the vector space consisting of the set of real-valued functions that have continuous second partial derivatives and satisfy Laplace's equation. (12.2.1)

2. Prove Lemma 1. (12.2.1)

3. Using Definition 3, prove that the set-theoretic intersection of two linear varieties is either empty or a linear variety. (12.2.2)

4. Using Definition 3, prove that the span of any $k + 1$ linearly independent points is a k-dimensional linear variety. (12.2.2)

5. Using Definition 3, is a mapping of $\mathscr{V}$ into itself that maps every line into one given line an affinity? (12.2.2)

6. Using Definition 5, prove the duality theorem described in the text for n-dimensional projective geometry.   (12.2.3)

7. Prove Theorem 3 for $n = 2$.   (12.2.3)

8. Using Definition 6, prove that any projectivity of an ordered projective geometry preserves the relation of separation.   (12.2.3)

9. Using Definition 7, give an example of a function from $V \times V$ into Re that has the properties of bilinearity and positive definiteness but not symmetry.   (12.2.4)

10. Prove Theorem 4.   (12.2.4)

11. Prove that the 3-parameter subgroup defined by (4) of Section 12.2.5 preserves scalar products of differences of vectors and therefore preserves the Euclidean metric.   (12.2.5)

12. With reference to Section 12.2.5, give an example to show affinities do not preserve the relation of perpendicularity.   (12.2.5)

13. Using Definition 9, give examples to show that the $\mathscr{L}^1$ norm and the supremum norm are not strongly convex.   (12.2.6)

14. Prove that a Minkowski geometry that is invariant under every rotation of the axes is Euclidean.   (12.2.6)

15. Confirm the example calculations for the first and second fundamental forms and Gaussian curvature for the sphere. Show how the fundamental forms change in various other coordinate systems (spherical, rectangular) and that the transformations of the first form follow the laws for a second degree covariant tensor.   (12.3.3)

16. Show that the scalar product in the tangent plane to a parametrized surface satisfied Holder's Inequality, i.e.,

$$|\langle v, w \rangle_u|^2 \leqslant |\langle v, v \rangle_u| \, |\langle w, w \rangle_u|. \qquad (12.3.3)$$

17. The mean curvature of a two-dimensional surface is defined as one-half of the sum of the principal curvatures. Prove that if the Gaussian and mean curvatures of a surface are everywhere zero, then the surface is a plane. (12.3.3)

18. Show directly (i.e., without using the power-function solution) that the functional equation for homogeneity $f(\alpha t) = g(\alpha)f(t)$ implies homogeneity of the distance from the origin $\delta_f(x, 0)$ defined by f, i.e., if $\delta_f(x, 0) =$

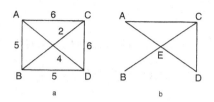

FIGURE 4. Two weighted graphs.

$f^{-1}[\Sigma_f(|x_i|]$, then $\delta_f(\alpha x, 0) = |\alpha|\delta_f(x, 0)$. This result has a valid converse, i.e., homogeneity of δ_f implies that the functional equation holds for f, but this is not so easy; see Theorem 15.10.

Solution: If $f(\alpha t) = g(\alpha)f(t)$, then $f^{-1}[g(\alpha)y] = \alpha f^{-1}(y)$; therefore,

$$\delta_f(\alpha x, 0) = f^{-1}[\Sigma f(|\alpha| |x_i|)]$$
$$= f^{-1}[g(|\alpha|)\Sigma f(|x_i|)]$$
$$= |\alpha|\delta_f(x, 0). \qquad (12.3.3)$$

19. Show that for $q, r > 0$ and $p \geqslant 1$, $f(t) = q(e^{rt} - 1)^p$ satisfies conditions (i)–(iii) of Mulholland's criterion, for solutions to inequality (14). (12.3.4)

20. Consider the two weighted graphs (a) and (b) shown in Figure 4. Show that it is impossible to assign positive weights to (b) such that the distance induced over $\{A, B, C, D\}$ coincides with the distance in (a). (12.3.4)

Chapter 13 Axiomatic Geometry
and Applications

13.1 INTRODUCTION

This chapter begins with a survey of the classical foundations of geometry, with two purposes in mind. First, we want to show that the standard problem of a representation theorem for a given classical, synthetic axiomatic geometry corresponds very closely to a representation theorem in the theory of measurement. The representation of the Euclidean plane by Cartesian coordinates is the most familiar and classical example. More general representations were discussed in Chapter 12—for example, n-dimensional analytic Euclidean space. Second, we want to provide a background in geometry that will be useful when, in the later sections of the chapter, we develop the foundations of physical and perceptual spaces. Chapter 14, on proximity measurement, also uses some results from this chapter.

An axiomatic approach to geometry is found in ancient Greek mathematics. Of course, numerical representation theorems were not then developed because the numerical relational structures were not yet known. Indeed, a major thrust of ancient geometry was to provide a theory of irrational quantities, for example, to incorporate the discovery that the diagonal and the side of a square are "incommensurable" (i.e., they cannot both be elements of the same standard sequence). On the other hand, it would be incorrect to hold that Greek mathematics was unconcerned with representation theorems. They simply took a different, nonnumerical form. A good example of such a representation theorem would be the first proposition of Archimedes' treatise, *Measurement of a Circle* (1897). This proposition asserts that the area of any circle is equal to that of a right triangle in which one leg is equal to the radius of the circle and the other leg to the circumference of the circle. (We can regard this as representing areas of circles in a two-factor, multiplicative conjoint measurement structure in which the objects are right triangles, the factors being their legs, and the ordering is based on the area of the corresponding triangle.)

A basic test of the adequacy of an axiom system for a particular branch of geometry is that it should lead to the desired numerical representation and the desired uniqueness theorem or automorphism group of the representation. This model view does not appear in clear and explicit form in mathematics until the end of the nineteenth century; it is found especially in the work of Hilbert (1899) and Klein (1893). (The Erlangen program of Klein has already been discussed in Section 12.2.5.) An influential and extraordinarily clear exposition of these basic ideas was given in the treatise of Veblen and Young (1910, 1918). One early use of representation theorems was to provide interpretations for the non-Euclidean systems discovered by Bolyai and Lobachevski at the beginning of the nineteenth century. For example, the Hilbert geometry in the interior of an ellipse (Section 12.2.7) provides a representation for the structures in which the Euclidean parallel postulate is replaced by the postulate that there exist at least two parallels through a given point outside a given line. (Ordinary Euclidean lines whose intersection is outside the ellipse are parallel in the Hilbert geometry.) Such interpretations established the consistency of the non-Euclidean geometries and provided more familiar representations for them.

In broad terms, the problem of proving such a representation theorem is the same for a system of geometry as for a system of measurement. The only real difference is that in geometry, in contrast to the one-dimensional representation that is characteristic of measurement, we ordinarily expect a point to be represented by an n-tuple of real numbers where n is the dimension of the space, though, in the important case of conjoint measurement developed in Chapter 6 of Volume I, we could already take a geometrical attitude toward the representation, as we did in Chapter 10. The problem of uniqueness also assumes the same conceptual form as it does in the case of measurement—finding the most general group of transformations under which the properties of the numerical representation are invariant.

We have organized the material in the first part of this chapter along the following lines. Section 2 considers affine and projective order on the line, and Section 3 contains proofs of the main theorems. Section 4 considers the projective plane, and Section 5 extends this to projective space. Section 6 develops the theory of order geometry and the important special cases of affine and absolute geometry. In addition, two extensions of absolute geometry, namely, Euclidean and hyperbolic geometry, are analyzed. Section 7 contains a brief treatment of elliptic geometry. In all of these various geometries, we begin with qualitative postulates and end with numerical representation and uniqueness theorems. Because the results are well known in the literature and the proofs are long, most of the proofs are omitted; but we have made it a point to state explicitly the representation and uniqueness theorems to show how close the theory of measurement (as developed in Volume I, in Chapter 16 of this volume and in Chapters 19 and 20 of Volume III) is to the classical foundations of geometry. It is somewhat curious that so much time elapsed between Hölder's 1901 paper on measurement, which was clearly an outgrowth of the then active work on foundations of geometry, and the modern revival of foundations of measurement around 1950.

Section 8 addresses the characterization of classical space-time structures in physics, and Section 9 does the same for the space-time structure of special relativity. We give a more detailed treatment of the qualitative axioms in this case because of the continuing development of the subject and the rather large recent literature on the axiomatic foundations. Section 10 is concerned with perceptual spaces, but only certain questions are examined; namely, those dealing with the geometrical structure of visual perception.

One general point about the axiomatizations of the several geometries given in this chapter is that choice of primitives for the relational structures is more arbitrary than in the case of most of the measurement structures of Volume I, in the following sense. A variety of choices, all rather natural in character, are available for the geometries. Take the familiar case of Euclidean geometry. In addition to the basic set of points, we have chosen as primitives the ternary relation of betweenness and the quaternary relation of segment congruence: $ab \sim cd$. Here are some of the better-known alternatives that were open to us: Pieri (1908) showed that a single ternary relation of a point being equally distant from two others will suffice. Beth and Tarski (1956) showed that the ternary relation of equilaterality is sufficient; and Scott (1956) showed the same for the ternary relation of orthogonality: a, b and c form a triangle with a right angle at b. On the other hand, Tarski (1956) showed that *no* binary relation of points could suffice. Yet, if we change from the basic set of points to the basic set being the set of lines, Schwabhäuser and Szczerba (1975) showed that the binary relation of perpendicularity suffices for dimensions greater than three and, in three dimensions, addition of the binary relation of intersection is sufficient. Of a different sort is Tarski's (1929) formulation of three-dimensional Euclidean geometry in which the basic set is, intuitively, the set of solids, and the two primitive concepts are that of a ball and the binary relation of one solid being a part of another.

Another important choice is whether to have just a single-sorted theory with, e.g., points as the basic objects or to have a many-sorted theory. There seems to be no overriding uniform choice. Both kinds of theories are exemplified below. Because of the fundamental duality principle of projective geometry it is natural to have a two-sorted theory of the projective plane, with both points and lines as basic. On the other hand, most of the recent axiomatizations of Euclidean geometry use a single-sorted theory with points as basic, as we do here.

Still another approach, not examined in detail here, is to include among the primitives a set of automorphisms, e.g., a set of motions (see Bachmann, 1959, which also discusses the earlier literature on these matters; and for a clear elementary axiomatic development, see Delessert, 1964).

This sample illustrates the possibilities. Results on some of the possible primitive concepts for hyperbolic and elliptic geometry are to be found in Robinson (1959), Royden (1959), and Kordos (1973).

13.2 ORDER ON THE LINE

We briefly develop affine and projective order on the line. The affine notion of order is the familiar ternary relation of betweenness for points on a line. The projective notion is the less familiar quaternary relation of separation. All the theorems of this section are ordinal in character.

13.2.1 Betweenness: Affine Order

The classical theory of order on the line is just the theory of the ternary relation of betweenness. The notion of order embodied in the relation of betweenness is a generalization of the notion of simple order introduced in Section 1.3.1. We recall that a simple order is a binary relation on a set that is transitive, connected, and antisymmetric. The generalization embodied in betweenness on the line is that the sense of direction is eliminated. The intuitive basis for this within classical Euclidean geometry is apparent. There is no preferred direction and consequently there is no meaningful way of fixing a direction; on the other hand, there is a natural intuitive order to points on the line as represented by betweenness. A variety of axioms that are necessary and sufficient for characterizing order on the line in terms of betweenness can be found in the literature. In the following set, the first five axioms are taken from Suppes (1972). The notation for betweenness is that introduced earlier in Section 4.10, namely, $a|b|c$ for b

being between a and c, but the separate ordinal theory of betweenness was not developed there. For simplicity of notation we shall refer to this ternary relation as B. Thus, explicitly, $B(a, b, c)$ if and only if $a|b|c$. We can define equivalence as follows: $a I c$ if and only if $a|b|c$ and $b|a|c$ and $a|c|b$; but already in Axiom 1 we require identity of a and b if $a|b|a$, in accordance with the usual geometrical view that a given spatial position can be occupied by only a single point. It is apparent that it would be easy to generalize this standard assumption to permit equivalence.

DEFINITION 1. *Let A be a nonempty set and B a ternary relation on A. Then $\mathfrak{A} = \langle A, B \rangle$ is a* one-dimensional betweenness structure *iff the following five axioms are satisfied for every a, b, c, and d in A:*

1. *If $a|b|a$, then $a = b$.*
2. *If $a|b|c$, then $c|b|a$.*
3. *If $a|b|c$ and $b|d|c$, then $a|b|d$.*
4. *If $a|b|c$ and $b|c|d$ and $b \neq c$, then $a|b|d$.*
5. *$a|b|c$ or $b|c|a$ or $c|a|b$.*

Moreover, $\mathfrak{A}$ is separable *iff there is a finite or countable order-dense subset of A, i.e., there is a finite or countable subset C of A such that for all a and c in A with $a \neq c$ there exists b in C such that $a|b|c$.*

The term *separable* refers to the fact that any two elements a and c of A can be separated by an element b of C with respect to the betweenness ordering. On the basis of these axioms, it is straightforward to prove the desired representation theorem using the approach followed in the proof of Theorem 2.2 in Volume I.

THEOREM 1. *Let $\mathfrak{A} = \langle A, B \rangle$ be a structure such that A is a nonempty set and B a ternary relation on A. Then $\mathfrak{A}$ is a separable one-dimensional betweenness structure iff there is a real-valued function φ such that for all a, b, and c in A*

$$a|b|c \quad iff \quad [\varphi(a) \leqslant \varphi(b) \leqslant \varphi(c) \quad or \quad \varphi(c) \leqslant \varphi(b) \leqslant \varphi(a)].[1]$$

---

[1] Notice that the following would be incorrect: either there exists a real-valued function φ on A such that for all a, b, c in A

$$a|b|c \quad iff \quad \varphi(a) \leqslant \varphi(b) \leqslant \varphi(c),$$

or there exists a real-valued function φ' on A such that for all a, b, c in A

$$a|b|c \quad iff \quad \varphi'(a) \geqslant \varphi'(b) \geqslant \varphi'(c),$$

because $a|b|c$ iff $c|b|a$.

The uniqueness theorem for the structures studied in this section is simple; it is similar to Theorem 2.3. A slight but familiar additional terminology is useful. A function h mapping a subset of Re into a subset of Re is *strictly monotone* iff it is either strictly increasing or strictly decreasing.

THEOREM 2. *Let* $\mathfrak{A} = \langle A, B \rangle$ *be a separable one-dimensional betweenness structure, and let* φ *and* φ' *be two real-valued functions satisfying Theorem 1. Let R and R' be the ranges of* φ *and* φ', *respectively. Then there exists a strictly monotone function h from R to R' such that for all* $a \in A$, $h[\varphi(a)] = \varphi'(a)$. *Moreover, if h is any strictly monotone function on R, then* $h \circ \varphi$ *defines another numerical representation of* $\mathfrak{A}$.

13.2.2 Separation: Projective Order

The projective theory of order on the line requires the four-place relation of separation. As is shown in many standard discussions of projective geometry, the axioms of separation for projective order on the line also characterize the axioms for order on the circle. The intuitive idea is that when we have $ab\,S\,cd$, then the points a and b separate the points c and d and conversely, as illustrated in Figure 1. Moreover, this order is preserved when the points are "projected" onto another line, as in Figure 2, from a point of perspective p that does not lie on either line. Note that, in Figure 2, betweenness is not preserved; for example, $a|c|b$ but $c'|a'|b'$.

The axioms for one-dimensional projective order are embodied in the following definition. As far as we know, this particular formulation is new, although Axioms 1–4 and 6 are similar to those in Crampe (1958). One minor point to note is this. The axioms for separation impose a strict order whereas those for betweenness impose an inclusive order, the first corresponding to numerical *less than* and the second to *less than or equal to*.

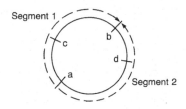

FIGURE 1. Separation on a circle.

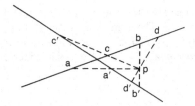

FIGURE 2. Invariance of separation under projection.

DEFINITION 2. *Let A be a nonempty set and S a quaternary relation on A. Then* $\mathfrak{A} = \langle A, S \rangle$ *is a* one-dimensional separation structure *if and only if the following six axioms are satisfied for every a, b, c, d, and e in A*:

1. *It is not the case* $ab\,S\,ac$ *or* $ab\,S\,cc$.

2. *If* $ab\,S\,cd$, *then* $cd\,S\,ab$.

3. *If* $ab\,S\,cd$, *then* $ab\,S\,dc$.

4. *If* $ab\,S\,cd$, *then it is not the case* $ac\,S\,bd$.

5. *If* $ab\,S\,cd$ *and* $ae\,S\,db$, *then* $ab\,S\,ce$.

6. *If* a, b, c, *and* d *are distinct, then* $ab\,S\,cd$ *or* $bc\,S\,ad$ *or* $ca\,S\,bd$.

Moreover, $\mathfrak{A}$ *is* separable *iff there is a finite or countable order-dense subset of A, i.e., there is a finite or countable subset B of A such that for all a and b in A with* $a \neq b$ *there exist c and d in B with* $ab\,S\,cd$.

The terminology introduced in this definition, "a separable one-dimensional separation structure," may sound a little strange, but in fact two distinct but related notions of separation are involved. Separability is just the standard mathematical term for having a finite or countable order-dense subset. As remarked earlier, the subset does the separating, modeled, of course, on the separation of any two distinct real numbers by a rational number. As is clear in Figures 1 and 2, separation in the sense of projective order is another related form of separation, required to keep the order projectively invariant.

To show the close relation of betweenness and separation, we first define strict betweenness in terms of the given weak betweenness,

$$B'(a, b, c) \quad \text{iff} \quad a|b|c \quad \text{and} \quad a \neq b \quad \text{and} \quad b \neq c.$$

Given a separation ordering S on A and x an element of A, then

$$B'_x(a, b, c) \quad \text{iff} \quad xb\,S\,ac$$

is a strict betweenness relation on $A - \{x\}$; also, for a, b, c and d in A, if $B_x'(a, b, d)$ and $B_x'(b, c, d)$, then $ac\,S\,bd$. Because the axioms of order embodied in Definition 2 are seldom studied, apart from more complex assumptions of projective geometry, several elementary theorems are given below as exercises.

The representation theorem for projective order is stated in terms of the concept of cross ratio introduced in Section 12.2.3; but in the one-dimensional case we shall not use homogeneous coordinates, to facilitate the most direct comparison with the theory of a binary ordering (developed in Chapter 2 of Volume I) and the immediately preceding theory of the ternary relation of betweenness. We return to homogeneous coordinates in Section 4 on the projective plane.

So, let a_i, $i = 1, 2, 3, 4$, be four distinct points on a line, and let the coordinates of a_i be x_i. The *cross ratio* of these four points in terms of the given coordinates is

$$\frac{x_1 - x_2}{x_1 - x_4} \cdot \frac{x_3 - x_4}{x_3 - x_2},$$

where in the canonical example of separation we have $a_1 a_3 \, S \, a_2 a_4$ and the numerical order $x_1 < x_2 < x_3 < x_4$.

We must deal with one complication in assigning coordinates to points. We must permit ∞ as well as any real number to be a coordinate (we comment on the reason for this soon). So our set of coordinates is

$$\mathrm{Re}^* = \mathrm{Re} \cup \{\infty\}.\,^2$$

For the algebra of ∞ and the cross ratios involving ∞ we assume, for $x \in \mathrm{Re}$:

(i) $x + \infty = \infty + x = x - \infty = \infty - x = \infty$.

(ii) If $x \neq 0$, then $x\infty = x/0 = \infty$ and $x/\infty = 0$.

(iii) If $x_4 = \infty$, then the cross ratio is

$$\frac{x_1 - x_2}{x_3 - x_2},$$

and similarly for any of the other three coordinates being ∞; and more generally, if a, b, c, and d are real numbers such that $ad - bc \neq 0$ and

---

^{2} Note that $-\infty$ is not required.

$x = \infty$, then

$$\varphi(x) = \frac{ax + b}{cx + d} = \frac{a}{c}.$$

(iv) As for order, $x < \infty$.

Note that, at most, one coordinate x_i, $i = 1, 2, 3, 4$, can be such that $x_i = \infty$, so that, e.g., we do not need to define $\infty - \infty$ or the cross ratio if two coordinates have the value ∞.

THEOREM 3. *Let $\mathfrak{A} = \langle A, S \rangle$ be a structure such that A is a nonempty set and S a quaternary relation on A. Then $\mathfrak{A}$ is a separable one-dimensional separation structure iff there is a function φ from A to Re^* such that for all a, b, c, and d in A*

$ab \, S \, cd$ *iff the cross ratio of a, b, c, d with respect to φ is negative.*

The uniqueness theorem for the numerical representation of the quaternary relation of separation requires a generalization of the concept of a monotone function used in the uniqueness conditions for the representation of simple order or linear betweenness. The intuitive idea is that the relation of separation is preserved, not only by monotone transformations of the numerical representations, but also by projective transformations that are not monotone. A function h from a subset R of Re^* to a subset of Re^* is *projectively monotone* iff h is strictly monotone or there exists a partition $(\underline{R}, \overline{R})$ of R such that every element of $\underline{R}$ is less than every element of $\overline{R}$ and either

(i) h is strictly increasing on $\underline{R}$ and on $\overline{R}$, and for every x in $\underline{R}$ and every x' in $\overline{R}$, $h(x') < h(x)$, or

(ii) h is strictly decreasing on $\underline{R}$ and on $\overline{R}$, and for every x in $\underline{R}$ and every x' in $\overline{R}$, $h(x) < h(x')$.

This general definition can be illustrated by linear fractional transformations and, at the same time, the need for ∞ can be further clarified. Consider, then, the projectively monotone function h_1 on Re^* defined by the equation

$$h_1(x) = \frac{x + 1}{2x + 1}.$$

(It is instructive to sketch the graph of this function.) $\underline{R}$ is the interval $(-\infty, -\frac{1}{2})$ and $\overline{R}$ is the interval $[-\frac{1}{2}, \infty]$, and it is easy to check that h_1 is strictly decreasing on $\underline{R}$ and $\overline{R}$; moreover, for x in $\underline{R}$ and x' in $\overline{R}$, $h_1(x) < h_1(x')$. Note that $h_1(-\frac{1}{2}) = \infty$ and $\cdot h_1(\infty) = \frac{1}{2}$. On the other hand, let

$$h_2(x) = \frac{-4x + 3}{-2x + 1}.$$

Then $\underline{R} = (-\infty, \frac{1}{2}]$ and $\overline{R} = (\frac{1}{2}, \infty]$, h_2 is strictly increasing on $\underline{R}$ and $\overline{R}$, and for x in $\underline{R}$ and x' in $\overline{R}$, $h_2(x) > h_2(x')$. In this case, $h_2(\frac{1}{2}) = \infty$ and $h_2(\infty) = 2$.

THEOREM 4. *Let* $\mathfrak{A} = \langle A, S \rangle$ *be a separable one-dimensional separation structure, and let* φ *and* φ' *be two representing functions satisfying Theorem 3. Let R and R' be the ranges of* φ *and* φ' *respectively. Then there exists a projectively monotone function h from R to R' such that for all* $a \in A$, $h[\varphi(a)] = \varphi'(a)$. *Moreover, if h is any projectively monotone function defined on R, then* $h \circ \varphi$ *is another numerical representation of* $\mathfrak{A}$.

13.3 PROOFS

THEOREM 3. *Let* $\mathfrak{A} = \langle A, S \rangle$ *be a structure such that A is a nonempty set and S a quaternary relation on A. Then* $\mathfrak{A}$ *is a separable one-dimensional separation structure iff there is a function* φ *from A to Re* such that for all a, b, c, and d in A, ab S cd iff the cross ratio of a, b, c, d with respect to* φ *is negative.*

PROOF. We give a sketch of the proof. Let a and b be distinct elements of A. Then it may be shown (see Exercise 3) that A is divided into two "segments" as shown in Figure 1. In each segment, define

$$x \prec_{ab} y \quad \text{iff} \quad ay \, S \, bx.$$

Then define one of the two orders (relative to a and b with a as first element) by

$$a \prec \text{segment 1 (under } \prec_{ab}) \prec b \prec \text{segment 2 (under } \prec_{ab}).$$

It can be shown that this ordering is a simple one satisfying Theorem 2 of Chapter 2, and hence there is an isomorphism φ of $\langle A, \prec_{ab} \rangle$ into $\langle \text{Re}, < \rangle$. [Note that $\langle \text{Re}^*, < \rangle$ is isomorphic to any interval (a, b), $a, b \in \text{Re}, a \neq b$.] It is straightforward but tedious to check that the cross ratio of any four distinct points is negative if and only if they form the appropriate separating pairs.   ◇

THEOREM 4. *Let $\mathfrak{A} = \langle A, S \rangle$ be a separable one-dimensional separation structure, and let φ and φ' be two representing functions satisfying Theorem 3. Let R and R' be the ranges of φ and φ', respectively. Then there exists a projectively monotone function h from R to R' such that for all $a \in A$, $h[\varphi(a)] = \varphi'(a)$. Moreover, if h is any projectively monotone function defined on R, then $h \circ \varphi$ is another numerical representation of $\mathfrak{A}$.*

PROOF. Let h be defined by $\varphi' = h \circ \varphi$. First, if A has less than four elements, or if h is monotone, there is nothing left to prove. So, assume there are four elements with respect to which h is not monotone. Without loss of generality we can assume them numerically ordered:

$$x_1 < x_2 < x_3 < x_4, \tag{1}$$

where $\varphi(a_i) = x_i$, with φ being one of the mappings from the empirical structure to the numerical structure, and $\varphi' = h \circ \varphi$. By hypothesis then from (1),

$$a_1 a_3 \, S \, a_2 a_4. \tag{2}$$

Excluding the monotone transformation of (1), one can easily see that to preserve (2) under $\varphi' = h \circ \varphi$, there are only six possibilities, for if transformation h preserves separation but not linear order, then h must induce one of three circular permutations of either $x_1 < x_2 < x_3 < x_4$ or of the reverse order.

$$h(x_2) < h(x_3) < h(x_4) < h(x_1)$$
$$h(x_3) < h(x_4) < h(x_1) < h(x_2)$$
$$h(x_4) < h(x_1) < h(x_2) < h(x_3)$$
$$h(x_1) < h(x_4) < h(x_3) < h(x_2)$$
$$h(x_2) < h(x_1) < h(x_4) < h(x_3)$$
$$h(x_3) < h(x_2) < h(x_1) < h(x_4) \tag{3}$$

Without essential loss of generality, we may consider four other distinct points b_i, where $\varphi(b_i) = y_i$, $i = 1, \ldots, 4$, with $y_1 < y_2 < y_3 < y_4$. We note at once that for these four points as well, h must satisfy one of the six possibilities of (3). Roughly speaking, what we must show is that the same type of six possibilities hold with the same partition $(\underline{R}, \overline{R})$, as described on p. 88. There are a large number of cases to consider. We give the details for two typical ones, labeled Case A and Case B.

Case A. Suppose, by way of contradiction, that the x_i satisfy the first inequality of (3) and the y_i the fourth, i.e.,

$$h(x_2) < h(x_3) < h(x_4) < h(x_1)$$
$$h(y_1) < h(y_4) < h(y_3) < h(y_2). \qquad (4)$$

(In the first inequality h is increasing on both $\underline{R}$ and $\overline{R}$, but in the second it is decreasing, which is the source of our contradiction.)

We also restrict ourselves by assuming that

$$x_3 < x_4 < y_3 < y_4. \qquad (5)$$

Given the restrictions imposed by the inequalities (4) and (5), there are six subcases to consider, each of which leads to a contradiction.

Subcase 1. $h(x_3) < h(x_4) < h(y_4) < h(y_3)$,

so h is not invariant since x_4 and y_4 are no longer separated.

Subcase 2. $h(x_3) < h(y_4) < h(x_4) < h(y_3)$.

Here again x_4 and y_4 are no longer separated under h.

Subcase 3. $h(x_3) < h(y_4) < h(y_3) < h(x_4)$.

To violate invariance we include x_2 whose position in this case is fixed without further assumption. So we have $x_2 < x_3 < y_3 < y_4$ and $h(x_2) < h(x_3) < h(y_4) < h(y_3)$, and h is not invariant, since x_3 and y_4 are no longer separated.

Subcase 4. $h(y_4) < h(y_3) < h(x_3) < h(x_4)$.

Here x_3 and y_3 are no longer separated under h.

Subcase 5. $h(y_4) < h(x_3) < h(y_3) < h(x_4)$.

Again x_3 and y_3 are no longer separated under h.

Subcase 6. $h(y_4) < h(x_3) < h(x_4) < h(y_3)$.
Subcase 6a. $x_4 < y_2$.

Then $x_3 < y_2 < y_3 < y_4$ and $h(y_4) < h(x_3) < h(y_3) < h(y_2)$, and x_3 and y_3 are no longer separated under h.

Subcase 6b. $y_2 < x_4$.

Then $y_2 < x_4 < y_3 < y_4$ and $h(y_4) < h(x_4) < h(y_3) < h(y_2)$, and x_4 and y_4 are no longer separated under h.

Case B. A similar but different analysis is required, when instead of (4), h is increasing on both the x_i and the y_i. Here a contradiction arises if the partitions implied by the x_i and y_i are different. We may show the basis of the contradiction without explicitly constructing the partitions. We replace the second line of inequalities of (4) by one representing an h increasing on $\underline{R}$ and $\overline{R}$. In particular we assume

$$h(y_3) < h(y_4) < h(y_1) < h(y_2), \tag{6}$$

the second line of (3), and assume as a typical ordering relating the x_i and y_i

$$y_1 < y_2 < y_3 < x_1 < x_2 \tag{7}$$

without commitment to the linear ordering of x_3, x_4, and y_4, which is not needed. We have from (3), (6), and (7)

$$h(y_3) < h(x_1) < h(y_1) < h(y_2),$$

where the position of $h(x_1)$ is fixed by (7), and the invariance of h given by (6). By similar invariant considerations,

$$h(y_3) < h(x_1) < h(x_2) < h(y_1) < h(y_2),$$

but this contradicts the first line of (4), which asserts $h(x_2) < h(x_1)$.

We now show that if h is projectively monotone, then $h \circ \varphi$ is another numerical representation of $\mathfrak{A}$.

First, if h is monotone, it is clear that a negative cross ratio is preserved. It is sufficient to consider one of the other two cases. Assume h is strictly increasing on $\underline{R}$ and on $\overline{R}$. As

$$x_1 < x_2 < x_3 < x_4,$$

then obviously the cross ratio

$$\frac{x_1 - x_2}{x_1 - x_4} \cdot \frac{x_3 - x_4}{x_3 - x_2} < 0,$$

for the signs of the numerical differences are

$$\frac{-}{-} \cdot \frac{-}{+} = -.$$

Now on the hypothesis on h, we have

$$\frac{h(x_1) - h(x_2)}{h(x_1) - h(x_4)} \cdot \frac{h(x_3) - h(x_4)}{h(x_3) - h(x_2)} = \frac{-}{+} \cdot \frac{-}{-} = -$$

as desired.

This completes the proof of Theorem 4.

13.4 PROJECTIVE PLANES

We now take up axioms for projective planes. Initially we shall consider the unordered plane and later add the concept of separation, analyzed earlier. To emphasize the duality of lines and points, one of the most beautiful features of projective geometry, we treat them symmetrically. We do not characterize lines as certain sets of points. Intuitively speaking, a projective plane represents the geometry that remains invariant when points and figures are viewed from various points of perspective. From another standpoint, projective geometry is the geometry of what can be constructed using only a straight edge.

Our basic set A of geometrical objects consists of lines and points. Hewing close to our earlier usage, we denote points by $a, b, c, d, e, \ldots$ and lines by $\alpha, \beta, \gamma, \epsilon, \ldots$. In addition to the set P of points and set L of lines, the relation of incidence I is a subset of $(P \times L) \cup (L \times P)$, that is, a point is incident to a line, and a line is incident to a point. If $a I \alpha$, we also say that a *lies on* α. We have two defined binary operations $P \times P \to L$ and $L \times L \to P$: any two points determine a line, and any two lines determine a point. It is this last property of any two lines determining a point that makes the axioms of the projective plane intrinsically much simpler than those of the Euclidean plane. These two binary operations we denote by juxtaposition, and when necessary by a dot, with parentheses as needed. Their definitions in terms of incidence, justified by Axioms 2 and 2′ below, are the following:

> If $a \neq b$, then $ab = \alpha$ iff $a I \alpha$ and $b I \alpha$.
> If $\alpha \neq \beta$, then $\alpha\beta = a$ iff $\alpha I a$ and $\beta I a$.

The notation ab for the line determined by points a and b is, of course,

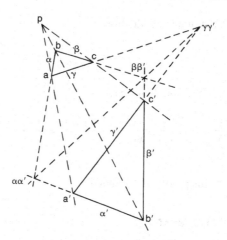

FIGURE 3. Desargues' proposition.

just the familiar notation of Euclidean geometry. The dual notation $\alpha\beta$ for the point determined by lines α and β is just as intuitive when, in the case of Euclidean geometry, α and β being parallel is excluded. That this exclusion need not be made in projective geometry is, as remarked, a chief reason for its simplicity.

To emphasize duality, the axioms are stated in pairs, the first unprimed member of the pair being mainly about points and its primed dual about lines. Under this arrangement some of the axioms are redundant but making the duality explicit seems worth it. We use informal geometrical language in the statement of the axioms, but the translation into the standard relational language is obvious. We say that two triangles are *perspective from a point* iff there is a one-to-one correspondence between the vertices of the two triangles such that the three lines passing through the three pairs of corresponding vertices are incident with a common point. Dually, two triangles are *perspective from a line* iff there is a one-to-one correspondence between the lines forming the sides of the two triangles such that pairs of corresponding lines meet at points that themselves lie on a common line. (See Figure 3, where a, b, c, and a', b', c' are corresponding vertices of triangles and α, β, γ and α', β', γ' are corresponding lines. Note that points $\alpha\alpha'$, $\beta\beta'$, and $\gamma\gamma'$ are collinear.)

DEFINITION 3. *Let A be a nonempty set, let P (the set of points) and L (the set of lines) be disjoint subsets of A, with $P \cup L = A$, and let I be a binary relation (of incidence) that is a subset of $(P \times L) \cup (L \times P)$. Then*

$\mathfrak{A} = \langle A, P, L, I \rangle$ *is a* projective plane *iff the following axioms are satisfied*:

1. *If a point is incident with a line, the line is incident with the point.*

1'. *If a line is incident with a point, the point is incident with the line.*

2. *Given two distinct points, there is a unique line with which both are incident.*

2'. *Given two distinct lines, there is a unique point with which both are incident.*

3. *There exist four distinct points of which no three are incident with the same line.*

3'. *There exist four distinct lines of which no three are incident with the same point.*[3]

Moreover, $\mathfrak{A}$ *is said to be* Desarguesian *iff the following holds*:

4. *If two distinct triangles are perspective from a point, they are perspective from a line.*

4'. *If two distinct triangles are perspective from a line they are perspective from a point.*[4]

The axioms all have a rather obvious intuitive content except for Axioms 4 and 4'. Axiom 4 is a classical form of Desargues' proposition,[5] as illustrated in Figure 3. Desargues' proposition asserts that if two triangles are perspective from a point, they are perspective from a line; Axiom 4' is its dual. Axiom 3 expresses the fact that the dimension of the space is at least two. That the space is of, at most, two dimensions is guaranteed by Axiom 2', which asserts that any two distinct lines have exactly one point in

---

[3] An algebraic formulation of Axioms 3 and 3' is the following: *There exist distinct points a_1, a_2, a_3, a_4 and distinct lines $\alpha_1, \alpha_2, \alpha_3, \alpha_4$ such that*

3. $a_1 a_2 = \alpha_1$, $a_2 a_3 = \alpha_2$, $a_3 a_4 = \alpha_3$, *and* $a_4 a_1 = \alpha_4$.

3'. $\alpha_1 \alpha_2 = a_1$, $\alpha_2 \alpha_3 = a_2$, $\alpha_3 \alpha_4 = a_3$, *and* $\alpha_4 \alpha_1 = a_4$.

[4] An algebraic formulation of Axioms 4 and 4' is this, where $NE(a_1, a_2, a_3)$ means a_1, a_2, and a_3 are distinct points:

4. *If for* $i, j = 1, 2, 3$, (i) $NE(a_1, a_2, a_3)$ *and* $NE(b_1, b_2, b_3)$, (ii) $a_i \neq b_i$, (iii) $a_i a_j \neq b_i b_j$, $i \neq j$, *and* (iv) $a_i b_i I c$, *then*

$$(a_1 a_2 \cdot b_1 b_2)(a_1 a_3 \cdot b_1 b_3) = (a_2 a_3 \cdot b_2 b_3)(a_1 a_3 \cdot b_1 b_3).$$

4'. *If for* $i, j = 1, 2, 3$, (i) $NE(\alpha_1, \alpha_2, \alpha_3)$ *and* $NE(\beta_1, \beta_2, \beta_3)$, (ii) $\alpha_i \neq \beta_i$, (iii) $\alpha_i \alpha_j \neq \beta_i \beta_j$, $i \neq j$, *and* (iv) $\alpha_i \beta_i I \gamma$, *then*

$$(\alpha_1 \alpha_2 \cdot \beta_1 \beta_2)(\alpha_1 \alpha_3 \cdot \beta_1 \beta_3) = (\alpha_2 \alpha_3 \cdot \beta_2 \beta_3)(\alpha_1 \alpha_3 \cdot \beta_1 \beta_3).$$

[5] In the literature of geometry, Axiom 4 is also called Desargues' *theorem* because he first proved it from other assumptions.

common. Axiom 3 can also be expressed by saying that there exists an incidence quadrangle, and Axiom 3′ that there exists an incidence quadrilateral.

We are mainly interested in axioms strong enough to establish a representation over the real numbers. For this purpose we add axioms of order and of completeness. Later we return to the general axioms of Definition 3. We extend the one-dimensional concepts of Section 12.2.2. The intuitive idea is that projective order is invariant under any perspective transformation or, indeed, any finite composition of perspective transformations that is a projective transformation. The appropriate axiom is formulated below [as (iii)] without introducing explicitly the concept of transformation.

DEFINITION 4. *Let* $\langle A, P, L, I \rangle$ *be a Desarguesian projective plane, and let S be a quaternary relation on A. Then $\mathfrak{A} = \langle A, P, L, I, S \rangle$ is a complete ordered Desarguesian projective plane iff*:

(i) *The relation S satisfies Axioms 1–5 of Definition 2.*

(ii) *The relation S satisfies Axiom 6 of Definition 2 iff a, b, c, and d are distinct and collinear.*

(iii) *If point e is not incident with lines α or β and if a, b, c, d are on α and $ab\,S\,cd$, then $(ae\beta)(be\beta)\,S\,(ce\beta)(de\beta)$.*

(iv) *(Axiom of Completeness) Given two arbitrary nonempty sets of points X and Y and a point c, if there exists a point a such that*

$$p \in X \quad and \quad q \in Y \quad implies \quad aq\,S\,pc,$$

then there exists a point b such that

$$p \in X - \{b\} \quad and \quad q \in Y - \{b\} \quad implies \quad pq\,S\,bc.$$

Condition (ii) expresses the limited connectivity on a line required for projective order in the plane. The meaning of (iii) is clarified by Figure 4.

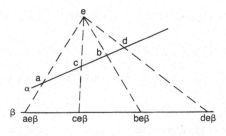

FIGURE 4. Perspective from a point.

Note that if $\alpha = \beta$, then $ae\beta = a$, etc. We next state the main representation theorem of this section, which shows that Definition 4 does give a qualitative axiomatization of analytic ordered plane projective geometry over the real numbers. For the reasons stated earlier, we omit the long classical proof.

THEOREM 5. *Let $\mathfrak{A}$ be a complete ordered Desarguesian projective plane in the sense of Definition 4. Then $\mathfrak{A}$ is isomorphic to a two-dimensional analytic ordered projective geometry over* Re *in the sense of Definition 5 of Section 12.2.3.*

As has already been suggested, the study of projective planes generalizes beyond the strong representation of Theorem 5. In particular, we survey informally and without proofs the natural representation of Desarguesian planes satisfying Definition 3, but not necessarily Definition 4. We also examine the nature of representations for non-Desarguesian planes. In this survey we are particularly focused on the interplay between geometric and algebraic conditions, a subject that has led to some of the most significant results in projective geometry.

The general representation theorem for Desarguesian projective planes in the sense of Definition 3 is formulated in terms of two-dimensional vector spaces over division rings. A division ring satisfies all the axioms of a field except that multiplication is not necessarily commutative.

We also need the concept of a *two-dimensional analytic geometry* $\mathcal{H}_2(\mathcal{F})$ over a division ring $\mathcal{F} = \langle F, +, \cdot \rangle$. Points of $\mathcal{H}_2(\mathcal{F})$ are certain equivalence classes of 3-tuples (x_0, x_1, x_2) that are elements of F^3, the Cartesian product of F three times. Two 3-tuples (x_0, x_1, x_2) and (y_0, y_1, y_2) are *left-proportional* if and only if there is a $c \neq 0$ in F such that $y_i = c \cdot x_i$ for $i = 0, 1, 2$. *Points* of $\mathcal{H}_2(\mathcal{F})$ are then left-proportional equivalence classes. Dually, *lines* are right-proportional equivalence classes: $y_i = x_i \cdot c$ for some $c \neq 0$ and $i = 0, 1, 2$.

A point $[x']$ is incident with a line $[y']$ if and only if for any $x \in [x']$ and $y \in [y']$ the "inner product" $x \cdot y = 0$, i.e.,

$$x_0 y_0 + x_1 y_1 + x_2 y_2 = 0.$$

The following result is then easy to check for any division ring $\mathcal{F}$. The two-dimensional homogeneous analytic geometry $\mathcal{H}_2(\mathcal{F})$ with the definitions of point, line, and incidence given above satisfies the axioms of Definition 3, that is, it is a Desarguesian projective plane. (Strictly speaking, if $\mathcal{F}$ is a field, then the set of points and the set of lines in the geometry are not disjoint but identical. We ignore this problem, which is easy to deal with set-theoretically.)

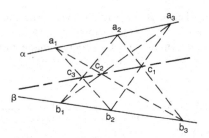

FIGURE 5. Pappus' proposition.

The representation theorem is then this. *For any Desarguesian projective plane there is a division* ring $\mathscr{F}$ *such that the plane is isomorphic to* $\mathscr{H}_2(\mathscr{F})$.

One of the most beautiful relations between geometry and algebra can be established by means of the theorem of Pappus, a geometer who lived in Alexandria in the fourth century A.D. We say that a Desarguesian projective plane *satisfies Pappus' proposition* if and only if, we are given two distinct lines α and β, distinct points a_1, a_2, and a_3, and distinct points b_1, b_2, and b_3 such that a_1, a_2, and a_3 are incident with α but not β, and b_1, b_2, b_3 are incident with β but not α, then the points $c_1 = (a_2 b_3) \cdot (a_3 b_2)$, $c_2 = (a_3 b_1) \cdot (a_1 b_3)$, and $c_3 = (a_1 b_2) \cdot (a_2 b_1)$ are collinear. (See Figure 5.) The fundamental result is this. *A Desarguesian projective plane satisfies Pappus' proposition iff the multiplication operation of "its" division ring is commutative.*

This result, together with Theorem 5, reveals some of the power hidden in order–completeness (Definition 4); a complete ordered Desarguesian projective plane must satisfy Pappus' proposition. It is also worth noting that if a projective plane satisfies all but Axioms 4 and 4′ of Definition 3 *and* Pappus' proposition, then it must be Desarguesian, i.e., it must satisfy Axioms 4 and 4′.

Hilbert gave an example of an ordered noncommutative division ring by modifying appropriately the definition of multiplication for the field of formal power series over the field of rationals. Details can be found in Heyting (1963, pp. 136–138). The ordered projective plane constructed over Hilbert's division ring can be used as a model to show that ordered Desarguesian projective planes do not necessarily satisfy Pappus' proposition.

The literature of projective geometry includes extensive consideration of finite spaces or finite geometries. Obviously, if a projective plane, for example, is finite, the division ring $\mathscr{F}$ underlying the isomorphic coordinate system $\mathscr{H}_2(\mathscr{F})$ is finite. One can use the famous theorem of Wedderburn that every finite division ring is a field, that is, multiplication is commutative, to infer that in any finite projective plane Pappus' proposition is satisfied.

Still weaker structures are characterized by using only Axioms 1, 1', 2, 2', 3, and 3' of Definition 3 for *projective planes*. The representation is in terms of ternary rings,[6] which are generalizations of division rings.

Let A be a nonempty set and t a ternary operation on A, i.e., t is a mapping from $A \times A \times A$ to A. Then $\langle A, t \rangle$ is a *ternary ring* iff the following axioms hold.

(i) There are distinct elements 0 and 1 in A such that

$$t(0, a, b) = t(a, 0, b) = b,$$
$$t(1, a, 0) = t(a, 1, 0) = a,$$

for all a, b in A;

(ii) For a, b, c, d in A with $a \neq c$, there exists a unique x in A such that

$$t(x, a, b) = t(x, c, d);$$

(iii) For a, b, c in A, there exists a unique x in A such that

$$t(a, b, x) = c;$$

(iv) For a, b, c, d in A with $a \neq c$, there exists a unique pair (x, y) in $A \times A$ such that

$$t(a, x, y) = b \quad \text{and} \quad t(c, x, y) = d.$$

A simple intuitive but special example of a ternary ring is this. Let A be the set of real numbers, and define

$$t(a, b, c) = a \cdot b + c.$$

To show how to get projective planes from ternary rings, we may define points as ordered pairs (x, y) of $A \times A$ and lines as sets of points

$$L(a, b) = \{(x, t(x, a, b)) | x \text{ in } A\},$$

---

[6] For discussion of the several different terminologies for ternary rings in the literature, see Dembowski (1968).

with the parameters a, b also in A, and also

$$L(c) = \{(c, y)|y \text{ in } A\}.$$

Incidence is now set-theoretical membership or its converse. Details of the construction are to be found in Dembowski (1968) or Stevenson (1972). Going the other way, we obtain the representation theorem: *Every projective plane is isomorphic to a projective plane constructed from a ternary ring.*
Details of the construction are to be found in many places, e.g., Blumenthal (1961) or Stevenson (1972).

A close relation also exists between Desarguesian planes and associativity, which is worth describing. We follow the exposition in Stevenson (1972). Let A be a nonempty set and ∘ a binary operation on A. Then $\langle A, \circ \rangle$ is a *loop* iff for every a and b in A:

   (i)   There is a unique c in A such that $a \circ c = b$;

   (ii) There is a unique c in A such that $c \circ a = b$;

   (iii) There is a unique e in A such that for all x in A,

$$e \circ x = x \circ e = x.$$

the element e is, as usual, called an *identity* element.

The following result relates ternary rings and loops. Let $\langle A, t \rangle$ be a ternary ring, and define the binary operations $+$ and ∘ as follows for a, b in A:

$$a + b = t(1, a, b),$$
$$a \circ b = t(a, b, 0).$$

Then $\langle A, + \rangle$ and $\langle A - \{0\}, \circ \rangle$ are loops, where 0 is the identity of $\langle A, + \rangle$ and 1 is the identity of $\langle A - \{0\}, \circ \rangle$. We say that the algebra $\langle A, +, \circ \rangle$, with $+$ and ∘ as just defined, is *generated* by the ternary ring $\langle A, t \rangle$. Note that in general neither of the binary operations defined is associative. We now describe conditions under which they are.

First, let $\langle A, t \rangle$ be a ternary ring, and let $\langle A, +, \circ \rangle$ be the algebra generated by the ternary ring. Then the ternary ring is *linear* iff for all a, b, c in A,

$$t(a, b, c) = (a \circ b) + c.$$

(Hereafter, when convenient, we let juxtaposition denote multiplication.) When the ternary ring is linear, we say that the algebra $\langle A, +, \circ \rangle$ is a *planar ring*. We leave as an exercise the proof that a planar ring has the following properties:

(i) $\langle A, + \rangle$ and $\langle A, \circ \rangle$ are loops;

(ii) $0 \circ a = a \circ 0 = 0$ for any a in A;

(iii) For any a, b, c, d in A with $a \neq c$ there is a unique x in A such that $xa + b = xc + d$;

(iv) For any a, b, c in A there is a unique x in A such that $ab + x = c$;

(v)   For any a, b, c, d in A with $a \neq c$ there is a unique pair (x, y) in $A \times A$ such that

$$ax + y = b \quad \text{and} \quad cx + y = d.$$

Now we come to the geometric equivalences, which are Desarguesian in character.

First, let $\mathfrak{A} = \langle A, t \rangle$ be a ternary ring with quadrangle (a, b, c, d), as guaranteed by Axiom 3 of Definition 3. Then $\mathfrak{A}$ is linear iff all triangles cef and ghi that are perspective from the point b are perspective from the line ab. Dually $\mathfrak{A}$ is linear iff all pairs of triangles that are perspective from the line bc are perspective from the point a.

To state detailed results about associativity, we tighten our characterization of Desarguesian planes. A plane is (a, α)-*Desarguesian* iff any two triangles that are perspective from point a are also perspective from line α. Let $\mathfrak{A} = \langle A, t \rangle$ be a ternary ring, $\mathfrak{A}' = \langle A, +, \circ \rangle$ its generated algebra, and P its generated projective plane. Then,

(i)   the projective plane P is (a, α)-Desarguesian for some a lying on α iff $\mathfrak{A}$ is a linear ternary ring and the operation of addition of $\mathfrak{A}'$ is associative;

(ii) the projective plane P is (a, α)-Desarguesian for some a not lying on α iff $\mathfrak{A}$ is a linear ternary ring and the operation of multiplication of $\mathfrak{A}'$ is associative. (In the literature, when the operation of addition of $\mathfrak{A}'$ is associative, $\mathfrak{A}'$ is often called a *Cartesian group*.)

Finally, we remark that it can be shown that the associativity of *both* binary operations of $\mathfrak{A}'$ does not imply any stronger geometrical condition (see Stevenson, 1972, pp. 308–309). The relation between Pappus' theorem and the commutativity of multiplication of the representing division ring for a projective plane is a classical and well-known result. The results just described about associativity are not as well known but provide some of the best geometrical examples of what a failure of associativity of a binary operation can mean. A much more general development of totally ordered nonassociative operations is given in Chapter 19 of Volume III.

13.5 PROJECTIVE SPACES

Without question the Desarguesian proposition (Axiom 4) is the most complicated axiom of Definition 3 for projective planes. The surprising and important fact is that in three or more dimensions this axiom can be proved from other axioms that correspond closely to the remaining ones given in Definition 3. This situation is reminiscent of the provability of Thomsen's or the stronger double cancellation axiom in conjoint measurement when the number of components exceeds two (see Section 6.11 of Volume I). Indeed, the highly special character of two-dimensional spaces is evident throughout geometry, as can be seen from several theorems in this chapter.

We need to extend in an obvious way the primitive concepts of Definition 3 by adding the set M of planes. Second, we add additional defined binary operations: if points a and b are distinct, then ab is a line; and if planes π_1 and π_2 are distinct, $\pi_1 \pi_2$ is a line. The definitions are these:

$$\text{If} \quad a \neq b, \quad \text{then} \quad ab = \alpha \quad \text{iff} \quad aI\alpha \quad \text{and} \quad bI\alpha.$$
$$\text{If} \quad \pi_1 \neq \pi_2, \quad \text{then} \quad \pi_1\pi_2 = \alpha \quad \text{iff} \quad \pi_1 I\alpha \quad \text{and} \quad \pi_2 I\alpha.$$

We also define the concept of a line lying on a plane:

$$\alpha \text{ lies on } \pi \quad \text{iff} \quad \text{for all points } a, \text{ if } aI\alpha, \text{ then } aI\pi.$$

This definition makes partially evident how close incidence is to set membership and lying on to set inclusion. It is also useful to define the intersection of two lines lying on a common plane:

$$\text{If} \quad \alpha \neq \beta \quad \text{and} \quad \alpha, \beta \quad \text{lie on } \pi,$$
$$\text{then} \quad \alpha\beta = a \quad \text{iff} \quad \alpha I a \quad \text{and} \quad \beta I a.$$

In three dimensions, points and planes are dual, as can be seen from the operations just defined. In defining three-dimensional projective space we use as many of the axioms of incidence of Definition 3 as possible. The axioms of order and the axiom of connectedness are just those given in the preceding section.

DEFINITION 5. *Let A be a nonempty set, let P, L, and M be disjoint subsets of A with $P \cup L \cup M = A$, let I be a binary relation that is a subset of*

$$(P \times L) \cup (L \times P) \cup (P \times M) \cup (M \times P) \cup (L \times M) \cup (M \times L),$$

and let S be a quaternary relation on A. Then $\mathfrak{A} = \langle A, P, L, M, I, S \rangle$ is a three-dimensional ordered projective space *iff the following axioms are satisfied for every a, b in P, α, β in L, and π, π₁, π₂ in M:*

Axioms of Incidence.

1. *The relation I is symmetric.*

2. *If $a \neq b$, then there is a unique line α such that $a\,I\,α$ and $b\,I\,α$.*

2′. *If $π_1 \neq π_2$, then there is a unique line α such that $π_1\,I\,α$ and $π_2\,I\,α$.*

3. *If $α \neq β$, $α\,I\,a$ and $β\,I\,a$, then there is a unique plane π such that α and β lie on π.*

3′. *If $α \neq β$, and α and β lie on π, then there is a unique point a such that $α\,I\,a$ and $β\,I\,a$.*

4. *If $a \neq b$, $a\,I\,π$ and $b\,I\,π$, then ab lies on π.*

4′. *If $π_1 \neq π_2$, $π_1\,I\,a$ and $π_2\,I\,a$, then $π_1π_2\,I\,a$.*

5. *If $a\,I\,α$ and $α\,I\,π$, then $a\,I\,π$.*

6. *At least three distinct points are incident with every line.*

6′. *Every line lies on at least three distinct planes.*

7. *All points are not incident with one common plane.*

7′. *All planes are not incident with one common point.*

8. *There are at least four noncoplanar points.*

Axioms of Order. The axioms are conditions (i)–(iii) of *Definition* 4, where (iii) *holds whenever lines α and β lie in the same plane.*

Axiom of Completeness. As stated in Definition 4.

The axioms of incidence can be reduced in number and, undoubtedly, some of them can be proved from others. However, each of them has an obvious and important intuitive meaning. It can be proved that every plane π in M, together with the lines and points incident with π, satisfies the axioms of Definition 3, that is, it is a Desarguesian projective plane.

The representation theorem for projective planes described in the previous section generalizes in a natural way to projective spaces.

THEOREM 6. *Every three-dimensional ordered projective space in the sense of Definition 5 is isomorphic to a three-dimensional analytic ordered projective geometry over Re in the sense of Definition 6 of Section 12.2.3.*

From the form in which we have stated the representation theorem it is clear that it embodies the classical result that the axioms for projective spaces we have given in Definition 5 are categorical. We recall that a theory is *categorical* if and only if any two models of the theory are isomorphic, and this follows at once from the representation theorem we have stated, because any two models are isomorphic to P_3. The uniqueness of the represention also follows from the results stated in Section 12.2.3.

THEOREM 7. *Any two analytic representations of a three-dimensional projective space are related by a projectivity, i.e., a mapping that takes points into points, lines into lines, planes into planes, and preserves incidence. Hence, any two representations are related by a nonsingular linear transformation of* Re^4.

13.6 AFFINE AND ABSOLUTE SPACES

We now return to the concept of order that is characteristic of affine geometry, namely, betweenness. We use the properties of this relation developed in Section 13.2.1. We begin by developing ordered geometry as a general foundation for affine or absolute geometry. By extending the axioms on the single primitive concept of betweenness, we get affine geometry. By adding the concept of congruence, but not the special affine axiom, we get absolute geometry, which itself specializes to either Euclidean or hyperbolic geometry.

There is a contrast between our approach to the primitives of projective and affine geometry that is more accidental than essential. We base affine geometry on the single ternary relation of betweenness. But we base ordered projective geometry on both the quaternary relation of separation and the binary relation of incidence. It should be apparent, however, that we could base ordered projective geometry on just the relation of separation. (We leave proof of this as an exercise.) Or, going in the other direction, we could have begun this section with unordered affine geometry, using the same primitive concepts as those for the unordered projective plane. Such affine axioms of incidence can be found in Pickert (1955, p. 10) or Lingenberg and Bauer (1960, p. 97). Another approach, in certain ways the most natural for affine geometry, is to take as primitive the quaternary relation of ab being parallel to cd, i.e., intuitively, the line determined by points a and b being parallel to the line determined by points c and d. A detailed development of this approach can be found in Szmielew (1983).

These remarks about alternative approaches to affine geometry help underscore the point that the particular arrangement of developments in this section is not in any sense mandatory. We begin with a characterization of ordered geometry, but we could have begun directly with either affine or absolute geometry, as will be clear in the sequel.

13.6.1 Ordered Geometric Spaces

For axioms of ordered geometry, we follow, with some modifications, Veblen (1904), Coxeter (1961), and Szczerba and Tarski (1965). Using the notation $a|c|b$ for point c lying between points a and b, as introduced in Section 13.2.1, we define first the *line* determined by two distinct points a and b:

$$ab = \{c|\ a|c|b \quad \text{or} \quad c|a|b \quad \text{or} \quad c|b|a\}.^7$$

The *segment* ab is just $\{c|\ a|c|b\}$. If $a = b$, then $ab = \{a\}$, but if $a \neq b$, then ab is the segment we expect it to be. Collinearity has a simple definition: points a, b, and c are *collinear* iff $c \in ab$ or $a = b$. A triple abc is a *triangle* iff points a, b, and c are not collinear. Let a, b, c be noncollinear; then the plane abc is defined as follows:

$$abc = \{d|(\exists e)(\exists f)(e \neq f \quad \text{and} \quad e, f \in ab \cup ac \cup bc$$
$$\text{and} \quad d, e, f \text{ are collinear})\}.$$

For 3-spaces, we use a *tetrahedron abcd* with the four points noncoplanar. Following the usual terminology, a, b, c, and d are the *vertices* of the tetrahedron, the segments ab, bc, ac, ad, bd, and cd are the *edges*, and the four triangular regions abc, bcd, cda, and dab are the *faces*. (If abc is a triangle, then the triangular region abc is

$$\{d|(\exists e)(\exists f)(e \neq f \quad \text{and} \quad e, f \in ab \cup ac \cup bc \quad \text{and} \quad e|d|f)\}.$$

The 3-space $abcd$ is then the set of all points collinear with pairs of points in one or two faces of the tetrahedron $abcd$.

We proceed in the same way for the 4-space, a concept that we need for the spaces of classical and special relativity. A quintuple $abcde$ is a *simplex* if and only if the five points do not all lie in a common 3-space. The *cells* of the simplex are the five tetrahedral regions determined by four points of the simplex. The *4-space abcde* is then the set of points collinear with pairs of points on one or two cells of the simplex.

Using the various concepts defined, we can now characterize ordered geometric spaces with axioms that have a rather obvious intuitive content.

DEFINITION 6.   *Let A be a nonempty set and* B *a ternary relation on A. Then* $\mathfrak{A} = \langle A, B \rangle$ *is an* ordered geometric space *iff the following axioms are*

---

[7] We have used the same notation for lines in this section as in Sections 13.4 and 13.5, although here lines are sets of points and in our development of projective geometry they are not.

satisfied for every a, b, c, d, e in A:

1. *Axioms 1–4 of Definition 1 are satisfied.*

2. *(Connectivity)   If a|b|c, a|b|d, and a ≠ b, then b|c|d or b|d|c.*

3. *(Extension)   There exists f in A such that f ≠ b and a|b|f.*

4. *(Pasch's Axiom)   If abc is a triangle, b|c|d, and c|e|a, then there is on line de a point f such that a|f|b.*

5. *(Axiom of Completeness)   For every partition of a line into two nonempty sets Y and Z such that*

     (i)   *no point b of Y lies between any a and c of Z and*

     (ii) *no point b' of Z lies between any a' and c' of Y,*

there is a point b of Y ∪ Z such that for every a in Y and c in Z, b lies between a and c.

Moreover, if all points of 𝔄 *lie on a plane, but not on a line,* 𝔄 *is* two-dimensional; *if all points of* 𝔄 *lie in a 3-space but not on a plane,* 𝔄 *is* three-dimensional; *if all points of* 𝔄 *lie in a 4-space but not in a 3-space,* 𝔄 *is* four-dimensional.

When 𝔄 is two-dimensional, a form of Desargues' proposition is sometimes added as an axiom, as for example, by Szczerba and Tarski (1965), who call the resulting two-dimensional spaces *restricted affine spaces.* We shall assume such an axiom here. The formulation simply in terms of betweenness is forbiddingly complex, as can be seen by using the form given by Szczerba and Tarski.

Desargues' axiom.   If d|a|a', d|b|b', d|c|c', a|b|e, a'|b'|e, a|c|f, a'|c'|f, b|c|g, b'|c'|g, not d|a|b, not a|b|d, not b|d|a, not d|b|c, not b|c|d, not c|d|b, not d|c|a, not c|a|d, not a|d|c, and a ≠ a', then e|f|g.

It should be be noted that this is a special form of Desargues' proposition because both the hypothesis and conclusion imply an ordering, which is not assumed in Axiom 4 of Definition 3. Coxeter (1961), on the other hand, does not include Desargues' proposition as one of the axioms of two-dimensional ordered geometry. It is provable for three- and four-dimensional ordered geometry. We explicitly include it as part of Definition 6 for the two-dimensional case.

We turn now to the representation theorem for n-dimensional ordered spaces $(n = 2, 3, 4)$. For completeness, we recall the definition of the relation of betweenness when the points are n-dimensional vectors, that is, ordered n-tuples of real numbers, in a Cartesian space. Let x, y, and z be three such points. Then the relation of betweenness holds if the following condition is satisfied:

$$B_R(x, y, z) \qquad \text{iff} \quad \text{there exists an } \alpha \in [0, 1] \text{ such that}$$
$$y = \alpha x + (1 - \alpha) z.$$

Let S be a nonempty, open, and convex set of n-dimensional Cartesian space. Then the structure $\langle S, B_R \rangle$ is an n-dimensional ordered space (over the real numbers) $n = 2, 3, 4$. It is easy to check that such an ordered space over the real numbers satisfies the axioms given in Definition 6. The following representation theorem can then be proved.

THEOREM 8. *Every n-dimensional ordered geometric space in the sense of Definition 6 (n = 2, 3, 4) is isomorphic to an n-dimensional ordered space over the field of real numbers.*

Appropriate methods of proof of this theorem are widely known in the literature and will not be considered here. We omit a uniqueness theorem for ordered spaces in general but state them for the various important special cases. We have restricted ourselves to $n = 2, 3, 4$ only to avoid formulating axioms of dimension for arbitrary $n > 4$. The other axioms remain unchanged.

13.6.2 Affine Spaces

We obtain ordered affine geometry from ordered geometry by adding an axiom equivalent to Euclid's famous fifth postulate. One classical form is Playfair's axiom (1795): *Given a point a and a line α on which a does not lie, there exists at most one line through a in the plane aα that does not meet α.* It is geometrically easy to see how this axiom strengthens the axioms of ordered geometry. If we take as a model of three-dimensional ordered geometry the interior of a sphere, then through point a as described above there would be many lines that lie in the plane $a\alpha$ but do not meet α. Playfair's axiom eliminates this possibility and yields for each dimension a categorical set of axioms. For our purposes we use an axiom, which we label as Euclidean but which is due in its present form to Szczerba and Tarski (1965). Its advantage is that it is formulated only in terms of betweenness. (Given the axioms of Definition 6, these various formulations are all equivalent.)

DEFINITION 7. *Let $\mathfrak{A} = \langle A, B \rangle$ be an ordered geometric space (in the sense of Definition 6). Then $\mathfrak{A}$ is an affine space iff the following Euclidean axiom E is also satisfied: for every a, b, c, d, and e there are points f and g such that if a|e|d, b|e|c, and a $\neq$ e, then a|b|f, a|c|g, and f|d|g.*

Using the numerical characterization of betweenness given in the previous section, the following representation theorem for affine spaces can be proved.

THEOREM 9. *Every n-dimensional affine space ($n = 2, 3, 4$) is isomorphic to the n-dimensional affine space over the field of real numbers. Moreover, for a given n, any two affine spaces are isomorphic.*

In other words, the theorem asserts the categorical character of the axioms for a given dimension. For $n = 2$ we get the affine Euclidean plane, etc. Generalization to $n > 4$ is immediate. As already remarked, we have not stated the general dimensionality axioms for arbitrary n in order to avoid further notation. A good discussion of these dimension axioms is to be found in Kordos (1969).

We now consider the uniqueness of the representation. In standard geometrical terms, the uniqueness is simply uniqueness up to affine transformations. An affine transformation is any linear transformation φ of n-dimensional Cartesian space such that

$$\varphi(x) = Ax + b,$$

where A is a nonsingular $n \times n$ matrix, b is any n-dimensional vector, that is, any n-tuple of real numbers, and x is also an n-dimensional vector. Intuitively, the matrix A represents a rotation together with stretches in different amounts along the n Cartesian coordinates. The vector b represents a translation of the origin of the Cartesian space. As mentioned in Section 12.2.6, there is an affine concept of congruence for collinear or parallel segments which is invariant under an affine transformation. Parallelograms can be used to define affine congruence. Opposite sides of a parallelogram are affine congruent.

THEOREM 10. *The representation of an affine space in terms of the Cartesian space over the field of real numbers is unique up to an affine transformation.*

Using the concepts of Section 12.2.2, if we introduce a defined concept of incidence, we can state that an affine space of n-dimensions is isomorphic to an n-dimensional analytic ordered affine geometry, as characterized by Definition 12.4.

We remark that an entirely different but equivalent approach to affine geometry is to begin with a projective space, say three-dimensional as an example, and then remove a plane, labeled the "plane at infinity." The resulting space is affine. In the two-dimensional case, a line at infinity is removed to obtain the affine plane.

13.6.3 Absolute Spaces

In this section we consider the so-called absolute spaces. Absolute geometry is a generalization of Euclidean and hyperbolic geometry in the following sense. To the axioms of absolute geometry, which we give in this subsection, we add one axiom to obtain the standard Euclidean geometry, and we add its negation to obtain hyperbolic geometry. These geometries are essential in the discussion of perceptual spaces (Section 13.10). Hyperbolic geometry is often called Bolyai–Lobachevskian, after its two creators, the Hungarian Janos Bolyai (1802–1860) and the Russian Nikolai Ivanovich Lobachevski (1793–1856).

The relational structure that we adopt for absolute geometry consists of the primitives of ordered geometry (a set of points and a ternary relation of betweenness) plus a quaternary relation of equidistance; $ab \sim cd$ is interpreted to mean that the distance between points a and b is the same as the distance between points c and d. The concept of equidistance is, of course, the concept of congruence for segments; it is also the concept of equivalence for absolute differences used in Section 4.10 of Volume I.

We continue to use the defined concepts of line, etc., introduced in ordered geometry. In addition, it is convenient to define the notion of a half-line or ray to use in the formulation of the axioms of congruence. Let a be any point of a line L. The set $L - \{a\}$ can be uniquely represented as the sum of two nonempty and disjoint sets,

$$L - \{a\} = A_1 \cup A_2,$$

such that for no two points b and c of A_i ($i = 1, 2$) do we have $b|a|c$. We call A_1 and A_2 half-lines or rays determined on line L by point a, and the point a is called the origin of the half-lines.

We turn now to the formal definition of absolute spaces. The dimensionality of the space depends on the dimensionality of the ordered space that is part of the definition.

DEFINITION 8.   Let $\langle A, B \rangle$ be an ordered geometric space (in the sense of Definition 6), and let $\sim$ be a quaternary relation on A. Then $\mathfrak{A} = \langle A, B, \sim \rangle$ is an absolute space iff the following axioms of congruence hold for every a, b, c, d, e, f, a', b', c', and d' in A:

1. If $aa \sim bc$, then $b = c$.
2. $ab \sim ba$.
3. If $ab \sim cd$ and $ab \sim ef$, then $cd \sim ef$.
4. If $a|b|c$, $a'|b'|c'$, $ab \sim a'b'$, and $bc \sim b'c'$, then $ac \sim a'c'$.

5. *For every half-line L with origin a and for every segment cd there exists just one point b in L such that ab ~ cd.*

6. *If abc and a'b'c' are two triangles with ab ~ a'b', bc ~ b'c', and ac ~ a'c', and if d and d' are two points such that b|c|d and b'|c'|d' with bd ~ b'd', then ad ~ a'd'.*

The first four axioms express properties postulated or proved for absolute differences in Section 4.10 of Volume I. It is left as an exercise to prove that Axioms 2 and 3 imply that congruence is an equivalence relation. Axiom 5 is close to Axiom 5 of Definition 4 of Section 4.10. Axiom 6 enables us to extend the relation of congruence from segments to angles.

Rather than state a representation theorem directly for absolute spaces, it is more convenent first to state the extensions to Euclidean and hyperbolic spaces.

13.6.4 Euclidean Spaces

DEFINITION 9. *An absolute space $\mathfrak{A} = \langle A, B, \sim \rangle$ is a Euclidean space iff Euclidean axiom E as formulated in Definition 7 is satisfied.*

It follows at once that a Euclidean space is affine.

It is worth noting that a large number of the standard theorems of Euclidean geometry are provable in absolute geometry. We shall not attempt to describe the situation in any detail, but if we consider the first book of Euclid's *Elements*, the first 26 propositions can be proved in absolute geometry. Propositions 27 and 28 can be proved if the concept of being parallel is replaced by that of not intersecting, since Euclid makes the first use of his fifth postulate in proving Proposition 29. A detailed and rigorous treatment of absolute geometry is to be found in the first three chapters of Borsuk and Szmielew (1960).

To formulate the representation theorem for Euclidean geometry, we add to the concepts introduced in Definition 7 of Chapter 12 the analytic or numerical notion of betweenness defined earlier in this section.

THEOREM 11. *Every n-dimensional Euclidean space is isomorphic to an n-dimensional Euclidean vector space over the field of real numbers in the sense of Definition 12.7. Moreover, the representation is unique up to the group of similarities (Definition 12.8).*

For later comparison with hyperbolic spaces it is worth noting that the following four propositions are all equivalent to Playfair's axiom or Axiom

E of Definition 7, once the other axioms of absolute geometry are assumed:

E1. *In every triangle the sum of the sizes of the angles is equal to π.*

E2. *There exists a triangle in which the sum of the sizes of the angles is equal to π.*

E3. *There exists a rectangle.*

E4. *There exist a line L and a point a not in L such that there is at most one line K passing through point a and parallel to line L.*

The proof of these equivalences is to be found in Borsuk and Szmielew (1960, p. 263). In formulating these propositions we make use of standard notions of absolute geometry, including the measure of angles; all of these concepts can be developed in straightforward fashion from the axioms of Definition 8. It is particularly important to note that the standard concept of angle measure is a concept of absolute, not just Euclidean, geometry.

13.6.5 Hyperbolic Spaces

We obtain hyperbolic spaces by negating the Euclidean axiom.

DEFINITION 10. *An absolute space $\mathfrak{A} = \langle A, B, \sim \rangle$ is a* hyperbolic *space iff axiom E is not satisfied.*

From Definitions 9 and 10 it follows immediately that any absolute space is either Euclidean or hyperbolic, but the sense in which this is true is stronger than these definitions directly indicate. For a given dimension, the Euclidean spaces are categorical and so are the hyperbolic spaces. Consequently, for a given dimension there are, up to isomorphism, only two absolute spaces: one Euclidean and the other hyperbolic.

The reason for using the term *hyperbolic* is worth noting. A central conic in classical Euclidean geometry is called an ellipse if it has no asymptote and is called a hyperbola if it has two. In analogous fashion, a non-Euclidean plane is said to be *elliptic* or *hyperbolic* if, as in projective geometry, each of its lines contains either no point at infinity or two points at infinity. Elliptic spaces can be derived from projective spaces, but since they do not satisfy the axioms of absolute geometry, they have not been considered in this section.

We state the representation theorem for hyperbolic spaces in terms of the Klein model. The Klein model takes as the space the interior of the unit sphere in n-dimensions. A line is any chord of the sphere, and the relations of betweenness and equidistance are defined in a manner to be described later in detail. It is obvious, of course, that the ordinary Euclidean notion of congruence cannot be the characterization of equidistance in this model. This is the intuitively simplest and most easily understood model of hyperbolic geometry. It has the defect, on the other hand, that it gives an interpretation of geometrical notions that makes them very distinct from those for Euclidean geometry.

The characterization we give of the n-dimensional Klein space over an arbitrary field follows Szmielew (1959) and expands upon the discussion in Section 12.2.7.

Let $\mathscr{F} = \langle F, +, \cdot, \leqslant \rangle$ be an arbitrary ordered field. By the n-dimensional Klein space $\mathscr{K}_n(\mathscr{F})$ over the field $\mathscr{F}$ we understand the system $\langle A_\mathscr{F}, B_\mathscr{F}, E_\mathscr{F} \rangle$ constructed in the following way: $A_\mathscr{F}$ is the set of all ordered n-tuples $x = (x_1, x_2, \ldots, x_n)$ in F^n for which

$$x_1^2 + x_2^2 + \cdots + x_n^2 < 1.$$

For any ordered n-tuples $x = (x_1, x_2, \ldots, x_n)$ and $y = (y_1, y_2, \ldots, y_n)$ in $A_\mathscr{F}$, let

$$x \cdot y = \sum_{i=1}^n x_i y_i$$

(thus $-1 < x \cdot y < 1$) and

$$\psi(x, y) = \frac{(1 - x \cdot x)(1 - y \cdot y)}{(1 - x \cdot y)^2}.$$

Note that $\psi(x, y) \leqslant 1$. The betweenness relation $B_\mathscr{F}$ among any three n-tuples x, y, z in $A_\mathscr{F}$ is characterized by the equivalence

$$B_\mathscr{F}(x, y, z) \text{ iff } \psi(x, z) = \frac{\psi(x, y)\psi(y, z)}{\{1 + [1 - \psi(x, y)]^{1/2}[1 - \psi(y, z)]^{1/2}\}^2}.$$

The equidistance relation $E_\mathscr{F}$ among any four n-tuples x, y, z, u in $A_\mathscr{F}$ is characterized by the equivalence

$$xy\,E_\mathscr{F}\,zu \qquad \text{iff} \qquad \psi(x, y) = \psi(z, u).$$

THEOREM 12. *An n-dimensional hyperbolic space is isomorphic to the n-dimensional Klein space over the field of real numbers.*

For later use in the discussion of concepts and experiments concerning perceptual space, it will be useful to have available a number of the elementary propositions of hyperbolic geometry. Anything like a detailed development with proofs of these propositions is beyond the scope of this chapter, however, so we shall have to be content with a rather sketchy outline.

First, in contrast to the four Euclidean theorems equivalent to Axiom E, in hyperbolic geometry one can prove the following theorems:

H1. *For any plane P, any line L, and any point $a \in P - L$, there exist at least two distinct lines $K_1 \subset P$ and $K_2 \subset P$ passing through point a and not intersecting line L.*

H2. *In every triangle the sum of the sizes of the angles is less than π.*

H3. *In every convex quadrangle the sum of the sizes of the angles is less than 2π.*

H4. *No quadrangle is a rectangle.*

We also have a weaker criterion of congruence for triangles in hyperbolic geometry. Recall that in Euclidean geometry we must have congruence of both angles and sides, and one of the important elementary topics is the characterization of the appropriate sides and angles to give congruence. In hyperbolic geometry we have a characterization of congruence just in terms of angles.

H5. *Two triangles are congruent iff the three angles of one triangle are congruent to the three angles of the other.*

We define the *defect* of a triangle as the number that is the difference between π and the sum of the triangle's interior angles. We then have the result:

H6. *If two triangles are congruent, then their defects are equal.*

Perhaps of greater interest is the following theorem about the addition of defects.

H7. *Given a triangle abc and $a|d|c$, then the defect of triangle abc is equal to the sum of the defects of triangles abd and bcd.*

By using the standard notion of the triangulation of a polygon, we can introduce the same concepts and define the defect of a polygon as the sum of the defects of triangles constituting any triangulation of the polygon. We can then prove the following theorem:

H8. *If two polygons are congruent, then their defects are equal.*

And, corresponding to H7 about sums of defects of triangles given above, we have the following:

H9.   *If a polygon is the union of two polygons with disjoint interiors, then the defect of the given polygon is equal to the sum of the defects of the two component polygons.*

We can develop a concept of parallel for absolute geometry by using the concept of half-lines. However, rather than enter into any detailed treatment, we shall introduce simply the intuitive notion for hyperbolic geometry. Given a line L and a point a that does not lie on L, we obtain in hyperbolic geometry two parallel lines that pass through a. These are constructed by constructing each of the two half-lines or rays that pass through a, each of which is parallel to the given line L.

As might be expected from this construction, it can then be proved that if K and L are two parallel lines in hyperbolic space, they approach indefinitely close to each other in one direction and recede indefinitely far from each other in the other direction. If two lines are not parallel and also do not intersect, then in hyperbolic geometry they are often called *divergent* or *hyperparallel* lines. We can show that divergent lines have a unique common perpendicular, but parallel lines do not have a common perpendicular. One can also show that two divergent and two intersecting lines recede indefinitely from one another in both directions. Some further developments are given in the exercises.

13.7   ELLIPTIC SPACES

The source of elliptic geometry is spherical geometry, whose earliest systematic development goes back at least to the fourth century BC. In fact, because of their importance in mathematical astronomy, spherical geometry and trigonometry were developed to a sophisticated level in ancient times. It was not, however, until the latter part of the nineteenth century that elliptic geometry, as we now think of it, was created by Felix Klein. He realized that spherical geometry could be brought closer to Euclidean and hyperbolic geometry if the most important deviant aspect of spherical geometry were removed. This deviant aspect is the fact that two great circles, which we interpret as lines, have not one but two points in common. Every point has associated with it a unique antipodal point. In the familiar geographical case, the North Pole has as its antipodal point the South Pole, etc. Klein's insight was that by identifying each pair of antipodal points as a single point, a geometry would be determined that is close to Euclidean and hyperbolic geometry. The familiar postulate that two points uniquely determine a line would hold, and in this geometry a line is still unbounded but finite because its length is half that of the great circle. We are of course not forced to take this model of elliptic geometry. Once the axioms are given, other models can be found. Another suggested model for the elliptic plane, which generalizes to higher dimensions, is to take in Euclidean three-dimensional space the lines and planes through a fixed point a. In the elliptic plane, a point is then a line through a, and a line in the elliptic plane is a plane through a. But it is clear that the first model just described in terms of the identification of antipodal points is the historical and conceptual source of elliptic geometry as we now think of it.[8]

The terminology of elliptic geometry can be a source of confusion, and therefore it is important to be explicit on the following distinctions. The reader will find in the literature references to both *single* and *double elliptic geometry* and also simply to *elliptic geometry*. The distinctions involved are these: First, double elliptic geometry is the same thing as spherical geometry. Second, single elliptic geometry is elliptic geometry with the identification of antipodal points as defined above. In almost all cases, the term *elliptic geometry* without any modifying *double* or *single* refers to single elliptic geometry. We first treat double elliptic geometry because of the greater familiarity of traditional spherical geometry.

13.7.1 Double Elliptic Spaces

Fortunately, a rather natural set of axioms for three-dimensional double elliptic geometry can be given by modifying the axioms for ordered geometry but retaining the same primitives—a set of points and the ternary relation of betweenness. Our exposition follows but modifies that of Kline (1916). The natural model of the axioms is the surface of the hypersphere $x^2 + y^2 + z^2 + w^2 = r^2$, and the betweenness relation $a|b|c$ holds iff the corresponding points $x|y|z$ on the hypersphere are such that y is on the shorter arc of the geodesic on the hypersphere from x to z.

---

[8] It is also possible to characterize, in a formal way, elliptic spaces in terms of projective spaces. Our brief discussion follows that of Veblen and Young (1918). The elliptic geometry of the plane is the geometry corresponding to the group of projective collineations that leave an imaginary ellipse invariant in the projective plane. Three-dimensional elliptic geometry is based on the notion of a polar system, a concept taken from projective geometry. We shall not give a rigorous definition of a polar system. Roughly speaking and referring just to the plane, we define in projective geometry a polarity as a mapping from points into lines and lines into points such that if it maps a point a into a line α it also maps α into the point a. Three-dimensional elliptic geometry is characterized by the properties that are invariant under projective collineations that leave invariant an arbitrary but fixed projective polar system, labeled the *absolute polar system*, in which no point is on its polar plane.

It is convenient in stating the axioms to introduce some defined concepts, including some modifications of earlier concepts. The important new concept is that of an opposite of a point. The point b is an *opposite* of point a iff for every point c, not $a|b|c$. (In the hypersphere model, b is the antipodal point of a.) On the basis of the axioms given below it can be proved that every point has a unique opposite. The definition of a line must include the opposite points, and so it generalizes the earlier definition for ordered geometry. Let a' be an opposite of a and b' an opposite of b. Then the line ab is the set of all points c such that $a|c|b$, $c|a|b$, $c|b|a$, $c = a'$, $c = b'$, or $a|c'|b$, where c' is an opposite of c. On the other hand, the definition of the segment ab is the same as for ordered geometry: $\{c|\ a|c|b\}$. The definition of collinear is the same as in ordered geometry, as are the definitions of triangle, tetrahedron, vertices, edges, and faces, as well as the definition of the 3-space. (The generalization to $n > 3$ dimensions is relatively straightforward, but we explicitly consider only $n = 3$.)

DEFINITION 11.   *Let A be a nonempty set and B a ternary relation on A. Then $\mathfrak{A} = \langle A, B \rangle$ is a* three-dimensional double elliptic space *iff the following axioms are satisfied*:

1.   *Axioms 1–3 of Definition 1 are satisfied.*

2.   *For every point a, there is a point b such that for every point c, not $a|b|c$.*

3.   *For points a, b, c, and a', if $a|b|c$ and a' is an opposite of a, then $a'|c|b$.*

4.   *There exist three distinct points a, b, and c such that a is not an opposite of b and c is not on the line ab.*

5.   *(Pasch's Axiom)   Same as Axiom 5 of Definition 6.*

6.   *(Completeness Axiom)   Same as Axiom 6 of Definition 6.*

7.   *All points lie in a 3-space but not on a plane.*

The changes from ordered geometry to double elliptic geometry are in Axioms 2, 3, and 4 above and in the dropping of Axiom 4 of Definition 1. The rest remain the same. The axiom of connectivity of Definition 6, which is Axiom 2 for ordered spaces, holds for elliptic spaces also but is provable from the other axioms of Definition 11. Axiom 2 above implies the negation of Axiom 3, the axiom of extension, of Definition 6. The representation theorem is most naturally stated in terms of the hypersphere mentioned earlier.

THEOREM 13. *Every three-dimensional double elliptic space is isomorphic to the hypersphere given by the equation $x^2 + y^2 + z^2 + w^2 = r^2$, and betweenness is characterized in the manner stated earlier, i.e., $x|y|z$ on the hypersphere are such that y is on the shorter arc of the geodesic from x to z.*

Compared with the axiomatizations of Euclidean and hyperbolic geometry, an obvious element of incompleteness exists in our treatment of double elliptic geometry. What about the concept of congruence? Unfortunately, the elliptic axioms for congruence are rather complicated. A standard procedure is to relativize the concept in the following fashion. A correspondence between the pairs of opposite points and planes of the double elliptic space constitutes a *polar system* if incidence is preserved, in particular, if the pair of opposite points incident with three noncollinear planes corresponds to the plane incident with the corresponding pairs of opposite points, and conversely. A polar system φ is *elliptic* if no pair of opposite points is on its corresponding plane. Congruence is now defined relative to an elliptic polar system φ. The segment *ab* is *congruent* to the segment *cd* iff *ab* can be carried into *cd* by a projective transformation that leaves φ invariant. Such a relativized notion of congruence could also have been introduced in ordered or affine spaces.

13.7.2 Single Elliptic Spaces

Single elliptic geometry derives its name from the identification of antipodal points. So, for example, in the familiar geographic case, the North and South Poles are treated as a single point.

An excellent intuitive axiomatic development of plane elliptic geometry is to be found in Podehl and Reidemeister (1934). Their axioms are based on three concepts: (i) incidence of lines and points, (ii) orthogonality of lines, and (iii) congruence of line segments. The set of nine congruence axioms is complicated, but as far as we know no simpler version exists.

It is therefore surprising that a relatively simple set of axioms for the elliptic plane has been given by Kordos (1973) in terms of a single binary relation as the only primitive concept. The relation may be interpreted as that of orthogonality for lines, as the only primitive objects, or dually as the relation of being at the maximal distance for points. Five of the seven axioms are relatively simple; for example, no line is orthogonal to itself, and the relation of orthogonality is symmetric. On the other hand, the special midpoint axiom requires 15 different variables to formulate, and the Pappus axiom requires 18. Kordos did not use a completeness axiom, so what he proved was a representation theorem in terms of Pythagorean fields, i.e., fields that have the closure property: if x and y are in F, then so is $(x^2 + y^2)^{1/2}$.

13.8 CLASSICAL SPACE-TIME

Our objective in this section is to use qualitative axioms, familiar in the theory of measurement, to characterize the space-time continuum of classical physics. The theory naturally comes under the general theory of affine spaces. The geometric points of this space are interpreted as four-dimensional point-events. (Wiener, 1914, and Russell, 1936, begin with extended events and derive point-events, but they do not develop a full theory of space-time.) The desired uniqueness result is formulated in terms of *Galilean* transformations, which are defined below (see also Section 10.14.1 of Volume I).

Linearity is introduced in terms of the ternary relation B of betweenness used as the conceptual basis of affine geometry in Section 13.6.2 (Definition 6). We add two additional primitives to this affine framework. The first is a distinguished set $\mathscr{I}$ of inertial lines, which may be thought of as the paths of particles acted on by no forces. The second is a Euclidean quaternary relation of congruence ∼ on A, which has a different meaning for segments of inertial lines from that for segments of simultaneous spatial lines (defined below). Our axioms are thus for structures $\mathfrak{A} = \langle A, B, \mathscr{I}, \sim \rangle$.

To simplify the statement of the axioms, some defined notions are convenient. Two distinct points a and b are *simultaneous* (in symbols, $a\,\sigma\,b$) iff either $a = b$ or $a \neq b$ and the line ab is not an inertial line. The set of all points simultaneous with a given point we denote by

$$[a] = \{ b \mid b \in A \quad \text{and} \quad b\,\sigma\,a \}.$$

Later it will also be useful to refer to *spatial* lines, i.e., lines connecting two simultaneous points.

DEFINITION 12.   *A structure* $\mathfrak{A} = \langle A, B, \mathscr{I}, \sim \rangle$ *is a* (four-dimensional) *classical space-time structure iff the following axioms are satisfied:*

1. *The structure* (A, B) *is a four-dimensional affine space.*

2. *The relation of congruence is an equivalence relation.*

3. *Opposite sides of any parallelogram are congruent.*

4. *The relation* σ *of simultaneity is transitive.*

5. *Given any point on an inertial line, there is on any other inertial line a point simultaneous with it.*

6. *If* $a\,\sigma\,b$, $c\,\sigma\,d$, *and not* $a\,\sigma\,c$, *then* $ac \sim bd$.

7. *If a σ b and not c σ d, then not ab ~ cd.*

8. *For every point a, the structure ⟨[a], B, ~ ⟩ is a three-dimensional Euclidean space.*

The intuitive meaning of each of the axioms should be clear, but some additional comments may be useful. Axiom 1, which in the analysis given here is common to both classical and restricted relativistic space-time, expresses the affine character of the spaces. (The relativistic case is the subject of the next section.) The flatness or linearity of both kinds of spaces is one of their most important shared characteristics. Axiom 2 states a standard elementary property expected of congruence. Axiom 3 plays the important role of connecting affine congruence as defined just in terms of betweenness to the general congruence relation. The fourth axiom expresses the transitivity of simultaneity. Reflexivity and symmetry of simultaneity follow at once from the definition. Axiom 5 describes a physically obvious property of simultaneity. Axioms 6 and 7 express elementary relations between simultaneity and congruence. Axiom 6 is the congruence axiom special to classical physics. It asserts the congruence of inertial segments that have the same elapsed time. Axiom 8 expresses the fundamental Euclidean spatial structure of all points simultaneous with a given point.

It should also be noted in connection with the technical formulation of Axiom 8 that the relations of betweenness and congruence referred to are restricted to the given domain, i.e., the set of points simultaneous with a given point.

The definition of a *four-dimensional classical space-time structure over the field of real numbers* should be evident. In particular, point-events in such a space are four-dimensional vectors, that is, ordered 4-tuples of real numbers, forming a Cartesian space; the relation of betweenness $B_R(x, y, z)$ holds for vectors x, y, and z iff there is an $\alpha \in [0, 1]$ such that

$$y = \alpha x + (1 - \alpha)z;$$

L is in $\mathscr{I}$ if L is a line and for any two vectors x and y in L, $x_4 \neq y_4$; the congruence or equidistance relation $xy \sim_R uv$ holds iff either (i)

$$x_4 = y_4 = u_4 = v_4,$$

and

$$\sum_{i=1}^{3} (x_i - y_i)^2 = \sum_{i=1}^{3} (u_i - v_i)^2;$$

(ii)

$$x_4 \neq y_4 \quad \text{and} \quad u_4 \neq v_4,$$

and

$$|x_4 - y_4| = |u_4 - v_4|;$$

or (iii) xy is parallel to uv and

$$\sum_{i=1}^{4} (x_i - y_i)^2 = \sum_{i=1}^{4} (u_i - v_i)^2.$$

For brevity, parallelism of two segments is interpreted to include their collinearity as a special case.

We have the following representation theorem in analogy with Theorems 9 and 11. The proof is given in Suppes (in press).

THEOREM 14. *Every four-dimensional classical space-time structure is isomorphic to the four-dimensional classical space-time structure over the field of real numbers. Moreover, any two four-dimensional classical space-time structures are isomorphic.*

The uniqueness theorem follows closely the formulation of Theorem 20 of Chapter 10 but is considerably simpler in form because only space and time rather than the concepts of mechanics are involved. As Galilean transformations are ordinarily defined, no changes of units of measurement either of space or of time are made by the transformations; but such changes are natural in our geometrical setting, and so we shall use *generalized* Galilean transformations, as in Chapter 10.

We recall from Section 13.6.2 that an affine transformation of four-dimensional Cartesian space is such that

$$\varphi(x) = Ax + b,$$

where A is a nonsingular four-dimensional matrix of real numbers and b is any four-dimensional vector. We say that a four-dimensional matrix A is a *generalized Galilean matrix* if and only if there exist real numbers α, β, and δ, a three-dimensional vector U, and an orthogonal matrix $\mathscr{E}$ of order 3 such that

$$A = \begin{pmatrix} \alpha\mathscr{E} & 0 \\ 0 & \beta\delta \end{pmatrix} \begin{pmatrix} I_3 & 0 \\ -U & 1 \end{pmatrix},$$

where $\alpha, \beta > 0$, $\delta^2 = 1$, and I_3 is the identity matrix of order 3.

The interpretation of the constants is close to that given for the constants of generalized Galilean carriers in Section 10.14.1. The positive number α represents a ratio transformation in the unit of spatial distance, β a ratio transformation in the unit of time, δ a possible transformation in the direction of time (if $\delta = 1$ there is no change, and if $\delta = -1$ the direction is reversed); U is the relative velocity of the new inertial frame of reference with respect to the old one; and the matrix $\mathscr{E}$ represents a rotation of the spatial coordinates. The vector b of the affine transformation also has a simple interpretation. The first three coordinates represent a translation of the origin of the spatial coordinates and the fourth a translation of the origin of the one-dimensional coordinate system for measurement of time.

We say that an affine transformation is *generalized Galilean* if and only if its four-dimensional matrix is generalized Galilean. It is easy to show that the generalized Galilean transformations form a group. Using these concepts, we then can state the following uniqueness theorem for classical space and time.

THEOREM 15. *The representation of a classical space-time structure in terms of a Cartesian space over the field of real numbers is unique up to the group of generalized Galilean transformations.*

13.9 SPACE-TIME OF SPECIAL RELATIVITY

The purpose of this section is to characterize by qualitative axioms the space-time continuum of special relativity. Just as in classical space-time, the geometrical points of the space are interpreted as four-dimensional point-events; and, also as in the classical case, linearity is introduced by means of the ternary relation of betweenness. The approach used here is to keep in the foreground the question of congruence of segments, i.e., equidistance. Of course, the simple classical space-time formulation is not possible because the simultaneity of distant events is not a meaningful concept in special relativity. However, as was pointed out many years ago by Robb (1914, 1921, 1936), there is a meaningful notion of congruence in special relativity for segments of inertial lines, i.e., possible paths of particles moving uniformly. There is also a separate notion of congruence for segments of separation (or spatial) lines, i.e., lines determined by two points that determine neither an inertial line nor an optical line. In contrast, there is no natural notion of congruence for segments of optical lines, i.e., lines representing possible paths of light rays, except affine congruence for parallel segments.

The intuitive content of these ideas is most easily explained by introducing the usual coordinate representation for special relativity. The four real coordinates (x_1, x_2, x_3, t) of a point-event are measured relative to an inertial space-time frame of reference. The coordinates x_1, x_2, and x_3 are the three orthogonal spatial coordinates, and t is the time coordinate. It is often notationally convenient to replace t by x_4. The relativistic distance between two points $x = (x_1, x_2, x_3, x_4)$ and $y = (y_1, y_2, y_3, y_4)$ is given by the indefinite quadratic metric, relative to frame f as:

$$I_f(x, y) = \left[\sum_{i=1}^{3} (x_i - y_i)^2 - c^2(x_4 - y_4)^2 \right]^{1/2},$$

where c is the numerical velocity of light. (The metric is called *indefinite* because $I_f^2(x, y)$ can be either positive or negative.) It is simpler to use the square of the relativistic distance. Keeping in mind that the velocity of a particle with mass or inertia must be less than c, it is easy to see that $I_f^2(x, y) < 0$ if x and y lie on an inertial line, $I^2(x, y) > 0$ if x and y lie on a separation line, and $I^2(x, y) = 0$ if x and y lie on an optical line. In these terms, two *inertial segments* xy and uv, i.e., segments of inertial lines, are congruent iff

$$I_f^2(x, y) = I_f^2(u, v),$$

and a similar condition holds for two segments of separation lines. On the other hand, this equality does not work at all for optical lines because the relativistic length of any optical segment is zero. Affine congruence holds for parallel segments of optical lines.

The desired uniqueness result is given in terms of the group of Lorentz transformations. There is also a generalization stated in terms of the Brandt groupoid of generalized Lorentz transformations. In the latter case, units of measurement are not held constant. It was shown in Suppes (1959) that invariance of relativistic length of inertial segments for different inertial frames is a sufficient condition to prove that the inertial frames must be related by a Lorentz transformation. The restriction of congruence to inertial segments is physically natural since such segments are more susceptible to direct measurement than are segments of separation lines, but we shall not impose this restriction here.

The axioms for special relativistic space-time are based on five primitive concepts. First, there is the basic set A of point-events, which we shall often refer to simply as points. Second, there is the affine ternary relation of betweenness. Then there are two distinguished families of affine lines, the inertial lines comprising the family $\mathscr{I}$ and the optical lines that are the elements of $\mathcal{O}$. The role of the optical lines is made clear by axioms given below. Finally, there is the quaternary relation $\sim$ of congruence or equidistance for inertial segments, i.e., segments of inertial lines, and also congruence for segments of separation lines and parallel segments of optical lines.

We need to define various concepts to state in parallel form the natural analogue of the classical space-time axiom of Section 13.8 that the set of points simultaneous to a given point forms a three-dimensional Euclidean space (with respect to betweenness and congruence). First, we define an affine parallelogram as an *optical* parallelogram if its sides are segments of optical lines. Second, let L be an inertial line. A line is *orthogonal* to L if the two lines form the diagonals of an optical parallelogram. This construction is the basis of Einstein's definition of simultaneity for noncontiguous events (notice that simultaneity is relative to an inertial line). More explicitly, let a be any point and L any inertial line through a. Then,

$$\mathscr{S}(a, L) = \{\, b \mid ab \text{ is orthogonal to } L \,\} \cup \{\, a \,\}.$$

In Einstein's terms $\mathscr{S}(a, L)$ is the set of points simultaneous to a relative to inertial line L.

DEFINITION 13. *Let A be a nonempty set, B a ternary relation on A, $\mathscr{I}$ and $\mathcal{O}$ sets of lines, where lines are subsets of A defined in terms of B (Section 13.6.1), and let $\sim$ be a quaternary relation on A. Then $\mathfrak{A} = \langle A, B, \mathscr{I}, \mathcal{O}, \sim \rangle$ is a* special relativistic space-time structure *iff the following axioms are satisfied for all points $a, a', a_1, a_2, b, b', c, c', d, e, f$ and all lines L, L_1, L_2:*

1. *$\langle A, B \rangle$ is a four-dimensional affine space.*

2. *There is an inertial line passing through a.*

3. *If L_1 is an inertial line and L_2 is parallel to L_1, then L_2 is an inertial line.*

4. *The relation of congruence is an equivalence relation.*

5. *Opposite sides of any parallelogram are congruent.*

6. *If cd is an inertial segment and L is any inertial half-line with origin a, then there exists exactly one point b on L such that $ab \sim cd$.*

7. *If L is an inertial line passing through a, then in any plane containing L there are exactly two optical lines passing through a.*

8. *If L_1 and L_2 are two distinct optical lines passing through a, a_1 lies on L_1, a_2 lies on L_2, with a, a_1, and a_2 distinct points, and a_1a_2 not an inertial line, then for any c on a_1a_2, the line ac is an inertial line iff c is strictly between a_1 and a_2.*

9. *If L is an inertial line, a lies on L, and b does not, then each of the optical lines passing through a in the plane of L and b is parallel to one of the optical lines passing through b in the plane of L and b.*

10. *For any point a and inertial line L the structure $\langle \mathscr{S}(a, L), B, \sim \rangle$ is a three-dimensional Euclidean space.*

11. *For any point a and inertial line L such that $a \in L$ the intersection of the space $\mathscr{S}(a, L)$ and the set of all optical lines through b on L with $b \neq a$ is an ellipsoid.*

Each of the axioms is intended to have an obvious intuitive meaning. Axioms 1–6 hold for classical space-time as well (see Exercise 12). Axiom 6 is a standard "transfer" axiom for congruence. Axioms 7 and 9 state strong restrictive conditions on optical lines. Axiom 8 states such a condition on inertial lines. Axiom 10 is the direct analogue of Axiom 8 for classical space-time. It shows exactly how Euclidean space is recovered in special relativity, i.e., by relativizing the space not just to a given point but also to an inertial line on which the point lies. Axiom 11 is a strong axiom to guarantee the quadric character of the light cone since under suitable affine coordinate change the ellipsoid postulated is a sphere with equation

$$x^2 + y^2 + z^2 = r^2.$$

Axiom 7 can be derived from Axiom 11 and the other axioms, but we keep it because of its clear intuitive content in terms of the usual two-dimensional space-time picture.

For the desired representation theorem we have already stated the intended analytic representation with the indefinite quadratic metric defined above. As before, the relation of betweenness B_R holds for vectors x, y, and z iff there is an $\alpha \in [0, 1]$ such that

$$y = \alpha x + (1 - \alpha)z.$$

An analytic affine line L is in the set $\mathscr{I}$ of inertial lines iff the square of the relativistic length of any segment xy, $x \neq y$, is negative, i.e., $I^2(x, y) < 0$, and for the set of optical lines the condition, as already indicated, is $I^2(x, y) = 0$. The analytic representation of congruence for inertial and separation segments was given above: $I^2(x, y) = I^2(u, v)$.

THEOREM 16. *Every four-dimensional special relativistic space-time structure is isomorphic to a four-dimensional special relativistic space-time structure over the field of real numbers with (positive) constant c. Moreover, any two such relativistic structures are isomorphic.*

The proof of this theorem is given in Suppes (in press).

The constant c representing the numerical velocity of light is a feature of relativistic space-time structures that has no analogue in the classical case and, as we shall see, creates certain problems in characterizing the set of transformations up to which a representation is unique. We remark that c is treated as a positive real number rather than as a physical quantity in the sense given in Section 10.14.2 of Volume I. This choice is motivated by the desire to keep the representation theorems in this section close to those in earlier sections of this chapter. Moreover, since the physics of the present chapter is very limited compared with that of Chapter 10, it seems reasonable to adopt here a purely geometrical viewpoint.

Lorentz transformations, like Galilean ones, are special cases of general affine transformations. We say that a four-dimensional nonsingular matrix A is a *generalized Lorentz matrix with constant c* if and only if there exist real numbers δ and γ, a positive real number λ, a three-dimensional vector U, and an orthogonal matrix $\mathscr{E}$ of order 3 such that

$$\delta^2 = 1$$

$$\gamma^2\left(1 - \frac{U^2}{c^2}\right) = 1$$

$$A = \lambda\begin{pmatrix} \mathscr{E} & 0 \\ 0 & \delta \end{pmatrix}\begin{pmatrix} I_3 + \dfrac{\gamma - 1}{U^2} & U^*U - \dfrac{\gamma U^*}{c^2} \\ -\gamma U & \gamma \end{pmatrix}.$$

Here, as in the classical case, δ represents a possible transformation in the direction of time; U is the relative velocity of the new inertial frame of reference with respect to the old one, and U^* is the transpose of U (thus U^* is a one-column matrix); and the matrix $\mathscr{E}$ represents a rotation of the spatial coordinates, or a rotation followed by a reflection. The two new constants in addition to c are γ (the famous *Lorentz contraction factor*, which is of course definable in terms of c and U) and λ (the parameter for change of scale).

An affine transformation φ, where

$$\varphi(x) = Ax + b,$$

is *a generalized Lorentz transformation* only when A is a generalized Lorentz matrix (with some constant c). It is easy to show that the generalized Lorentz transformations with the same constant c form a group.

THEOREM 17. *The representation of a special relativistic space-time structure in terms of a four-dimensional Cartesian space over the field of real numbers with constant c is unique up to the group of generalized Lorentz transformations with constant c.*

As already implied, the proof of Theorem 17 is to be found in Suppes (1959).

From a measurement standpoint it is not natural to fix the numerical constant c in advance, but rather to let it vary with changing units of measurement. The complications thereby introduced are not negligible, however, because the set of generalized Lorentz transformations is now no longer a group. In particular, the composition of two generalized Lorentz transformations is not such a transformation itself. Because this kind of complication is relatively rare in the theory of measurement, though not unknown, we develop the details. We follow in considerable part Rubin and Suppes (1954).

First, we extend the definition of generalized Lorentz matrices. Let c, c', and λ be positive real numbers. Then a four-dimensional matrix A is a *generalized Lorentz matrix with respect to* (c, c', λ) if and only if there exist real numbers δ and γ, a three-dimensional vector U, and an orthogonal matrix $\mathscr{E}$ of order 3 such that

$$\delta^2 = 1$$

$$\gamma^2\left(1 - \frac{U^2}{c'^2}\right) = 1$$

$$A = \lambda \begin{pmatrix} I_3 & 0 \\ 0 & \dfrac{c}{c'} \end{pmatrix} \begin{pmatrix} \mathscr{E} & 0 \\ 0 & \delta \end{pmatrix} \begin{pmatrix} I_3 + \dfrac{\gamma - 1}{U^2} U^*U & -\dfrac{\gamma U^*}{c'^2} \\ -\gamma U & \gamma \end{pmatrix}.$$

The constants δ, γ, U, and $\mathscr{E}$ have the same interpretation as before. The constants c and c' are the numerical velocities of light in the old and new frame, respectively, and λ is the change in measurement scale. Note that λ must apply uniformly to all four dimensions, unlike the independence of spatial and temporal units characteristic of the Galilean case.

The following lemma about the matrices just defined helps clarify their structure and is useful in proving additional facts about generalized Lorentz transformations.

LEMMA 1. *As before, let c, c', and λ be positive real numbers. Then a four-dimensional matrix A is a generalized Lorentz matrix with respect to* (c, c', λ) *iff*

$$A \begin{pmatrix} I_3 & 0 \\ 0 & -c'^2 \end{pmatrix} A^* = \lambda^2 \begin{pmatrix} I_3 & 0 \\ 0 & -c^2 \end{pmatrix}.$$

As should be obvious, a *generalized Lorentz transformation* is an affine transformation whose matrix is a generalized Lorentz one.

Although the set of generalized Lorentz transformations is not a group under function composition, it is a *Brandt groupoid*, which is not necessarily closed under the group operation, in this case the operation of function composition. Various algebraic structures with operations that were only partial were introduced throughout Volume I, beginning with Chapter 2, but Brandt groupoids, which we now define, did not arise naturally in any of these structures. (We follow the definition given in Jónsson and Tarski, 1952.) A Brandt groupoid differs from a group mainly in the fact that closure under the binary operation is not assumed.

DEFINITION 14. *Let A be a nonempty set, $*$ a partial binary function from $A \times A$ to A, J a subset of A, and $^{-1}$ a unary operation from A to A. Then the structure $\mathfrak{A} = \langle A, *, J, ^{-1} \rangle$ is a* Brandt groupoid *iff the following axioms are satisfied for every a, b and c in A*:

1. *If $a * b$, $b * c \in A$, then $(a * b) * c \in A$ and $(a * b) * c = a * (b * c)$.*
2. *If $a * b \in A$ and $a * b = a * c$, then $b = c$.*
3. *If $a * c \in A$ and $a * c = b * c$, then $a = b$.*
4. *If $a \in J$, $a * a = a$.*
5. *$a^{-1} * a \in J$ and $a * a^{-1} \in J$.*
6. *If $a, c \in J$, there exists a b in A such that $a * b$, $b * c \in A$.*

If φ_1 is a generalized Lorentz transformation with respect to (c_1, c_1', λ_1), and φ_2 is so with respect to (c_2, c_2', λ_2), then their composition $\varphi_2 \circ \varphi_1$ is a generalized Lorentz transformation iff $c_1' = c_2$, and the resulting transformation is with respect to $(c_1, c_2', \lambda_1, \lambda_2)$. More generally, we have the result:

THEOREM 18. *The set of generalized Lorentz transformations is a Brandt groupoid under function composition, where J is the set of generalized Lorentz transformations with positive constants $(c, c, 1)$, and the inverse of a transformation φ with positive constants (c, c', λ) is the affine inverse with positive constants $(c', c, 1/\lambda)$.*

This theorem shows how the uniqueness result of Theorem 17 is generalized to take account of changes in the units of measurement.

13.9.1 Other Axiomatic Approaches

What is surprising and important is that, in the case of special relativity, the primitive binary relation of temporal order is a sufficient structural basis for the entire geometry. This was shown many years ago by Robb (1911, 1914, 1921, 1928, 1930, 1936). The most extensive exposition of his ideas is to be found in his 1936 book. If we assume the affine structure that is imposed by the ternary relation of betweenness, some considerable simplification of Robb's axioms can be made, but we shall not pursue this idea here. A detailed discussion and some extensions of Robb's work can be found in Winnie (1977). Alexandrov (1950, 1975) showed that invariance of the numerical representation of Robb's single binary relation is sufficient to imply the Lorentz transformations. This was also shown later and independently by Zeeman (1964, 1967).

Since the time of Robb's early work, a number of papers have been published on qualitative axiomatic approaches to the geometry of special relativity. We survey here in somewhat cursory fashion the examples known to us. The literature is large and scattered through a variety of periodicals published in many countries; consequently, we have in all likelihood omitted some significant developments. We do emphasize that we survey qualitative axiomatic approaches and do not consider those approaches relevant to Theorem 17 which already assume a coordinate system or a vector space and then proceed to derive the Lorentz transformations.

An early approach was that of Wilson and Lewis (1912), who gave an incomplete axiomatization of the two-dimensional case with an emphasis on parallel transformations. Reichenbach (1924) gave a philosophically interesting and intuitively appealing axiomatic discussion, but it was more in the spirit of physics than of an explicit and precise geometrical axiomatization. He emphasized the analysis of simultaneity and philosophical issues of conventionalism. A work of Schnell (1937), which has not been widely read and suffers from being too much in the logistical and logical positivist tradition of the Vienna Circle, nevertheless represents an interesting qualitative approach. Walker (1948, 1959), building on the physical approach to relativity of the British physicist E. A. Milne (1948), took as undefined, in addition to space-time points, particles as certain classes of such points (intuitively the classes are particle paths) and Robb's binary relation of *after*, with one particle being distinguished as the particle-observer. In addition, he assumed a primitive relation of signal correspondence between the space-time points of any two particles. Walker's detailed treatment was developed independently of Robb's and has considerably less of a geometrical flavor. Perhaps the most unsatisfactory feature of Walker's approach is that the signal-mappings are complex functions that, at one stroke, do work that should be done by a painstaking buildup of more elementary and more intuitively characterizable operations or relations. An independent but related approach to that of Walker was given by Hudgin (1973), who began with timelike geodesics and signal correspondence represented by a communication function. Like Walker, Hudgin used the physical ideas of Milne. Domotor (1972) published a detailed discussion of Robb and provided another viewpoint on Robb's approach by giving a vector-space characterization. This characterization is, of course, in the spirit of the analytic representations given in Chapter 12, but Domotor discussed a number of aspects of Robb's axioms and thereby contributed to their further clarification. Latzer (1972) showed that it is possible to take a binary relation of light signaling as primitive. He did not give a new axiomatic analysis but showed how Robb's ordering relation can be defined in terms of his symmetric light relation. From a purely geometrical standpoint, there is a good argument for adopting such a symmetric relation because there is no nonarbitrary way of distinguishing a direction of time on purely geometrical grounds. It actually follows from much earlier work of Alexandrov (1950, 1975) that Latzer's binary primitive is adequate.

One of the most detailed recent efforts, roughly comparable to that of Robb, is the work of Schutz (1973), which continued and developed the work of Walker. Schutz's axiom system goes beyond Walker's in two respects. First, Walker did not state axioms strong enough to characterize completely the space-time of restricted relativity; and, second, more than half of the axioms have a form that is not closely related to any of those given by Walker, though in several cases they have a rather strong appeal in terms of physical intuition. As Schutz indicated, the most important immediate predecessor of his book is the article by Szekeres (1968), whose approach resembled that of Walker, but Szekeres treated both particles and light signals as objects rather than treating light signals as formally being characterized by a binary relation of signaling.

Schutz's formulation is based upon eleven axioms, the first five of which are similar to axioms already given by Walker. The final six axioms are more powerful and depend primarily upon a concept that is geometrically very natural but that has not ordinarily been adequately exploited in the axiomatizations of special relativity. This is the concept which Schutz called a "spray" of the set of inertial paths that pass through a given point, in other words, the set of inertial paths that are contained in the light cone whose vertex is the given point. Schutz refers to a spray as the set of inertial particles whose paths have the properties mentioned. Obviously, in both his and Walker's development the concept of an inertial particle is not really a physical but a geometrical concept, and one could just as well speak of inertial paths as of inertial particles. The four essential properties of sprays expressed in the axioms are the following: (i) between any two distinct particles of a spray there is a particle that is distinct from both; (ii) each spray is isotropic; (iii) there is a spray with a maximal symmetric subspray of four distinct particles (this is the dimension axiom); and (iv) each bounded infinite subspray is compact.

The remaining two axioms (of the six) postulate that space-time can be "connected" by particles and that given any two distinct particles coinciding at some event, that is, intersecting at some space-time point, there is another particle that forms a third side of the "triangle."

As this informal discussion should make clear, each of Schutz's axioms has a clear intuitive content. On the other hand, his axiomatization is not entirely satisfactory from the standpoint of the formal simplicity and ease of comprehension that seem desirable or possible. For instance, the axiom formulating the isotropy of sprays is quite complicated in its explicit form. More recently, Schutz (1979) has simplified his axioms and cast them in better form but with essentially the same approach.

The axiomatization given here has been most influenced by the recent work of Mundy (1986a, 1986b). In (1986b) Mundy gave a detailed analysis of the physical content of Minkowski geometry. He also included a survey of other axiomatizations more oriented toward physics, such as that of Mehlberg (1935, 1937), as well as detailed remarks on the as yet unpublished work of Dorling (1989) that obtains Minkowski geometry by a minimal modification of the Euclidean axioms of congruence. Mundy emphasized the physical content of Minkowski geometry, but his emphasis is different from ours. He stressed in his axiomatization the constancy of the speed of light as the physical core of Minkowski geometry. We have emphasized, first of all, the way in which classical ideas play a role. This is seen especially in Axiom 10, which postulates ordinary three-dimensional Euclidean space relative to a space-time point and an inertial line. In addition, specific elementary properties of optical lines and inertial lines are postulated whereas Mundy uses more complicated, and fewer, axioms. All the same, the axioms of Definition 13 are closer to Mundy's than to any others.

In (1986a) Mundy gave an optical axiomatization that is close in spirit to that of Robb, although the single binary primitive is close to Latzer's (1972). Mundy's formulation is certainly an improvement on Robb's and is the best set of axioms thus far that depend on a single binary relation; in his case the relation is that of optical signaling, but it is given such an elementary framework that some of the axioms are inevitably rather complicated.

In a recent work, Goldblatt (1987) emphasizes, probably for the first time in the literature, the central role that can be played by the concept of orthogonality generalized to relativistic space-time. In a Euclidean vector space with the standard inner product, two vectors x and y are orthogonal iff $x \cdot y = 0$. Using the indefinite quadratic metric $I^2(x, y)$, we obtain the following: when two distinct lines are orthogonal, one must be timelike (an inertial line) and the other spacelike (a separation line); an optical line is orthogonal to itself. Axiomatically, Goldblatt adds to the affine axioms based on betweenness, additional axioms based on the quaternary relation of ab being orthogonal to cd.

13.10 PERCEPTUAL SPACES

The literature on perception is vast. Our analysis of perceptual spaces is restricted to a highly focused part of the psychological literature, that concerned with the geometrical character of visual perception. We ignore even so closely related matters as the role of texture cues in the perception of depth.

13.10.1 Historical Survey through the Nineteenth Century

The natural place to begin is with Euclid's *Optics*, the oldest extant treatise on mathematical optics. (This survey follows Suppes, 1977.) It is important to emphasize that Euclid's *Optics* is really a theory of vision and not a treatise on physical optics. A large number of the propositions are concerned with vision from the standpoint of perspective in monocular vision. Indeed, Euclid's *Optics* could be characterized as a treatise on perspective within Euclidean geometry. The tone of Euclid's treatise (1945) can be seen from quoting the initial part, which consists of seven "definitions." (The translation is taken from that given by Burton.)

1. Let it be assumed that lines drawn directly from the eye pass through a space of great extent;

2. and that the form of the space included within our vision is a cone, with its apex in the eye and its base at the limits of our vision;

3. and that those things upon which the vision falls are seen, and that those things upon which the vision does not fall are not seen;

4. and that those things seen within a larger angle appear larger, and those seen within a smaller angle appear smaller, and those seen within equal angles appear to be of the same size;

5. and that things seen within the higher visual range appear higher, while those within the lower range appear lower;

6. and, similarly, that those seen within the visual range on the right appear on the right, while those within that on the left appear on the left;

7. but that things seen within several angles appear to be more clear.

The development of Euclid's *Optics* is mathematical in character, but it is not axiomatic in the way that the *Elements* are. For example, at one point Euclid proves two propositions: "to know how great is a given elevation when the sun is shining" and "to know how great is a given elevation when the sun is not shining." As would be expected, there is no serious introduction of the concept of the sun or of shining, but they are treated in an informal, commonsense, physical way with the essential thing for the proof being rays from the sun falling upon the end of a line. Visual space is of course treated by Euclid as Euclidean in character.

The restriction to monocular vision is one that we shall meet repeatedly in this survey. However, it should be noted that Euclid proves several propositions involving both eyes; for example, "If the distance between the eyes is greater than the diameter of the sphere, more than the hemispheres will be seen." Euclid is not restricted to some simple geometrical optics but is indeed concerned with the theory of vision, as is evident from the proposition that "if an arc of a circle is placed on the same plane as the eye, the arc appears to be a straight line." This kind of proposition is a precursor of later theories that emphasize the non-Euclidean character of visual space.

We skip the period after Euclid to the eighteenth century, not because there are no matters of interest in this long period but because there do not seem to be salient changes of opinion about the character of visual space, or at least if there were, they are not known to us. We looked, for example, at the recent translation by David C. Lindberg of the thirteenth-century treatise *Perspectiva Communis* of John Pecham (1970) and found nothing to report in the present context, although the treatise itself and Lindberg's comments on it are full of interesting matter of great importance concerning other questions in optics, as, for example, theories about the causes of light. Another important continuous tradition is that of the theory of perspective that extends from Alhazen in the eleventh century to Kepler in the seventeenth century, but the representation of perspective is entirely within Euclidean geometry (for a good history of theories of vision in this period, see Lindberg, 1976).

Newton's *Opticks* (1704/1730/1931) is in marked contrast to Euclid's. The initial definitions do not make any mention of the eye until Axiom VIII, and then only in very restrained fashion. Almost without exception, Newton's propositions are concerned with geometrical and especially physical properties of light. Only in several of the Queries at the end are there any conjectures about the mechanisms of the eye, and these conjectures do not bear on the topic at hand.

Five years after the publication of the first edition of Newton's *Opticks*, Berkeley's *An Essay Towards a New Theory of Vision* (1709/1901) appeared. Berkeley does not have much of interest to say about the geometry of visual space, except in a negative way. He makes the point that distance cannot be seen directly and, in fact, seems to categorize the perception of distance as a matter of tactile rather than visual sensation because the muscular convergence of the eyes is tactile in character. He emphatically makes the point that we are not able geometrically to observe or compute the optical angle generated by a remote point as a vertex with sides pointing toward the centers of the two eyes. Here is what he says about the perception of optical angles. "Since therefore those angles and lines are not themselves perceived by sight, it follows, ... that the mind does not by them judge the distance of objects" (paragraph 13). What he says about distance he also says about magnitude not being directly perceived visually. In this passage (paragraph 53), he is especially negative about trying to use the geometry of the visual world as a basis for visual perception.

It is clear from these and other passages that for Berkeley visual space is not Euclidean because there is no proper perception of distance or magnitude; at least, visual space is not a three-dimensional Euclidean space. What he seems to say is ambiguous as to whether it is a two-dimensional Euclidean space. Our own inclination is to judge that his views on this are more negative than positive. Perhaps a sound negative argument can be made up from his insistence on there being a minimum visible, that is, a visual threshold. As he puts it, "It is certain sensible extension is not infinitely divisible. There is a minimum tangible, and a minimum visible, beyond which sense cannot perceive. This everyone's experience will inform him" (paragraph 54). In fact, toward the end of the essay, Berkeley makes it clear that even two-dimensional geometry is not a proper part of visual space or, as we might say, the visual field. As he says in the final paragraph of the essay, "By this time, I suppose, it is clear that neither abstract nor visible extension makes the object of geometry."

Of much greater interest here is Thomas Reid's *Inquiry into the Human Mind*, first published in 1764 (1764/1967). Chapter 6 deals with seeing, and Section 9 is the celebrated one entitled "Of the geometry of visibles." It is sometimes said that this section is a proper precursor of non-Euclidean geometry, but if so, it must be regarded as an implicit precursor because the geometry explicitly discussed by Reid as the geometry of visibles is wholly formulated in terms of spherical geometry, which had of course been recognized as a proper part of geometry since ancient times, especially because of its use in astronomy. [For example, Daniels (1972) has argued that Reid's geometry of visibles is not simply a use of spherical geometry but is an introduction by Reid of an elliptic space as a non-Euclidean geometry. A similar argument is made by Angell (1974).] The viewpoint of Reid's development is clearly set forth at the beginning of the section:

"Supposing the eye placed in the centre of a sphere, every great circle of the sphere will have the same appearance to the eye as if it was a straight line; for the curvature of the circle being turned directly toward the eye, is not perceived by it. And, for the same reason, any line which is drawn in the plane of a great circle of the sphere, whether it be in reality straight or curved, will appear straight to the eye." It is important to note that Reid's geometry of visibles is a geometry of monocular vision. He mentions in other places binocular vision, but the detailed geometrical development is restricted to the geometry of a single eye. The important contrast between Berkeley and Reid is that Reid develops in some detail the geometry in a straightforward mathematical fashion. No comparable development occurs in Berkeley.

Single elliptic geometry, as created by Felix Klein at the end of the nineteenth century, was not even hinted at by Reid. Klein recognized that a natural geometry very similar to Euclidean geometry or hyperbolic geometry could be obtained from spherical geometry by identifying antipodal points as a single point (see Section 13.7.2). The difficulty with spherical geometry as a geometry having a development closely parallel to that of Euclidean geometry is that two great circles, which correspond to lines, have two points of intersection, not one. However, by identifying the two antipodal points as a single point, a number of standard Euclidean postulates remain valid. It is clear that no such identification of antipodal points was made by Reid, for he says quite clearly in the fifth of his propositions, "Any two right lines being produced will meet in two points, and mutually bisect each other." This property of meeting in two points rather than one is what keeps his geometry of visibles from being a single elliptic geometry and forces us to continue to think of it in terms of the spherical model used directly by Reid himself.

In spite of the extensive empirical and theoretical work of Helmholtz on vision, he does not have a great deal to say that bears directly on the Euclidean or non-Euclidean nature of visual space; thus, after an analysis of some methodological issues, we move on to experiments and relevant psychological theory in the twentieth century.

13.10.2   General Considerations Concerning Perceptual Spaces

What perceptual domains have useful geometrical representations? This is not a question that can be given a definitive answer, either mathematically or empirically. We raise it in part to get the benefit of its presuppositions: first, that perception can be subdivided into domains that may require rather different theoretical formulations; second, that a perceptual geometry is, at most, a useful representation of the phenomena within some particular domain. The latter point suggests that one should not be overly disturbed if rather different geometries are used to represent distinct and loosely related types of perceptual judgment.

The most obvious domains for geometrical representation are those in which perceptual judgments yield, more or less directly, empirical relations that match familiar geometrical primitives. For example, any perceptual judgment concerning three distinct objects can be abstracted as a ternary empirical relation and thus is a candidate to be represented geometrically by collinearity of three points. In particular, people readily classify lines by whether they appear straight or not, and by extension, they classify triples of points with regard to apparent collinearity. Such judgments of *perceptual collinearity* can be abstracted as a ternary empirical relation; it is at least plausible that, if perceptually distinct points a, b, c are judged collinear, and b, c, d likewise, then, in separate tests, b and c will each be judged collinear with a and d. Thus, judgments of apparent collinearity plausibly satisfy at least some of the crucial axioms for collinearity relations in standard geometries. To give another example, any perceptual judgment concerning four distinct objects can be abstracted as a quaternary relation, and is a candidate to be represented geometrically by congruence of two pairs of points; in particular, judgments of *perceived equidistance* are particularly favored, since they plausibly satisfy axioms such as symmetry and transitivity, which are necessary for the usual geometrical representations. Certain other geometrical properties likewise have perceptual counterparts that satisfy some of the same axioms. Thus, people readily judge parallelism, perpendicularity, and equality or inequality of angles. Other salient perceptual judgments, however, violate the fundamental symmetries common to many geometries, most notably judgement of verticality.

It is not hard to see the difficulty of encompassing even all "perceptual geometry" judgments by a single geometrical representation. The geometrical visual illusions provide clear evidence that this cannot be done. Based on the well-known Mueller–Lyer illusion, for example, it is not hard to construct a quadrilateral (embedded in additional lines) whose angles are all perceived as right angles, and whose opposite sides are perceived as parallel, but such that one pair of opposite sides appears far from congruent. It is sometimes suggested that the geometrical visual illusions can be explained by using a three-dimensional geometrical representation rather than the obvious two-dimensional one. However it seems doubtful that the detailed phenomena that arise in quantitative studies of illusions can be subsumed by any standard geometry.

In fact, normal perception, as well as the perception of two-dimensional drawings or paintings, is organized in terms of scenes and objects. There are a number of very interesting models of this organization and, though many of them use geometry or geometrical objects, they have not taken the form of representations in a standard geometrical space. We mention in this connection work on picture grammars (Fu, 1974; Suppes and Rottmayer, 1974), development of computer systems for object recognition (Besl and Jain, 1985), attempts to understand perceived shape and slant in terms of visual invariants, and models of human object recognition (Hochberg, 1988).

In order to separate out a domain of perceptual relations that might plausibly be isomorphic to a standard geometry, investigators have worked to suppress scenes and objects, by using small, unmoving point sources of light that provide minimal illumination on the surfaces of an otherwise darkened room. One hopes thus to study perception of collinearity, equidistance, parallelism, etc., under conditions in which those perceptual relations are determined by nothing more than the visual angles between different points of light and the convergence angles of the two eyes as they fixate various points.

Perhaps the most important method used to obtain data on these perceptual relations is the method of adjustment. One of the points of light is put under the observer's control, and its position in three dimensions is adjusted to meet an experimentally imposed requirement—collinearity with other points, particular equidistance relations with other points, or parallelism of a line determined by the movable point and another with some other line. Other methods sometimes used include "yes/no" judgments of whether some particular relation obtains among a set of visually presented points, or numerical judgments of distance for pairs of such points.

Once such judgments are obtained, the mathematical question is whether or not the finite relational structure representing the experimental data can be embedded in a two- or three-dimensional space of a given type (Euclidean, hyperbolic, etc.). The dimensionality depends on the character of the experiment. In many cases the points are restricted to a plane, and therefore embedding in two dimensions is required; in other cases, embedding in three dimensions is appropriate. For example, suppose that A is a finite set of points and that the judgments we have asked for are judgments of equidistance of points. Let $\sim$ be the quaternary relation of perceived equidistance. Then to say that the finite relational structure[9] $\mathfrak{A} = \langle A, \sim \rangle$ can be embedded in three-dimensional Euclidean space is to say that there exists a function φ defined on A such that φ maps A into Re^3 and such that

$$ab \sim cd \quad \text{iff} \quad \delta[\varphi(a), \varphi(b)] = \delta[\varphi(c), \varphi(d)],$$

[9] By a *finite relational structure* we mean, as usual, a relational structure whose domain is finite.

where δ is the Euclidean distance in Re^3. In principle, it is straightforward to answer the embedding question: Given a set of data from an individual's visual judgments of equidistance between pairs of points, can we determine in a definite and constructive mathematical manner whether such a Euclidean embedding is possible?

Immediately, however, a problem arises. It can be grasped by considering the analogous physical situation. Suppose we are making observations of the stars and want to test either a similar proposition or some more complex proposition of celestial mechanics. We are faced with the problem recognized early in the history of astronomy, and also in the history of geodetic surveys, that the data are bound not to fit the theoretical model exactly. The classical way of putting this is that errors of measurement arise, and our problem is to determine if the model fits the data within the limits of the error of measurement. Because of the complexity and subtlety of the statistical questions concerning errors of measurement in the present setting, for purposes of simplification we shall ignore them; but it is absolutely essential to recognize that they must be dealt with in any detailed analysis of experimental data.

Returning to the formal problem of embedding relations among a finite set of points into a given space, it is surprising to find that the results of the kind that we need for this perceptual problem are not to be found in the enormous mathematical literature on finite geometries.[10] The tradition of considering finite geometries goes back at least to the beginning of this century. Construction of such geometries by Veblen and others was a fruitful source of models for proving independence of axioms, etc. On the other hand, the literature that culminates in Dembowski's magisterial survey consists almost entirely of projective and affine geometries that have relatively weak structures. From a mathematical standpoint, such structures have been of considerable interest in connection with a variety of problems in abstract algebra. Some general theorems on the embedding of finite structures in projective and affine planes are given in Szczerba and Tarski (1979) and Szczerba (1984).

The corresponding theory of finite geometries of a stronger type, such as finite Euclidean, finite elliptic, or finite hyperbolic geometries, is not as developed. As a result, the experimental literature does not deal directly with such finite geometries, although they are a natural extension of the weaker finite geometries on the one hand and finite measurement structures on the other.

Another basic methodological approach to the geometrical character of visual space is to assume that a standard metric representation already exists and then to determine which kind of space best fits the data. We

---

[10] For example, Dembowski (1968) contains over 1200 references.

consider this approach in some detail in the next section, which is devoted to Luneburg's theory of binocular vision. Of especial relevance here is multidimensional scaling, some results of which are reported.

A third approach is to go back to the well-known Helmholtz–Lie problem of the nature of space and to replace finiteness by considerations of continuity and motion. This problem and its modern treatment and solution were discussed in Chapter 12. Some of the basic mathematical results are used in the next section to give a rigorous and explicit treatment of Luneburg's theory.

13.10.3 Experimental Work before Luneburg's Theory

Hillebrand (1902) and Blumenfeld (1913) were among the first to perform specific experiments to show that, in one sense, phenomenological visual judgments do not satisfy all Euclidean properties. Blumenfeld performed experiments with so-called parallel and equidistance alleys that improved on those of Hillebrand. In a darkened room, a subject sitting at a table and looking straight ahead was required to adjust two rows of point sources of light placed on either side of the normal plane, i.e., the vertical plane that bisects the horizontal segment joining the centers of the two eyes. The first task was to construct a parallel alley. The two furthest lights were fixed and were placed symmetrically and equidistant from the normal plane. The subject was then asked to arrange the other lights so that they formed a parallel alley extending toward him from the fixed lights, that is, to arrange the lights so that he perceived them as straight and parallel to each other in his visual space. The second task was to construct an equidistance alley. In this case, all the lights except the two fixed lights were turned off and a pair of lights was presented, which the subject adjusted so as to seem the same physical distance apart as the fixed lights (the kind of equidistance judgment discussed earlier). That pair of lights was then turned off and another pair of lights closer to the subject was presented for adjustment, and so forth.

The experimental results are that the two physical configurations do not coincide, as they should in Euclidean geometry where straight lines are parellel if and only if they are equidistant from each other along any mutual perpendiculars. The discrepancies observed in Blumenfeld's experiment are taken to be evidence that visual space is not Euclidean. In both the parallel-alley and equidistance-alley judgments the lines diverge in the direction away from the subject, but the angle of divergence tends to be greater in the case of parallel than in the case of equidistance alleys. Since the most distant pair is the same for both alleys, this means the equidistance alley lies outside the parallel alley. These results were later taken by Luneburg to support his hypothesis that visual space is hyperbolic.

There is one obvious reservation to be made about Luneburg's inference that visual space is hyperbolic. As is evident from Definition 7 there is no unique concept of lines being parallel in hyperbolic space. Indow (1979) discussed Luneburg's choice rather carefully and showed that it has some justification. Essentially he used orthogonality to characterize being parallel. The situation is worse when visual space's being elliptic is tested by alley data, for no two lines can be parallel in such a space. A local concept must be used; for any standard choice it can be shown that in elliptic space the parallel alley lies outside the equidistance alley, contrary to Blumenfeld's data. Blank (1961) made a suggestion for an alternative experiment that is better from a geometrical viewpoint for deciding between the three geometries because it does not use the concept of being parallel. Consider any triangle. The segment joining the midpoints of any two sides is half the length of the third side (Euclidean case), less than half the length (hyperbolic case), or more than half the length (elliptic case). Six of Blank's seven subjects made judgments supporting the hyperbolic hypothesis; the seventh supported the Euclidean hypothesis. This setup, which is excellent from a theoretical standpoint, does not seem to have been used by other experimenters.

13.10.4 Luneburg Theory of Binocular Vision

The theory of binocular vision developed by R. K. Luneburg and his collaborators beginning in the 1940s is still the most detailed and sophisticated viewpoint to receive both mathematical and experimental attention. The object of much of the experimental work we report in the next section (13.10.5) is direct testing of the Luneburg theory or some modification of it; this is certainly true of the extensive experimental work of Tarow Indow and his collaborators. (The exposition in this and the next section expands upon that in Suppes, 1989.)

Essentially, Luneburg wanted to postulate that the space of binocular vision must be a Riemannian space of constant curvature K in order to have free mobility (connections with the Helmholtz–Lie problem are set forth in more detail below). It is well known that there are just three types of Riemannian spaces of constant curvature: If $K = 0$, the space is Euclidean; if $K < 0$, hyperbolic; and if $K > 0$, elliptic. (These spaces were analyzed from a qualitative viewpoint earlier.) Moreover, Luneburg felt the evidence to be extremely strong in support of the conclusion that the space of binocular vision of most persons is hyperbolic. Luneburg and his collaborators adopted a metric viewpoint rather than a synthetic one toward hyperbolic space. We recapitulate some of the main lines of development here. In particular, we begin with the Luneburg (1950) axioms for determining a metric on visual space that is unique up to an isometry, that is, a similarity transformation. Some preliminary definitions are useful. Let $\mathfrak{A} = \langle A, \delta \rangle$ be a metric space. We define a betweenness relation B_δ (relative to δ) and an equidistance relation E_δ in the obvious way:

$$B_\delta = \{(a, b, c) | \delta(a, b) + \delta(b, c) = \delta(a, c), \quad \text{for} \quad a, b, c \in A\}$$
$$E_\delta = \{(a, b, c, d) | \delta(a, b) = \delta(c, d), \quad \text{for} \quad a, b, c, d \in A\}.$$

We also use from Section 12.3.1 on G-spaces the concept of a metric space being *metrically convex* and that of its being complete (Axioms 1 and 4, respectively, for G-spaces in Definition 12 of Chapter 12).

If we think of B_δ and E_δ as the (idealized) observed betweenness and equidistance relations in visual space, then roughly speaking any two metrics for which they are the same are related by an isometry. More explicitly and precisely, we have the following theorem (Luneburg, 1950).

THEOREM 19. *Let $\mathfrak{A} = \langle A, \delta \rangle$ and $\mathfrak{A}' = \langle A, \delta' \rangle$ be metric spaces that are complete and metrically convex, and let the betweenness and equidistance relations be the same for the two spaces, i.e., let $B_\delta = B_{\delta'}$ and $E_\delta = E_{\delta'}$. Then there is a positive real number c such that for all a and b in A*

$$\delta'(a, b) = c\delta(a, b).$$

This theorem is, of course, quite general, and we need a stronger postulate to determine that visual space must be a Riemannian space of constant curvature. (The general concept of curvature in Riemannian spaces was developed in Section 12.3.3.) This is contained in the condition of free mobility as formulated in Wang's theorem (Theorem 12.7) for G-spaces of dimension 2, 3, or any odd number. This theorem may be regarded as a precise expression of Luneburg's intuitive idea, except for the possibility that the space is spherical. This may be excluded by a direct appeal to the elementary global requirement that two distinct points determine a unique line, which is true for Euclidean, hyperbolic, and single elliptic spaces but not spherical (i.e., double elliptic) spaces. Recall that antipodal points on a sphere have an infinity of distinct "lines," i.e., geodesics, passing through them. We summarize these matters in the following theorem, whose proof is obvious from what has been said.

THEOREM 20. *Let* $\mathfrak{A} = \langle A, \delta \rangle$ *be a G-space of dimension 2 or 3 satisfying the hypothesis of free mobility of Theorem 12.7, and let* $\mathfrak{A}$ *be such that two distinct points always determine a unique geodesic. Then* $\mathfrak{A}$ *is Euclidean, hyperbolic, or elliptic, i.e., is a Riemannian space of constant curvature.*

The geometry of hyperbolic and elliptic spaces needs to be developed in some detail, especially for the case of binocular vision, in order to have appropriate tools for experimental investigation of the nature of visual space. A good recent development is Indow (1979), and we refer the reader to his treatment. We give here a sketch of Luneburg's original approach because of its historical importance and its intuitive clarity. The development is closest to Luneburg (1948).

First, on the basis of Theorem 20, we can use the differential expression for a line element in Riemannian spaces of constant curvature,

$$ds^2 = \frac{d\xi^2 + d\eta^2 + d\zeta^2}{\left[1 + \frac{1}{4}K(\xi^2 + \eta^2 + \zeta^2)\right]^2}, \tag{1}$$

where the sensory coordinates ξ, η, ζ are ordinary Cartesian coordinates in a three-dimensional Euclidean space when $K = 0$. The origin $\xi = \eta = \zeta = 0$ is selected to represent the apparent center of observation of the observer. Second, we introduce polar coordinates ρ, φ, θ by the relations

$$\begin{aligned}
\xi &= \rho \cos \varphi \cos \theta \\
\eta &= \rho \sin \varphi \\
\zeta &= \rho \cos \varphi \sin \theta.
\end{aligned} \tag{2}$$

The angles φ, θ determine the apparent direction of a sensed point relative to the observer, and the parameter ρ is the radius (in visual space) of a sphere around the observer centered as stated above.

It would be possible to pursue a research program of determining, for a given class of experimental configurations, the intrinsic structure of visual space without reference to physical space, but Luneburg takes a more psychophysical approach. As would be expected, physical space is assumed to be Euclidean. We assume next that in physical space the centers of rotation R and L of the observer's eyes are at the point $x = z = 0$, $y = \pm 1$ of a Cartesian coordinate system with the x-axis horizontal and perpendicular to the line joining R and L, the z-axis defined similarly in the vertical direction, and the y-axis being the horizontal line joining R and L.

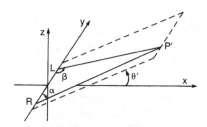

FIGURE 6.   Convergence of optical axes of the eyes.

Let P' be a luminous physical point focused on by both eyes. (We use primes for physical points.) Then the optical axes of the eyes converge at P' and have directions RP' and LP' in the physical space. These directions are characterized by angles α, β, and θ' (see Figure 6): angle α is the angle formed by the y-axis and RP', angle β is the angle formed by the y-axis and LP', and angle θ' is the angle of elevation, that is, the angle between the horizontal x, y plane and the plane RLP'. The physically luminous point P' is seen as a sensory point P in the visual space of the observer, so the construction just described gives a three-dimensional Euclidean map of the observer's visual space.

Instead of using physical angles α and β, we can get a somewhat more suggestive Euclidean map by defining two other angles in terms of α and β (see Figure 7):

$$\gamma = \pi - \alpha - \beta$$
$$\varphi' = \tfrac{1}{2}(\beta - \alpha). \tag{3}$$

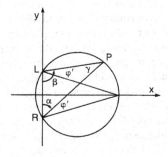

FIGURE 7.   Bipolar parallax γ and bipolar latitude φ'.

Luneburg calls the angle γ the *bipolar parallax* arising from spatial separation of the two eyes and φ' the *bipolar latitude*. In any plane of elevation it is obvious from the construction that the curves γ = constant are circles passing through the eyes, the Vieth-Muller circles as they are known in the literature. If a Vieth-Muller circle is rotated about the y-axis, we obtain a Vieth-Muller torus. A fixed φ' gives a circle of latitude on the torus (note that meridians are horizontal and the latitudes vertical, contrary to our conventional pictorial representation of latitudes and longitudes on the surface of the earth). The coordinates θ' and φ' are thus visually interpreted as spherical coordinates.

As is intuitively obvious, the apparent visual distance of a point is a monotonically decreasing function of γ for fixed spherical coordinates. Based on the facts described, Luneburg thus postulates the following equations to relate the sensory coordinates ξ, η, ζ and the physical coordinates γ, φ', θ':

$$
\begin{aligned}
\xi &= f(\gamma)\cos\varphi'\cos\theta' \\
\eta &= f(\gamma)\sin\varphi' \\
\zeta &= f(\gamma)\cos\varphi'\sin\theta',
\end{aligned}
\tag{4}
$$

where the function f is not yet fixed. On the basis of the Blumenfeld alley experiments and related ones, Luneburg concluded that it is reasonable to combine Equations (2) and (4) to assume:

$$
\begin{aligned}
f(\gamma) &= \rho \\
\varphi' &= \varphi \\
\theta' &= \theta.
\end{aligned}
\tag{5}
$$

The intuitive basis for Equation (5) is clear, but a more detailed argument would be desirable. In any case, we proceed with Luneburg's development of ideas and reserve general critical remarks for the next section.

The line element ds of visual space can now be written in terms of physical coordinates as (Luneburg, 1948, p. 232):

$$
ds^2 = \frac{f(\gamma)^2}{\left[1 + \frac{1}{4}Kf(\gamma)^2\right]^2}\left[\frac{f'(\gamma)^2}{f(\gamma)^2}d\gamma^2 + d\varphi^2 + \cos^2\varphi\, d\theta^2\right],
\tag{6}
$$

where f' is the derivative of f and where, in view of Equation (5), the primes are dropped from φ and θ. Luneburg gives different methods in different publications for determining the function $f(\gamma)$. In the 1948 publication the following intuitive but rather sketchy argument is given. Transform the physical space by

$$\begin{aligned}
\gamma_1 &= \gamma + \tau \\
\varphi_1 &= \varphi + \delta \\
\theta_1 &= \theta + \epsilon.
\end{aligned} \tag{7}$$

Then any configuration is observed by the same sequence of retinal images before and after the transformation. As can be seen immediately, under the transformation $\Delta\gamma$, $\Delta\varphi$, and $\Delta\theta$ remain unchanged. Such transformations are called *iseikonic*, and two configurations that can be transformed into each other by a transformation such as (7) are called *iseikonically congruent*. Observations of two such congruent configurations show that corresponding angles in the visual configuration are equal. Thus, iseikonic transformations, i.e., angle-preserving ones, are conformal transformations of the visual space.

The critical hypothesis here is that iseikonically congruent configurations produce the same sequence of retinal images. In Luneburg (1948) this hypothesis was treated as a fact; in Luneburg (1947) a detailed qualitative argument was given based on experiments by A. Ames, Jr., at the Dartmouth Eye Institute, but the experiments are not described in any detail and do not seem to have been published elsewhere. Some related references are Ames (1946) and Ames, Ogle, and Gliddon (1933).

Consider now the special iseikonic transformation

$$\begin{aligned}
\gamma_1 &= \gamma + \tau \\
\varphi_1 &= \varphi \\
\theta_1 &= \theta.
\end{aligned} \tag{8}$$

Since angles are preserved under iseikonic transformation, it can be shown that in Equation (6) for the line element ds^2, $f'(\gamma)/f(\gamma)^2$ must be a constant, say σ^2. We then have at once that

$$f(\gamma) = ce^{-\sigma\gamma} \tag{9}$$

and without loss of generality we can choose $c = 2$ so that we now have as the expression for the line element,

$$ds^2 = \frac{4}{(e^{\sigma\gamma} + Ke^{-\sigma\gamma})^2}(\sigma^2 d\gamma^2 + d\varphi^2 + \cos^2\varphi \, d\theta^2). \tag{10}$$

We also then have by substituting Equation (9) into Equation (4),

$$\xi = 2e^{-\sigma\gamma}\cos\varphi\cos\theta$$
$$\eta = 2e^{-\sigma\gamma}\sin\varphi \qquad (11)$$
$$\zeta = 2e^{-\sigma\gamma}\cos\varphi\sin\theta$$

as a suitable expression relating the physical coordinates γ, φ, θ to the sensory coordinates ξ, η, ζ.

Note that in the basic Equation (10) for the line element, the parameters K and σ are individual parameters that can only be determined experimentally.

The concept of iseikonic congruence was crucial in the argument just given for determining the form of the function $f(\gamma)$. In Luneburg (1950) a different and more global argument is given, using the results of the alley experiments of Hillebrand (1902) and Blumenfeld (1913); although the argument provides in a certain sense a route that is more firmly based experimentally for determining $f(\gamma)$, we shall omit it here.

13.10.5   Experiments Relevant to Luneburg's Theory

In Luneburg (1947, 1948, 1950), a number of experimental applications of the theory were sketched; for example, determination of the parameters K and σ for a given observer, quantitative analysis of observational data for equidistance and parallel alleys, analysis and prediction of visually congruent configurations, and analysis of what is visually congruent to infinite horizons in physical space. After Luneburg's premature death, detailed analytic suggestions for experiments, quantitative analysis of the data, or determination of parameters were made by his associate A. A. Blank (1953, 1957).

Hardy et al. The most extensive early test of Luneburg's theory is found in the report of L. H. Hardy, Rand, Rittler, Blank, and Boeder (1953) of the experiments carried out at the Knapp Memorial Laboratories, Institute of Ophthamology, Columbia University. Without entering into a detailed description of the experiments, we summarize the experimental setup and their main conclusions. All experiments were carried out in a dark room with configurations made up of a small number of low intensity point sources of light. The intensities were adjusted to appear equal to the observer but low enough not to permit any perceptible surrounding illumination.[11] Observers' heads were fixed in a headrest, and they always viewed static configurations; no perception of motion was investigated. All observations were made binocularly, and observers were permitted to let their point of regard vary over the entire physical configuration until stable judgments about the visual geometry of the configuration were reached. An important condition was that all experiments were restricted to the horizontal plane, that is, in the notation of the previous section, $\theta = 0$.

Their main conclusions were these:

(i) There is considerable experimental evidence to support Luneburg's hypothesis that two configurations related by an iseikonic transformation are visually congruent. Consistent with the notation used in the last section, in one experiment the iseikonic transformation was

$$\gamma' = \gamma$$
$$\varphi' = \varphi + \mu,$$

with $\mu = \pm 10°$. The observer was first asked to arrange a set of nine lights so that they appeared to lie on a straight line between two pre-set fixed lights symmetrically placed about the median. Then the pair of fixed lights was reset by adding μ to φ for one of the lights, and the observer was asked again to set some subset of the nine lights to lie on a line determined by the new (oblique) location of the two fixed lights. The nine lights were placed so as to be adjustable along the φ-lines with $\varphi_n = 5n°$ ($n = 0, \pm 1, \pm 2, \pm 3, \pm 4$) and with the fixed lights at $\varphi = \pm 20°$ initially symmetric to the median with $x = 330$ cm. These were the instructions to the observers: "The two end lights are fixed. Adjust the remaining lights by having them moved toward you or away from you until they appear to lie on a straight line between the end points."

Not all of the details of the experiment were ideal, as the authors explicitly state, since it had been originally designed to test a different hypothesis. For example, on the resetting of the fixed points not all nine points were available to the observer for resetting but only a subset. However, the qualitative agreement of data and theory for five observers was good, even if not within experimental error.

(ii) The experiments on parallel and equidistance alleys confirmed the classical results of Blumenfeld.

(iii) The efforts to determine the individual observer constants K and σ were not quantitatively successful. (Hardy et al. used a somewhat different

---

[11] This adjustment of physical intensity with varying distance of the light points from the subject so as to appear equal in intensity to the subject is standard procedure in the various experiments of this type described in this section.

formulation from that of Luneburg given above, but determination of their function $r(\Gamma)$ comes to the same thing.) The main problem was drift in the values of the constants through a sequence of experiments. The values of K obtained here and in related experiments supported Luneburg's hypothesis that, for most persons, visual space is hyperbolic, that is, $K < 0$.

Some closely related data and analysis are given in Blank (1958, 1961); in the main the results support the hypothesis that the curvature of visual space is negative. Other closely related experiments are those of Zajaczkowska (1956a, 1956b).

Indow and collaborators. The main group to continue in a direct way the theoretical and experimental work of Luneburg, Blank, and the Knapp Memorial Laboratories at Columbia has been the group centered around Tarow Indow, first at Keio University in Japan and later at the University of California at Irvine. The list of publications extends over a period of more than two decades, and the references we give here are far from complete. Indow, Inoue, and Matsushima (1962a, 1962b) reported extensive experiments conducted over a period of three years to test Luneburg's theory and, in particular, to estimate the individual parameters K and σ of Equation (10) of the previous section. In the 3-point experiment, three points of light Q_0, Q_1, and Q_2 were presented in the horizontal plane relative to the subject, but both horizontally and vertically relative to the darkened room. Q_0 and Q_1 were fixed, and it was the task of the subject to move Q_2 so that the segment Q_0Q_1 was visually congruent to the segment Q_0Q_2, with Q_2 constrained to a Vieth-Muller circle on which Q_0 and Q_1 do not lie. Conditions were similar in the 4-point experiment except that points Q_2 and Q_3 were to be adjusted so that Q_2Q_3 was visually congruent to Q_0Q_1. Of the 26 experimental runs with six subjects reported in Indow *et al.* (1962a), for 23 runs the estimated value of K was in the range $-1 < K < 0$ with a satisfactory goodness of fit, which directly supports Luneburg's theory that visual space is hyperbolic. It should be mentioned that repeated runs with the same subjects showed considerable fluctuation in the value of K. In Indow *et al.* (1962b) the same experimental setup and subjects were used to replicate the alley experiments of Hillebrand (1902) and Blumenfeld (1913) mentioned earlier. The equidistant and parallel alleys were in the relation observed in the earlier investigations and thus supported Luneburg's theory. But one aspect was not theoretically satisfactory. The values of K and σ estimated for individual subjects in Indow *et al.* (1962a) did not satisfactorily predict the alley data. Quite different estimated values were needed to fit these data.

Indow, Inoue, and Matsushima (1963) repeated the experiments of Indow *et al.* (1962a, 1962b), but with the points of light located in a spacious field.

In the earlier experiments the most distant point of light was 300 cm from the subject. In this study it was 1610 cm, made possible by conducting the experiment in a darkened gymnasium. Qualitatively, the results agreed with the earlier experiments, but the quantitative aspects, as reflected in the estimated parameters K and σ, did not.

To meet the criticism that the results of the alley experiments arise not from the hyperbolic nature of visual space but from a difference of scanning direction in constructing the two kinds of alleys, Indow and Watanabe (1984a) performed what is probably the first alley experiment with moving points. In their study, all stimulus points were presented on the horizontal plane at eye level. As the authors remark, it is technically complicated to produce both the perception of moving points and also have the points adjustable on the y-axis so that the subject can construct a parallel or equidistance alley. The solution was that of apparent movement. Discrete points were used with their presentations controlled in order and duration to give the subjects a clear impression of movement. The experimental data predominantly supported the hypothesis that visual space is hyperbolic: $K < 0$ for 79 of the 89 sets of data; these sets of data contained both stationary and moving points for all subjects. The authors also give a detailed analysis of the 10 cases in which $K > 0$: 8 of the 10 cases occurred with two subjects who produced data that did not fit well. In contrast, for 77 cases the root mean square of the deviations in fit were $\leqslant 0.3$. The actual details of the method of fit are too intricate to give here but are detailed in the article.

In contrast to these results, Indow and Watanabe (1984b) report alley experiments in the frontoparallel plane, which is perpendicular to the horizontal plane used in all earlier studies. They performed the experiments with the plane at distances of 98, 186, and 276 cm from the subject. In Experiment I the subject adjusted the position of point-light stimuli as described earlier. In Experiment II ratios of perceptual distances between stimuli were assessed by the subject. In contrast to the many alley experiments on the horizontal plane, the results of neither experiment unequivocally supported the hyperbolic hypothesis. In fact, the data were about as well fit by the Euclidean hypothesis that $K = 0$.

Starting in 1967 and extending over a number of years, Indow and associates have applied multidimensional scaling methods (MDS) to the direct investigation of the geometrical character of visual space. However, there are several points about MDS to keep in mind. First, the results would be difficult to interpret if the number of scaling dimensions exceeded the number of physical dimensions. Second, MDS is most often used when there are not strong structural constraints given in advance. This is not so in the present context since visual space is approximately Euclidean. Is the accuracy of MDS sufficient to pick up the sorts of discrepancies found in the alley experiments?

Matsushima and Noguchi (1967) used data from experiments of a Luneburg type—small light points in a dark room—and observation of stars in the night sky to obtain good fits to the Euclidean metric using MDS with the appropriate dimensionality. (Subjects were asked to judge ratios of interpoint distances.) However, the mapping between the physical space and the visual space determined by MDS was much more complicated than that proposed by Luneburg [Equation (5) of Section 13.10.4] and, in fact, was too complicated to describe in any straightforward mathematical fashion. Nishikawa (1967) continued the same line of investigation by arranging the light stimuli in ways to test the standard alley results and also standard horopter results.[12] He also suggested a theoretical approach to explain the Luneburg-type results, which he replicated, on the MDS assumption that visual space is Euclidean. The essence of the approach is to assume that the mapping function between visual and physical space changes substantially with a change in task and instruction, and is different from the Luneburg mapping function defined by Equation (4). The existence of such an effect seems likely, but Nishikawa's theoretical analysis does not get very far. Similar theoretical arguments were advanced by Indow (1967), but he expressed appropriate skepticism about the Euclidean solution. Closely related empirical results and theoretical ideas are also analyzed with care by Indow (1968, 1974, 1975a), who gives a particularly good quantitative account of the accuracy of the Euclidean model for various subjects when MDS methods are used.

Both methodologically and conceptually, it is natural to be somewhat skeptical that verbal estimates of ratios of distances, the MDS method used in the studies just cited, were sensitive and accurate enough to discriminate the Euclidean or hyperbolic nature of visual space. Indow expresses similar skepticism about nonmetric MDS, whose lack of sensitivity to details is well known. A thorough discussion of these matters is given in Indow (1982).

Indow (1982) extended in a detailed way the methods of MDS to using a hyperbolic or elliptic metric as well as a Euclidean one. Although the quantitative fit is not much better than that of the Euclidean metric, the hyperbolic metric does give a better account of the standard alley data.

---

[12] A horopter curve is, in this context, a physical curve symmetrical about the x-axis (the depth axis) that appears to the subject to be visually straight. The concept was introduced and studied extensively already in the nineteenth century by Helmholtz and others.

13.10.6 Other Studies

We restrict ourselves to a few other studies that are especially pertinent to the viewpoint of this chapter. Blank (1978) provided an overview of the geometrical theory and some experiments as of 1978, although he omits all reference to the many studies of Indow and his collaborators. Eschenburg (1980) is a useful mathematical study of the Luneburg theory. Foley (1964a, 1964b, 1972) undertook the important task of studying the qualitative properties of visual space with an emphasis on whether or not it is Desarguesian. In the first two papers his answer is tentatively affirmative, and in the last one, negative. Foley's work represents a line of attack that has not been followed up by other investigators as much as it should. The importance of Foley's careful experiments needs to be underscored. The two studies of 1964 will not be described here since the significant negative outcome of the 1972 study needs to be emphasized. A subject sat in a dark room with both eyes open, and a light A was fixed in the subject's visual field in the horizontal plane at the eye level. The subject was asked to set light B so that the line OB (O is position of the subject) is perceptually perpendicular to line AB, and segment OB is apparently equal in length to segment AB. The subject was next asked to set light C so that OC was perceived to be perpendicular to OB with OC equal to OB. The subject was then asked to judge the relative lengths of OA and BC. (See Figure 8.) For a homogeneous Riemannian space, i.e., a space of constant curvature (Euclidean, hyperbolic, or elliptic), by the construction right-angled isosceles triangles OBA and BOC should be congruent, and thus OA and BC should be judged equal in length. The results were that 24 subjects in 40 of 48 trials judged BC to be significantly longer than OA. From two other closely related experiments reported in the same article, Foley plausibly argued that these negative results may arise from the process of locating

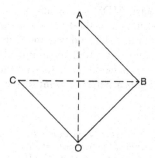

FIGURE 8. Foley's homogeneity experiment (1972).

line segments, as reflected in angle judgments, being independent of the process of judging length or distance as reflected in the comparison of the length of OA and of BC. Regardless of whether this explanation stands up, the results of the experiment raise serious questions about our being able to construct any simple but complete geometrical models of visual space. Foley's additional work on the binocular theory of perceived distance (1978, 1980) is also important in several respects, e.g., in presenting data that argue against the Luneburg hypothesis that iseikonically congruent configurations produce the same sequence of retinal images. We refer the reader to the two articles for the detailed developments.

A significant and careful experimental study that reaches some different conclusions about visual space is that of Wagner (1985). The methodology of the work is notable for two reasons. First, the experiments were conducted outdoors in full daylight in a large field with judgments about the geometrical relations of 13 white stakes. Second, four procedures were used for judging distances, angles and areas: magnitude (ratio) estimation, category estimation, mapping, and perceptual matching, where *mapping* means constructing a simple scale map of what is seen. Only the results for distance will be discussed here. In this case, perceptual matching was not feasible in the experimental setup and was not used.

The results for distance are surprising and interesting. The Luneburg model of hyperbolic space did not fit well at all. What did fit reasonably well is a Euclidean model of visual space, but the Euclidean visual space is a nontrivial affine transformation of Euclidean physical space. We can use the x and y axes of Figure 7 to discuss the results. The x-axis is the one perpendicular to the vertical plane through the eyes; it is the depth axis. The y-axis is the frontal axis passing through the two eyes. Let $(x, 0)$ and $(0, y)$ be two physical points such that $x = y$, i.e., along their respective axes the two points are equidistant from the origin (the point midway between the two eyes, but in visual space $x' = 0.5y'$) approximately, i.e., visual foreshortening amounts to the perceived distance along the depth axis being half of the physical distance when perceived frontal distances are equated to the physical distances. Call this foreshortening factor c; then $x' = cx$ and $y' = y$, with c varying with subjects but being approximately 0.5. This very strong effect is surprising, but it has also been found by Foley (1972) in the experiments in a dark room previously discussed. Given the very different types of experimental conditions, the foreshortening phenomenon seems quite robust. In an earlier study under somewhat similar conditions of full illumination, Battro, Netto, and Rozestraten (1976) also got results strongly at variance with the Luneburg predictions.

From a variety of viewpoints, the notable omission is experimental study of projective geometry since the essence of vision is the projection of objects in three-dimensional space onto the retina. Fortunately, Cutting (1986) has recently published a book on these matters. We shall not attempt a resumé of the many experiments he reports but concentrate on one fundamental point. The most important quantitative invariant of projective geometry is the cross ratio of four collinear points (defined in Section 12.2.3). In perceiving lines in motion, i.e., from a continuously changing perspective, does our perception of the cross ratio as invariant provide the visual evidence of rigidity in the actual relative spatial positions of given lines? Cutting provides evidence that the answer is by and large affirmative. This result also solves La Gournerie's paradox (1859), described in Pirenne (1975): linear perspective is mathematically correct for just one fixed point of view, but almost any position in front of a painting will not disturb our perception of it. As Cutting points out, an explanation of the paradox is that when the cross ratio of points that are projected onto a plane surface is preserved, it is preserved from any viewer position. Further pursuit of this kind of projective analysis should throw more light on the Euclidean or non-Euclidean nature of visual space.

Some comments. We close this section with some general observations about visual space. First, we have not presented detailed numerical data on the various parameters estimated or statistical goodness of fits because most investigators have commented on the instability of the quantitative results, even for repetition of a particular stimulus display with the same subjects. The fluctuating nature of the results is in great contrast to the precision and stability of the measurements of physical space. Second, changes in stimulus configurations or other conditions such as illumination seem to change drastically the underlying geometry of visual space, but the way the visual geometry apparently changes from one scene to another has as yet scarcely been studied in any systematic generality. Third, size constancy and depth perception—to mention just two important concepts—have not entered into many of the geometrical studies of visual space we have surveyed and, we believe, have not really been dealt with systematically in any studies aimed at characterizing the overall nature of visual space. (Foley, 1980, already referred to, is excellent on binocular depth perception but does not focus on the full geometry.) Fourth and finally, it seems that the kinds of *qualitative* structures we have focused on in this work could be used more effectively than they yet have been to understand the fundamental properties of visual space. For example, do the axioms of projective order or incidence, with the possible exception of Desaurgues' proposition, hold in all or almost all of the visual configurations that have been studied thus far? Certainly Foley's work should be pursued further. There are also a number of different ways to study qualitative axioms for visual spaces, as exemplified in Cutting's study of projective invariants. Perhaps more important, we need to develop qualitative axioms that are more directly suited to visual space since the favored position and orientation of the viewer is contrary to the homogeneous character of classical axiomatic geometry in which there are no distinguished points or orientation.

EXERCISES

Let $\langle A, S \rangle$ be a one-dimensional separation structure. Exercises 1–4 concern such structures.

1. Prove the following elementary results:

(i)   If $ab\,S\,cd$, then $ba\,S\,cd$.

(ii) If $ab\,S\,cd$, then a, b, c, and d are distinct.

(iii) If $ab\,S\,cd$ and $bc\,S\,de$, then $cd\,S\,ea$.

(iv) If $ab\,S\,cd$ and $ac\,S\,be$, then $ab\,S\,de$.

(v)   If $ab\,S\,cd$ and e is distinct from a, b, c, and d, then $ab\,S\,ce$ or $ab\,S\,de$.   (13.2.2)

2. Using a five-element set, prove the independence of Axiom 5 of Definition 2.   (13.2.2)

3. Define $xE_{a,b}y$ iff a and b are distinct from x and y, and it is not the case $ab\,S\,xy$.

(i) Prove that for all a, b in A, $E_{a,b}$ is an equivalence relation on $A - \{a, b\}$.

(ii) Prove that in $A - \{a, b\}$ there are at most two equivalence classes with respect to $E_{a,b}$.

(iii) Call each of the two equivalence classes of (ii) a segment. For each segment define $x <_{ab} y$ iff $ay\,S\,bx$. Prove that this binary relation is a strict simple order on each segment, i.e., asymmetric, transitive, and connected.

(iv) Let $<$ be a strict simple order of a nonempty set A. Define S' such that $xy\,S'\,uv$ iff x, y, u, and v are distinct points and exactly one of the points u, v is between x and y, where betweenness is defined in the obvious way in terms of $<$. Prove that S' satisfies Axioms 1–5 of Definition 2.   (13.2.2)

4. Define $a|b|c$ in terms of S, and prove that this defined relation satisfies the axioms of Definition 1.   (13.2.2)

5. Prove from the axioms of Definition 3 for Desarguesian projective planes the following elementary theorems, where $NE(a, b, c)$ means a, b, and c are distinct:

    (i) If $a \neq b$, then $a I ab$.

    (ii) If $\alpha \neq \beta$, then $\alpha I \alpha\beta$.

    (iii) If $a \neq b$, $a I \alpha$ and $b I \alpha$, then $\alpha = ab$.

    (iv) If $\alpha \neq \beta$, $\alpha I a$ and $\beta I a$, then $a = \alpha\beta$.

    (v) If $NE(a, b, c)$ and $ab = ac$, then $ab = bc$.

    (vi) We define the important concept of perspective. Triangles abc and $a'b'c'$ are *perspective from point e* iff

        (a) $NE(a, b, c, a', b', c', e)$

        (b) $NE(ab, ac, bc, a'b', a'c', b'c')$,

        (c) $ea = ea'$, $eb = eb'$, $ec = ec'$.

Correspondingly, trilaterals $\alpha\beta\gamma$ and $\alpha'\beta'\gamma'$ are perspective from line ϵ iff

        (a) $NE(\alpha, \beta, \gamma, \alpha', \beta', \gamma', \epsilon)$,

        (b) $NE(\alpha\beta, \alpha\gamma, \beta\gamma, \alpha'\beta', \alpha'\gamma', \beta'\gamma')$,

        (c) $\epsilon\alpha = \epsilon\alpha'$, $\epsilon\beta = \epsilon\beta'$, $\epsilon\gamma = \epsilon\gamma'$.

Show that if two triangles are perspective from a point the corresponding trilaterals are perspective from a line. (This is another formulation of Desargues' theorem.)   (13.4)

6. With reference to the definition of three-dimensional ordered projective spaces in Section 13.4, prove the following elementary theorems.

    (i) If three points are noncollinear, then there exists exactly one plane with which they are incident.

    (ii) For any planes π_1 and π_2, if there exists a point a such that $a I \pi_1$ and $a I \pi_2$, then there exists a point b distinct from a such that $b I \pi_1$ and $b I \pi_2$.

    (iii) Given a line and a point not on the line, there is a unique plane such that the line lies on the plane and the point is incident with the plane.

    (iv) If a and b are distinct points and π_1 and π_2 are distinct planes and if both points are incident with both planes, then

$$ab = \pi_1\pi_2.   (13.5)$$

7. With reference to ordered spaces, prove that the axioms of Definition 6 imply the following proposition:

If $a \neq b$, $c \neq d$, and c, d are on line ab, then a is on line cd. (13.6.1)

8. Define collinearity in terms of betweenness, and using the axioms on betweenness, prove the following properties, where $L(abc)$ means points a, b, and c are collinear:

(i) $L(aba)$.

(ii) If $a \neq b$, $L(abc)$, and $L(abd)$, then $L(acd)$.

(iii) If $L(abc)$, then $L(bac)$.

(iv) If $a = b$ or $a = c$ or $b = c$, then $L(abc)$.

(v)   If $a \neq b$, $L(abp)$, $L(abq)$, and $L(abr)$, then $L(pqr)$. (13.6.2)

9. We define for a plane the concept of the segment determined by points a and b being parallel to the segment determined by c and d as follows:

$$ab\|cd \quad \text{iff} \quad a = b \quad \text{or} \quad \forall p \text{ if } L(abp), \text{ then } L(cdp)$$

or
$$\forall p \text{ if } L(abp), \text{ then not } L(cdp).$$

Using the properties of collinearity stated in the preceding exercise, prove:

(i) $ab\|ba$.

(ii) $ab\|cc$.

(iii) If $a \neq b$, $ab\|cd$, $ab\|ef$, then $cd\|ef$.

(iv) If $ab\|ac$, then $ba\|bc$. (13.6.2)

10. The definition of collinearity in terms of parallelism is much simpler than the converse:

$$L(abc) \quad \text{iff} \quad ab\|ac.$$

Using properties (i)–(iv) of the preceding exercise, prove collinear properties (i)–(v) of Exercise 8. (13.6.2)

11. Use Axioms 2 and 3 of Definition 8 to prove that congruence is an equivalence relation. (13.6.3)

12. We define perpendicularity in absolute geometry as follows: Line K is *perpendicular* to line L iff K and L intersect at some point p such that there exist half-lines $A \subset K$ and $B \subset L$ both with origin p such that the angle AB is a right angle. Prove the following in three-dimensional absolute geometry:

(i) If L is a line on plane P and $a \in L$, there is a unique line $K \subset P$ such that $a \in K$ and K is perpendicular to L.

(ii) If $a \notin L$, then there is a unique line K such that $a \in K$ and K is perpendicular to L.

(iii) Given three lines K, L, and M in a plane P such that $K \neq L$ and K and L are both perpendicular to M, then $K \cap L = \varnothing$.

(iv) There is a unique line through a given point perpendicular to a given plane.

(v) If a plane P contains a line K perpendicular to a plane R, then R contains a line L perpendicular to P.

(vi) If P and R are perpendicular planes, any line in P that is perpendicular to the line of intersection of P and R is also perpendicular to R.

(vii) Two lines that are perpendicular to the same plane are coplanar. (13.6.3)

13. Let $\langle A, B, \sim \rangle$ be an n-dimensional Euclidean space in the sense of Definition 9. For this exercise, individual line segments ab are called *bound* segments. A *free* segment is an equivalence class of bound segments:

$$\alpha = [ab] = \{ cd \mid cd \sim ab \}.$$

Let F be the set of free segments, let $\alpha = [a_1 b_1]$, $\beta = [a_2 b_2]$, and define in the obvious fashion $a_1 b_1 \succsim a_2 b_2$ in terms of betweenness and congruence. Now define for free segments:

$$\alpha \succsim \beta \qquad \text{iff} \qquad a_1 b_1 \succsim a_2 b_2;$$

$\alpha + \beta = \gamma$ if $\exists a, b, c$ such that $B(abc)$, $\alpha = [ab]$, $\beta = [bc]$, and $\gamma = [ac]$. Prove from the axioms of Definition 8 that $\langle F, \succsim, + \rangle$ is a closed extensive structure in the sense of Definition 3.1 of Volume I.   (13.6.4)

14. Prove H1–9 of Section 6 for hyperbolic geometry. (Recall that the angle measure used in hyperbolic geometry is just that of absolute geometry.)   (13.6.5)

15. Prove that Axioms 2, 3, and 6 of special relativity (Definition 13) hold also for classical space-time (Definition 12).   (13.9)

16. This exercise relates to Theorem 17. Let A be the set of physical space-time points, and let $\mathscr{F}$ be a set of inertial frames or coordinate systems for four-dimensional space-time in the sense of special relativity. The relativistic distance $I_f(ab) = I_f[f(a)f(b)]$ is defined on page 122. Recall that a and b lie on an inertial path if $I_f^2(ab) < 0$. Let ab be an inertial path. The *slope* of ab relative to f is the three-dimensional vector W such that

$$W = \frac{[f_1(a), f_2(a), f_3(a)] - [f_1(b), f_2(b), f_3(b)]}{f_4(a) - f_4(b)},$$

where $f_i(a)$ is the ith coordinate of a in the frame f. The *speed* of ab is $|W|$. The speed of a line ab such that $I_f^2[f(a), f(b)] = 0$ is c for any $f \in \mathscr{F}$, i.e., is an optical line. A system $\langle A, \mathscr{F}, c \rangle$ is a *collection of inertial frames* if and only if for any a and b that lie on an inertial path with respect to some f in $\mathscr{F}$, then for all f and f' in $\mathscr{F}$,

$$I_f(ab) = I_{f'}(ab).$$

Note that this limited condition does not guarantee that arbitrary lines are carried into lines by an affine transformation from one frame to another. This must be proved.

LEMMA 1.   *If $k \geqslant 0$ and $f(a) - f(b) = k[f(c) - f(d)]$, then $I_f(ab) = kI_f(cd)$.*

LEMMA 2.   *If the points $f(a)$, $f(b)$, and $f(c)$ are collinear and $f(b)$ is between $f(a)$ and $f(c)$, then*

$$I_f(ab) + I_f(bc) = I_f(ac).$$

LEMMA 3.   *If any two of the three points a, b, and c are distinct and lie on an inertial path with respect to f and if $I_f(ab) + I_f(bc) = I_f(ac)$, then the points $f(a)$, $f(b)$, and $f(c)$ are collinear, and $f(b)$ is between $f(a)$ and $f(c)$.*

LEMMA 4.   *The property of being the midpoint of a finite segment of an inertial path is invariant in $\mathscr{F}$, i.e., the property holds uniformly for every f in $\mathscr{F}$, or for none.*

LEMMA 5.   *The property of being the midpoint of an arbitrary finite segment is invariant in $\mathscr{F}$.*

LEMMA 6. *The property of two finite segments of inertial paths being parallel and in a fixed ratio is invariant in $\mathscr{F}$.*

LEMMA 7. *The property of two arbitrary finite segments being parallel and in a fixed ratio is invariant in $\mathscr{F}$.*

LEMMA 8. *Any two frames in $\mathscr{F}$ are related by a nonsingular affine transformation.*

LEMMA 9. *Any two frames in $\mathscr{F}$ are related by a Lorentz transformation.*
(13.9)

17. Prove Theorem 19.

Chapter 14   Proximity Measurement

14.1   INTRODUCTION

Much work in the social and the behavioral sciences is concerned with the analysis and representation of proximity relations among entities such as colors, faces, emotions, consumer products, and political parties. Proximity relations can be inferred from various types of observations, such as rating of similarity or dissimilarity, classification of objects, correlation among variables, frequency of co-occurrences, and errors of substitution. Each of these procedures gives rise to a proximity ordering of the form either *b is closer to a than to c* or *the similarity of a to b exceeds that of c to d*.

Proximity data are commonly used to infer the structure of the entities under study and to embed them in a suitable formal structure. Formal representations serve two related functions: they provide convenient methods for describing, summarizing, and displaying proximity data, and they are also treated as theories about the data-generating process. Representations of proximity relations can be divided into two general classes: geometrical (spatial) models and set-theoretical (feature) models.

Geometrical models represent the objects under study as points in a space so that the proximity ordering of the objects is represented by the ordering of the metric distances among the respective points. The most familiar example of such a representation is the embedding of objects as points in an n-dimensional Euclidean space. Earlier formulations of this model (Torgerson, 1952; G. Young and Householder, 1938) assumed interval scale measurement of proximity, but much modern work (e.g., Coombs, 1964; Guttman, 1968; Kruskal, 1964; Shepard, 1962a, 1962b, 1980) has relied primarily on ordinal data.

Set-theoretical, or feature, models of proximity represent each object as a collection of features and express the proximity between objects in terms of the measures of their common and their distinctive features. Because a feature can be represented by the set of objects that share this feature, it is possible to express feature models in terms of the clusters they induce. The most familiar clustering model is the ultrametric tree (see Section 12.3.4), also called the hierarchical clustering scheme (S. C, Johnson, 1967; Sokal and Sneath, 1963). For other classificatory models, see Carroll (1976), Corter and Tversky (1986), and Shepard and Arabie (1979).

Spatial and clustering representations of proximity relations have been used extensively in many fields including biology, ecology, psychology, anthropology, linguistics, political science, and marketing. Indeed, there is an extensive literature on multidimensional scaling and cluster analysis dealing with scaling algorithms for spatial representations (mostly Euclidean) and clustering solutions (mostly trees) (see, e.g., Carroll and Arabie, 1980; Davies and Coxon, 1982; Davison, 1983; Hartigan, 1975; Jardine and Sibson, 1971; Kruskal and Wish, 1978; Schiffman, Reynolds, and Young, 1981). In this chapter we do not address the scaling literature. Instead, we analyze metric, dimensional, and set-theoretical representations of proximity from the standpoint of measurement theory.

We first investigate a general metric model with additive segments (Sections 14.2 and 14.3) using the theory of extensive measurement developed in Chapter 3. We then turn to the analysis of dimensional metric models (Sections 14.4 and 14.5) using the theory of difference and conjoint measurement developed in Chapters 4 and 6. Experimental investigations of the basic axioms underlying dimensional metric representations of proximity data are reviewed in Section 14.6. Finally, set-theoretical models of proximity based on feature matching are discussed in Sections 14.7 and 14.8 from both theoretical and empirical perspectives.

The basic qualitative structure is characterized by the following definition, in which $(a, b) \succeq (c, d)$ denotes the observation that the dissimilarity between a and b is at least as great as that between c and d.

DEFINITION 1.   *Suppose A is a set and $\succeq$ is a quaternary relation on A, i.e., a binary relation on $A \times A$. The pair $\langle A, \succeq \rangle$ is a proximity structure iff the following axioms hold for all $a, b \in A$:*

1. $\succsim$ is a weak ordering of $A \times A$, i.e., it is connected and transitive.
2. $(a, b) \succ (a, a)$ whenever $a \neq b$.
3. $(a, a) \sim (b, b)$ (minimality).
4. $(a, b) \sim (b, a)$ (symmetry).

The structure is factorial iff

$$A = \overset{n}{\underset{i=1}{\textbf{X}}} A_i .$$

The definition of a proximity structure imposes natural, though not trivial, constraints on the observed relation of dissimilarity. Axiom 1 is the familiar ordering assumption applied to proximity. Axiom 2 imposes the natural constraint that other-dissimilarity exceeds self-dissimilarity. Axiom 3 requires all self-dissimilarities to be equivalent, and Axiom 4 requires dissimilarity to be symmetric.

Several comments regarding Definition 1 are in order. First, Axiom 2 may fail when a and b are physically distinct but indistinguishable to the observer. In this case, the analysis should be applied to the appropriate equivalence classes provided they are well defined (see Chapter 16). Second, both minimality and symmetry are violated in some situations. To illustrate the problem let $P(a, b)$ be the probability that a is (erroneously) identified as b in a forced-choice recognition experiment. Naturally, the greater the similarity between a and b, the more likely they are to be confused. Thus confusion probabilities are commonly used to define the proximity order between stimuli. In this case, however, both minimality and symmetry are often violated. The probability of correct identification is generally not the same for all stimuli, and the probability of confusing b with a, $P(b, a)$, is not always the same as the probability of confusing a with b, $P(a, b)$ (see, e.g., G. A. Miller and Nicely, 1955).

There are several ways of treating violations of minimality or of symmetry. One can define the proximity order in a manner that assures minimality and symmetry, for example, by defining self-dissimilarities to be zero and by averaging $P(a, b)$ and $P(b, a)$. Alternatively, one can attribute failures of minimality and symmetry to a response bias that can be estimated within an appropriate model (see, e.g., Holman, 1979; Luce, 1963). Finally, one can reject metric models altogether and employ more general models that do not impose minimality and symmetry (see Section 14.7). The assumptions of Definition 1 are assumed to hold in Sections 14.2 through 14.6, which are concerned with the representation of a proximity structure in a metric space.

DEFINITION 2. *Suppose A is a set and $\succsim$ is a quaternary relation on A. The structure $\langle A, \succsim \rangle$ is representable by a metric iff there exists a real-valued function δ on A^2 such that*

1. $\langle A, \delta \rangle$ *is a metric space, i.e., δ is a metric on A (Definition 12.10).*
2. δ *preserves $\succsim$, i.e., for all $a, b, c, d \in A$*

$$(a, b) \succsim (c, d) \qquad iff \qquad \delta(a, b) \geqslant \delta(c, d).$$

Axioms 1 through 4 of Definition 1 are clearly necessary for representing $\langle A, \succsim \rangle$ by a metric. Axiom 1 follows at once from Part (2) of Definition 2, and Axioms 2, 3, and 4 follow from the fact that any metric δ (Definition 12.11) satisfies

$$\delta(a, b) = \delta(b, a) > \delta(a, a) = \delta(b, b) \qquad \text{for all } a \neq b.$$

An additional necessary condition for $\langle A, \succsim \rangle$ to be representable by a metric is that $A^2/\sim$ must have a finite or countable order-dense subset (Definition 2.1, p. 40). If a proximity structure $\langle A, \succsim \rangle$ satisfies this condition, then by Theorem 2.2 there exists a real-valued function δ on A^2 that preserves $\succsim$. Furthermore, since the representation is an ordinal scale, we can select δ so that $\delta(a, a) = 0$ and $1 < \delta(a, b) < 2$ for all $a \neq b$. It is easy to verify that δ is a metric. In particular, the triangle inequality follows from the fact that for any three numbers x, y, z between 1 and 2, $x + y \geqslant z$.

THEOREM 1. *Suppose A is a set and $\succsim$ is a quaternary relation on A. Then the structure $\langle A, \succsim \rangle$ is representable by a metric (Definition 2) iff $\langle A, \succsim \rangle$ is a proximity structure (Definition 1) and $A^2/\sim$ has a finite or countable order-dense subset.*

Theorem 1 shows that, in the absence of additional restrictions, the triangle inequality does not impose any empirical constraint. Thus representability by an arbitrary metric exists under very general conditions, and the resulting representation is essentially an ordinal scale. For a metric representation to carry more interesting information, the class of metric spaces must be restricted, and corresponding stronger empirical conditions must be imposed. The following sections consider various classes of metric models and formulate ordinal conditions that are necessary and/or sufficient for the representation of a proximity structure by a model of each of these classes.

14.2 METRICS WITH ADDITIVE SEGMENTS

14.2.1 Collinearity

In this section we develop a theory due to Beals and Krantz (1967) for the representation of a proximity structure by a metric space with additive segments, i.e., a metric space in which every pair of points is joined by a segment along which distances are additive; see Definition 12.12. Some important examples of such metrics are discussed in Section 12.3.

One major reason for considering such structures is that practically all metric models that have been studied in any depth in mathematics are metrics with additive segments. A measurement axiomatization for this class is, therefore, a natural starting point for a foundational treatment of any of the more specific types, discussed in later sections.

To derive necessary conditions for a metric with additive segments, we first need to develop an ordinal criterion for a point b to lie on an additive segment from a to c.

Suppose that $\langle A, \delta \rangle$ is a metric space with additive segments and that b lies on an additive segment from a to c. Then

$$\delta(a, b) + \delta(b, c) = \delta(a, c). \tag{1}$$

Suppose that a', b', c' are any three points such that

$$\delta(a', b') \leqslant \delta(a, b) \tag{2}$$

and

$$\delta(a, c) \leqslant \delta(a', c'). \tag{3}$$

Using the triangle inequality and Equations (1) and (3), we have

$$\delta(a', b') + \delta(b', c') \geqslant \delta(a', c') \geqslant \delta(a, c) = \delta(a, b) + \delta(b, c).$$

This inequality, together with Equation (2), implies

$$\delta(b', c') \geqslant \delta(b, c). \tag{4}$$

That is, if b lies on a segment from a to c, Equations (2) and (3) imply Equation (4). Furthermore, if either antecedent inequality is strict, so is the conclusion. This implication involves only inequalities in its statement, though addition of numbers is involved in its proof. It can be translated, therefore, into the following, purely ordinal, criterion.

DEFINITION 3. *For $a, b, c \in A$, the ternary relation (abc) is said to hold iff for all $a', b', c' \in A$:*

1. *If $(a, b) \succsim (a', b')$ and $(a', c') \succsim (a, c)$, then $(b', c') \succsim (b, c)$.*

2. *If either of the preceding antecedent inequalities is strict, then the consequent inequality is also strict.*

The relation $\langle abc \rangle$ is said to hold iff both (abc) and (cba) hold. Definition 3 introduces a *collinear betweenness* relation (much like the one used in Section 13.6.2 on Euclidean geometry) in terms of the proximity relation $\succsim$. It is instructive to examine the defined relation in terms of concentric spheres.

The spherical boundary, or isosimilarity contour, $B(a, cd)$ can be defined in terms of $\succsim$ as follows:

$$B(a, cd) = \{ b \in A | (a, b) \sim (c, d) \}.$$

Intuitively, $B(a, cd)$ is a sphere with center a and radius (c, d). Accordingly, we speak of b as being outside, on, or inside the spherical boundary depending on whether $(a, b) \succ (c, d)$, $(a, b) \sim (c, d)$, or $(a, b) \prec (c, d)$.

Suppose that $B(a, ae)$ and $B(a, af)$ are two concentric spherical boundaries and that c is a point on the outer one; then a point b on the inner one lies on a segment from a to c if the (b, c) distance is minimal, i.e., equal to the *difference in the radii* of the spheres. If $B(a', ae)$ and $B(a', af)$ are any two other concentric spherical boundaries with center a', which may differ from a, but with the same radii as the spheres around a, and if b' is on or inside the inner one and c' is on or outside the outer one, then (b', c') cannot be smaller than the difference in radii; so $(b', c') \succsim (b, c)$. Furthermore, if b' is strictly inside the inner boundary or c' is strictly outside the outer boundary, then $(b', c') \succ (b, c)$. A graphical illustration of this relation is presented in Figure 1.

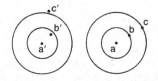

FIGURE 1. Illustration of the collinear betweenness relation in terms of isosimilarity contours (see text for explanation).

If $\langle A, \succsim \rangle$ is representable by a metric with additive segments, then for any distinct $a, c \in A$ and for any nonnegative $\alpha \leqslant \delta(a, c)$, there is a point b on a segment from a to c whose distance from a is exactly α. Thus a necessary condition for a representation as a metric with additive segments is the following *segmental solvability* condition:

> For any $a, c, e, f \in A$ with $(a, c) \succsim (e, f)$, there exists $b \in A$
> such that $(a, b) \sim (e, f)$ and $\langle abc \rangle$. (5)

To see the content of this condition in a different way, consider two concentric boundaries $B(a, ae)$ and $B(a, af)$ and a point c on the outer contour. If one can always construct a point b on the inner contour that minimizes the distance to c, then segmental solvability merely asserts that the minimized distance is independent of the *direction* (choice of c) and of the *location* of the spheres (choice of a); it depends only on the radii (a, e) and (a, f). Put differently, the space must be homogeneous in the sense that the spacing between concentric isosimilarity contours is independent of direction and location. Equation (5) excludes, for example, the exponential metric (Section 14.4.5) for which distances are additive along the major coordinates but not elsewhere.

Unlike transitivity and symmetry, which are subject to straightforward experimental tests, the empirical test of segmental solvability requires a rather elaborate construction. Given two concentric isosimilarity contours and a point c on the outer contour, one finds a point b on the inner contour that is closest to c. If this is done for different directions (choices of c) and for other concentric pairs of contours with the same radii, one can test whether the minimized distances are, in fact, equal. Thus the first problem is to construct isosimilarity contours (i.e., multiple solutions to an equation); then one must repeatedly search within one of them to minimize the distance to a point on another contour. With substantial experimental errors, these operations can be difficult to perform. Thus segmental solvability is an unlikely target for a direct experimental test. It may, nevertheless, prove useful in the analysis of isosimilarity contours and in the testing of specific hypotheses concerning preferred spatial directions.

14.2.2 Constructive Methods

To gain further insight into the representation of a proximity structure by a metric with additive segments, we examine next the constructions required to measure distances according to such a model. Suppose $\langle A, \succsim \rangle$ is a proximity structure (Definition 1) satisfying segmental solvability [Equation (5)]. Fundamentally, two constructions are required to estimate the ratio of two distances, say $\delta(a, b)$ and $\delta(c, d)$.

First, one must construct segments from a to b and from c to d; or, at least, find many closely spaced points along such segments. Second, one must partition each segment into approximately equal subsegments, with the same size subsegment for both. If p such subsegments are needed for the segment from a to b and q for the segment from c to d, then the additivity of distances along the segments implies that the ratio of $\delta(a, b)$ to $\delta(c, d)$ is approximately p/q.

Points along a segment can be constructed by repeated applications of segmental solvability. This can be done in many ways. We illustrate one that has the advantage that the points so constructed partition the segment into subsegments, all but one of which are equal. Thus both constructions follow the same process.

Start by choosing a distance (e, f), with $e \neq f$, which will serve as the common length of the subsegments. To obtain points along a segment from a to b, let $a_0 = a$ and find a_1 such that $\langle a_0 a_1 b \rangle$ and $(a_0, a_1) \sim (e, f)$. Assuming that $(a, b) \succsim (e, f)$, one can do this by constructing the isosimilarity contour $B(a, ef)$ and finding an a_1 on it that is as close as possible to b. Next, if $(a_1, b) \succ (e, f)$, find a_2 such that $\langle a_1 a_2 b \rangle$ and $(a_1 a_2) \sim (e, f)$. One continues this process until one finds a_{p-1} with $(e, f) \succsim (a_{p-1}, b)$. Then let $a_p = b$. The points $a_0, a_1, \ldots, a_p$ lie on a segment from a to b and partition it into p subsegments, the first $p - 1$ of which are equal in distance to (e, f), and the pth one is equal to or smaller than the distance between e and f. In the same way, one constructs $c_0, \ldots, c_q$ along a segment from c to d, with $(c_{i-1}, c_i) \sim (e, f)$ for $i = 1, \ldots, q - 1$ and $(e, f) \succsim (c_{q-1}, c_q)$. Since the distances along segments are additive, we can then assert that

$$\frac{p-1}{q} < \frac{\delta(a, b)}{\delta(c, d)} < \frac{p}{q-1}. \tag{6}$$

A number of comments about this construction are in order. The role of isosimilarity contours is fundamental. The cornerstone of the repetitive construction of a_i is the construction of an isosimilarity contour with center a_{i-1} and radius (e, f). From the preceding constructions and segmental solvability, we know that $\langle a_{i-1} a_i b \rangle$ for $i = 1, \ldots, p - 1$ and that $\langle c_{i-1} c_i d \rangle$ for $i = 1, \ldots, q - 1$. However, to have $a_0, \ldots, a_n$ all lying on one segment, we must have $\langle a_i a_j a_k \rangle$ for $i \leqslant j \leqslant k$. Similar considerations apply to $c_1, \ldots, c_q$. These statements all follow from the following proposition.

THEOREM 2. *Suppose* $\langle A, \succsim \rangle$ *is a proximity structure (Definition 1) that satisfies segmental solvability [Equation (5)]. Then for any a, b, c, $d \in A$*

$$if \quad \langle abc \rangle \quad and \quad \langle acd \rangle, \quad then \quad \langle abd \rangle \quad and \quad \langle bcd \rangle.$$

The proof of this theorem is given in Section 14.3.1. It establishes the fundamental property that must be satisfied by a *collinear betweenness* relation (see Definition 9, Chapter 4, and the following remarks, p. 172). The fact that our ordinal criterion for collinear betweenness (Definition 3) is strong enough to prove Theorem 2 makes segmental solvability the key property for the construction of a metric with additive segments.

Our assumptions do not guarantee that the above construction of $a_0, a_1, \ldots$ eventually terminates in the construction of a_{p-1} such that $(e, f) \succsim (a_{p-1}, b)$. Thus one needs an Archimedean assumption to guarantee that (e, f) cannot be "infinitely small" compared with (a, b), provided $e \neq f$. Once such a technical assumption is introduced, the ratio $\delta(a, b)/\delta(c, d)$ is uniquely determined by the preceding construction, and so distances are measured on a ratio scale.

The main limitation of this method lies in the repeated use of constructions involving $\sim$. If determinations of $\sim$ could be observed directly, then the error in estimating $\delta(a, b)/\delta(c, d)$ would only be a function of the size of (e, f). However, if the $\sim$ relations are only approximate, additional unknown errors are introduced. Hence, the measurement procedure would be improved if we could base the estimation of $\delta(a, b)/\delta(c, d)$ on inequality rather than on equivalence observations.

An obvious way would be to consider all sequences $a_0, \ldots, a_p$, with $a_0 = a$ and $a_p = b$, such that $(e, f) \succ (a_{i-1}, a_i)$ for $i = 1, \ldots, p$. The sequence $a_0, \ldots, a_p$ is not constrained to lie on an additive segment. Find such a sequence with the minimal p. This exists because p cannot be smaller than the value of $\delta(a, b)/\delta(e, f)$ in a metric representation. This construction does not involve observations of equal similarity or $\sim$. The difficulty is that we have no procedure for finding such a sequence or for knowing that one of minimal length has been attained. A constructive method leading to an approximation for $\delta(a, b)/\delta(c, d)$, with bound of error specified on the basis of a finite number of observations, would be desirable.

As will be shown in the next section, the assumption that $\langle A, \succsim \rangle$ is a proximity structure satisfying segmental solvability and an Archimedean axiom implies the existence of an essentially unique representation by a metric that is additive for any points a, b, c satisfying $\langle abc \rangle$. To obtain a metric with complete additive segments joining any two points, however, additional technical assumptions must be added. In the next section, these assumptions are introduced, and the main representation and uniqueness theorem are stated.

14.2.3 Representation and Uniqueness Theorems

The conditions needed to establish a representation by a metric with additive segments are summarized in the following definition.

DEFINITION 4. *A proximity structure* $\langle A, \succsim \rangle$ *(Definitions 1), with a trinary relation* $\langle\ \rangle$ *defined in terms of* $\succsim$ *(Definition 3), is segmentally additive iff the following two axioms hold for all* $a, c, e, f \in A$:

1. *If* $(a, c) \succsim (e, f)$, *then there exists* $b \in A$ *such that* $(a, b) \sim (e, f)$ *and* $\langle abc \rangle$.

2. *If* $e \neq f$, *then there exist* $b_0, \ldots, b_n \in A$ *such that* $b_0 = a$, $b_n = c$ *and* $(e, f) \succsim (b_{i-1}, b_i)$, $i = 1, \ldots, n$.

A segmentally additive proximity structure is complete iff the following two axioms hold for all $a, c, e, f \in A$:

3. *If* $e \neq f$, *then there exist* $b, d \in A$, *with* $b \neq d$, *such that* $(e, f) \succ (b, d)$.

4. *Let* $a_1, \ldots, a_n, \ldots$ *be a sequence of elements of* A *such that for all* $e, f \in A$, *if* $e \neq f$, *then* $(e, f) \succsim (a_i, a_j)$ *for all but finitely many pairs* (i, j). *Then there exists* $b \in A$ *such that for all* $e, f \in A$, *if* $e \neq f$, *then* $(e, f) \succsim (a_i, b)$ *for all but finitely many* i.

Axiom 1 is the segmental solvability condition, Equation (5), discussed in Section 14.2.1. Axiom 2 is the Archimedean (or connectedness) property needed for the construction leading to the evaluation of distance. Axiom 3 is a nondiscreteness condition. It specifies that there is no smallest positive distance. Axiom 4 expresses the usual completeness property of metric spaces: every Cauchy sequence has a limit. The class of models for which all the axioms of Definition 4 are both necessary and sufficient is the class of complete metric spaces with additive segments.

THEOREM 3. *Suppose* $\langle A, \succsim \rangle$ *is a segmentally additive proximity structure (Definition 4). Then there exists a real-valued function* δ *on* A^2 *such that, for any* $a, b, c, d \in A$,

(i) $\langle A, \delta \rangle$ *is a metric space (Definition 12.10).*

(ii) $(a, b) \succsim (c, d)$ *iff* $\delta(a, b) \geq \delta(c, d)$.

(iii) $\langle abc \rangle$ *iff* $\delta(a, b) + \delta(b, c) = \delta(a, c)$.

(iv) *If* δ' *is another metric on* A *satisfying the above conditions, then there exist* $\alpha > 0$ *such that* $\delta' = \alpha\delta$.

THEOREM 4. *Suppose* $\langle A, \succsim \rangle$ *is a complete segmentally additive proximity structure (Definition 4). Then there exists a real-valued function* δ *on* A^2 *such that* (i)–(iv) *of Theorem 3 hold, and* $\langle A, \delta \rangle$ *is a complete metric space with additive segments.*

Theorems 3 and 4 are proved in Sections 14.3.3 and 14.3.4, respectively. The proofs are based on the reduction of a segmentally additive proximity structure to an extensive-measurement structure. A longer proof, based directly on the construction outlined in Section 14.2.2, is presented in Beals and Krantz (1967).

14.3 PROOFS

All the numbered axioms in this section refer to Definition 4 unless otherwise stated.

14.3.1 Theorem 2 (p. 167)

Suppose $\langle A, \succsim \rangle$ *is a proximity structure satisfying segmental solvability. For any* $a, b, c, d \in A$ *if* $\langle abc \rangle$ *and* $\langle acd \rangle$, *then* $\langle abd \rangle$ *and* $\langle bcd \rangle$.

The proof of Theorem 2 is based on the following two lemmas.

LEMMA 1. $\langle abc \rangle$ *implies* $(a, c) \succsim (a, b)$ *and* $(a, c) \succsim (b, c)$.

PROOF. Since by Definition 1 $(a, b) \succsim (a, a)$ and $(a, c) \succsim (a, c)$, $\langle abc \rangle$ implies $(a, c) \succsim (b, c)$. Similarly, from $\langle cba \rangle$ we obtain $(a, c) \succsim (a, b)$. ◇

LEMMA 2. *If* $\langle abc \rangle$, $(a, b) \succsim (a', b')$, $(b, c) \succsim (b', c')$, *and* $(a', c') \succsim (a, c)$, *then* $\langle a'b'c' \rangle$.

PROOF. If any of the three inequalities in the hypothesis were strict, then (ii) of Definition 3 would yield a contradiction of one of the other two hypothesized inequalities. Thus $(a', b') \sim (a, b)$, $(b', c') \sim (b, c)$, and $(a, c) \sim (a', c')$. From these inequalities and Definition 3 $\langle a'b'c' \rangle$ is immediate. ◇

We turn now to Theorem 2, i.e., given $\langle abc \rangle$ and $\langle acd \rangle$, we prove $\langle bcd \rangle$ and $\langle abd \rangle$.

Proof of $\langle bcd \rangle$. By Lemma 1, $(a, d) \succsim (a, c)$. Since $\langle abc \rangle$, we have $(b, d) \succsim (b, c)$. By Axiom 1, construct c' with $(b, c') \sim (b, c)$ and $\langle bc'd \rangle$. From $\langle abc \rangle$ and $(b, c') \sim (b, c)$, we have $(a, c) \succsim (a, c')$. From this and

(acd), $(c', d) \gtrsim (c, d)$. From $(b, c') \sim (b, c)$, $(c', d) \gtrsim (c, d)$, and Lemma 2, we have $\langle bcd \rangle$.

Proof of $\langle abd \rangle$. First we show $(d, a) \gtrsim (d, b)$. Otherwise, (dca) and $(d, b) \succ (d, a)$ imply $(c, b) \succ (c, a)$, contradicting $\langle cba \rangle$ (by Lemma 1). By Axiom 1 construct b' with $(d, b') \sim (d, b)$ and $\langle db'a \rangle$. Next, we show $(a, b') \gtrsim (a, b)$. By $\langle bcd \rangle$, shown above, $(d, b') \gtrsim (d, c)$. By Axiom 1 construct c' with $(d, c') \sim (d, c)$ and $\langle dc'b' \rangle$. If $(a, b) \succ (a, b')$, then by $\langle abc \rangle$ either $(a, c) \succ (a, c')$ or $(b', c') \succ (b, c)$. But the former is false because $\langle acd \rangle$ and $(a, c) \succ (a, c')$ would imply $(c', d) \succ (c, d)$, contrary to construction; and the latter is false because $\langle dc'b' \rangle$, $(d, c) \sim (d, c')$, and $(d, b) \sim (d, b')$ imply $(b, c) \gtrsim (b', c')$. Thus $(a, b') \gtrsim (a, b)$. From $(d, b') \sim (d, b)$, $(a, b') \gtrsim (a, b)$, $\langle db'a \rangle$, and Lemma 2, we have $\langle abd \rangle$.   ◇

14.3.2 Reduction to Extensive Measurement

We show in this section that the construction of the desired metric representation is reducible to the construction of extensive measurement. To apply extensive measurement to the present problem, let

$$N = \{(a, a) | a \in A\} \quad \text{and let} \quad S = (A^2 - N)/\sim .$$

Thus S is the set of all equivalence classes of A^2 except the class consisting of the null pairs (a, a), (b, b), etc. To denote elements of S, we simply omit parentheses and commas. Thus, for any $a, b \in A$ with $a \neq b$, $ab = \{(e, f) | e, f \in A, (e, f) \sim (a, b)\}$. For simplicity, we use $\gtrsim$ to denote the natural (induced) order on S.

Next, a concatenation operation $\circ$ on S is defined as follows. For any ab, $a'b'$, $ed \in S$, $ab \circ a'b' = ed$ iff there exists $f \in A$, such that $ab = ef$, $a'b' = fd$, and $\langle efd \rangle$. Let T be the set of all pairs $(ab, a'b')$ for which $ab \circ a'b'$ is defined.

LEMMA 3. *The quadruple* $\langle S, T, \gtrsim, \circ \rangle$ *is an extensive structure with no essential maximum* (*Definition 3, Chapter 3*).

PROOF. First note that $\circ$ is well defined and commutative, since if $\langle efd \rangle$, $\langle e'f'd' \rangle$, $ef \sim e'f'$, and $fd \sim f'd'$, then $ed \sim e'd'$. Next, we verify the six axioms of Definition 3.3.

1. $\gtrsim$ is a weak ordering of S. Obvious.

2. We show that if $(ab, a'b') \in T$ and $(ab \circ a'b', a''b'') \in T$, then $(a'b', a''b'') \in T$, $(ab, a'b' \circ a''b'') \in T$, and $(ab \circ a'b') \circ a''b'' = ab \circ (a'b' \circ a''b'')$.

Let $(ab \circ a'b') \circ a''b'' = ed$, and let f be such that $\langle efd \rangle$ and $ab \circ a'b' = ef$. Such an f can be found by Axiom 1. Use Axiom 1 to construct f' such that $\langle ef'f \rangle$ and $ef' = ab$. Then by definitions of $\circ$ and $\langle \ \rangle$, $f'f = a'b'$. By Theorem 2, $\langle ef'd \rangle$ and $\langle f'fd \rangle$. The latter relation shows that $(a'b', a''b'') \in T$ and $a'b' \circ a''b'' = f'd$. Then from $\langle ef'd \rangle$ we have $(ab, a'b' \circ a''b'') \in T$ and $ab \circ (a'b' \circ a''b'') = ed$, as required.

3. It follows from Axiom 1 that if $(ab, a''b'') \in T$ and $ab \succsim a'b'$, then $(a''b'', a'b') \in T$ and $ab \circ a''b'' \succsim a''b'' \circ a'b'$.

4. It also follows from Axiom 1 that if $ed \succ ab$, then there exists $a'b'$ such that $(ab, a'b') \in T$ and $ab \circ a'b' = ed$.

5. If $(ab, a'b') \in T$, then since $a'b' \succ bb$, $ab \circ a'b' \succ ab$. Note that this property guarantees that there is no essential maximum (Definition 3.4).

6. Let $1ab = ab$, and if $(n-1)ab$ is defined and $[(n-1)ab, ab] \in T$, let $nab = (n-1)ab \circ ab$. Then for any $ab, ef \in S$, the set

$$\{ n \mid nef \text{ defined}, \quad ab \succsim nef \}$$

is finite. Here, for the first time, we use Axiom 2. Let $b_0, \ldots, b_p$ be such that $b_0 = a$, $b_p = b$ and $b_{i-1}b_i \precsim ef$ for $i = 1, \ldots, p$. This is possible because $e \neq f$, since $ef \in S$. We show that for $n > p$, either nef is undefined or $nef \succ ab$. Suppose that for some $n > p$, $ab \succsim nef$. Then by repeated application of Axiom 1 we can construct $a_0, \ldots, a_n$ such that $a_0 = a$, $\langle a_0 a_i b \rangle$, and $a_{i-1} a_i \sim ef$, $i = 1, \ldots, n$. We have $b_0 b_1 \precsim a_0 a_1$. Also, if $b_0 b_k \precsim a_0 a_k$ for $1 \leqslant k \leqslant p-1$, then from $\langle a_0 a_k a_{k+1} \rangle$ and $b_k b_{k+1} \precsim a_k a_{k+1}$, we conclude $b_0 b_{k+1} \precsim a_0 a_{k+1}$. Thus $ab = b_0 b_p \precsim a_0 a_p \prec a_0 a_n \precsim ab$, a contradiction. Hence, for $n > p$, $ab \succsim nef$ cannot hold. $\Diamond$

Note that the key step in this proof involves applying Theorem 2 to establish the associativity of $\circ$.

14.3.3 Theorem 3 (p. 168)

If $\langle A, \succsim \rangle$ is a segmentally additive proximity structure, then there exists a function δ on A^2 such that for any $a, b, c, d \in A$

(i) $\langle A, \delta \rangle$ *is a metric space.*

(ii) $(a, b) \succsim (c, d)$ *iff* $\delta(a, b) \geqslant \delta(c, d)$.

(iii) $\langle abc \rangle$ *iff* $\delta(a, b) + \delta(b, c) = \delta(a, c)$.

(iv) δ *is a ratio scale.*

PROOF. By Lemma 3, $\langle S, T, \succsim, \circ \rangle$ is an extensive structure with no essential maximum. Hence, by Theorem 3.3, there exists a function φ, unique up to multiplication by a positive constant, from S into Re^+ such that for all $ab, cd \in S$,

$$ab \succsim cd \quad \text{iff} \quad \varphi(ab) \geqslant \varphi(cd),$$

and

$$\text{if } (ab, cd) \in T, \text{ then } \varphi(ab \circ cd) = \varphi(ab) + \varphi(cd).$$

(Note that in Theorem 3, Chapter 3, φ need not be defined on the maximal element of S, if any exists, but the discussion following the theorem (Volume I, p. 85) shows that in this case, φ is defined on all of S.)

Define δ on A^2 by letting

$$\delta(a, a) = 0 \quad \text{and} \quad \delta(a, b) = \varphi(ab) \quad \text{for } a \neq b.$$

Hence, properties (ii) and (iv) of Theorem 3 follow from the corresponding properties of the extensive measurement representation. To establish (i), we have to show that δ satisfies the triangle inequality. The positivity and the symmetry of δ are immediate.

Select arbitrary $a, b, c \in A$, and suppose $(a, c) \succsim (a, b)$ (otherwise, $\delta(a, b) + \delta(b, c) > \delta(a, c)$ is trivial). By Axiom 1, construct e such that $(a, b) \sim (a, e)$ and $\langle aec \rangle$. By definition of $\langle \ \rangle$, $(b, c) \succsim (e, c)$. Consequently,

$$\delta(a, b) + \delta(b, c) \geqslant \delta(a, e) + \delta(e, c) = \delta(a, c)$$

as required.

Finally, to establish (iii), suppose $\delta(a, b) + \delta(b, c) = \delta(a, c)$ for some $a, b, c \in A$. By constructing e such that $(a, b) \sim (a, e)$ and $\langle aec \rangle$ and applying the above argument, we obtain $(b, c) \sim (e, c)$ and, by Lemma 2, $\langle abc \rangle$. The fact that $\langle abc \rangle$ implies $\delta(a, b) + \delta(b, c) = \delta(a, c)$ follows readily from the definition of $\circ$ and the properties of φ. $\diamond$

14.3.4 Theorem 4 (p. 169)

If $\langle A, \succsim \rangle$ is a complete segmentally additive proximity structure, then there exists a ratio scale δ on A^2 such that $\langle A, \delta \rangle$ is a complete metric with additive segments, and δ preserves $\succsim$.

PROOF. We first show that a segmentally additive proximity structure that violates Axioms 3 or 4 need not have a representation by a metric with additive segments. For a finite example, which violates Axiom 3, let $A = \{a, b, c\}$ and $(a, c) \succ (b, c) \sim (a, b)$. Axiom 1 is satisfied because

$\langle abc \rangle$ holds, and Axiom 2 is true in any finite structure. The representation of Theorem 3 is given by

$$\delta(a, b) = \delta(b, c) = \alpha \quad \text{and} \quad \delta(a, c) = 2\alpha,$$

where $\alpha > 0$ is arbitrary. But, of course, additive segments do not exist in a finite structure. To prove Theorem 4, it is sufficient to show that any two points in A are joined by an additive segment. The other parts of the theorem were established in Theorem 3. To demonstrate the existence of complete additive segments, Axioms 3 and 4 are used. By Axiom 3 we can construct additional distinct points between any two points, and Axiom 4 guarantees that there are no "holes" in the dense set of points thus shown to exist.

DEFINITION 5. *For $a, b \in A$, a subset B of A is a* partial segment *from a to b iff*

(i) $a, b \in B$;

(ii) *for any $e \in B$, $\langle aeb \rangle$;*

(iii) *for any $e, f \in B$ either $\langle aef \rangle$ or $\langle afe \rangle$.*

The set B is called a segment *from a to b if it is a maximal partial segment, i.e., a partial segment such that for any partial segment C from a to b, if $B \subseteq C$, then $B = C$.*

Note that if B is a partial segment from a to b, then by Theorem 2, for any $e, f, d \in B$, either $\langle efd \rangle$ or $\langle fed \rangle$ or $\langle edf \rangle$. For example, if $\langle aef \rangle$ and $\langle afd \rangle$, then $\langle efd \rangle$, etc.

By the principle of transfinite induction, one can show that for any $a, b \in A$ there exists a segment from a to b. In fact, the partial segments from a to b can be partially ordered by set-inclusion; and any union of a family of partial segments, any two of which are comparable by inclusion, is also a partial segment. Stated more formally, if Γ is a family of partial segments from a to b such that, for $B, C \in \Gamma$, either $B \subseteq C$ or $C \subseteq B$, then $\overline{B} = \bigcup_{B \in \Gamma} B$ is a partial segment from a to b. To verify this assertion, suppose that e, f are in $\overline{B}$, then $e \in B$ and $f \in C$ for some $B, C \in \Gamma$, but then either $e, f \in B$ or $e, f \in C$. In either case, $\langle aef \rangle$ or $\langle afe \rangle$. By Zorn's lemma, any partially ordered set that includes all upper bounds of all its totally ordered subsets has a maximal element.

Let B be a segment from a to c, and let δ be the metric constructed in Theorem 3. We wish to show that, for any real α satisfying $0 < \alpha < \delta(a, c)$, there is a unique element $b \in B$ such that $\delta(a, b) = \alpha$. This will show that B is an additive segment in $\langle A, \delta \rangle$ in the sense of Definition 12.12.

Let $\alpha_1 = \sup\{\beta | \beta \leqslant \alpha$ and, for some $e \in B, \ \delta(a, e) = \beta\}$. We show that for some $e_1 \in B, \ \delta(a, e_1) = \alpha_1$. In fact, if there were no $e_1 \in B$ such that $\delta(a, e_1) = \alpha_1$, then for every integer $n > 0$, there would be $d_n \in B$ such that $\alpha_1 - 1/n < \delta(a, d_n) < \alpha_1$. For $m > n$, we have $\langle ad_n d_m \rangle$ or $\langle ad_m d_n \rangle$; thus $\delta(d_m, d_n) < 1/n$. This shows that the sequence $d_1, d_2, \ldots$ satisfies the hypotheses of Axiom 4: for any $e \neq f$, there exists n_0 with $1/n_0 \leqslant \delta(e, f)$, so for $m, n \geqslant n_0, \ (e, f) \succsim (d_m, d_n)$. Therefore, by Axiom 4, there exists e_1 such that for any $e \neq f, \ (e, f) \succsim (d_m, e_1)$ for all but finitely many m. In particular, for any $n, \ \delta(d_m, e_1) \leqslant 1/n$ for all but finitely many m, i.e., $\lim_{m \to \infty} \delta(d_m, e_1) = 0$. We now show that $\delta(a, e_1) = \alpha_1$. On one hand,

$$\delta(a, e_1) \leqslant \delta(a, d_m) + \delta(d_m, e_1) < \alpha_1 + \delta(d_m, e_1),$$

so that $\delta(a, e_1) \leqslant \alpha_1$. On the other hand,

$$\alpha_1 - 1/n < \delta(a, d_n) \leqslant \delta(a, e_1) + \delta(e_1, d_n),$$

so that $\alpha_1 \leqslant \delta(a, e_1)$. Hence, $\delta(a, e_1) = \alpha_1$. It remains only to show that $e_1 \in B$. To do this, we use the maximality of B. If $e_1 \notin B$, let $C = B \cup \{e_1\}$. We show that C is a partial segment from a to c. The only problem is to show that for $f \in B$, either $\langle ae_1 f \rangle$ or $\langle afe_1 \rangle$. If $(a, e_1) \precsim (a, f)$, then $(a, d_n) \prec (a, f)$ for all n, so that $\langle ad_n f \rangle$, since B is a partial segment. Thus $\delta(a, f) = \delta(a, d_n) + \delta(d_n, f)$ for all n. Since $\delta(d_n, e_1) \to 0$, it follows that $\delta(a, d_n) \to \delta(a, e_1)$ and $\delta(d_n, f) \to \delta(e_1, f)$. Thus $\delta(a, f) = \delta(a, e_1) + \delta(e_1, f)$ and $\langle ae_1 f \rangle$ follows by Theorem 3. If $(a, f) \prec (a, e_1)$, then for all but finitely many $n, \ (a, f) \prec (a, d_n)$, so that $\langle afd_n \rangle$. It follows in the same way that $\delta(a, e_1) = \delta(a, f) + \delta(f, e_1)$, hence, $\langle afe_1 \rangle$. This proves that $B \cup \{e_1\}$ is a partial segment, so that $e_1 \in B$.

Let $\alpha_2 = \inf\{\beta | \beta \geqslant \alpha$ and for some $e \in B, \delta(a, e) = \beta\}$. By a similar argument, there exists $e_2 \in B$ such that $\delta(a, e_2) = \alpha_2$. We show that $\alpha_1 = \alpha_2 = \alpha$. Otherwise, $\alpha_1 < \alpha_2$, i.e., $\delta(a, e_1) < \delta(a, e_2)$. Since $\delta(e_1, e_2) > 0$, we can use Axiom 3 to obtain e', f', with $e' \neq f'$ and $(e_1, e_2) \succ (e', f')$. Choose e such that $(e_1, e) \sim (e', f')$ and $\langle e_1 e e_2 \rangle$. By Theorem 2, we have $\langle ae_1 e \rangle$ and $\langle aee_2 \rangle$. We show that $B \cup \{e\}$ is a partial segment. In fact, if $f \in B$, then either $\delta(a, f) \leqslant \delta(a, e_1)$ or $\delta(a, e_2) \leqslant \delta(a, f)$. In the former case, $\langle afe_1 \rangle$ and $\langle ae_1 e \rangle$ imply $\langle afe \rangle$; in the latter case, $\langle aee_2 \rangle$ and $\langle ae_2 f \rangle$ imply $\langle aef \rangle$. By maximality of B we have $e \in B$. However, $\delta(a, e_1) < \delta(a, e) < \delta(a, e_2)$, which is impossible. Thus $\alpha_1 = \alpha_2 = \alpha$. This shows that B is an additive segment in $\langle A, \delta \rangle$ from a to c. $\diamondsuit$

14.4 MULTIDIMENSIONAL REPRESENTATIONS

Much of the research on the representation of proximity is motivated by the attempt to embed the objects in some metric space so that the distance between them is expressible in terms of their respective coordinates. Such a representation not only quantifies the ordinal concept of proximity but also accounts for the observed proximities in terms of the underlying structure of the objects. In most applications of multidimensional scaling, the dimensional structure of the objects, or even the dimensionality of the space, is not known in advance. One usually assumes a general class of (dimensionally organized) metric spaces, e.g., Euclidean, and searches for a space in this class with minimal dimensionality, in which the objects under study can be embedded so that the dissimilarity ordering coincides (approximately) with the ordering of the respective metric distances.

Among the many possible metric models, almost all the research on multidimensional scaling has concentrated on one family called the power metric (or Minkowski r-metric) in the scaling literature and the $\mathcal{L}^p$ metric in the mathematical literature (see Section 12.2.6). According to this model, there is a real-valued mapping φ on A^n and a constant $r \geqslant 1$ such that the distance between $a = a_1 \cdots a_n$ and $b = b_1 \cdots b_n$ is given by

$$\delta(a, b) = \left[\sum_{i=1}^{n} |\varphi(a_i) - \varphi(b_i)|^r \right]^{1/r}. \tag{7}$$

Note that the components a_i, b_i, $i = 1, \ldots, n$, can be nominal scale values. The familiar Euclidean and city-block models are obtained from this equation by letting r equal 2 or 1, respectively. The power metric embodies four fundamental assumptions:

(a) *Decomposability.* The distance between stimuli is a function of componentwise contributions.

(b) *Intradimensional subtractivity.* Each componentwise contribution is the absolute value of an appropriate scale difference.

(c) *Interdimensional additivity.* The distance is a function of the sum of componentwise contributions.

(d) *Homogeneity.* Affine (straight) lines are additive segments.

By imposing these assumptions separately, the following generalizations of the power metric are obtained.

The general form of the equation that captures the decomposability (a) assumption is

$$\delta(a, b) = F[\psi_1(a_1, b_1), \ldots, \psi_n(a_n, b_n)], \qquad (8)$$

where F is an increasing[1] function in each of its n real arguments, and each ψ_i, $i = 1, \ldots, n$, is a real-valued symmetric function in two (nominal scale) arguments a_i, b_i satisfying $\psi_i(a_i, b_i) > \psi_i(a_i, a_i)$ whenever $a_i \neq b_i$.

If subtractivity (b) is assumed, then the power metric is generalized to

$$\delta(a, b) = F[|\varphi_1(a_1) - \varphi_1(b_1)|, \ldots, |\varphi_n(a_n) - \varphi_n(b_n)|], \qquad (9)$$

where F is as in Equation (8), but $\psi_i(a_i, b_i)$ is replaced by the specialized form $|\varphi_i(a_i) - \varphi_i(b_i)|$, for some real-valued function φ_i defined on A_i, $i = 1, \ldots, n$. Thus $|\varphi_i(a_i) - \varphi_i(b_i)|$ is the coordinatewise difference between a and b along the ith dimension. (Note Section 12.2.6; if F is convex and homogeneous of degree one, we have a general Minkowski metric.)

If additivity (c) alone is imposed, then Equation (8) is specialized to

$$\delta(a, b) = G\left[\sum_{i=1}^{n} \psi_i(a_i, b_i) \right], \qquad (10)$$

where the ψ_i are as in Equation (8), and G is an increasing function in one argument. Note that subtractivity is an intradimensional property since it refers to differences along the same dimension, whereas additivity is an interdimensional property since it refers to a summation across different dimensions.

If both additivity and subtractivity are assumed, we obtain the additive-difference model defined by

$$\delta(a, b) = G\left\{ \sum_{i=1}^{n} F_i[|\varphi_i(a_i) - \varphi_i(b_i)|] \right\}, \qquad (11)$$

where G and F_i, $i = 1, \ldots, n$ are all increasing functions in one argument. Clearly, the additive-difference model, Equation (11), is a special case of both Equations (9) and (10). The power metric, Equation (7), on the other hand, is a special case of the additive-difference model where all F_i are the same convex power function and G is its inverse.

---

[1] We use the terms *increasing* and *decreasing* functions to denote strictly monotonically increasing and decreasing real-valued functions, respectively.

The present section develops ordinal properties that are necessary and/or sufficient for the representation of a proximity structure by each of the above models. More specifically, we investigate the conditions imposed on a proximity structure $\langle A, \succsim \rangle$ under which there exists a real-valued function δ such that, for all $a, b, c, d \in A$,

$$(a, b) \succsim (c, d) \qquad \text{iff} \qquad \delta(a, b) \geqslant \delta(c, d),$$

where δ is defined by one of Equations (7)–(11).

The key structural assumption underlying this development is that the object set A possesses a product structure, i.e., $A = \mathbf{X}_{i=1}^{n} A_i$ and that each object $a \in A$ can be adequately characterized as an n-tuple,[2] denoted $a_1 \cdots a_n$. Thus A may be a set of color patches characterized in terms of their brightness, hue, and saturation; a set of facial expressions described in terms of intensity and pleasure; or a set of rectangles characterized by their heights and widths.

Some comments concerning the characterization of the objects, or stimuli, are in order. First, the components of each stimulus can be any nominal scale values, not necessarily real numbers. Second, the factorial representation of the stimuli is, in general, not unique. Rectangles, for example, can be characterized by any pair of the four attributes: height, width, area (height × width) and shape (height/width). Since the desired representation depends in an essential way on the initial characterization of the objects, the selection of an appropriate set of factors is an essential aspect of the analysis. In particular, it is important to select factors so that equivalence classes with respect to any one factor are well defined. If rectangles, for instance, are characterized by their physical height and width, and if perceived height depends on width, then two rectangles with the same physical height (but different width) could have different perceived heights. If this is the case, none of the above representations [Equations (7)–(11)] hold relative to the characterization of the objects in terms of their (physical) height and width, although they might hold relative to a characterization in terms of perceived height and perceived width. The problem of identifying an appropriate product structure is of special importance in the analysis of proximity, but it also exists in conjoint measurement (see Section 6.5.7).

Ordinal assumptions leading to decomposability [Equation (8)], subtractivity [Equation (9)], additivity [Equation (10)], and the additive-difference model [Equation (11)] are presented, respectively, in the following four subsections. These representation theorems are all reduced to results on

---

[2] Throughout this chapter we assume that $n \geqslant 2$.

difference measurement and additive conjoint measurement. The conditions under which the additive-difference model is a metric and an ordinal characterization of the power metric [Equation (7)] are presented in the last subsection (14.4.5). These developments are due to Tversky and Krantz (1970).

14.4.1 Decomposability

Consider a factorial proximity structure $\langle A, \succsim \rangle$, with $A = \mathbf{X}_{i=1}^{n} A_i$, that satisfies decomposability, Equation (8). It can be recast as a *decomposable conjoint structure* (Definition 7.1) by regrouping each pair of n-tuples, $(a, b) = (a_1 \cdots a_n, b_1 \cdots b_n)$, as an n-tuple of pairs, denoted as $ab = (a_1 b_1) \cdots (a_n b_n)$. In formulating conditions leading to Equations (7) to (11), there are several occasions in which this regrouping is convenient, so we present it as a definition for later reference.

DEFINITION 6. *Let* $\langle A = \mathbf{X}_{i=1}^{n} A_i, \succsim \rangle$ *be a factorial proximity structure (Definition 1). Let* A_i^2 *be the Cartesian product* $A_i \times A_i$, *and let* A^* *be the Cartesian product* $\mathbf{X}_{i=1}^{n} A_i^2$. *For* (a, b) *in* $A \times A$, *denote by* ab *the corresponding n-tuple* $(a_1 b_1) \cdots (a_n b_n)$ *of* A^*; *continue to denote by* $\succsim$ *the induced order on* A^*, *that is,*

$$ab \succsim a'b' \quad iff \quad (a, b) \succsim (a', b').$$

$\langle A^*, \succsim \rangle$ *is called the* induced conjoint structure *corresponding to the proximity structure* $\langle A = \mathbf{X}_{i=1}^{n} A_i, \succsim \rangle$.

The analysis of monotonic decomposability in Chapter 7 can be transferred in an obvious way to factorial proximity structures. Theorem 7.1 showed that the necessary and sufficient conditions to be satisfied in the induced conjoint structure are weak ordering, existence of a countable order-dense subset, and independence of the induced ordering on each single factor. This last condition can be restated directly, if somewhat awkwardly, in terms of the factorial proximity structure.

DEFINITION 7. *A factorial proximity structure* $\langle A, \succsim \rangle$ *satisfies* one-factor independence *iff the following holds for all* a, a', b, b', c, c', d, $d' \in A$:
If the two elements in each of the pairs (a, a'), (b, b'), (c, c'), (d, d') *have identical components on all but one factor, and the two elements in each of the pairs* (a, c), (a', c'), (b, d), (b', d') *have identical components on the remaining factor, then*

$$(a, b) \succsim (a', b') \quad iff \quad (c, d) \succsim (c', d').$$

One-factor independence is closely related to the most general form of the independence condition, which plays a central role in the theory of conjoint measurement. (See Definitions 6.1, 6.11, and 7.3.) In fact, if one considers all pairs whose elements differ with respect to the ith factor only, then one-factor independence asserts that for each $i = 1, \ldots, n$ the induced weak ordering of A_i^2 is independent of the fixed components of the remaining A_j^2 for $j \neq i$. The following theorem, which corresponds to Theorem 7.1, shows that this property is not only necessary for decomposability, but is also sufficient provided $\langle A, \succeq \rangle$ is representable in $\langle \mathrm{Re}, \geqslant \rangle$.

THEOREM 5. *Suppose* $\langle A, \succeq \rangle$ *is a factorial proximity structure and* $A^2/\sim$ *has a finite or countable order-dense subset (Definition 2.1). Then the following two conditions are equivalent:*

(i) $\langle A, \succeq \rangle$ *satisfies one-factor independence (Definition 7).*

(ii) *There exists a real-valued function F of n arguments, and real-valued functions* ψ_i *defined on* A_i^2, $i = 1, \ldots, n$, *such that, for all* $a, b, c, d \in A$,

$$(a_1 \cdots a_n, b_1 \cdots b_n) \succeq (c_1 \cdots c_n, d_1 \cdots d_n)$$

iff

$$F[\psi_1(a_1, b_1), \ldots, \psi_n(a_n, b_n)] \geqslant F[\psi_1(c_1, d_1), \ldots, \psi_n(c_n, d_n)],$$

where F is increasing in each of its arguments and

$$\psi_i(a_i, b_i) = \psi_i(b_i, a_i) > \psi_i(a_i, a_i) = \psi_i(b_i, b_i)$$
$$\text{for} \quad a_i \neq b_i, i = 1, \ldots, n.$$

The theorem is proved in Section 14.5.1.

14.4.2 Intradimensional Subtractivity

As suggested by its name, intradimensional subtractivity is a property that depends on each of the factors, or dimensions, separately. In fact, Equation (9) is a straightforward extension of the absolute-difference model, developed in Section 4.10. To establish intradimensional subtractivity, therefore, we first have to establish decomposability for the entire structure and then impose an absolute-difference structure (Definition 4.8) on each factor $\langle A_i^2, \succeq_i \rangle$, where $\succeq_i$ is the induced ordering defined by one-factor independence.

An important fact, incorporated in the statement of Theorem 5, is that one-factor independence permits transfer of the minimality and symmetry properties (Axioms 2, 3, and 4 of Definition 1) to each factorial structure $\langle A_i^2, \succsim_i \rangle$.

We introduce a ternary relation of betweenness, denoted $a|b|c$, as a natural extension of the unidimensional concept of Definition 4.9. Unidimensional betweenness is required for each factor separately. In particular, the minimality property for each factor assures us that $a|b|c$ and $a_i = c_i$ imply that b_i also coincides with a_i and c_i. The following definition also introduces the axioms for betweenness that follow naturally from intradimensional subtractivity and lead to construction of absolute difference structures (Definition 4.8) for each factor.

DEFINITION 8. *Let* $\langle A = \mathbf{X}_{i=1}^n A_i, \succsim \rangle$ *be a factorial proximity structure that satisfies one-factor independence. Let* $\succsim_i$ *denote the induced order on* A_i^2.

We say that b is between *a and c, denoted* $a|b|c$, *iff*

$$(a_i, c_i) \succsim_i (a_i, b_i), (b_i, c_i) \qquad \text{for each } i.$$

We say that $\langle A, \succsim \rangle$ *satisfies the* betweenness *axioms iff the following hold for all* $a, b, c, d, a', b', c' \in A$:

1. *Suppose* a, b, c, d *differ on at most one factor, and* $b \neq c$, *then*
    (i) *if* $a|b|c$ *and* $b|c|d$, *then* $a|b|d$ *and* $a|c|d$;
    (ii) *if* $a|b|c$ *and* $a|c|d$, *then* $a|b|d$ *and* $b|c|d$.

2. *Suppose* a, b, c, a', b', c' *differ on at most one factor,* $a|b|c$, $a'|b'|c'$, *and* $(b, c) \sim (b', c')$, *then*

$$(a, b) \succsim (a', b') \qquad \text{iff} \qquad (a, c) \succsim (a', c').$$

Axioms 1 and 2 of Definition 8 are both unidimensional properties that apply to elements that coincide on all but one factor. Indeed, they are essentially identical to Axioms 3 and 4 of Definition 4.8. (See also Definition 13.1.) An experimental test of these properties is described in Section 14.6.1.

To establish intradimensional subtractivity (as well as additivity), however, two additional axioms of a more technical nature are introduced.

DEFINITION 9. *A factorial proximity structure* $\langle A, \succsim \rangle$ *satisfies* restricted solvability *iff, for all* $e, f, d, a, c \in A$, *if* $(d, a) \succsim (e, f) \succsim (d, c)$, *then there exists* $b \in A$ *such that* $a|b|c$ *and* $(d, b) \sim (e, f)$.

DEFINITION 10. *A proximity structure* $\langle A, \succsim \rangle$ *satisfies the* Archimedean axiom *iff, for all* $a, b, c, d \in A$ *with* $a \neq b$, *any sequence*

$e^{(k)} \in A$, $k = 0, 1, \ldots$ *that varies on at most one factor, such that* $e^{(0)} = c, (e^k, e^{(k+1)}) \succ (a, b)$ *and* $(c, d) \succ (c, e^{(k+1)}) \succ (c, e^{(k)})$ *for all k is finite.*

Restricted solvability and the Archimedean axiom defined above extend, respectively, Axioms 5 and 6 of Definition 4.8. The Archimedean axiom, of course, is a necessary condition for the desired representation. Restricted solvability is not a necessary condition; rather it describes a class of proximity structures for which the following representation can be established.

THEOREM 6. *Suppose* $\langle A, \succsim \rangle$ *is a factorial proximity structure that satisfies the following axioms:*

1. *One-factor independence* (*Definition 7*)
2. *Betweenness axioms* (*Definition 8*)
3. *Restricted solvability* (*Definition 9*)
4. *Archimedean axiom* (*Definition 10*)

Then there exist real-valued functions φ_i *defined on* A_i, $i = 1, \ldots, n$ *and a real-valued function F that increases in each of its n real arguments such that*

$$\delta(a, b) = F[|\varphi_1(a_1) - \varphi_1(b_1)|, \ldots, |\varphi_n(a_n) - \varphi_n(b_n)|]$$

and

$$(a, b) \succsim (c, d) \quad \text{iff} \quad \delta(a, b) \geqslant \delta(c, d).$$

Furthermore, $\varphi_1, \ldots, \varphi_n$ *are unique up to an increasing linear transformation of the form* $\alpha_i \varphi_i + \beta_i$, $\alpha_i > 0$; *F is unique up to an arbitrary order-preserving transformation.*

The proof of this theorem is given in Section 14.5.2.

14.4.3 Interdimensional Additivity

Interdimensional additivity, Equation (10), assumes that the perceived distance between stimuli is expressible as an additive combination of n functions of the components of the stimuli on each of the n factors. It is closely related, therefore, to the theory of additive conjoint measurement discussed in Chapter 6; in fact, the present development is based directly on the results established there.

The essential difference between additive conjoint measurement and the present development is that the former is based on an ordering of objects with respect to some evaluative attribute (e.g., preference or intensity) whereas the latter is based on an ordering of pairs of objects with respect to proximity or dissimilarity. Consequently, the present representation has the form of a sum of functions in *two* arguments that correspond to the respective values of the two stimuli on each of the dimensions.

It is instructive to investigate directly some consequences of interdimensional additivity. Consider a factorial proximity structure $\langle A, \succsim \rangle$. Let a and a' be two elements of A that coincide on the ith factor, and let b and b' be two other elements of A that coincide on the ith factor. Suppose Equation (10) holds. Since, by assumption, $a_i = a_i'$ and $b_i = b_i'$, we obtain $\psi_i(a_i, b_i) = \psi_i(a_i', b_i')$. Consequently, the relation $(a, b) \succsim (a', b')$ must be independent of the ith components of the elements, provided that $a_i = a_i'$ and $b_i = b_i'$. Specifically, the above inequality should be preserved if we change the common value of a and a' and the common value of b and b' on the ith factor. This consequence of interdimensional additivity is just the independence property for the induced conjoint structure (see Definitions 6.11 and 7.3). Stated formally, we have the following.

DEFINITION 11. *A factorial proximity structure* $\langle A, \succsim \rangle$ *satisfies* independence *iff the following holds for all* $a, a', b, b', c, c', d, d' \in A$. *If the two elements in each of the pairs* (a, a'), (b, b'), (c, c'), (d, d') *have identical components on one factor, and the two elements in each of the pairs* (a, c), (a', c'), (b, d), (b', d') *have identical components on all the remaining factors, then*

$$(a, b) \succsim (a', b') \qquad iff \qquad (c, d) \succsim (c', d').$$

The reader can verify that independence implies one-factor independence (Definition 7) but not conversely (Exercise 2). Recall that in the two-factor case of additive conjoint measurement an additional property, called the Thomsen condition (Definition 6.3, p. 250) is assumed. The corresponding property for a proximity structure is introduced as follows.

DEFINITION 12. *A factorial proximity structure* $\langle A, \succsim \rangle$, *with* $A = A_1 \times A_2$, *satisfies the* Thomsen condition *iff, for all* $a_i, b_i, c_i, d_i, e_i, f_i$, $i = 1, 2,$

$$(a_1 e_2, b_1 f_2) \sim (e_1 c_2, f_1 d_2)$$

and

$$(e_1 a_2, f_1 b_2) \sim (c_1 e_2, d_1 f_2)$$

imply

$$(a_1 a_2, b_1 b_2) \sim (c_1 c_2, d_1 d_2).$$

As in additive conjoint measurement, the formulation of the theory depends on the number of essential (i.e., nontrivial) factors, defined as follows.

DEFINITION 13. *Suppose* $\langle A, \succsim \rangle$ *is a factorial proximity structure. A factor* A_i $(i = 1, \ldots, n)$ *is essential iff there exist* $a, b \in A$ *that coincide on all but the* i^{th} *factor and that satisfy* $(a, b) \succ (a, a)$.

Thus A_i is inessential whenever all a and b that differ on A_i alone satisfy $(a, b) \sim (a, a)$. When independence also holds, inessential factors can be omitted. We assume this has been done so that n denotes the number of essential factors. The representation and uniqueness theorem can now be stated. The proof is given in Section 14.5.3.

THEOREM 7. *Suppose* $\langle A, \succsim \rangle$ *is a factorial proximity structure that satisfies independence* (*Definition* 11) *and restricted solvability* (*Definition* 9), *and suppose that in the induced conjoint structure* (*Definition* 6) *each pair of factors satisfies the Archimedean axiom* (*Volume I, Definition* 6.4). *If* $n \geqslant 3$, *then there exist real-valued functions* ψ_i *defined on* A_i^2, $i = 1, \ldots, n$, *such that, for all* $a, b, c, d \in A$,

$$(a_1 \cdots a_n, b_1 \cdots b_n) \succsim (c_1 \cdots c_n, d_1 \cdots d_n)$$

iff

$$\sum_{i=1}^{n} \psi_i(a_i, b_i) \geqslant \sum_{i=1}^{n} \psi_i(c_i, d_i).$$

Furthermore, if ψ_i', $i = 1, \ldots, n$ *is another set of functions satisfying the above relation, then there exist constants* $\alpha > 0$, β_i *such that*

$$\psi_i' = \alpha \psi_i + \beta_i, \qquad i = 1, \ldots, n.$$

If $n = 2$, *the above assertions hold provided the Thomsen condition* (*Definition* 12) *is also satisfied.*

For $n \geqslant 3$ the Thomsen condition follows from the other axioms, as is the case in additive conjoint measurement. Note that this theorem uses an Archimedean axiom different from Definition 10. We do not know whether, given the other assumptions of Theorem 7, the condition defined in Definition 10 is sufficient to ensure that the induced conjoint structure is Archimedean.

14.4.4   The Additive-Difference Model

The theory developed in Section 14.4.2 assumes that the contribution of each factor to the dissimilarity between stimuli depends only on the absolute difference between the scale values along that factor. This theory does not specify how these absolute differences along the various factors are combined to generate the dissimilarity ordering, except for the assumption that dissimilarity increases with any of the absolute differences between the scale values. The theory developed in Section 14.4.3, on the other hand, assumes that the various factors contribute additively to the overall dissimilarity between stimuli but does not specify how the contribution of each of the factors is formed. If both theories are assumed, we obtain the additive-difference model [Equation (11)], where intradimensional subtractivity and interdimensional additivity are satisfied simultaneously. The required assumptions are summarized in the following definition.

DEFINITION 14.   *A factorial proximity structure* $\langle A, \succsim \rangle$ *is an* additive-difference structure *iff the following axioms hold*:

1. *Betweenness axioms* (*Definition* 8)
2. *Independence* (*Definition* 11)
3. *Thomsen condition* (*Definition* 12)
4. *Restricted solvability* (*Definition* 9)
5. *Archimedean axiom* (*Definition* 10)

One-factor independence is not included in this definition because it follows from independence. The Thomsen condition must be assumed only for $n = 2$; for $n \geqslant 3$ it follows from the other axioms. Note that all the axioms, except restricted solvability, are necessary conditions for Equation (11). In the presence of the betweenness axioms, independence, and restricted solvability, the Archimedean condition of Definition 10 is sufficient; the extra Archimedean assumption of Theorem 7 can be dropped. This is proved in Section 14.5.3.

Suppose $\langle A, \succsim \rangle$ is an additive-difference proximity structure. By Theorems 6 and 7, therefore, there exists a real-valued function δ on A^2 that preserves $\succsim$ and that satisfies

$$\delta(a, b) = F[|\varphi_1(a_1) - \varphi_1(b_1)|, \ldots, |\varphi_n(a_n) - \varphi_n(b_n)|]$$
$$= G\left[\sum_{i=1}^{n} \psi_i(a_i, b_i)\right],$$

where F, G, φ_i, ψ_i, $i = 1, \ldots, n$, are as in Equations (9) and (10).

Consequently, there exist increasing functions $F_1, \ldots, F_n$ on Re such that

$$F_i[|\varphi_i(a_i) - \varphi_i(b_i)|] = \psi_i(a_i, b_i), \qquad i = 1, \ldots, n.$$

Substituting F_i for ψ_i yields the additive-difference model. Thus, by combining Theorems 6 and 7, we obtain the following.

THEOREM 8. *If $\langle A, \succsim \rangle$ is an additive-difference factorial proximity structure (Definition 14), then there exist real-valued functions $\varphi_1, \ldots, \varphi_n$ defined on $A_1, \ldots, A_n$, respectively, and increasing functions $F_1, \ldots, F_n$ each defined on Re, such that for all $a, b, c, d \in A$,*

$$(a_1 \cdots a_n, b_1 \cdots b_n) \succsim (c_1 \cdots c_n, d_1 \cdots d_n)$$

iff

$$\sum_{i=1}^{n} F_i[|\varphi_i(a_i) - \varphi_i(b_i)|] \geq \sum_{i=1}^{n} F_i[|\varphi_i(c_i) - \varphi_i(d_i)|].$$

Furthermore, the φ_i are interval scales and the F_i are interval scales with a common unit, $i = 1, \ldots, n$.

The additive-difference model, established in Theorem 8, expresses the similarity between stimuli in terms of two sets of scales, or transformations. The transformations from one set $\{\varphi_i\}$ apply to the physically characterized components of the stimuli and describe their psychological counterparts along each of the factors. The transformations from the second set $\{F_i\}$ apply to the absolute values of componentwise differences and describe their contribution to the overall similarity between the respective stimuli. Thus similarity judgments can be interpreted according to this model as composed of two independent processes, a "perceptual" process satisfying intradimensional subtractivity and an "evaluative" process satisfying interdimensional additivity.

14.4.5 Additive-Difference Metrics

The additive-difference model established in Theorem 8 yields a measurement of proximity between stimuli that differ along several factors. The resulting proximity measure, however, is not necessarily a metric because the triangle inequality need not hold. [It is easy to see that by setting $G(0) = F_i(0) = 0$, $i = 1, \ldots, n$, in Equation (11), $\delta(a, a) = 0$. Symmetry and minimality of δ (Definition 1) follow at once from the respective properties of $\succsim$.] Furthermore, the additive-difference model need not have additive segments.

In this section we investigate the conditions under which the additive-difference model is compatible with a metric. We consider first a class of metrics in which $\delta(a, b) + \delta(b, c) = \delta(a, c)$ is required to hold for points that differ on one factor only. The more restrictive case of a metric with additive segments is considered later.

DEFINITION 15. *A factorial proximity structure $\langle A, \succsim \rangle$ is compatible with a metric iff there exists a real-valued function δ on A^2 such that the following conditions hold for all $a, b, c, d \in A$.*

(i) $(a, b) \succsim (c, d)$ *iff $\delta(a, b) \geqslant \delta(c, d)$.*

(ii) *$\langle A, \delta \rangle$ is a metric space.*

(iii) *If a, b and c differ on one factor only and $a|b|c$, then $\delta(a, b) + \delta(b, c) = \delta(a, c)$.*

THEOREM 9. *Suppose $\langle A, \succsim \rangle$ is an additive-difference proximity structure with the additive-difference representation of the form (Theorem 8)*

$$f(a, b) = \sum_{i=1}^{n} F_i[|\varphi_i(a_i) - \varphi_i(b_i)|],$$

where f preserves $\succsim$ on A^2. Suppose further that the domain of each F_i is a real interval, and $F_i(0) = 0$, $i = 1, \ldots, n$.

For $\langle A, \succsim \rangle$ to be compatible with a metric, (i) it is necessary that there exists a continuous, convex function F such that $F_i(x) = F(t_i x)$, $t_i > 0$, for all x in the domain of F_i, $1 \leqslant i \leqslant n$; and (ii) it is sufficient that, in addition, the function $\log F(e^x)$ is convex.

Theorem 9 is proved in Section 14.5.4. Part (i) extends a result of Tversky and Krantz (1970), and part (ii) is an application of a theorem of Mulholland (1947) that generalizes Minkowski's inequality in the form of a triangle inequality. Part (i) shows that Definition 15 imposes strong constraints on the measurement scales. In particular, the functions F_i must all be identical except for a change of unit of their domains. Moreover, they must all be continuous and convex. If we incorporate the t_i into the φ_i, we see that any metric compatible with an additive-difference model is of the form

$$\delta(a, b) = F^{-1}\left\{ \sum_{i=1}^{n} F[|\varphi_i(a_i) - \varphi_i(b_i)|] \right\}, \quad (12)$$

where F is continuous and convex. Not every function F satisfying these conditions, however, yields a metric distance function. To demonstrate this, consider the following example. Let

$$F(x) = \begin{cases} x & \text{for} \quad 0 \leqslant x \leqslant 1 \\ 1 + p(x - 1) & \text{for} \quad x > 1, \ p > 1. \end{cases}$$

This function is clearly continuous and convex. Substituting this form in Equation (12), however, does not yield a metric distance function because the triangle inequality is violated. The verification of this assertion is left as Exercise 3.

The power metric [Equation (7)] provides one example of a metric that is an additive-difference model. It is obtained from Equation (12) by letting $F(x) = x^r$ for some real $r \geqslant 1$. Another example is the *exponential metric* defined by letting $F(x) = e^{qx} - 1$ for some $q > 0$. Other functions that yield compatible metrics are sinh x and $x^p e^{qx^r}$, $p \geqslant 1$, $q, r \geqslant 0$. The reader is invited to check that these functions satisfy part (ii) of Theorem 9 (see Exercise 4).

Thus there are several families of additive-difference proximity models with compatible metrics. The following theorem shows that the power family is the only one that yields a metric with additive segments.

THEOREM 10. *Suppose* $\langle A, \succsim \rangle$ *is a complete, segmentally additive, factorial proximity structure* (*Definition* 4) *that is also an additive-difference structure* (*Definition* 14). *Then there exist a unique* $r \geqslant 1$ *and real-valued functions* φ_i *defined on* A_i, $i = 1, \ldots, n$, *that are unique up to a positive linear transformation, such that, for all* $a, b, c, d \in A$,

$$(a, b) \succsim (c, d) \qquad \text{iff} \qquad \delta(a, b) \geqslant \delta(c, d),$$

where

$$\delta(a, b) = \left[\sum_{i=1}^{n} |\varphi_i(a_i) - \varphi_i(b_i)|^r \right]^{1/r}.$$

The proof of Theorem 10 is given in Sections 14.5.5 and 14.5.6. It consists of showing first that a metric with additive segments satisfying intradimensional subtractivity is homogeneous (Lemma 4). That is, Euclidean (straight) lines are additive segments. From homogeneity, together with interdimensional additivity, we then derive the power metric using standard results on homogeneous means. Theorem 10 shows that an additive difference model that is compatible with a metric with additive segments gives rise to the power metric, thereby providing an ordinal characterization of this model.

14.5 PROOFS

14.5.1 Theorem 5 (p. 179)

A factorial proximity structure $\langle A, \succsim \rangle$ *satisfies decomposability iff* (i) $A^2/\sim$ *has a finite or countable order-dense subset and* (ii) *one-factor independence is satisfied.*

PROOF. The necessity of (i) is obvious; the necessity of (ii) was established in Section 14.4.1. To prove sufficiency, note that by Theorem 2, Chapter 2 (p. 40), there is a homomorphism H of $\langle A^2/\sim, \succsim \rangle$ into $\langle \mathrm{Re}, \geqslant \rangle$. We define the function ψ_i by holding all components constant except the ith and setting $\psi_i = H$. More precisely, let e be some fixed element of A. Let $E_i(a)$ be the element of A whose jth component is e_j for $j \neq i$ and whose ith component is a_i. Define

$$\psi_i(a_i, b_i) = H[E_i(a), E_i(b)].$$

By Axiom 2 of Definition 1,

$$\psi_i(a_i, b_i) = \psi_i(b_i, a_i) > \psi_i(a_i, a_i) = \psi_i(b_i, b_i), \quad \text{if} \quad a_i \neq b_i.$$

Let T_i be the range of the function ψ_i, $i = 1, \ldots, n$. We define a function F of n real variables on $T_1 \times \cdots \times T_n$ by

$$F[\psi_1(a_1, b_1), \ldots, \psi_n(a_n, b_n)] = H(a, b).$$

To prove decomposability, we employ one-factor independence to prove that F is well defined and increasing in each variable. Suppose that

$$\psi_i(a_i, b_i) = \psi_i(a_i', b_i'), i = 1, \ldots, n.$$

To show that F is well defined, we have to show that $(a, b) \sim (a', b')$. For $k = 0, \ldots, n$, let $a^{(k)}$ be the element of A whose ith component is a_i for $i > k$ and a_i' for $i \leqslant k$; define $b^{(k)}$ similarly. Then for $i = 1, \ldots, n$, we have, for $j \neq i$,

$$a_j^{(i-1)} = a_j^{(i)}, \qquad b_j^{(i-1)} = b_j^{(i)},$$
$$E_i(a)_j = E_i(a')_j, \qquad E_i(b)_j = E_i(b')_j,$$

and for $j = i$,

$$a_i^{(i-1)} = E_i(a)_i, \qquad b_i^{(i-1)} = E_i(b)_i,$$
$$a_i^{(i)} = E_i(a')_i, \qquad b_i^{(i)} = E_i(b')_i.$$

We can now apply one-factor independence (Definition 6), with $a^{(i-1)}$, $b^{(i-1)}$, $a^{(i)}$, $b^{(i)}$ playing the roles of a, b, a', b' and with $E_i(a)$, $E_i(b)$, $E_i(a')$, $E_i(b')$ playing the roles of c, d, c', d', to obtain

$$[a^{(i-1)}, b^{(i-1)}] \succsim [a^{(i)}, b^{(i)}] \quad \text{iff} \quad [E_i(a), E_i(b)] \succsim [E_i(a'), E_i(b')].$$

On the other hand, by definition of ψ_i,

$$[E_i(a), E_i(b)] \sim [E_i(a'), E_i(b')];$$

hence, for $i = 1, \ldots, n$,

$$[a^{(i-1)}, b^{(i-1)}] \sim [a^{(i)}, b^{(i)}].$$

Finally, since $a^{(0)} = a$, $b^{(0)} = b$, $a^{(n)} = a'$, $b^{(n)} = b'$, we have $(a, b) \sim (a', b')$ as required.

The proof that F is increasing in each variable is exactly parallel to the above, except that one of the equalities $\psi_i(a_i, b_i) = \psi_i(a_i', b_i')$ is replaced by a strict inequality, and we obtain correspondingly a strict inequality between $(a^{(i-1)}, b^{(i-1)})$ and $(a^{(i)}, b^{(i)})$ and, hence, between (a, b) and (a', b'), as required.

14.5.2 Theorem 6 (p. 181)

If $\langle A, \succsim \rangle$ is a factorial proximity structure satisfying (1) one-factor independence, (2) betweenness axioms, (3) restricted solvability, and (4) Archimedean axiom, then intradimensional subtractivity holds and the φ_i, $i = 1, \ldots, n$, are interval scales.

PROOF. We first show that for any A_i^2, $i = 1, \ldots, n$, the structure $\langle A_i^2, \succsim_i \rangle$ is an absolute-difference structure in the sense of Definition 4.8 (Volume I, p. 172).

Axioms 1 and 2 of Definition 4.8 follow from Definition 1; Axioms 3 and 4 follow from Definition 8; Axiom 5 of Definition 4.8 follows from Definition 9; and Axiom 6 of Definition 4.8 follows from Definition 10.

Since each $\langle A_i^2, \succsim_i \rangle$ is an absolute-difference structure, then, according to Theorem 4.6 (Volume I, p. 173), there exist real-valued functions φ_i

defined on A_i, $i = 1, \ldots, n$, such that for all $a_i, b_i, c_i, d_i \in A_i$

$$(a_i, b_i) \succsim (c_i, d_i) \quad \text{iff} \quad |\varphi_i(a_i) - \varphi_i(b_i)| \geqslant |\varphi_i(c_i) - \varphi_i(d_i)|.$$

Furthermore, the φ_i, $i = 1, \ldots, n$, are all interval scales.

To construct the function F, define $H_i(a_i, b_i) = |\varphi_i(a_i) - \varphi_i(b_i)|$, and for each $i = 1, \ldots, n$ select a countable order-dense subset of the range of H_i. Let B_i be the inverse image of the selected order-dense subset. Clearly, B_i is countable and order-dense in $A_i^* = A_i^2/\sim_i$. Let $B = B_1 \times \cdots \times B_n$; hence, B is countable. To show that B is order-dense in $A^* = A_1^* \times \cdots \times A_n^*$, suppose that $a^* \succ b^*$ for some $a^*, b^* \in A^*$. (Recall that we use $\succsim$ to denote the induced order on A^*.) By restricted solvability, therefore, there exists some $c^* \in A^*$ satisfying $c^* \sim b^*$ and $a_i^* \succsim_i c_i^*$ for all i, where $\succsim_i$ is the ordering induced on A_i^* by fixing the values of A_j^* for all $j \neq i$. Since B_i is order-dense in A_i^*, there exists $d_i^* \in B_i$ satisfying $a_i^* \succsim_i d_i^* \succsim_i c_i^*$, $i = 1, \ldots, n$. Consequently, $a^* \succ d^* \succsim b^* \sim c^*$ and B is order-dense in A^* as required.

By Theorem 2, Chapter 2 (p. 40), therefore, there exists a real-valued function H satisfying $H(a, b) \geqslant H(c, d)$ iff $(a, b) \succsim (c, d)$. Define a real-valued function F in n real arguments by

$$F[|\varphi_1(a_1) - \varphi_1(b_1)|, \ldots, |\varphi_n(a_n) - \varphi_n(b_n)|] = H(a, b).$$

It is left to be shown that F is well defined, i.e., that if

$$|\varphi_i(a_i) - \varphi_i(b_i)| = |\varphi_i(c_i) - \varphi_i(d_i)| \quad \text{for} \quad i = 1, \ldots, n,$$

then $(a, b) \sim (c, d)$, and that F is increasing in each of its arguments. Both propositions follow using one-factor independence, and the proof is identical to that of Theorem 5 (see Section 14.5.1). $\diamond$

14.5.3 Theorem 7 (p. 183)

If $\langle A, \succsim \rangle$ is a factorial proximity structure, satisfying (1) independence, (2) the Thomsen condition, (3) restricted solvability, and (4) the Archimedean property for the induced conjoint structure, then interdimensional additivity holds and the ψ_i, $i = 1, \ldots, n$, are interval scales with a common unit. If $n \geqslant 3$, the above follows without the Thomsen condition.

PROOF. The proof of Theorem 7 consists of showing that the induced conjoint structure is an additive conjoint-measurement structure. Let $A^* = A_1^2 \times \cdots \times A_n^2$ be the induced conjoint structure (Definition 6).

We show next that for $n \geqslant 3$, $\langle A_1^2, \ldots, A_n^2, \succsim \rangle$ is an n-component structure of additive conjoint measurement by verifying the five axioms of Definition 13, Chapter 6 (Volume I, p. 301).

1. Weak ordering. Immediate.

2. Independence follows from Definition 11. Specifically, consider some ab, $a'b'$, cd, $c'd' \in A^*$ such that for some $i = 1, \ldots, n$, $a_i b_i = a_i' b_i'$, $c_i d_i = c_i' d_i'$, and for all $j \neq i$, $a_j b_j = c_j d_j$, $a_j' b_j' = c_j' d_j'$. Each of the elements in the pairs (a, a'), (b, b'), (c, c'), (d, d') coincides on the ith factor, whereas each of the elements in the pairs (a, b), (c, d), (a', b'), (c', d') coincides on all other factors. By the independence condition of Definition 11 and the definition of $\succsim$ on A^*, therefore, $ab \succsim a'b'$ iff $cd \succsim c'd'$.

3. Restricted solvability. Follows from Definition 9.

4. Archimedean axiom. This is assumed as a hypothesis of the theorem.

5. $n \geqslant 3$. By hypothesis.

Hence, for $n \geqslant 3$, Theorem 7 follows from Theorem 6.13 (Volume I, p. 302), without assuming the Thomsen condition of Definition 12. For $n = 2$, it is readily shown that by including this condition, all the axioms of Definition 6.7 (p. 256) are satisfied by $\langle A_1^2, A_2^2, \succsim \rangle$; and Theorem 7 follows, in this case, from Theorem 6.2 (p. 257). $\Diamond$

In Theorem 8 (p. 184), we drop the Archimedean assumption on the induced conjoint structure; instead, we incorporate the Archimedean property given in Definition 10, which was used also for Theorem 6. We point out here why the induced conjoint structure must be Archimedean in this case.

Since factors are considered only pairwise, we assume $n = 2$. By restricted solvability and independence, a standard sequence in A_1 can be written as a sequence of pairs $a_1^{(0)} a_1^{(k)}$, $k = 1, 2, \ldots$ satisfying an equation of form

$$(a_1^{(0)} a_1^{(k+1)})(a_2 a_2) \sim (a_1^{(0)} a_1^{(k)})(b_2 c_2).$$

For standard sequences of sufficiently small mesh, we can also construct $b|c|d$ with $b_2 c_2 \sim_2 c_2 d_2$; we can then use independence, betweenness, and restricted solvability to show that $a_1^{(k)} a_1^{(k+1)} \sim b_2 c_2$ (in the obvious sense). Thus, the sequence $a_1^{(k)}$ has mesh bounded below in the sense of Definition 10. The Archimedean property of Definition 10 now assures that any bounded standard sequence is finite.

14.5.4 Theorem 9 (p. 186)

Assume an additive-difference proximity model, in which the domain of each F_i is a real interval and $F_i(0) = 0$, $i = 1, \ldots, n$. For the model to be compatible with a metric, (i) it is necessary that there exists a continuous, convex function F such that $F_i(x) = F(t_i x)$, $t_i > 0$, $i = 1, \ldots, n$; and (ii) it is sufficient that, in addition, the function $\log F(e^x)$ is convex.

PROOF OF (i). By the hypotheses of the theorem there exists a metric δ on A and a strictly increasing function G such that for all $a, b \in A$,

$$\delta(a, b) = G\left\{ \sum_{i=1}^{n} F_i[|\varphi_i(a_i) - \varphi_i(b_i)|] \right\}.$$

Suppose a, b, and c differ on the ith factor only and $a|b|c$. Hence, by Definition 15,

$$G\{ F_i[|\varphi_i(a_i) - \varphi_i(b_i)|] \} + G\{ F_i[|\varphi_i(b_i) - \varphi_i(c_i)|] \}$$
$$= G\{ F_i[|\varphi_i(a_i) - \varphi_i(c_i)|] \}.$$

Letting $|\varphi_i(a_i) - \varphi_i(b_i)| = x$ and $|\varphi_i(b_i) - \varphi_i(c_i)| = y$ yields

$$G[F_i(x)] + G[F_i(y)] = G[F_i(x + y)].$$

Define $H_i(x) = G[F_i(x)]$. Hence, the above equation reduces to $H_i(x) + H_i(y) = H_i(x + y)$ for all x, y in the domain of F_i, $1 \leqslant i \leqslant n$. The only monotonic solution of this equation is $H_i(x) = t_i x$, $t_i > 0$. Since G is strictly increasing, it has an inverse G^{-1} satisfying

$$G^{-1}[H_i(x)] = G^{-1}[t_i x] = F_i(x),$$

whenever $F_i(x)$ is defined. But since G^{-1} is independent of i, the F_i can differ only in range and in the unit of the domain. With no loss of generality, we can assume that $\sup F_n \geqslant \sup F_i$, $1 \leqslant i \leqslant n$. Because the range of F_n includes the range of F_i, we can let $F = F_n$ since $F_i(x) = F_n(t_i x)$ for any x in the domain of F_i.

To establish continuity and convexity, we first prove the following lemmas.

LEMMA 4. $F(\alpha + \beta + \gamma) + F(\gamma) \geqslant F(\alpha + \gamma) + F(\beta + \gamma)$.

PROOF. In the triangle inequality,

$$F^{-1}\left[\sum_{i=1}^{n} F(\alpha_i)\right] + F^{-1}\left[\sum_{i=1}^{n} F(\beta_i)\right] \geqslant F^{-1}\left[\sum F(\alpha_i + \beta_i)\right],$$

let $\alpha_1, = \gamma, \alpha_2 = F^{-1}[F(\alpha + \gamma) - F(\gamma)]$, $\alpha_j = 0$ for $3 \leqslant j \leqslant n$, $\beta_1 = \beta$, $\beta_j = 0$ for $2 \leqslant j \leqslant n$. Hence,

$$\begin{aligned}
F(\alpha + \beta + \gamma) &= F\left\{F^{-1}\left[\sum_{i=1}^{n} F(\alpha_i)\right] + F^{-1}\left[\sum_{i=1}^{n} F(\beta_i)\right]\right\} \\
&\geqslant \sum_{i=1}^{n} F(\alpha_i + \beta_i) \qquad \text{(by the triangle inequality)} \\
&= F(\beta + \gamma) + F(\alpha + \gamma) - F(\gamma). \qquad \diamond
\end{aligned}$$

LEMMA 5. F is continuous.

PROOF. Since F is increasing, any point of discontinuity is a gap. Suppose $F(\alpha)$ is the lower boundary of a gap, then there exists $\epsilon > 0$ such that for all $\delta > 0$, $F(\alpha + \delta) - F(\alpha) > \epsilon$. By Lemma 4, however, for any $\beta > 0$,

$$F(\alpha + \beta + \delta) - F(\alpha + \beta) \geqslant F(\alpha + \delta) - F(\alpha) > \epsilon,$$

and so there is a gap at any $\alpha + \beta$, and F must be discontinuous everywhere. This is impossible, so F must be continuous. $\diamond$

The following lemma completes the proof of part (i) of Theorem 9.

LEMMA 6. Suppose F is a strictly increasing function defined on the nonnegative reals. Then F is convex iff for all $\alpha, \beta, \gamma \geqslant 0$,

$$F(\alpha + \beta + \gamma) + F(\gamma) \geqslant F(\alpha + \gamma) + F(\beta + \gamma).$$

PROOF. Suppose F is convex, and let $p = \alpha/(\alpha + \beta)$, then

$$\begin{aligned}
p(\alpha + \beta + \gamma) + (1 - p)\gamma &= \alpha + \gamma \\
(1 - p)(\alpha + \beta + \gamma) + p\gamma &= \beta + \gamma.
\end{aligned}$$

Hence, by convexity,

$$
\begin{aligned}
F(\alpha + \gamma) + F(\beta + \gamma) &= F[p(\alpha + \beta + \gamma) + (1 - p)\gamma] \\
&\quad + F[(1 - p)(\alpha + \beta + \gamma) + p\gamma] \\
&\leqslant pF(\alpha + \beta + \gamma) + (1 - p)F(\gamma) \\
&\quad + (1 - p)F(\alpha + \beta + \gamma) + pF(\gamma) \\
&= F(\alpha + \beta + \gamma) + F(\gamma).
\end{aligned}
$$

To prove the converse, recall that by Lemma 5 F is continuous. Suppose $\alpha > \beta$, then

$$
\begin{aligned}
F(\alpha) + F(\beta) &= F\left(\frac{\alpha - \beta}{2} + \frac{\alpha - \beta}{2} + \beta \right) + F(\beta) \\
&\geqslant 2F\left(\frac{\alpha - \beta}{2} + \beta \right) \qquad \text{(by Lemma 4)} \\
&= 2F\left(\frac{\alpha + \beta}{2} \right),
\end{aligned}
$$

which is equivalent to convexity for a continuous F. $\diamondsuit$

PROOF OF (ii). Sufficiency is established by the following result of Mulholland (1947, p. 300); see Section 12.3.4.

Suppose F is a continuous, convex, and strictly increasing function with $F(0) = 0$. If $\log F(e^x)$ is convex, then

$$
F^{-1}\left[\sum_{i=1}^{n} F(x_i) \right] + F^{-1}\left[\sum_{i=1}^{n} F(y_i) \right] \geqslant F^{-1}\left[\sum_{i=1}^{n} F(x_i + y_i) \right],
$$

for any x_i, $y_i \geqslant 0$. $\diamondsuit$

14.5.5 Preliminary Lemma

The proof of Theorem 10 is based on the following geometric lemma due to M. Perles (personal communication).

LEMMA 7. *Let S be an open convex subset of n-dimensional Euclidean space. Let δ be a metric on S that satisfies intradimensional subtractivity [Equation (9)] with respect to linear coordinates in S. If $\langle S, \delta \rangle$ is a metric with additive segments, then δ is homogeneous, i.e., for any $x, z \in S$ and any real $0 \leqslant t \leqslant 1$,*

$$
\delta[x, x + t(z - x)] = t\delta(x, z).
$$

We remark that, as a corollary, if S equals all of Euclidean n-space, the only metrics with additive segments satisfying intradimensional subtractivity are the general Minkowski metrics (see Busemann, 1955, pp. 94–104). The homogeneity of the metric in Theorem 10 follows from Lemma 7 without use of interdimensional additivity; then, using homogeneity and interdimensional additivity, we obtain the power metric from a theorem of G. H. Hardy, Littlewood, and Polya (1952) on homogeneous means.

Note that in Lemma 7, and in the following proof, the terms *open*, *closed*, and *continuous* refer exclusively to the natural topology of Euclidean n-space, *not* to the relative topology induced by S nor to the metric topology induced by δ.

PROOF OF LEMMA 7. First consider the case in which $t = 1/m$, where m is an integer ≥ 1. Suppose $x, z \in S$. For $k = 0, 1, \ldots, m$, define $z^{(k)} = x + k(z - x)/m$. By convexity of S, $z^{(k)} \in S$. By intradimensional subtractivity,

$$\delta[z^{(k-1)}, z^{(k)}] = \delta[x, x + (z - x)/m] \quad \text{for} \quad k = 1, \ldots, m.$$

From the triangle inequality, we have

$$\delta(x, z) \leq \sum_{k=1}^{m} \delta[z^{(k-1)}, z^{(k)}] = m\delta[x, x + (z - x)/m].$$

Thus $\delta[x, x + (z - x)/m] \geq \delta(x, z)/m$. We need to obtain the opposite inequality to establish the lemma for $t = 1/m$.

By segmental additivity, we can choose $y^{(0)}, y^{(1)}, \ldots, y^{(m)} \in S$ equally spaced on the segment from x to z, i.e.,

$$\delta[y^{(i-1)}, y^{(i)}] = \delta(x, z)/m, \quad i = 1, \ldots, m,$$

where $y^{(0)} = x$, $y^{(m)} = z$. Let $x^{(i)} = y^{(i)} - y^{(i-1)}$, $i = 1, \ldots, m$. Since S is open, there is a number $\alpha > 0$ such that $x + u$ is in S for every u such that $|u_j| < \alpha$, $j = 1, \ldots, n$. If we choose m large enough, $\delta(x, z)$ is arbitrarily small; since δ is monotone in each coordinate, we can obtain $|x_j^{(i)}| = |y_j^{(i)} - y_j^{(i-1)}| \leq \alpha$, $j = 1, \ldots, n$. Thus $\delta[x, x + x^{(i)}]$ is defined and $= \delta[y^{(i-1)}, y^{(i)}] = \delta(x, z)/m$ for $i = 1, \ldots, m$. Now $x + (z - x)/m$ is a convex combination of the vectors $x + x^{(i)}$, namely:

$$\sum_{i=1}^{m} [x + x^{(i)}]/m = x + \sum_{i=1}^{m} [y^{(i)} - y^{(i-1)}]/m$$
$$= x + (z - x)/m.$$

Hence, the inequalities $\delta[x, x + x^{(i)}] \leqslant \delta(x, z)/m$ imply the inequality $\delta[x, x + (z - x)/m] \leqslant \delta(x, z)/m$ provided that spheres are convex: if m points are within a given distance of x, then any convex combination of them is within that same distance of x.

Note that once this convexity result is established so that $\delta[x, x + (z - x)/m] = \delta(x, z)/m$ for sufficiently large integers m, the lemma follows. For by segmental additivity, the same equation holds for $t = k/m$ wherever $0 \leqslant k \leqslant m$ and m is sufficiently large; and by continuity it follows for $0 \leqslant t \leqslant 1$. So we turn to the matter of convexity of spheres. To establish this, we use the Krein-Milman theorem (1940), which asserts that a closed convex set in Euclidean n-space is the closed convex hull of the set of its extreme points, i.e., points that are not proper convex combinations of any other points.

Let x be fixed and let $B(\alpha)$ denote the sphere with center x and radius α, i.e., $\{y | \delta(x, y) \leqslant \alpha\}$. Let $C(\alpha)$ be the convex hull of $B(\alpha)$, i.e., the smallest convex set containing $B(\alpha)$, which is the set of all convex combinations of elements of $B(\alpha)$. By continuity of δ, for α sufficiently small, $B(\alpha)$ is a closed set entirely contained in S. The convex hull of a closed set is closed; so $C(\alpha)$ is a closed convex set. By the Krein-Milman theorem, the set of extreme points of $C(\alpha)$, denoted $E(\alpha)$, is the minimal set whose convex hull is dense in $C(\alpha)$. In particular, $B(\alpha)$ contains $E(\alpha)$. Thus, to prove that $B(\alpha)$ is convex [equal to $C(\alpha)$], it suffices to show that any convex combination of elements of $E(\alpha)$ is in $B(\alpha)$. It even suffices to show this for convex combinations of form $\sum_{i=1}^{r} y^{(i)}/r$, where $y^{(i)}$ is in $E(\alpha)$ because such combinations are dense in $C(\alpha)$. Any convex combination with rational coefficients can be written in such a form, by letting r be a common denominator.

We showed above, using segmental additivity and intradimensional subtractivity, that $y \in B(\alpha)$ implies that y/r is in $C(\alpha/r)$. (There, we used a vector of form $y = z - x$, but this is irrelevant.) But if y is in $E(\alpha)$, then y/r must be in $E(\alpha/r)$. Suppose the contrary, i.e., $y/m = \sum_{i=1}^{p} \lambda_i w^{(i)}$, where $\lambda_i \geqslant 0$, $\sum_{i=1}^{p} \lambda_i = 1$, and $w^{(i)}$ is in $C(\alpha/r)$. Since $C(\alpha/r)$ is the set of convex combinations of $B(\alpha/r)$, we can suppose without loss of generality that $w^{(i)}$ is in $B(\alpha/r)$. By the triangle inequality, $rw^{(i)}$ is in $B(\alpha)$; hence, $y = \sum_{i=1}^{p} \lambda_i rw^{(i)}$ is not in $E(\alpha)$, a contradiction. Hence, a convex combination $\sum_{i=1}^{r} y^{(i)}/r$ with $y^{(i)}$ in $E(\alpha)$ has each $y^{(i)}/r$ in $B(\alpha/r)$. By translation-invariance and the triangle inequality, $\sum_{i=1}^{r} y^{(i)}/r$ is in $B(\alpha)$ as required. Thus we have shown that $B(\alpha)$ is convex for α small enough so that $B(\alpha)$ is a closed set in S. Since we apply the result only for $\alpha = \delta(x, z)/m$, for sufficiently large m, we can assume that m is large enough to obtain $B(\alpha)$ convex. $\Diamond$

14.5.6 Theorem 10 (p. 187).

The power metric is the only additive-difference model that is also a metric with additive segments.

PROOF. By Theorem 9, δ must be of the form

$$\delta(a, b) = F^{-1}\left[\sum_{i=1}^{n} F(|\varphi_i(a_i) - \varphi_i(b_i)|)\right].$$

Since the metric δ is defined on a (convex) cube, the Cartesian product of the ranges of $|\varphi_i(a_i) - \varphi_i(b_i)|$, $i = 1, \ldots, n$, Lemma 7 applies. If we let $\alpha_i = |\varphi_i(c_i) - \varphi_i(a_i)|$, the equation $\delta[x, x + t(z - x)] = t\delta(x, y)$ implies the following functional equation:

$$F^{-1}\left[\sum_{i=1}^{n} F(t\alpha_i)\right] = tF^{-1}\left[\sum_{i=1}^{n} F(\alpha_i)\right]. \tag{13}$$

We shall show that the only solutions to this functional equation are of the form $F(\alpha) = k\alpha^r$. To do this, two steps are required. First, we make some remarks about the domains of α_i and t, for which Equation (13) has been shown to hold, and show that F can be extended to map $[0, \infty)$ onto itself, with Equation (13) valid for $0 \leqslant \alpha_1, \ldots, \alpha_n$, $t < \infty$. Second, we show that Equation (13) implies that F satisfies another functional equation, for which the only solutions are $F(\alpha) = k\alpha^r$, according to a result of G. H. Hardy *et al.* (1952, Theorem 84).

By Theorem 9, the additive difference model

$$\delta(a, b) = G\left[\sum_{i=1}^{n} F_i[|\varphi_i(a_i) - \varphi_i(b_i)|]\right]$$

yields a proper metric only if each $F_i(\alpha) = F(t_i\alpha)$, where $F = G^{-1}$. Thus F^{-1} is defined for all numbers $\sum_{i=1}^{n} F_i[|\varphi_i(a_i) - \varphi_i(b_i)|]$, and each F_i coincides with F (except for the factor t_i multiplying the argument) in its own domain. Moreover, since the domains of F_i are intervals (δ is a metric with additive segments), we can choose $\omega > 0$ such that for each $i = 1, \ldots, n$ and each α, $0 \leqslant \alpha \leqslant \omega/t_i$, α is in the domain of F_i. Note that in particular, if $\omega_i = \omega/t_i$, $F^{-1}[\sum_{i=1}^{n} F_i(\omega_i)] = F^{-1}[nF(\omega)]$ is defined. We denote this quantity by Ω; note that $F = G^{-1}$ is defined at least on the interval $[0, \Omega]$. We can now extend F as follows.

For $\alpha < \omega$, we have by Equation (13),

$$F^{-1}[nF(\alpha)] = F^{-1}[nF(\omega\alpha/\omega)]$$
$$= (\alpha/\omega)F^{-1}[nF(\omega)]$$
$$= \alpha\Omega/\omega.$$

Thus F satisfies

$$F(\alpha) = nF(\alpha\omega/\Omega), \qquad 0 \leqslant \alpha \leqslant \Omega; \tag{14}$$
$$F^{-1}(a) = (\Omega/\omega)F^{-1}(a/n), \qquad 0 \leqslant a \leqslant nF(\omega). \tag{15}$$

We can use Equation (14) to define F for α such that the right side of this equation is defined, i.e., for $0 \leqslant \alpha \leqslant \Omega^2/\omega$. Note that $\Omega^2/\omega > \Omega$; so this really extends F beyond the interval $[0, \Omega]$. Correspondingly, F^{-1} satisfies Equation (15) for the extended interval $0 \leqslant a \leqslant nF(\Omega) = n^2F(\omega)$. Moreover, Equation (13) is now valid for $0 \leqslant \alpha_i \leqslant \Omega$ and $0 \leqslant t \leqslant 1$:

$$F^{-1}\left[\sum_{i=1}^{n} F(t\alpha_i)\right] = F^{-1}\left[\sum_{i=1}^{n} nF(t\alpha_i\omega/\Omega)\right] \qquad \text{(by Equation 14)}$$

$$= F^{-1}\left[n\sum_{i=1}^{n} F(t\alpha_i\omega/\Omega)\right]$$

$$= (\Omega/\omega)F^{-1}\left[\sum_{i=1}^{n} F(t\alpha_i\omega/\Omega)\right] \qquad \text{(by Equation 15)}$$

$$= t(\Omega/\omega)F^{-1}\left[\sum_{i=1}^{n} F(\alpha_i\omega/\Omega)\right] \qquad \begin{array}{l}\text{(by Equation 13 applied} \\ \text{for } \beta_i = \alpha_i\omega/\Omega \leqslant \omega)\end{array}$$

$$= tF^{-1}\left[n\sum_{i=1}^{n} F(\alpha_i\omega/\Omega)\right] \qquad \text{(by Equation 15)}$$

$$= tF^{-1}\left[\sum_{i=1}^{n} nF(\alpha_i\omega/\Omega)\right]$$

$$= tF^{-1}\left[\sum_{i=1}^{n} F(\alpha_i)\right]. \qquad \text{(by Equation 14)}$$

Repeatedly applying the arguments extends the intervals in which F is defined and in which Equation (13) is valid by a factor Ω/ω each time. Thus we can extend F to $[0, \infty)$ and have the equation valid for $0 \leqslant t \leqslant 1$.

Under these circumstances, it follows immediately that Equation (13) is valid for $0 \leqslant t < \infty$. The function F, so extended, coincides with the original F at least on $[0, \Omega]$.

Now by Equation (15) we have, for $0 \leqslant \alpha_i < \infty$,

$$F^{-1}\left[\sum_{i=1}^{n} F(\alpha_i)\right] = (\Omega/\omega)F^{-1}\left[\frac{1}{n}\sum_{i=1}^{n} F(\alpha_i)\right].$$

It follows immediately that Equation (13) applies to the mean-value function $F^{-1}[(1/n)\sum_{i=1}^{n} F(\alpha_i)]$:

$$F^{-1}\left[\frac{1}{n}\sum_{i=1}^{n} F(t\alpha_i)\right] = tF^{-1}\left[\frac{1}{n}\sum_{i=1}^{n} F(\alpha_i)\right]. \tag{16}$$

According to G. H. Hardy *et al.* (1952, Theorem 84), the only monotonic functions satisfying Equation (16) for $0 \leqslant \alpha_i < \infty$ are functions of form $k\alpha^r$ (see also Aczél, 1966, p. 153).

Going back to the original function F, we see that $F(\alpha) = k\alpha^r$, at least for $0 \leqslant \alpha \leqslant \Omega$. But now, for any $|\varphi_i(a_i) - \varphi_i(b_i)|$, $i = 1, \ldots, n$, we can find $t > 0$ so small that $t|\varphi_i(a_i) - \varphi_i(b_i)| \leqslant \omega$, $i = 1, \ldots, n$. Thus, using Equation (13) (since we can assume $t < 1$), we have

$$
\begin{aligned}
\delta(a, b) &= F^{-1}\left[\sum_{i=1}^{n} F(|\varphi_i(a_i) - \varphi_i(b_i)|)\right] \\
&= \frac{1}{t}F^{-1}\left[\sum_{i=1}^{n} F(t|\varphi_i(a_i) - \varphi_i(b_i)|)\right] \\
&= \frac{1}{t}\left[\sum_{i=1}^{n} (t|\varphi_i(a_i) - \varphi_i(b_i)|)^r\right]^{1/r} \\
&\qquad\qquad\qquad \text{(since } t|\varphi_i(a_i) - \varphi_i(b_i)| \leqslant \omega) \\
&= \left[\sum_{i=1}^{n} (|\varphi_i(a_i) - \varphi_i(b_i)|)^r\right]^{1/r}
\end{aligned}
$$

Thus δ is a power metric, and by the uniqueness part of Theorem 8, $F_i(\alpha) = k\alpha^r$.

Finally, the restriction $1 \leqslant r < \infty$ follows from the triangle inequality or, more specifically, from the Minkowski inequality (see Section 12.2.6). ◇

14.6   EXPERIMENTAL TESTS OF MULTIDIMENSIONAL REPRESENTATIONS

The axiomatic approach developed in the preceding sections assumes that the appropriate dimensional structure of the objects under study is specified in advance and provides necessary and/or sufficient conditions for the representation of proximity by a metric distance function. Because the dimensional structure is usually unknown, most work on the representation of proximity *assumes* a particular spatial model that is used to *infer* the structure of the domain. Thus measurement-theoretical analysis and scaling research proceed in different directions: the former assumes the structure and investigates the model whereas the latter assumes the model and uses it to derive the structure.

The need to summarize complex data, the intuitive appeal of geometric representations, and the availability of effective scaling programs have produced numerous applications of multidimensional scaling in the social sciences. For examples, see the contributions in the collections edited by Davies and Coxon (1982), Romney, Shepard, and Nerlove (1972), Shepard, Romney, and Nerlove (1972), and the reviews by Carroll and Arabie (1980), Shepard (1980), and F. W. Young (1984). Multidimensional scaling provides convenient methods for summarizing and displaying proximity data; it is also used by many authors, notably Shepard (1984, 1987), as a psychological theory about mental representations. In particular, Shepard (1987) has proposed an essentially Bayesian theory of generalization based on the prior expectation of the organism and showed that it can give rise to the Euclidean and the city-block metrics, under different assumptions concerning the variations of the dimensions of the space. Because of the predominance of the descriptive approach, most applications have not been concerned with testing the basic assumptions of multidimensional scaling. Rather, the models are generally treated as methods of data reduction whose usefulness depends primarily on the interpretability of the spatial configuration derived from the data.

The present section, in contrast, treats multidimensional proximity models as psychological theories about mental representations and reviews several attempts to test their underlying assumptions. The testing of these models is complicated by the fact that, unless the dimensional structure of the objects is known, any finite symmetric data set can always be embedded in a Euclidean space with a sufficiently large number of dimensions. To overcome this difficulty, researchers have investigated the proximity between objects whose dimensional structure, or at least its physical counterpart, can be specified in advance. Following the work of Attneave (1950), researchers constructed multidimensional stimuli and investigated the psychological counterparts of physical dimensions and the rule for combining psychological attributes.

Perhaps the most interesting finding that emerged from this research on multidimensional psychophysics concerns the distinction between integral and separable stimuli (see, Garner, 1974; Shepard, 1964; Torgerson, 1958). Integral stimuli (e.g., colors varying in lightness and chromaticness) are perceived as wholes whereas separable stimuli (e.g., circles varying in brightness and size) are perceived in terms of their separate attributes. Although the distinction between separable and integral attributes is not always clear, they appear to be governed by different rules of combination. The proximity between separable stimuli is better described by the city-block model (i.e., a power metric with $r = 1$) than by the Euclidean model whereas the opposite holds for integral stimuli. For further discussion of this issue see Section 14.6.3.

The axiomatic analysis of multidimensional scaling (Beals, Krantz, and Tversky, 1968; Tversky and Krantz, 1970) has motivated several studies that were designed to test the psychological validity of decomposability [Equation (8)], intradimensional subtractivity [Equation (9)], and interdimensional additivity [Equation (10)]. These models are of special interest because they can be viewed as defining properties of psychological dimensions. The axiomatic analysis of these models isolates the key ordinal properties that can be used to test whether an objectively specified attribute acts like a subjective dimension.

The investigation of critical ordinal properties could provide more powerful tests than those based on global measures of goodness-of-fit, which are often insensitive to small, albeit significant, deviations. The flat-earth model, for example, provides a good fit to the distances between all major cities in the United States, although systematic departures from the model can be detected by a properly designed test. Experimental tests of the axioms can also show in detail which aspect of the theory, if any, is wrong and can suggest alternative representations. In this section we discuss, in turn, some experimental tests of intradimensional subtractivity, interdimensional additivity, and the triangle inequality.

14.6.1 Relative Curvature

Consider two triples of stimuli a, b, c and a', b', c' that vary along a single dimension. Suppose that b lies between a and c whereas b' lies between a' and c', i.e., $a|b|c$ and $a'|b'|c'$. Suppose further that $(a, b) \succsim (a', b')$ and $(b, c) \succsim (b', c')$; hence, the betweenness axioms (Part 2 of Definition 8) implies $(a, c) \succsim (a', c')$. This property, which is necessary for intradimensional subtractivity, requires constant curvature. It is satisfied by the points on a straight line or on the circumference of a semicircle with distances measured along the chord. It is violated, however, by the points on an arc of an ellipse, which does not have constant curvature.

The same formal property can also be used to compare two different attributes. Suppose a, b, and c differ only on Attribute 1 whereas a', b', and c' differ only on Attribute 2. As before, we assume that $a|b|c$ and $a'|b'|c'$. If intradimensional subtractivity [Equation (9)] holds, then the above implication should be satisfied, even when a, b, c differ only on Attribute 1 and a', b', c' differ only on Attribute 2. A systematic failure of this property could be revealing. Specifically, we say that Attribute 1 is *curved relative* to Attribute 2 whenever $(a, b) \sim (a', b')$ and $(b, c) \sim (b', c')$, but $(a, c) \prec (a', c')$. By examining this relation for many pairs of triples that vary along the two attributes, one can test whether one attribute is curved relative to the other. Note that the relation of relative curvature is asymmetric.

Relative curvature was investigated in studies of dissimilarity between rectangles and between house plants. Krantz and Tversky (1975) devised a set of rectangles designed to test which pair of attributes—height H and width W, or area $A = HW$ and shape $S = H/W$—behave like subjective dimensions. A detailed ordinal analysis of the dissimilarity ratings of individual subjects showed that shape was curved relative to area for most respondents, but no consistent pattern of relative curvature was observed for height and width.

Following some ideas of Restle (1959) and Stevens (1957), Gati and Tversky (1982) proposed a representation of quantitative and qualitative attributes, respectively, as nested and nonnested feature sets (see Section 14.7.1). This representation suggests the hypothesis that qualitative attributes (e.g., shape) are curved relative to quantitative attributes (e.g., size). Gati and Tversky (1982) confirmed this hypothesis in a study of dissimilarity between house plants that vary in two qualitative attributes (shape of the leaves and form of the pot) and one quantitative attribute (size of the plant). Comparison of the quantitative attribute (size) with a qualitative attribute (shape) showed that shape was curved relative to size for most subjects. The comparison of the two qualitative attributes did not reveal a consistent pattern. Because relative curvature violates intradimensional subtractivity, spatial representations of objects that vary along both qualitative and quantitative attributes should be interpreted with special care.

14.6.2 Translation Invariance

Recall that both interdimensional additivity [Equation (10)] and decomposability [Equation (8)] imply (one-factor) independence (Definition 6).

Suppose $A = A_1 \times A_2$, where $a, b \in A_1$ and $p, q \in A_2$. It follows from (one-factor) independence that

$$\delta(ap, bp) = \delta(aq, bq) \quad \text{and} \quad \delta(ap, aq) = \delta(bp, bq). \quad (17)$$

This is an ordinal analogue of translation invariance; the "distance" between a and b is unaffected by the "translation" from p to q. Equation (17) is a necessary condition for any symmetric proximity model of the form

$$\delta(ap, bq) = F[\varphi_1(a, b), \varphi_2(p, q)],$$

where F is a strictly increasing function in two real arguments.

The equality prediction [Equation (17)] has been tested in the two studies described in the previous section. The analysis of the dissimilarity between rectangles casts serious doubt on the validity of the above decomposability assumption for either the height $\times$ width or the area $\times$ shape design. The most striking failure of Equation (17) can be described as an augmentation effect: the same shape difference produced greater dissimilarity when the (common) area of the rectangles was large than when it was small. Thus area augments the difference in shape. The augmentation effect for height and width was considerably smaller, but area differences were more pronounced for the more extreme (i.e., elongated) rectangles. Similar failures of independence between area and shape of rectangles were observed by Wender (1971) and by Wiener-Ehrlich (1978).

Figure 2 displays a multidimensional scaling solution of the dissimilarity of the rectangle, derived from INDSCAL (Carroll and Chang, 1970). This program constructs a common (Euclidean) space for all subjects, and it permits individual differences in the weighting of the dimensions. Each numbered point in the figure represents the location of the respective rectangle in the plane. The solid and the broken lines represent, respectively, the $H \times W$ and the $A \times S$ factorial designs. Figure 2 shows the substantial augmentation of shape by area, e.g., $\delta(1, 2) < \delta(8, 17) < \delta(7, 6)$, but not the equally pronounced finding that shape is curved relative to area (Section 14.6.1). This observation illustrates a general point: multidimensional scaling provides a useful method for discovering and describing empirical regularities, but some significant patterns (e.g., relative curvature) may not be captured by the spatial solution (see Section 14.7.3).

Gati and Tversky (1982) tested independence [Equation (17)] in their study of dissimilarity of house plants. They found that the same difference in the shape of the leaves produced greater dissimilarity for large plants than for small plants, but they found no systematic violations of indepen-

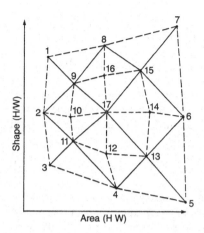

FIGURE 2. Two-dimensional Euclidean representation (INDSCAL) of the dissimilarity between rectangles (Krantz and Tversky, 1975). The solid and the broken lines represent, respectively, the $H \times W$ and the $A \times S$ factorial designs.

dence between the two qualitative attributes: shape of leaves and form of pot. Gati and Tversky (1982) hypothesized that a quantitative attribute (e.g., size) augments the difference in a qualitative attribute (e.g., shape) to which it applies. A further example of augmentation was obtained by Burns, Shepp, McDonough, and Wiener-Ehrlich (1978) who found that hue differences were more pronounced for high chroma levels than for low ones. In contrast, these authors found that (one-factor) independence was satisfied for dial-like figures that vary in size of circle and angle of radial line and for squares that vary in size and brightness.

Another source of systematic violations of independence and decomposability is the presence of additive binary attributes. To illustrate this phenomenon, we compare two studies of dissimilarity between schematic faces conducted by Tversky and Krantz (1969) and by Gati and Tversky (1984). In both studies, subjects rated the dissimilarity between all pairs of eight schematic faces generated by a $2 \times 2 \times 2$ design. The critical difference between the studies concerns the nature of the attributes. The schematic faces in the first study vary on three *substitutive* attributes: long face versus wide face, empty eyes versus filled eyes, and straight mouth versus curved mouth. The schematic faces in the second study vary on three *additive* attributes that correspond to the presence or absence of eyeglasses, a hat, and a beard.

The two studies produced very different results. The study of Tversky and Krantz (1969) that used substitutive features provided strong support for independence and interdimensional additivity. The dissimilarity between a pair of faces was practically independent of the attributes shared by the two faces. In contrast, Gati and Tversky (1984, Studies 8 and 9) found that independence was violated in a highly predictable manner. Letting 1 and 0 denote, respectively, the presence and absence of each attribute, the data implied that

$$\delta(100, 010) > \delta(101, 011), \qquad \delta(100, 110) > \delta(101, 111), \qquad \text{etc.}$$

That is, adding a common attribute (e.g., eyeglasses) to a pair of faces increases their similarity, contrary to independence. Hence, traditional multidimensional representations that assume translation invariance, or decomposability, cannot accommodate additive attributes. For further evidence and discussion, see Gati and Tversky (1984). Indeed, the failure of traditional models to account for the effect of common features has led to an alternative representation of proximity, discussed in Section 14.7.1.

14.6.3 The Triangle Inequality

The use of ordinal data to test the triangle inequality is beset by several difficulties. As was noted earlier (Section 14.1), the triangle inequality can be satisfied vacuously in the absence of additional constraints on the space or the metric. A segmentally additive proximity structure (Definition 4) permits a test of the triangle inequality, but the test requires a fairly elaborate construction that poses nontrivial experimental difficulties (see Section 14.2.1). However, a much simpler test of the triangle inequality can be devised by exploiting the dimensional structure of the objects under study.

Suppose $A = A_1 \times A_2$, $a, b, c \in A_1$, $p, q, r \in A_2$, $a|b|c$, and $p|q|r$. Assume that $\langle A, \succsim \rangle$ satisfies an additive difference metric (Section 14.4.5), and consider the five points displayed in Figure 3. (The point bq is the *center*, and the other four points are the *corners*.) In an additive-difference metric, the *corner path* $\delta(ap, cp) + \delta(cp, cr)$ is greater than or equal to the

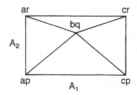

FIGURE 3. The design for the test of the corner inequality.

center path $\delta(ap, bq) + \delta(bq, cr)$. To demonstrate this assertion note that

$$\delta(ap, bq) + \delta(bq, cr)$$
$$= F^{-1}[F(|\varphi_1(a) - \varphi_1(b)|) + F(|\varphi_2(p) - \varphi_2(q)|)]$$
$$+ F^{-1}[F(|\varphi_1(b) - \varphi_1(c)|) + F(|\varphi_2(q) - \varphi_2(r)|)]$$
$$\text{(for some convex } F; \text{ see Equation 12)},$$
$$\leqslant |\varphi_1(a) - \varphi_1(b)| + |\varphi_2(p) - \varphi_2(q)| + |\varphi_1(b) - \varphi_1(c)|$$
$$+ |\varphi_2(q) - \varphi_2(r)|$$
$$\text{(because } F^{-1}[F(x) + F(y)] \leqslant x + y, \text{ by convexity)},$$
$$= |\varphi_1(a) - \varphi_1(c)| + |\varphi_2(p) - \varphi_2(r)|$$
$$\text{(since } a|b|c \text{ and } p|q|r),$$
$$= F^{-1}[F(|\varphi_1(a) - \varphi_1(c)|) + F(0)]$$
$$+ F^{-1}[F(0) + F(|\varphi_2(p) - \varphi_2(r)|)],$$
$$= \delta(ap, cp) + \delta(cp, cr).$$

This condition, called the *corner inequality*, states that a path from ap to cr via an internal point bq cannot be longer than a path through a corner cp. The two paths are equal only if F is linear.

Given an additive-difference proximity structure with two attributes, therefore, one can test the corner inequality and hence the triangle inequality from which it is derived. Although the corner inequality is defined in terms of sums of distances, it can be easily converted to an ordinal test. Given the configuration of points in Figure 3, the corner path is said to be shorter than the center path (in an ordinal sense) whenever

$$(ap, bq) \succsim (ap, cp) \quad \text{and} \quad (bq, cr) \succsim (cp, cr),$$

or

$$(bq, cr) \succsim (ap, cp) \quad \text{and} \quad (ap, bq) \succsim (cp, cr),$$

provided at least one inequality is strict; such a pattern clearly violates the corner inequality. The center path is said to be shorter than the corner path (in an ordinal sense) if the opposite pattern of inequalities holds. Note that neither pattern may hold, in which case the data do not provide ordinal information for the test of the corner inequality. This procedure provides an ordinal test for the convexity of F that is implied by the triangle inequality (Theorem 9).

Tversky and Gati (1982) tested the corner inequality in a series of studies of two-dimensional stimuli with separable attributes (including frequency histograms, schematic faces, parallelograms, and house plants), using judgments of similarity and dissimilarity, classification, and recognition errors. They found that corner stimuli, which coincide on one attribute, were judged closer to each other (and were confused more often) than the corner and the center. For separable stimuli, therefore, the corner path was shorter than the center path, contrary to the implication of the triangle inequality. This conclusion accords with earlier findings of Shepard (1964) and of Eisler and Knöppel (1970). The size of the deviations from the corner inequality increases with the separability of the attributes (integral stimuli satisfy the corner inequality), the transparency of the dimensional structure, and the discriminability of the levels within each dimension.

These results suggest that the triangle inequality holds for integral but not for separable stimuli. Because the city-block metric ($r = 1$) corresponds to the limiting case of the corner inequality, it is not surprising that it fits separable attributes better than the Euclidean model ($r = 2$). However, the experimental tests of the corner inequality indicate that, for separable stimuli, the best estimate of the exponent of the power metric is actually less than one. The estimates of r obtained by Tversky and Gati (1982) range from 0.9 for squares that vary in size and brightness to 0.66 for house plants that vary in the shape of the leaves and the form of the pot.

Altogether, the experimental tests of the metric and the dimensional axioms underlying multidimensional representations of proximity have led to several interesting observations. These include the curvature of qualitative relative to quantitative attributes (14.6.1), the augmentation of qualitative by quantitative attributes (14.6.2), and the systematic violations of the triangle inequality for separable attributes (14.6.3). These conclusions cannot be easily reached by examining spatial configurations or by evaluating overall goodness-of-fit. They help delineate the conditions under which the dimensional metric model is likely to be valid, as a psychological theory, and they suggest ways in which it could be modified to account for the observed data.

14.7 FEATURE REPRESENTATIONS

The dimensional metric models discussed in the preceding sections are based on two general notions: the representation of objects as points in a coordinate space and the use of metric distance to represent proximity. These assumptions have led to fruitful developments, but they are not always valid. In particular, the systematic failures of subtractivity, additivity, and the triangle inequality, described in the previous sections, motivate alternative representations of proximity that are neither metric nor dimensional in the conventional sense.

In this section we investigate a model of proximity that represents each single object as a set (interpreted as a set of features) and expresses the proximity of two objects in terms of their set intersection (the shared features) and their set differences (distinctive features for each object relative to the other). Our primary goal is to obtain a dissimilarity scale of the form

$$\delta(a, b) = \alpha\varphi(a - b) + \beta\varphi(b - a) - \theta\varphi(a \cap b), \qquad (18)$$

where the set function φ evaluates the weight or salience of each set difference, $a - b$ and $b - a$, and of the set intersection $a \cap b$; the coefficients α, β, and θ describe the relative contribution of the respective arguments.

Equation (18) is the *contrast model* introduced by Tversky (1977). It encompasses two common conceptions of proximity, based exclusively on common and on distinctive features. If we set $\alpha = \beta = 0$, the contrast model reduces to the common-features model, in which the proximity of objects is given by the measure of their common features. That is, $\delta(a, b)$ depends only on $\varphi(a \cap b)$. This model serves as a basis for many clustering methods (see, e.g., Hartigan, 1975; Shepard and Arabie, 1979), and it has been used to assess the commonality and the prototypicality of concepts (Rosch and Mervis, 1975; E. E. Smith and Medin, 1981). An alternative model of proximity, based exclusively on distinctive features, arises when we set $\theta = 0$ and $\alpha = \beta = 1$ so that $\delta(a, b) = \varphi(a - b) + \varphi(b - a)$. If φ is additive, we obtain the symmetric-difference model, which defines a metric distance between sets; see Restle (1961) for a general discussion and Corter and Tversky (1986) for scaling applications. The contrast model can also accommodate asymmetric proximities that arise when α and β are unequal; see Section 14.7.2.

In this section we develop an axiomatic analysis of the contrast model. Along the way we investigate a more general representation with different salience functions, rather than different coefficients for common and for distinctive features, and a more restrictive representation in which ϕ is required to be an additive measure. In the present development we treat the two differences, $a - b$ and $b - a$ and the intersection $a \cap b$ as three factors contributing to the ordering $\succsim$, and we impose the usual axioms of conjoint measurement to obtain decomposability with respect to a general rule of combination (Section 7.2.1), general additivity (Section 6.11.1), or additivity for identical factors (Section 6.11.2).

Section 14.7.1 presents an axiomatic analysis of the contrast model. The main departure from the standard treatment of conjoint measurement (see, e.g., Theorem 7) is the replacement of a factorial structure by a more general structure satisfying factorial connectedness (Definition 18). Empirical applications of the contrast model are discussed in Section 14.7.2. The simplest spatial and feature representations of proximity data, respectively, are two-dimensional maps and weighted trees. These representations are illustrated and compared in Section 14.7.3.

14.7.1 The Contrast Model

Let S be a nonempty set, and let A be a family of subsets of S. We interpret A as the set of objects under study and S as the set of all possible relevant features of the objects of A. Thus, each object $a \in A$ corresponds to a subset of S, and for each pair $(a, b) \in A^2$, we can form the subsets $a - b$, $b - a$, and $a \cap b$.

We do not assume that $A = 2^S$ so that every nonempty subset of S corresponds to a possible object. Likewise, we do not assume that every subset can appear as a set of common or of distinctive features. Rather, A is a proper subset of 2^S, and the possible sets of distinctive features are defined via A. The family S_D of *difference* subsets is $\{a - b \mid a, b \in A\}$, and the set S_C of *common* subsets is $\{a \cap b \mid a, b \in A\}$.

We assume that the sets S_D and S_C are ordered with respect to their contributions to dissimilarity. Because proximities may be asymmetric, we assume two weak orders of S_D. We include among the primitives not only an ordering of pairs of objects (dissimilarity ordering) but also three orderings of feature sets (salience orderings). These orderings can be inferred directly from the dissimilarity ordering or derived from other data, such as direct judgments. Thus we conclude that $a - b$ exceeds $c - d$ whenever $b - a = d - c$, $a \cap b = c \cap d$, and $(a, b) \succeq (c, d)$. Similarly, we conclude that $a \cap b$ exceeds $c \cap d$ whenever $c - d = a - b$, $d - c = b - a$, and $(c, d) \succeq (a, b)$. Given a pair (a, b), $\succeq_1$ represents the contribution of $a - b$, $\succeq_2$ represents the contribution of $b - a$, and $\succeq_3$ is the ordering on S_C. All three weak orders are assumed to be monotone with respect to set inclusion, that is, they are extensions of the partial order $\supseteq$ on S_D or on S_C.

The dissimilarity ordering $\succeq$ is a weak order on A^2. In line with the intended representation of Equation (18), we say that the pair (a, b) *dominates* (a', b') if $a - b \succeq_1 a' - b'$, $b - a \succeq_2 b' - a'$, and $a' \cap b' \succeq_3 a \cap b$. We assume that in any such case of dominance, $(a, b) \succeq (a', b')$. In particular, suppose b is between a and c in the sense that $a \cup c \supseteq b \supseteq a \cap c$, denoted $a|b|c$. By monotonicity of $\succeq_3$ for set inclusion, together with dominance, we conclude that $(a, c) \succeq (a, b)$ and $(a, c) \succeq (b, c)$.

The primitives and properties assumed so far are collected in the following definition:

DEFINITION 16. *Suppose S is a set, A is a subset of 2^S, and let $S_D = \{a - b \mid a, b \in A\}$ and $S_C = \{a \cap b \mid a, b \in A\}$. Let $\succsim$ be a binary relation on A^2, and let $\succsim_1$, $\succsim_2$, and $\succsim_3$ be binary relations on S_D, S_D and S_C respectively. The structure $\langle S, A, \succsim, \succsim_1, \succsim_2, \succsim_3 \rangle$ is a proximity feature-structure provided that the following axioms are satisfied:*

1. *$\succsim$ is a weak order on A^2.*

2. *$\succsim_1$ and $\succsim_2$ are weak orders on S_D, and $\succsim_3$ is a weak order on S_C.*

3. *Each $\succsim_i$ is an extension of the partial order by inclusion on S_D or S_C.*

4. *Each $\succsim_i$ is nondegenerate, i.e., contains at least two equivalence classes.*

5. *If $a - b \succsim_1 a' - b'$, $b - a \succsim_2 b' - a'$, and $a' \cap b' \succsim_3 a \cap b$, then $(a, b) \succsim (a', b')$.*

In the remainder of this section, except where otherwise stated, we assume that $\langle S, A, \succsim, \succsim_1, \succsim_2, \succsim_3 \rangle$ is a proximity feature-structure. To map this structure into an additive conjoint measurement model, we take the additive factors to be the sets of equivalence classes induced by the weak orders $\succsim_i$. Thus we let $X_1 = S_D/\sim_1$, $X_2 = S_D/\sim_2$, and $X_3 = S_C/\sim_3$. We map A^2 onto a subset Y of $X_1 \times X_2 \times X_3$ by the correspondence

$$(a, b) \to ([a - b], [b - a], [a \cap b]),$$

where [ ] designates the equivalence class determined by the set inside the brackets. We refer to $[a - b]$, $[b - a]$, and $[a \cap b]$ as the first, second, and third *components* of (a, b).

By the dominance axiom (Definition 16, Axiom 5) we can define a relation, denoted $\succsim$, on Y that is homomorphic to $\succsim$ on A^2:

$$([a - b], [b - a], [a \cap b]) \succsim ([c - d], [d - c], [c \cap d])$$
$$\text{iff} \quad (a, b) \succsim (c, d).$$

This relation is well defined because (a, b) and (a', b') map to the same element of Y if and only if their components are equivalent, that is,

$a - b \sim_1 a' - b'$, $b - a \sim_2 b' - a'$, and $a \cap b \sim_3 a' \cap b'$; hence, by dominance, $(a, b) \sim (a', b')$.

If $\langle A^2, \succsim \rangle$ has a countable order-dense subset, then we can map $\langle Y, \succsim \rangle$ and each $\langle X_i, \succsim_i \rangle$ homomorphically into $\langle \text{Re}, \geqslant \rangle$, thereby obtaining an order-preserving function δ on A^2 of form

$$\delta(a, b) = F[\varphi_1(a - b), \varphi_2(b - a), \varphi_3(a \cap b)],$$

with F strictly increasing in its first two variables and decreasing in its third. That is, a proximity feature-structure that is ordinally scalable is *decomposable* in a sense similar to that of factorial decomposability discussed in Section 4.1.

We turn now to the question of when the general function of $\varphi_1, \varphi_2, \varphi_3$ can be replaced by $\varphi_1 + \varphi_2 + \varphi_3$. The problem in applying conjoint measurement to this question is that in the present context it is desirable to avoid the assumption that Y is a Cartesian product. The following definitions are introduced for this purpose.

DEFINITION 17.   *A proximity feature-structure is* partially factorial *if for any* a, b, c, d, e, f, g, h *in* A *such that* $a - b \succsim_1 c - d$, $b - a \succsim_2 f - e$, $a \cap b \succsim_3 g \cap h$, *there exist* a', b' *in* A *such that* $a' - b' \sim_1 c - d$, $b' - a'$ $\sim_2 f - e$, *and* $a' \cap b' \sim_3 g \cap h$.

The structure introduced in Definition 17 is weaker than the usual factorial structure in two important respects. First, it requires equivalence ($\sim_i$) rather than equality. Thus, we do not assume that any component of a pair of objects (e.g., $a - b$) can be juxtaposed with any component of another pair (e.g., $c \cap d$). We only assume juxtaposition for an appropriately chosen element of the respective equivalence classes $[a - b]$ and $[c \cap d]$. Second, the factorial assumption is restricted to feature sets that are bounded by the components of some pair of objects in A. Specifically, for any a, b in A, define

$$Y_1 = \{ x_1 \,|\, [a - b] \succsim_1 x_1 \}$$
$$Y_2 = \{ x_2 \,|\, [b - a] \succsim_2 x_2 \}$$
$$Y_3 = \{ x_3 \,|\, [a \cap b] \succsim_3 x_3 \}$$

Then $Y(a, b) = Y_1 \times Y_2 \times Y_3$ is the Cartesian product *bounded by* (a, b).

DEFINITION 18.   *Let* $\langle S, A, \succsim, \succsim_1, \succsim_2, \succsim_3 \rangle$ *be a proximity feature-structure.*

1. *The pairs* (c, d), (e, f) *are factorially comparable if their components lie in a common Cartesian product bounded by some* (a, b), *that is,*

$$a - b \succsim_1 c - d, e - f$$
$$b - a \succsim_2 d - c, f - e$$
$$a \cap b \succsim_3 c \cap d, e \cap f.$$

2. *A proximity feature-structure is factorially connected if for any two pairs* $(c, d) \succsim (e, f)$ *there exist* $(a_0, b_0), \ldots, (a_n, b_n)$ *with* $(a_0, b_0) = (c, d)$ *and* $(a_n, b_n) = (e, f)$ *such that* $(a_{i-1}, b_{i-1}) \succsim (a_i, b_i)$, $i = 1, \ldots, n$, *and all adjacent pairs in the sequence are factorially comparable.*

Factorial connectedness assumes that any two pairs of elements are connected by a finite order-preserving chain in which successive pairs belong to a common Cartesian product. It assumes, in effect, that the image of the representation is a convex set (see Figure 6 on p. 275 of Volume I). This assumption permits the extension of conjoint measurement to a proper subset of a Cartesian product. Although we have developed the notion of factorial connectedness in the context of proximity measurement, it is equally applicable to other instances of additive conjoint measurement. Thus it provides a simple answer to the question posed in Section 6.5.3 of Volume I regarding the extension of conjoint measurement to nonfactorial structures (see also Miyamoto, 1988).

We say that (a, b) and (c, d) *agree* on the first component if $a - b \sim_1 c - d$. Agreement on the second and third components is defined similarly. With this definition we can introduce the remaining conjoint-measurement axioms.

DEFINITION 19. *Let* $\langle S, A, \succsim, \succsim_1, \succsim_2, \succsim_3 \rangle$ *be a proximity feature-structure.*

1. *It is* independent *iff the following condition is satisfied. Suppose the pairs* (a, b) *and* (c, d) *agree on two components, and the pairs* (a', b') *and* (c', d') *also agree on the same two components. Furthermore, suppose the pairs* (a, b) *and* (a', b') *agree on the remaining (third) component, and the pairs* (c, d) *and* (c', d') *also agree on the remaining (third) component. Then,*

$$(a, b) \succsim (a', b') \qquad iff \qquad (c, d) \succsim (c', d').$$

2. *It satisfies* restricted solvability *iff* $(a, b) \succsim (c, d) \succsim (e, f)$ *implies that there exist* (p, q) *in* A^2 *such that* $(p, q) \sim (c, d)$, *and* (p, q) *agrees with* (a, b) *and* (e, f) *on all components on which they agree with each other.*

3. *It is* Archimedean *iff for all* $a, b, c, d \in A$, $a \neq b$, *any sequence* $e^{(k)} \in A$, $k = 0, 1, \ldots$, *that agrees throughout on two components, such that* $e^{(0)} = c$, $(e^{(k)}, e^{(k+1)}) \succ (a, b)$ *and* $(c, d) \succ (c, e^{(k+1)}) \succ (c, e^{(k)})$ *for all* k *is finite.*

As we shall see below, a proximity feature-structure that is factorially connected and satisfies independence, restricted solvability, and the Archimedean condition has a representation of form $\varphi_1 + \varphi_2 + \varphi_3$. The following definitions impose additional constraints on the relations between these scales.

DEFINITION 20. *A proximity feature-structure is* uniform *iff* $S_D = S_C$ *and* $\succsim_1$, $\succsim_2$, $\succsim_3$ *are identical.*

Uniformity says that the salience ordering of feature sets is the same for all three components. This assumption can be tested, provided the relevant features can be identified and controlled. Suppose, for example, that adding beards to a pair of faces increases their similarity more than adding moustaches. Uniformity implies, then, that adding a beard to one face only should decrease the similarity between them more than adding a moustache to one face only (see Gati and Tversky, 1984). Naturally, uniformity implies that all φ_i must be monotonically related. The next definition introduces a stronger assumption that ensures that all φ_i are essentially proportional.

DEFINITION 21. *Consider an independent proximity feature-structure. For distinct* $i, j \in \{1, 2, 3\}$, *let* $\succsim_{ij}$ *be the ordering induced on a subset of* $X_i \times X_j$ *by holding the remaining (third) component fixed. The structure is* invariant *iff it is uniform and for any distinct* i, j *and any distinct* k, l *in* $\{1, 2, 3\}$

$$\text{if} \quad xy \sim_{ij} x'y' \quad \text{and} \quad x'y \sim_{ij} x''y',$$
$$\text{then} \quad xz \sim_{kl} x'z' \quad \text{iff} \quad x'z \sim_{kl} x''z'.$$

The assumption of invariance asserts, in effect, that if x, x', x'' is a standard sequence in component i relative to component j, then it should also form a standard sequence in component k relative to component l. A similar condition was used in Theorem 6.15, Section 6.11.2, to obtain linearly related scales in an additive conjoint measurement model.

Finally, we introduce a condition that ensures that φ is finitely additive.

DEFINITION 22. *A proximity feature-structure is* extensive *iff each of the following is true whenever* $x \subseteq S$ *is disjoint from* $a \cup b \cup c \cup d$ *and the required unions with x are elements of A.*

$$(a, b) \succsim (c, d) \quad \textit{iff} \quad (a \cup x, b) \succsim (c \cup x, d),$$
$$\textit{iff} \quad (a, b \cup x) \succsim (c, d \cup x),$$
$$\textit{iff} \quad (a \cup x, b \cup x) \succsim (c \cup x, d \cup x).$$

The consequences of the preceding assumptions are summarized below.

THEOREM 11. *Suppose* $\langle S, A, \succsim, \succsim_1, \succsim_2, \succsim_3 \rangle$ *is a proximity feature-structure (Def. 16) that is partially factorial (Def. 17), factorially connected (Def. 18), and satisfies independence, restricted solvability, and the Archimedean axiom (Def. 19).*

(i) *Then there exist nonnegative functions* φ_i *on* X_i, $i = 1, 2, 3$ *such that* $(a, b) \succsim (c, d)$ *iff*

$$\varphi_1(a - b) + \varphi_2(b - a) - \varphi_3(a \cap b)$$
$$\geq \varphi_1(c - d) + \varphi_2(d - c) - \varphi_3(c \cap d).$$

Furthermore, the φ_i *are interval scales with a common unit.*

(ii) *If the structure is uniform (Def. 20), then* $\varphi_i = H_i(\varphi)$ *for some function* φ *on* $X_1 = X_2 = X_3$, *and for some increasing functions* H_1, H_2, H_3.

(iii) *If the structure is invariant (Def. 21), then* $\varphi_i = \alpha_i \varphi$, $\alpha_i \geq 0$, $i = 1, 2, 3$.

(iv) *If the structure is extensive (Def. 22), then each* φ_i *is finitely additive, and the* φ_i *are ratio scales.*

The proof of Theorem 11 is presented in Section 14.8.1. It establishes a hierarchy of models which imposes increasing constraints on the representation of a proximity order between objects in terms of their common and distinctive features. The general additive model (i) imposes no constraints on the relation among the scales whereas in (ii) and (iii) all φ_i are related to each other via a monotone or a linear transformation, respectively. Finally, (iv) imposes feature additivity, that is, $\varphi_i(x \cup y) = \varphi_i(x) + \varphi_i(y) - \varphi_i(x \cap y)$. An alternative axiomatization of the latter model was developed by Osherson (1987).

The identifiability of the coefficients was investigated by Sattath and Tversky (1987) in the case for which φ is an additive measure. They observed that, given a particular feature-structure, different coefficients give rise to different proximity orderings. However, if there exists an additive measure φ such that for all $a, b \in A$

$$\delta(a, b) = \alpha\varphi(a - b) + \alpha\varphi(b - a) - \theta\varphi(a \cap b)$$

for some $\alpha, \theta \geq 0$, then for any $\alpha', \theta' \geq 0$, there exists an additive measure φ' such that

$$\delta(a, b) = \alpha'\varphi'(a - b) + \alpha'\varphi'(b - a) - \theta'\varphi'(a \cap b).$$

The two measures, φ and φ', are not linearly related, and they do not have the same support. (The support of a measure is the set of elements for which it is nonzero.) For proof and further discussion, see Sattath and Tversky (1987).

14.7.2 Empirical Applications

In this section we discuss briefly some applications of the contrast model to the analysis of psychological phenomena that are not readily accommodated by the traditional metric models. In particular, we use the contrast model to analyze (a) asymmetric similarities, (b) the relation between similarity and dissimilarity judgments, and (c) the weighting of common and distinctive features.

The application of the contrast model to empirical data requires some assumptions about the feature-structure under study. The theory developed in the previous section assumed a known feature-structure. In practice, however, the feature-structure (like the dimensional structure in multidimensional models) is not fully specified in advance, which limits the applicability of the model. However, as will be illustrated, the contrast model can be tested under fairly weak assumptions about the underlying feature-structure. For further discussion of feature representations, see Tversky (1977) as well as Gati and Tversky (1984).

Directionality and asymmetry. According to the contrast model, similarity (and dissimilarity) are not necessarily symmetric. It follows readily from (18) that $(a, b) \sim (b, a)$ if either $\varphi(a - b) = \varphi(b - a)$ or $\alpha = \beta$. Thus $(a, b) \succ (b, a)$ whenever (i) $\varphi(b - a) > \varphi(a - b)$ and (ii) $\alpha > \beta$. We interpret (i) as a characteristic of the objects and (ii) as a characteristic of the task. In particular, $\varphi(b - a) > \varphi(a - b)$ (which is equivalent to $\varphi(b) > \varphi(a)$ in the additive case) implies that b is more prominent or salient than a. The inequality $\alpha > \beta$ asserts that the distinctive features of the first argument, called the subject or the topic, loom larger than the distinctive features of the second argument, called the referent. (In the statement "a is more similar to b than c is to d," denoted $(a, b) \prec (c, d)$, a and c serve as subjects whereas b and d serve as referents.)

Naturally, the asymmetry depends on the directionality of the task that is reflected in the discrepancy between α and β. Compare the following instructions: (i) assess the degree to which a and b are similar to each other, and (ii) assess the degree to which a is similar to b. In (i) the task is nondirectional; hence, it essentially imposes symmetry. In (ii), on the other hand, the task is directional; hence, asymmetry might be observed.

If we assume that the subject of a directional comparison looms larger than the referent (i.e., $\alpha > \beta$), then the contrast model implies that the less prominent object should be closer to the more prominent object than vice versa. This prediction is consistent with the observation that in normal discourse we tend to select the more salient object (or the prototype) as a referent and the less salient object (or the variant) as the subject. Thus we say "the son resembles the father" rather than "the father resembles the son," and we say "an ellipse is like a circle" rather than "a circle is like an ellipse."

Moreover, studies of similarity have shown that the asymmetry in the *choice* of similarity statements is associated with asymmetry in *judgments* of similarity. Rosch (1975) showed that the prototype is generally less similar to a variant than vice versa. For example, the perceived distance of a focal red to an off-red was greater than the perceived distance of an off red to a focal red. Tversky and Gati (1978) obtained the same pattern in both perceptual and conceptual comparisons. In a study of similarity between countries they observed, for example, that the judged similarity of North Korea to China exceeded the judged similarity of China to North Korea; and in a study of similarity of geometric forms they observed that an ellipse was judged more similar to a circle than a circle to an ellipse. The direction of the observed asymmetry, therefore, is predicted by the relative prominence of the objects as established by an independent empirical procedure, such as the choice of subject and referent in similarity statements (see Tversky, 1977).

As was noted earlier, the contrast model expresses the dissimilarity of objects as a linear combination of the measures of their common and their distinctive features, with parameters α, β, and θ. The preceding discussion of asymmetry focuses on the relation between α and β that reflects the directionality of the comparison. In the remainder of this section we confine our attention to the symmetric case by assuming $\alpha = \beta = 1$ and focus on θ, which reflects the weight of the common relative to the distinctive features.

Similarity versus dissimilarity. Because dissimilarity is generally viewed as the opposite of similarity, it has been commonly assumed that the two orderings are mirror images. However, semantic considerations suggest that, in judging the similarity of a to b, the respondent attends primarily to the features they share whereas the dissimilarity of a to b is influenced primarily by the features that distinguish them. As a consequence, common features loom larger in similarity than in dissimilarity judgments whereas distinctive features exhibit the opposite pattern.

Let $\succsim_S$ and $\succsim_D$ denote, respectively, the similarity and the dissimilarity orders defined on the same feature-structure. If both orders can be represented by the contrast model (18) with the same scale φ, and $\alpha = \beta = 1$, then the orders can differ only by θ. The preceding discussion suggests the hypothesis that $\theta_S > \theta_D$, where each subscript refers to the respective order. Note that if θ_S is much larger than one, $\succsim_S$ is determined primarily by the common features, and if θ_D is much smaller than one, $\succsim_D$ is determined primarily by the distinctive features. (Recall that $\alpha = \beta = 1$.) Therefore, if the discrepancy between θ_S and θ_D is sufficiently large, it should be possible to identify a pair of prominent objects (a, b) and a pair of nonprominent objects (c, d) such that $(a, b) \succsim_D (c, d)$ and $(a, b) \succsim_S (c, d)$.

Tversky and Gati (1978) confirmed this pattern in the comparison of pairs of well-known countries with pairs of countries that were less known to the respondents. For example, most respondents in the similarity group selected East Germany and West Germany as more similar to each other than Sri Lanka and Nepal whereas most respondents in the dissimilarity group selected East Germany and West Germany as more different from each other than Sri Lanka and Nepal. Evidently, the relative weight of the common features is larger in judgment of similarity than in judgment of difference, as evidenced by the nonmonotonicity of the two orders.

Weighting common and distinctive features. Consider a pair of stimuli a, b (e.g., schematic faces) and a separable feature x (e.g., glasses) that could be added to them. Let $s(a, b)$ denote the similarity between a and b. Adding x to both a and b tends to increase similarity, and this increment provides an estimate of the impact of x as a common feature, denoted $C(x)$. That is, $C(x) = s(a \cup x, b \cup x) - s(a, b)$, where $a \cup x$ is the concatenation of a and x. (Recall that, under the standard interpretation of x as a binary additive dimension, the common observation $s(a \cup x, b \cup x) > s(a, b)$ violates translation invariance; see Section 14.6.2.) Adding x to one object only tends to decrease similarity, and this decrement provides an estimate of the impact of x as a distinctive feature, denoted $D(x)$. That is, $D(x) = s(a, b) - s(a \cup x, b)$.

Gati and Tversky (1984) constructed several sets of verbal stimuli (e.g., descriptions of persons, meals, or trips) and pictorial stimuli (e.g., schematic

faces or landscapes) with separable additive components and estimated
C and D for more than 50 features. The data revealed a highly consistent pattern: C was greater than D for almost all the verbal stimuli whereas
D was greater than C for almost all the pictorial stimuli. Thus common
features loom larger in the verbal stimuli whereas the distinctive features
loom larger in the pictorial stimuli. Evidently, the relative weight of these
components depends on the nature of the task (similarity vs dissimilarity)
as well as on the nature of the comparison (verbal vs pictorial).

14.7.3 Comparing Alternative Representations

Attempts to test and evaluate proximity models have led several investigators to compare alternative representations of the same data set, particularly a two-dimensional Euclidean solution and an additive, or an ultrametric tree (see, e.g., Fillenbaum and Rapoport, 1971; E. Johnson and Tversky,
1984; Pruzansky, Tversky, and Carroll, 1982; Shepard, 1974, 1980). These
models have been selected because they are the simplest nontrivial examples, respectively, of the spatial and the clustering approaches to the
representation of proximity, because they are easy to display and interpret,
and because they have a similar number of free parameters. In this section
we illustrate the two representations, discuss their interpretation, and
explore the use of some statistical indices to diagnose which representation,
if any, is appropriate for a particular data set.

As an example, we have selected a study by Henley (1969) who reported
mean ratings of dissimilarity between 30 animals for a homogeneous group
of 18 respondents. Figure 4 displays the two-dimensional Euclidean representation of these data in which the order of the interpoint distances in the
plane represents the dissimilarity order between the respective animals.
Although the precise conditions for the existence of such a representation
are not known, there are several effective computer programs (see, e.g.,
Carroll and Arabie, 1980; Coxon, 1982; Davison, 1983; Schiffman et al.,
1981) that construct a spatial solution of a given dimensionality (typically 2
or 3) by maximizing some index of goodness-of-fit. The resulting solution is
commonly used to infer the spatial relations among the objects and assess
their dimensional structure. For example, Henley (1969) interpreted the
horizontal dimension in Figure 4 as size, with elephant and mouse at the
two extremes, and the vertical dimension as ferocity with wild animals at
the top and domesticated animals at the bottom. This interpretation can be
tested, e.g., by obtaining independent ratings of the animals on the hypothesized dimensions.

Figure 5 represents the same data set as a (rooted) additive tree. Recall
that a tree is a connected graph without cycles. In an additive tree, the
objects appear as terminal nodes, and the dissimilarity between objects is

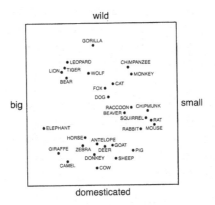

FIGURE 4. Two-dimensional Euclidean representation (SSA) of the dissimilarity between animals (Henley, 1969).

represented by the horizontal length of the path that joins them; the vertical lines in Figure 5 are introduced for graphical convenience. A rooted additive tree can be interpreted as a feature tree in which each arc corresponds to the set of features shared by the objects that follow from this arc, and the length of each arc represents the measure of the respective set of features (Sattath and Tversky, 1977; Tversky, 1977). The terminal arcs correspond to the unique features of the objects under study (see Figure 5), and the length of the arc labeled *rodents*, for example, corresponds to the measure of this feature or cluster. Thus the additive tree is a special case of the additive contrast model (Section 14.7.1), in which $\theta = 0$ and the feature-structure is nested in the sense that any two clusters of objects (i.e., branches of the tree) are either disjoint or one includes the other. Note that the choice of a root in an additive tree, like the choice of a coordinate system in the Euclidean model, does not affect the distances though it is often important for the interpretation of the structure. The labeling of the dimensions and the clusters, therefore, is merely suggestive. Figure 5, for example, suggests the presence of four distinct clusters labeled *rodents, apes, carnivores,* and *herbivores,* which are further subdivided to smaller clusters.

It is easy to verify (Exercise 8) that any four points a, b, c, d in an additive tree satisfy the following *tree inequality* (Section 12.3.4),

$$\delta(a, b) + \delta(c, d) \leqslant \max\{\delta(a, c) + \delta(b, d), \delta(a, d) + \delta(b, c)\},$$

where δ is the (path-length) distance between the respective nodes. Because

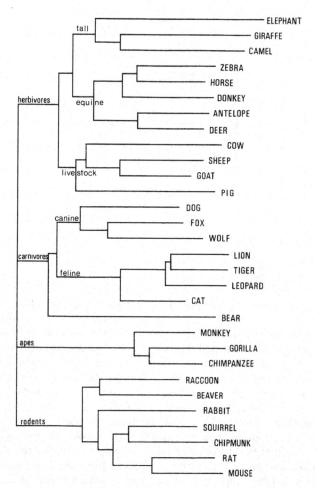

FIGURE 5. Additive tree representation (ADDTREE) of the dissimilarity between animals (Henley, 1969).

a tree is a connected graph without cycles, each pair of nodes is joined by a unique path. The tree inequality implies that the nodes of a tree can be labeled so that

$$\delta(a, b) + \delta(c, d) \leqslant \delta(a, c) + \delta(b, d) = \delta(a, d) + \delta(b, c).$$

As was demonstrated by several authors (see, e.g., Buneman, 1971, 1974;

Dobson, 1974; Hakimi and Yau, 1964; Patrinos and Hakimi, 1972), the tree inequality (also called the four-point condition) is not only necessary but also sufficient for the representation of a symmetric distance matrix by an additive tree. Note that the tree inequality is not ordinal because it is formulated in terms of the ordering of sums. Because the equations are linear, however, an ordinal analogue to the tree inequality can be developed using the methods of Chapter 9.

The tree inequality generalizes the *ultrametric inequality* (Section 12.3.4),

$$\delta(a, b) \leqslant \max\{\delta(a, c), \delta(b, c)\},$$

that is both necessary and sufficient for an ultrametric tree in which all terminal nodes are equidistant from the root (see, e.g., Hartigan, 1975; Jardine and Sibson, 1971; S. C. Johnson, 1967). This model, however, is quite restrictive because it implies that for any three points, the two larger distances are equal; that is, $\delta(a, b) \leqslant \delta(b, c) = \delta(a, c)$ for some labeling of the nodes.

The comparison of tree and planar representations is often instructive. In many cases, they fit the proximity order about equally well, but they highlight different aspects of the data. For example, the spatial solution of Henley's data (Figure 4) emphasizes the global dimensions (e.g., size) that do not emerge from the tree (Figure 5) whereas the tree provides a sharper view of the clustering of the animals. In other cases, the data appear to favor one representation over another (Holman, 1972). Pruzansky *et al.* (1982) compared the two models using many different data sets and found that perceptual data (e.g., similarity of colors and confusion between letters) tend to favor the spatial model whereas the conceptual data (e.g., similarity of animals and occupations) tend to favor the tree model. E. Johnson and Tversky (1984) investigated lay perceptions of risks (e.g., diseases, accidents, and natural hazards) using judgments of similarity and the ratings of each risk on nine evaluative dimensions (e.g., controllability, dread, voluntary). The analysis of these data showed that an additive tree accounts for the similarity judgments better than a two- (and even a three-) dimensional Euclidean model. The spatial model, on the other hand, fits the correlations between the risks (computed across the nine evaluative dimensions) better than the additive tree. These observations suggest that the appropriateness of a representation depends both on the nature of the task and on the definition of proximity.

Besides the evaluation of goodness-of-fit and the test of critical axioms, some authors have explored statistical properties of the observed and recovered proximities that can be used to diagnose which representation is more appropriate for a given data set. For example, Sattath and Tversky (1977) observed that a "typical" configuration of points in the plane tends to produce many small distances and fewer large distances whereas a "typical" tree tends to produce many large distances and fewer small distances. Using both real and synthetic data, Pruzansky *et al.* (1982) showed that this property can help diagnose whether the data are better represented by a tree or by a plane.

Another class of diagnostics for multidimensional scaling is based on the relation of nearest neighbor (Tversky, Rinott, and Newman, 1983). Given a symmetric dissimilarity order without ties, b is the *nearest neighbor* of a iff $(a, b) \prec (a, c)$ for all $c \neq b$ in A. Suppose A is finite, and let N_a be the number of elements in A whose nearest neighbor is a. Thus N_a measures the "centrality" of a with respect to A. Note that the average value of N_a, for all $a \in A$, equals 1.

Spatial models impose an upper bound on the number of points that can share the same nearest neighbor. For example, it is easy to see that a point in the plane can serve as the nearest neighbor of at most five points corresponding to the vertices of a regular (or nearly regular) pentagon. In a regular hexagon, the distance between the center and a vertex is not smaller than the distance between adjacent vertices. Thus N_a cannot exceed 5 for any $a \in A$. In the three-dimensional Euclidean model, N_a must be less than 12. High values of N_a, therefore, can be used to rule out low-dimensional representations. Tversky and Hutchinson (1986) have analyzed 100 proximity matrices reported in the literature and found that most perceptual data satisfy the geometric bound, but many conceptual data sets do not. The most striking discrepancies between the data and their multidimensional solutions arise in semantic fields when the object set includes a focal element (e.g., a superordinate category) that is the nearest neighbor of many of its instances. For example, the category name *bird* was the most frequent associate (i.e., the nearest neighbor) of all its instances. Such data cannot be represented in a low-dimensional space, but they can be readily accommodated by a tree because a (short) branch can be the nearest neighbor of any number of branches. A more restrictive bound (that is essentially independent of the dimensionality of the space) is implied by the assumption that the data points represent a sample from some continuous distribution in a multidimensional Euclidean space. For further details and discussion of the use of nearest neighbor analysis, see Tversky and Hutchinson (1986) and Tversky *et al.* (1983).

14.8 PROOFS

14.8.1 Theorem 11

A proximity feature-structure (Def. 16) that is partially factorial (Def. 17) and factorially connected (Def. 18), and satisfies independence, restricted

solvability, and the Archimedean axiom (Def. 19), has a representation of the form $\delta(a, b) = \varphi_1(a - b) + \varphi_2(b - a) - \varphi_3(a \cap b)$, *where the* φ_i *are interval scales.*
If the structure is uniform (Def. 20), the φ_i *are monotonically related; if the structure is invariant (Def. 21), the* φ_i *are linearly related; if the structure is extensive (Def. 22), each* φ_i *is finitely additive, and the* φ_i *are ratio scales.*

PROOF. For any $(a, b) \in A^2$, let $Y(a, b)$ be the Cartesian product bounded by (a, b). The structure $\langle Y(a, b), \succsim \rangle$ is an additive factorial proximity structure, in the sense of Theorem 7, with three essential components. We review the hypotheses of this theorem in turn.

Weak order: transitivity follows from Definition 16 and the definition of the induced ordering whereas connectedness follows from Definition 17, which guarantees that any two elements of $Y(a, b)$ in the image of A^2.

Independence, restricted solvability, and the Archimedean axiom follow, respectively, from Parts 1, 2, and 3 of Definition 19. The requirement of three essential factors follows from Part 4 of Definition 16.

By Theorem 7, therefore, there exists a real-valued function φ_i defined on Y_i, $i = 1, 2, 3$, such that for all pairs $(c, d), (e, f) \in Y(a, b)$,

$$(c, d) \succsim (e, f) \qquad \text{iff}$$
$$\varphi_1(c - d) + \varphi_2(d - c) - \varphi_3(c \cap d)$$
$$\geq \varphi_1(e - f) + \varphi_2(f - e) - \varphi_3(e \cap f),$$

where the φ_i are interval scales with a common unit.

The uniqueness of the representations for each $Y(a, b)$ ensures the agreement among the scales of the overlapping factorial structure (see Section 6.5.3). To extend the representation beyond this domain, suppose $(c, d) \succsim (e, f)$ but the two pairs do not belong to the same Cartesian product. By factorial connectedness (Definition 18) there exists $(a_0, b_0), \ldots, (a_n, b_n)$ with $(a_0, b_0) = (c, d)$ and $(a_n, b_n) = (e, f)$ such that $(a_{i-1}, b_{i-1}) \succsim (a_i, b_i)$, $i = 1, \ldots, n$, and all adjacent pairs are factorially comparable. That is, (a_{i-1}, b_{i-1}) and (a_i, b_i) lie in a common Cartesian product bounded by some (a, b). Since the representation of each $Y(a, b)$ is order-preserving, it is extendable to all of Y.

If the structure is uniform (Definition 20) so that the $\succsim_i$ are identical, $i = 1, 2, 3$, it follows readily that the φ_i are strictly monotonic. If the structure is invariant (Definition 21), then

$$\varphi_i(x) - \varphi_i(y) = \varphi_i(u) - \varphi_i(v) \qquad \text{implies}$$
$$\varphi_j(x) - \varphi_j(y) = \varphi_j(u) - \varphi_j(v).$$

By the uniqueness of the representation, therefore, the φ_i are linearly related.

Finally, if the structure is extensive (Definition 22), then

$$\varphi_1(x) + \varphi_2(y) - \varphi_3(z) \geqslant \varphi_1(x') + \varphi_2(y') - \varphi_3(z') \qquad \text{iff}$$
$$\varphi_1(x \cup w) + \varphi_2(y) - \varphi_3(z) \geqslant \varphi_1(x' \cup w) + \varphi_2(y') - \varphi_3(z'), \qquad \text{etc.,}$$

for all $w, x, x', y, y', z, z' \subseteq S$ for which all φ_i are defined and the sets $w \cap (x \cup y \cup z)$ and $w \cap (x' \cup y' \cup z')$ are empty. It follows readily from the above that $\varphi_i(x) = \varphi_i(y)$ iff $\varphi_i(x \cup w) = \varphi_i(y \cup w)$, $i = 1, 2, 3$, provided $w \cap (x \cup y)$ is empty. Consequently, $\varphi_i(x \cup w)$ can be expressed as $F_i[\varphi_i(x), \varphi_i(w)]$, where F_i is symmetric and increasing in each argument. Letting $z = z'$ in the above equation yields

$$\varphi_1(x) - \varphi_1(x') = \varphi_2(y') - \varphi_2(y) \qquad \text{iff}$$
$$\varphi_1(x \cup w) - \varphi_1(x' \cup w) = \varphi_2(y') - \varphi_2(y)$$

for all w disjoint from $x \cup x'$. Hence,

$$F_1[\varphi_1(x), \varphi_1(w)] - F_1[\varphi_1(x'), \varphi_1(w)] = \varphi_1(x) - \varphi_1(x'),$$

or in a simplified form, $F(\alpha, \gamma) - F(\beta, \gamma) = \alpha - \beta, \alpha, \beta, \gamma \geqslant 0$. Letting $\beta = \gamma$ yields $F(\alpha, \beta) - F(\beta, \beta) = \alpha - \beta$. Similarly $F(\beta, \alpha) - F(\alpha, \alpha) = \beta - \alpha$; but since $F(\alpha, \beta) = F(\beta, \alpha)$ by the commutativity of $\cup$, $F(\beta, \beta) + \alpha - \beta = F(\alpha, \alpha) + \beta - \alpha$. Hence, $F(\beta, \beta) - 2\beta = F(\alpha, \alpha) - 2\alpha = k$ since it is independent of α and β. Consequently,

$$F(\alpha, \beta) = F(\beta, \beta) + \alpha - \beta = (k + 2\beta) + \alpha - \beta = \alpha + \beta + k.$$

Setting $k = 0$ yields $\varphi_1(x \cup w) = \varphi_1(x) + \varphi_1(w)$, provided $x \cap w$ is empty. The same argument applies to the other two components, which completes the proof of the theorem. $\diamondsuit$

EXERCISES

1.   Let A be the set of all intervals on the real line. For any $x = (x_1, x_2)$ and $y = (y_1, y_2)$ in A with $x_1 < x_2$ and $y_1 < y_2$, define

$$\delta(x, y) = \begin{cases} |x_1 - x_2| + |y_1 - y_2| & \text{if either } x_2 \leqslant y_1 \text{ or } y_2 \leqslant x_1 \\ |x_1 - y_1| + |x_2 - y_2| & \text{otherwise.} \end{cases}$$

Show that $\langle A^2, \delta \rangle$ is a metric space with additive segments.   (14.2)

2. Verify that independence (Definition 11) implies one-factor independence (Definition 6), and show that the converse is not true.   (14.4.3)

3. For any $x, y \in \mathrm{Re}^n$, let

$$\sigma(x, y) = F^{-1}\left[\sum_{i=1}^{n} F(|x_i - y_i|) \right],$$

where

$$F(\alpha) = \begin{cases} \alpha & \text{for } 1 \geqslant \alpha \geqslant 0 \\ 1 + p(\alpha - 1) & \text{for } \alpha > 1 \text{ for some } p > 1. \end{cases}$$

Show that σ violates the triangle inequality.   (14.4.5)

4. Let $F_1(\alpha) = \alpha^p e^{q\alpha^r}$, $p \geqslant 1$, $q, r \geqslant 0$, $F_2(\alpha) = \sinh \alpha$, and $F_3(\alpha) = p(e^{q\alpha} - 1)^r$, $p, q > 0$, $r \geqslant 1$. Show that substituting F_i, $i = 1, 2, 3$, in the preceding expression for $\sigma(x, y)$ (see Exercise 3) yields a metric distance function.   (14.4.5)

5. Show that if the function $\log F(e^\alpha)$ has no points of inflection, then its convexity is necessary for the compatibility of the respective additive–difference model with a metric; see Theorem 9.   (14.4.5)

6. Show that (one-factor) independence [Equation (17)] is a necessary condition for decomposability.   (14.6.2)

7. Show that if $\varphi(a) = \varphi(b)$ for all $a, b \in A$ and φ is additive, then the common-features model can be expressed as a distinctive-features model and vice versa.   (14.7.1)

8. Show that any additive tree satisfies the tree inequality.   (14.7.3)

9. Let N_a be the number of elements in A whose nearest neighbor is a. Show that $\sum_{a \in A} N_a$ equals the cardinality of A.   (14.7.3)

10. Show that in the two-dimensional city-block model $\max N_a = 7$.   (14.7.3)

Chapter 15   Color and Force
Measurement

15.1 INTRODUCTION

This chapter concerns homomorphisms that map a set of infinite-dimensional vectors into a low-dimensional vector space. Such representations are quite different from those in the preceding two chapters. Previously, the objects in the empirical relational structure either had no internal structure (e.g., points in the foundations of geometry) or at most a Cartesian product structure; the vector representation was introduced on the basis of qualitative geometric properties. Here we begin with objects that have an extremely rich internal structure, namely, infinite-dimensional vectors. The representation theorem does not introduce an algebraic structure, but rather reduces such a structure to a manageable, low-dimensional size.

The infinite-dimensional objects with which we are concerned can be thought of as input functions to an operator. The reduction to low dimensionality is based on an equivalence relation: two input functions are equivalent if the operator acts on them to produce the same output. The two principal empirical examples are force and color. In the case of force, an input function specifies all the various force-magnitudes acting on various directions at a point. In the two-dimensional case, the domain of the input function is angle, $0 \leqslant \theta < 2\pi$ radians, and the value $a(\theta)$ of the input function a is the magnitude of the force acting in direction θ; in the three-dimensional case, the domain of the input function consists of directions specified by points on a sphere rather than a circle. The output is the resultant force. Two input configurations of force-magnitudes are deemed equivalent if they yield the same resultant, i.e., if they are in equilibrium with the same force. In the case of color, an input function specifies the radiant energy at different wavelengths; the corresponding output is perceived color, i.e., two inputs are equivalent—*metameric* is the technical term—if they match exactly in color. For both color and force, the usual homomorphism is onto vectors in three dimensions. This corresponds to resolution of forces into components in three independent spatial directions and, for color, to tristimulus coordinates or to three independent physiological codes for color.

It is important to realize that this reduction of infinite-dimensional vectors to a space of low dimensionability is a fact of nature, not a mathematical trick. For example, in the case of complex tones, an input function specifies the acoustic energy at various frequencies, and the corresponding output is the perceived tone. Two inputs can be said to be equivalent if they match exactly in loudness, pitch, and timbre (or tone quality). Formally, these definitions are parallel to those for color, but empirically there is an enormous difference. As far as anyone knows, there do not exist two drastically different distributions of acoustic energy that are judged as indistinguishable tones. The perception of tones is not reducible to a low-dimensional space in a way analogous to the representation of color.

Likewise in the case of force, the reduction of an infinite dimensional "force configuration" to a finite-dimensional resultant vector is an important law of physics. In this chapter we develop the precise statements of these laws for color and force.

Another difference between this and the previous chapter is that we use algebraic rather than metric properties of the representing vector space. This corresponds to the fact that the homomorphism simply preserves algebraic structure and carries the equivalence relation in the empirical structure into vector equality in the numerical representation.

The organization of this chapter is as follows. In the next section, we present two slightly different but related axiom systems characterizing *Grassmann structures* and *equilibrium structures*. The latter turn out (Theorem 2) to be equivalent to a subtype of the Grassmann structures. Theorem 3 gives a representation for Grassmann structures in an abstract vector space over the reals, and Theorem 4 deals with finite-dimensional representations. Proofs are provided in Section 15.3.

We offer in Section 15.4 a rather detailed interpretation for the axiomatic structure of color matching. Color matching is often misunderstood, and we make some effort to confront the most common confusions.

For both color and force, the usual vector representations are three-dimensional. In the case of color, this fact has been the basis of major theoretical interpretations dating to Thomas Young. By contrast, the fact that force vectors are three-dimensional has generally been regarded as an obvious consequence of the three-dimensionality of the physical world. In Section 15.5 we analyze the three-dimensionality of both color and force. In both cases, three-dimensionality derives from an underlying structure of one-dimensional components. The analogies and disanalogies between these two domains turn out to be surprising and useful. They lead us to formulate, at the end of Section 15.5, a measurement-theoretic description of the traditional program of color theory.

Finally, in Section 15.6, we axiomatize two approaches to color theory, the Koenig theory based on reduction dichromacy and the Hering/Hurvich/Jameson theory based on opponent colors. We cover in detail those portions of the theories in which the extensions of the measurement representation for trichromatic color matching are sufficiently simple and natural to permit straightforward axiomatic analysis. In particular, the analysis of opponent-colors theory uses one-dimensional equilibrium structures just as in the analysis of components of force in Section 15.5.3.

Section 15.7 collects together the proofs of the theorems of Sections 15.5 and 15.6.

The name *Grassmann structures* commemorates the original formulation of the axioms satisfied by metameric color matching (Grassmann, 1854). However, trichromacy of color matching was known before Grassmann—Brindley (1970) discusses the history—and was the basis for theoretical interpretation beginning with T. Young (1802). The effort to give a systematic account of color vision in terms of three underlying one-dimensional structures was begun by Helmholtz (1867, 1896) and Hering (1878, 1920). Jameson and Hurvich (1955) initiated the systematic measurement and theoretical use of one-dimensional color equilibria. Krantz (1975a, 1975b) gave the axiomatic formulation of Grassmann and equilibrium structures in color measurement.

In *Principia*, Newton used the three-dimensional vector representation for force. He recognized that an empirical law was involved but did not give it a precise formulation. An axiomatic analysis of one-dimensional resultant force was sketched by Euler (1736, 1912). Some subsequent analyses attempted to treat force as a defined concept (especially, Kirchhoff, 1876) or, at the opposite extreme, took a three-dimensional vector representation of force as a primitive concept (McKinsey, Sugar, and Suppes, 1953). Jammer (1957) recognized force configuration to be an important concept. Krantz (1973) adopted this concept as a primitive, treating resultant force as neither primitive nor defined but as a representation of a structure closely analogous to the structures of color measurement.

15.2 GRASSMANN STRUCTURES

15.2.1 Formulation of the Axioms

The primitive concepts of a Grassmann structure are a set A, a binary operation $\oplus$ on A, an operation $*$ from $\mathrm{Re}^+ \times A$ into A, and a binary relation $\sim$ on A. In the case of an equilibrium structure, the primitives are A, $\oplus$, $*$ as before and a subset E of A (i.e., a unary rather than a binary relation on A).

We discuss these primitives, and the axioms that follow, in terms of a particular interpretation, in which elements of A are two-dimension force configurations, the operation $\oplus$ concatenates two force configurations, and the operation $t*$ scales a force configuration by the factor t. The interpretation of the primitives in color vision will be presented in Section 15.4.

A device known as a *force table* (Figure 1) is sometimes used to demonstrate the conditions for static equilibrium in two dimensions. A number of pulleys are attached to the edge of a circular platform, and a small ring is placed at the center of the platform. To the ring are tied several light cords; each cord passes over a pulley and its free end hangs

FIGURE 1. A force table with an equilibrium configuration of five weights.

downward and is attached to a weight. If the weights, or the angles between the pulleys, are suitably adjusted, then the forces on the small ring are in equilibrium and the ring remains motionless above the center of the platform, without touching the center post.

The positions of the pulleys can be designated by angles relative to some arbitrary radius. A configuration of forces in this apparatus can be described by a function a whose values $a(\theta)$ are the magnitudes of the weights attached to the cords passing over the pulleys placed at various angles θ. For the force table, $a(\theta)$ is zero except for finitely many values of θ. A configuration a can be considered a vector with infinitely many components, the θth component being $a(\theta)$; only finitely many components are nonzero. Two configurations a, b can be concatenated by attaching both weights to the cord at angle θ, for each θ. If one of the configurations has zero weight at angle θ, then the addition involves only the weight from the other configuration at that angle. Thus $(a \oplus b)(\theta) = a(\theta) + b(\theta)$, or vectors are added componentwise. Likewise, scalar multiplication is defined by $(t * a)(\theta) = t[a(\theta)]$, for t in Re^+. Note that this interpretation uses only a nominal scale of angle, since θ is just the label of a position, but requires a ratio scale of weight based on extensive measurement.

For the force table, the subset E consists of just those configurations that are in equilibrium, leaving the ring motionless without touching the center post. The binary relation $\sim$ is defined in terms of E: configurations a and b are equivalent if for every configuration c, $a \oplus c$ is in E if and only if $b \oplus c$ is in E.

Some of the axioms involve only the triple $\langle A, \oplus, * \rangle$. These of course are exactly the same for both Grassmann and equilibrium structures. They can be summarized by saying that the triple $\langle A, \oplus, * \rangle$ is an abstract *convex cone of vectors*: $\oplus$ behaves like vector addition, $*$ like multiplication by positive scalars, and A is closed under both operations (see Definition 1). The term *convex cone* arises from the closure properties. The set A is *convex* because if a, b are in A, then for any $t, 0 < t < 1, (t * a) \oplus [(1 - t) * b]$ is also in A; it is a *cone* because it contains, with any vector a, the whole ray $\{t * a \mid t > 0\}$.

These axioms are valid for the (idealized) force table, as a consequence of the properties of extensive measurement of weight. This point will be elaborated following Definition 1.

The most interesting axioms from an empirical standpoint are those that involve $\sim$ or E as well as $\langle A, \oplus, * \rangle$. The relation $\sim$ is assumed to be an equivalence relation on A; and the $\oplus$, $*$ are assumed to be well defined for equivalence classes. (See Definition 2.)

For the force table or for equilibrium structures more generally, we require axioms for E that lead to these properties of the binary relation $\sim$ defined in terms of E. These are introduced in Definition 3. The precise statement of the axioms is contained in Definitions 1–3.

DEFINITION 1. *Let A be a nonempty set, $\oplus$ a binary function from $A \times A$ into A, and $*$ a binary function from $\mathrm{Re}^+ \times A$ into A. The triple $\langle A, \oplus, * \rangle$ is a convex cone iff the following two axioms hold:*

1. *$\langle A, \oplus \rangle$ is a commutative cancellation semigroup, i.e., $\oplus$ is associative and commutative and for all $a, b, c \in A$, if $a \oplus c = b \oplus c$, then $a = b$.*

2. *$*$ is a scalar multiplication on $\langle A, \oplus \rangle$, i.e., for all $a, b, \in A$ and $t, u \in \mathrm{Re}^+$,*
   (i) *$t*(a \oplus b) = (t * a) \oplus (t * b)$;*
   (ii) *$(t + u)*a = (t * a) \oplus (u * a)$;*
   (iii) *$t*(u * a) = (tu)* a$;*
   (iv) *$1 * a = a$.*

These axioms are approximately true for actual force tables and can be assumed to be idealizations. In Axiom 1, associativity, commutativity, and the cancellation property follow from the componentwise definition of $\oplus$ and from the corresponding properties for extensive measurement of weight. (Closure of concatenation can be difficult if the attached weights at a point become too heavy or if the angular separation between points at which weights should be attached becomes too small for the size of the pulleys.) Axiom 2 relies heavily on the numerical measurement of weight based on extensive measurement *onto* the positive reals. That is, $t * a(\theta)$ is defined in terms of the extensive scales as being the weight whose numerical value is t times the numerical value of $a(\theta)$. The actual properties in the axiom then follow, for each θ, from the properties of addition and multiplication of real numbers.

Note that these axioms are the same as those for abstract vector spaces over Re (Definition 12.1), with the following exceptions: instead of assuming that $\langle A, \oplus \rangle$ has an identity element and that inverses exist, we substitute the cancellation property in Axiom 1; and Axiom 2 is assumed only for positive scalars, rather than for all of Re. This definition characterizes convex cones in abstract vector spaces; for any such convex cone satisfies Axioms 1 and 2, and conversely, we have the following embedding theorem.

THEOREM 1. *If $\langle A, \oplus, * \rangle$ satisfies Axioms 1 and 2 of Definition 1, then there exists a vector space $\langle V, +, \cdot \rangle$ over Re, a convex cone C in V, and a* *function φ from A onto C, such that for all a, $b \in A$, $t \in \mathrm{Re}^+$, and $x \in V$, the following properties hold*:

(i) $\varphi(a \oplus b) = \varphi(a) + \varphi(b)$;

(ii) $\varphi(t * a) = t \cdot \varphi(a)$;

(iii) $x = \varphi(c) - \varphi(d)$ *for some* c, $d \in A$.

Moreover, if $\langle V', +', \cdot' \rangle$, $C' \subset V'$, *and* φ' *satisfy all the same properties, then there is a nonsingular linear transformation* ψ *of V onto V', which carries C onto C' such that* $\varphi' = \psi\varphi$.

Properties (i) and (ii) express that φ is an isomorphic embedding of A onto a convex cone in V; property (iii) guarantees that V is as small as possible—containing no irrelevant vectors—and thus assures uniqueness up to isomorphisms.

Theorem 1 is a simple corollary of the main theorem for Grassmann structures (Theorem 3) and is closely related to the classic embedding theorems of algebra (see discussion in Section 15.2.3).

DEFINITION 2. $\langle A, \oplus, *, \sim \rangle$ *is a* Grassmann structure *iff* $\langle A, \oplus, * \rangle$ *is a convex cone* (*Definition* 1), $\sim$ *is a binary relation on A, and for all a, b, $c \in A$ and $t \in \mathrm{Re}^+$, the following three axioms hold*:

1. $\sim$ *is an equivalence relation.*

2. $a \sim b$ *iff* $a \oplus c \sim b \oplus c$.

3.   *If $a \sim b$, then $t * a \sim t * b$.*

A Grassmann structure *is m-chromatic if m is a positive integer and if*

4.   (i)   *For any* $a_0, a_1, \ldots, a_m \in A$, *there exist* $t_i, u_i \in \mathrm{Re}^+$ *for* $0 \leqslant i \leqslant m$ *such that* $t_i \neq u_i$ *for at least one i and such that*

$$(t_0 * a_0) \oplus \cdots \oplus (t_m * a_m) \sim (u_0 * a_0) \oplus \cdots \oplus (u_m * a_m).$$

(ii) *There exist* $a_1, \ldots, a_m \in A$ *such that if* $t_i, u_i \in \mathrm{Re}^+$ *for* $1 \leqslant i \leqslant m$, *and*

$$(t_1 * a_1) \oplus \cdots \oplus (t_m * a_m) \sim (u_1 * a_1) \oplus \cdots \oplus (u_m * a_m),$$

then $t_i = u_i$ for all i.

In effect, Axiom 4(i) asserts that any $m + 1$ equivalence classes of vectors in A are linearly dependent, and 4(ii) asserts that there exist m linearly independent equivalence classes. It is shown in Section 15.4.3 that, for $m = 3$, Axiom 4 is equivalent to the usual statement of the law of trichromacy of color matching. Hence the name *m-chromacy* for the general property. In Section 15.6, 1-chromatic and 2-chromatic structures play major roles. We now define the second kind of structure dealt with in this section.

DEFINITION 3. $\langle A, \oplus, *, E \rangle$ *is an equilibrium structure iff* $\langle A, \oplus, * \rangle$ *is a convex cone* (*Definition* 1), *E is a nonempty subset of A, and for all* $a, b \in A$ *and* $t \in \text{Re}^+$, *the following three axioms hold*:

1. *There exists* $a' \in A$ *such that* $a \oplus a' \in E$.
2. *If* $a \in E$, *then* $b \in E$ *iff* $a \oplus b \in E$.
3. *If* $a \in E$, *then* $t * a \in E$.

The three axioms of Definition 3 are a set of physical laws satisfied by the idealized force table (with frictionless pulleys, weightless strings, etc.). For this interpretation, configuration a' in Axiom 1 can always be chosen to consist of just one pulley and weight, with direction opposite and magnitude equal, respectively, to the "resultant force" for configuration a. However, this is not a general feature of equilibrium structures. Axioms 2 and 3 are important qualitative laws of statics.

Note that in a three-dimensional analogue of the force table, in which pulleys are attached at different points on a sphere, Axioms 2 and 3 are false unless the central object that is pulled by the various weights is counterweighted to remove the effects of gravity. Without such counterweighting, the earth itself is an element of every configuration but cannot be concatenated with itself very easily! Counterweighting can easily be incorporated into the formal definitions by letting E' be the set of configurations for which actual equilibrium is observed (that is, the set of counterweighting configurations) and then defining E in terms of E' by

$$a \in E \text{ iff, for all } b \in A, a \oplus b \in E' \text{ iff } b \in E'.$$

This idea is developed in Exercise 4. Counterweighting is unnecessary on the force table because slight vertical components in the string tensions compensate for the gravitational force on the central ring.

The connection between Grassmann and equilibrium structures is given by the following definition and theorem.

DEFINITION 4. *In a Grassmann structure* $\langle A, \oplus, *, \sim \rangle$ *define a subset* E_- *of A by*

$$E_- = \{ a \mid \text{for all } b \in A, \quad a \oplus b \sim b \}.$$

A Grassmann structure is called proper *iff* $E_- = \varnothing$, *otherwise*, improper. *At the other extreme, the structure satisfies* solvability *iff for any* $a \in A$, *there exists* $a' \in A$ *such that* $a \oplus a' \in E_\sim$.

In an equilibrium structure $\langle A, \oplus, *, E \rangle$, *define a binary relation* $\sim_E$ *on* A *by*

$$a \sim_E b \text{ iff, for all } c \in A, \quad a \oplus c \in E \text{ iff } b \oplus c \in E.$$

THEOREM 2. *If* $\langle A, \oplus, *, \sim \rangle$ *is a Grassmann structure that satisfies* solvability, *then* $\langle A, \ominus, *, E_\sim \rangle$ *is an equilibrium structure. Conversely, if* $\langle A, \oplus, *, E \rangle$ *is an equilibrium structure, then* $\langle A, \oplus, *, \sim_E \rangle$ *is a Grassmann structure that satisfies solvability. The two constructions are inverses, i.e.*,

$$\sim_{(E_\sim)} = \sim \quad \text{and} \quad E_{(\sim_E)} = E.$$

The proof of this theorem is left as Exercise 3.

We say that an equilibrium structure is *m*-dimensional if and only if the associated Grassmann structure is *m*-chromatic. It is not difficult, but it is unnecessary, to translate *m*-dimensionality into an axiom stated directly in terms of $\oplus$, $*$, and E. We do so for the specific case of a one-dimensional equilibrium structure in Section 15.6; see also Exercise 2.

The force table (Figure 1) is two-dimensional. This is an empirical law that forms the basis for the usual statement that the configuration is an equilibrium structure if and only if the components of resultant force in any two independent directions are zero. The force sphere just described, with counterweighting, would be a three-dimensional equilibrium structure. In Section 15.5.3 we relate the dimensionality of the equilibrium structure to that of the kinematic space resulting from measurement of position and acceleration.

15.2.2 Representation and Uniqueness Theorems

THEOREM 3. *Let* $\langle A, \oplus, *, \sim \rangle$ *be a Grassmann structure. There exists a vector space* $\langle V, +, \cdot \rangle$ *over the real numbers, a convex cone* C *in* V, *and a function* φ *from* A *onto* C, *such that for all* a, $b \in A$, $t \in \text{Re}^+$, *and* $x \in V$, *the following properties hold*:

   (i) $\varphi(a \oplus b) = \varphi(a) + \varphi(b)$;

   (ii) $\varphi(t * a) = t \cdot \varphi(a)$;

   (iii) $x = \varphi(c) - \varphi(d)$ *for some* c, $d \in A$;

   (iv) $a \sim b$ *iff* $\varphi(a) = \varphi(b)$.

Moreover, if $\langle V', +', \cdot' \rangle$, C' in V', and φ' from A onto C' satisfy the same properties, then there is a nonsingular linear transformation ψ from V onto V' such that $\psi(C) = C'$ and $\varphi' = \psi\varphi$.

THEOREM 4. *Under the hypotheses of Theorem 3, the vector space $\langle V, +, \cdot \rangle$ is m-dimensional iff $\langle A, \oplus, *, \sim \rangle$ is m-chromatic.*

We note that Theorem 3 generalizes Theorem 1 by addition of the empirical relation $\sim$ and its representation by $=$ in V [condition (iv)].

We remark also that the Grassmann structure is proper if and only if the convex cone C does not contain the zero vector; it satisfies solvability (and hence, is equivalent to an equilibrium structure) if and only if $C = V$. An equilibrium structure is homomorphic to a structure $\langle V, +, \cdot, \{0\} \rangle$, where 0 is the identity of the vector space V.

Theorems 3 and 4, applied to two- or three-dimensional equilibrium structures defined by forces, justify the usual procedure of treating a configuration of forces by means of the calculated resultant vector in two or three dimensions, with concatenation of configurations corresponding to vector addition of their resultants (the parallelogram law).

15.2.3 Discussion of Proofs of Theorems 3 and 4

We start by giving a brief outline of the proof of a combined theorem: that every m-chromatic Grassmann structure is homomorphic to a convex cone in Re^m. This can be proved by using m-chromacy from the outset to introduce coordinates.

Let $a_1, \ldots, a_m$ be a set of elements of A that are linearly independent in the sense stated in Axiom 4(ii) of Definition 2. Now, for any a in A, let a play the role of a_0 in Axiom 4(i); therefore, there exist numbers t, u and $t_i, u_i, 1 \leqslant i \leqslant m$ in Re^+, with not all the t's equal to the u's, such that

$$(t * a) \oplus \sum_{i=1}^{m} (t_i * a_i) \sim (u * a) \oplus \sum_{i=1}^{m} (u_i * a_i). \tag{1}$$

(For clarity, we use the Σ notation for sums involving $\oplus$.) It is impossible that $t = u$; for in the contrary case, we could cancel $t * a$ from both sides of Equation (1) (Axiom 2 of Definition 2) obtaining

$$\sum_{i=1}^{m} (t_i * a_i) \sim \sum_{i=1}^{m} (u_i * a_i')$$

and by choice of the a_i, this implies $t_i = u_i$ for $i = 1, \ldots, m$, making the entire Equation (1) trivial. Since $t \neq u$, we introduce coordinates for a from Equation (1).

$$\varphi_i(a) = \frac{u_i - t_i}{t - u}, \quad i = 1, \ldots, m;$$
$$\varphi(a) = (\varphi_1(a), \ldots, \varphi_m(a)). \tag{2}$$

Thus, φ is a mapping of A into Re^m. It is straightforward to show that φ satisfies (i)–(iv) of Theorem 3. The definition of φ, in Equation (2), is based on "subtracting" $\Sigma(t_i * a_i)$ from both sides of Equation (1), and similarly "subtracting" $u * a$, to get an "equation" for a in terms of the a_i:

$$(t - u) * a \sim \sum_{i=1}^{m} (u_i - t_i) * a_i.$$

This equation is, of course, meaningless unless $t > u$ and $u_i > t_i$, $i = 1, \ldots, m$; but it is a heuristic that generates a meaningful definition of coordinates, by Equation (2). Note that Axioms 1–3 of Definition 2 are not used at all to define the representation φ; but, of course, they must be used heavily to show that it has the desired properties.

In Section 15.3, we take a quite opposite course: using more abstract methods, we define the representation in a general (possibly infinite-dimensional) vector space using only the axioms of a Grassmann structure (1–3 of Definition 2, plus, of course, the properties of a convex cone, Definition 1). The m-chromacy is not introduced until the proof of Theorem 4, to fix the dimensionality of the representation at m.

The basic idea of this latter proof is to perform "subtractions" like those used above in the heuristic manipulation of Equation (1) to justify Equation (2). We introduce the set of ordered pairs $A \times A$, and treat (a, b) as a formal difference between a and b. Two pairs (a, b) and (a', b') define the same vector in the representing space if and only if $a \oplus b' \sim a' \oplus b$.

This procedure is like the construction of rational numbers from integers: (p, q) and (p', q') define the same rational number $p/q = p'/q'$ if and only if the products pq' and $p'q$ are equal as integers. Here we have the additional feature of *equivalence*, rather than *equality*, of sums $a \oplus b'$, $a' \oplus b$; this means that $\sim$ is mapped into $=$ in the representing space.

Despite the fact that the proof in Section 15.3 is similar to the proofs of classical theorems, e.g., the embedding of the (ring of) integers in the (field of) rational numbers, we give it in detail. Those readers who are familiar with the classical embedding theorems (e.g., Jacobson, 1951, Chap. 4) can scan the proof quickly. For other readers, it may be instructive to follow in detail how the abstract vector-space structure is defined using just transitivity and linearity of color matching. The details of this proof are not to be found elsewhere. (A similar proof is found in Section 3.3, but it cannot easily be adapted to give a detailed proof of the present theorems.)

15.3 PROOFS

15.3.1 Theorem 3 (p. 234)

Any Grassmann structure is homomorphic to a structure $\langle C, +, \cdot, = \rangle$, where C is a convex cone in a vector space $\langle V, +, \cdot \rangle$ over Re and C generates V. The homomorphism is unique up to nonsingular linear transformations.

Let $\langle A, \oplus, *, \sim \rangle$ be a Grassmann structure. Define $\approx$ on $A \times A$ by

$$(a, b) \approx (c, d) \; if \; a \oplus d \sim c \oplus b.$$

PROOF. Reflexivity and symmetry of $\approx$ follow respectively from the corresponding properties of $\sim$. Transitivity of $\approx$ follows from Axioms 1 and 2 of Definition 2 and the associativity and commutativity of $\oplus$: if $(a, b) \approx (c, d)$ and $(c, d) \approx (e, f)$, then

$$
\begin{aligned}
(a \oplus f) \oplus (c \oplus d) &\sim (a \oplus d) \oplus (c \oplus f) \\
&\sim (c \oplus b) \oplus (c \oplus f) \\
&\sim (c \oplus f) \oplus (c \oplus b) \\
&\sim (e \oplus d) \oplus (c \oplus b) \\
&\sim (e \oplus b) \oplus (c \oplus d).
\end{aligned}
$$

Therefore $a \oplus f \sim e \oplus b$, or $(a, b) \approx (e, f)$. Thus $\approx$ is an equivalence relation on $A \times A$.

Let $V = A \times A / \approx$. Let $[a, b]$ denote the equivalence class in V determined by (a, b). Define $+$ on $V \times V$ by

$$[a, b] + [c, d] = [a \oplus c, b \oplus d].$$

This is well defined because $(a, b) \approx (a', b')$ and $(c, d) \approx (c', d')$ yield $(a \oplus c, b \oplus d) \approx (a' \oplus c', b' \oplus d')$ by an argument very similar to that given above for transitivity of $\approx$ (the last step, cancellation, relying on the *if* part of Axiom 2 of Definition 2, is unneeded here). It is easy to show that $\langle V, + \rangle$ is a commutative group with identity $[a, a] = [b, b]$ and with $-[a, b] = [b, a]$.

Define $\cdot$ on $Re \times V$ by:

$$t \cdot [a, b] = \begin{cases} [t*a, t*b], & \text{if} \quad t \in Re^+; \\ [a, a], & \text{if} \quad t = 0; \\ [|t|*b, |t|*a], & \text{if} \quad -t \in Re^+. \end{cases}$$

It can easily be shown that $\cdot$ is well defined and satisfies the properties required for scalar multiplication, i.e., 2(i)–2(iv) of Definition 1, for all $t, u \in Re$. For example, 2(i) is $t \cdot (u \cdot [a, b]) = (tu) \cdot [a, b]$. For $-t, -u \in Re^+$, $tu \in Re^+$, so

$$\begin{aligned} t \cdot (u \cdot [a, b]) &= t \cdot [|u|*b, |u|*a] \\ &= [|t|*(|u|*a), |t|*(|u|*b)] \\ &= [(tu)*a, (tu)*b] = (tu) \cdot [a, b]. \end{aligned}$$

The other cases are similar. To take one more example, we prove 2(ii) for $t \in Re^+$, $-u$, $-(t + u) \in Re^+$. We have

$$\begin{aligned} (t + u) \cdot [a, b] &= [|t + u|*b, |t + u|*a] \\ &= [(|u| - t)*b, (|u| - t)*a] \\ &= [(|u| - t)*b, (|u| - t)*a] + [t*a, t*a] \\ &\quad + [t*b, t*b] \\ &= [(|u|*b) \oplus (t*a), (|u|*a) \oplus (t*b)] \\ &= [t*a, t*b] + [|u|*b, |u|*a] \\ &= t \cdot [a, b] + u \cdot [a, b]. \end{aligned}$$

Therefore, $\langle V, +, \cdot \rangle$ is a vector space over Re.

Let $\varphi: A \to V$ be defined by

$$\varphi(a) = [a \oplus b, b].$$

This definition is independent of b, a fact that we shall use extensively.
For a, b in A, $t \in Re^+$,

$$\begin{aligned} \varphi(a \oplus b) &= [(a \oplus b) \oplus (c \oplus c), c \oplus c] \\ &= [a \oplus c, c] + [b \oplus c, c] \\ &= \varphi(a) + \varphi(b); \\ \varphi(t*a) &= [(t*a) \oplus (t*b), t*b] \\ &= t \cdot [a \oplus b, b] \\ &= t \cdot \varphi(a). \end{aligned}$$

For any a, b in A,

$$\varphi(a) - \varphi(b) = [a \oplus c, c] - [b \oplus c, c]$$
$$= [a \oplus c, c] + [c, b \oplus c]$$
$$= [a \oplus c \oplus c, b \oplus c \oplus c]$$
$$= [a, b].$$

Finally,

$$a \sim b \quad \text{iff} \quad (a \oplus c) \oplus c \sim (b \oplus c) \oplus c$$
$$\text{iff} \quad [a \oplus c, c] = [b \oplus c, c]$$
$$\text{iff} \quad \varphi(a) = \varphi(b).$$

Thus, φ satisfies all the required properties, letting $C = \{\varphi(a) | a \in A\}$.

To prove uniqueness, let φ' be another homomorphism of $\langle A, \oplus, *, \sim \rangle$ onto $\langle C', +', \cdot', = \rangle$, where C' is a convex cone that generates a vector space $\langle V', +', \cdot' \rangle$ over Re. Define a function ψ from V into V' by

$$\psi[a, b] = \varphi'(a) - '\varphi'(b).$$

To show ψ well-defined and one-to-one, note that

$$[a, b] = [a', b'] \quad \text{iff} \quad a \oplus b' \sim a' \oplus b$$
$$\text{iff} \quad \varphi'(a \oplus b') = \varphi'(a' \oplus b)$$
$$\text{iff} \quad \varphi'(a) + '\varphi(b') = \varphi'(a') + '\varphi(b)$$
$$\text{iff} \quad \psi[a, b] = \psi[a', b'].$$

It is also easy to show that ψ is a vector-space homomorphism and that ψ is onto all of V' [since every vector in V' has the form $\varphi'(a) - '\varphi'(b)$ for some $a, b \in A$]. Clearly $\psi[a \oplus b, b] = \varphi'(a)$, and so ψ maps C onto C' and $\psi\varphi = \varphi'$. ◇

15.3.2 Theorem 4 (p. 235)

A Grassmann structure is m-chromatic iff its representation is m-dimensional.

PROOF. Let φ be a homomorphism of the Grassmann structure $\langle A, \oplus, *, \sim \rangle$ onto $\langle C, \oplus, *, \sim \rangle$, where C is a convex cone that generates a vector space V over Re. If $b_1, \ldots, b_n \in A$ satisfy

$$\sum_{i=1}^{n} t_i * b_i \sim \sum_{i=1}^{n} u_i * b_i,$$

then by the homomorphism property,

$$\sum_{i=1}^{n} (t_i - u_i) \cdot \varphi(b_i) = 0,$$

where 0 is the vector-space identity. Conversely, if $\varphi(b_1), \ldots, \varphi(b_n)$ satisfy a linear equation,

$$\sum_{i=1}^{n} \lambda_i \varphi(b_i) = 0,$$

choose $\mu > \sup\{|\lambda_i| \,|\, i = 1, \ldots, n\}$. We have

$$\sum_{i=1}^{n} (\lambda_i + \mu) \varphi(b_i) = \sum_{i=1}^{n} \mu \varphi(b_i),$$

and since $\lambda_i + \mu, \mu \in \mathrm{Re}^+$,

$$\sum_{i=1}^{n} (\lambda_i + \mu) * b_i \sim \sum_{i=1}^{n} \mu * b_i.$$

This shows that $b_1, \ldots, b_n$ satisfy a nontrivial equation in $\langle A, \oplus, *, \sim \rangle$ iff $\varphi(b_1), \ldots, \varphi(b_n)$ are linearly dependent. Therefore, if $\langle A, \oplus, *, \sim \rangle$ is m-chromatic, then Axiom 4(i) implies that any $m + 1$ elements of C are linearly dependent, and since C generates V, any $m + 1$ elements of V are also linearly dependent. Moreover, Axiom 4(ii) implies that there exist m elements of C that are not linearly dependent. Therefore V is m-dimensional. Conversely, if V is m-dimensional, then any $m + 1$ elements of C are linearly dependent, and so 4(i) holds. There must be m linearly independent elements of C because, otherwise, V would be n-dimensional with $n < m$; hence, Axiom 4(ii) also holds.   ◇

15.4 COLOR INTERPRETATIONS

15.4.1 Metameric Color Matching

In experiments on color, the overall stimulus consists of a *focal* stimulus element (Jameson and Hurvich, 1959) together with nonfocal stimulus elements. By focal element we mean the spatiotemporal part of the stimulus that is judged or matched in color. A nonfocal element differs from the focal one in space (region of the visual field) or time or both. The nonfocal elements can affect the perceived color of the focal stimulus; these effects on perceived color are studied in the literature under various headings, such as adaptation, contrast, assimilation, and masking.

In a color-matching experiment, the focal element of one stimulus is compared with the focal element of another, and the two are judged the same or different in color. Only the focal elements are judged to match, and they are required only to match in *color*, i.e., in hue, saturation, and brightness, not necessarily in other perceptual attributes such as size, duration, and glossiness.

The focal element is partly characterized by a radiant power spectral distribution, i.e., a function that gives the radiant power at any wavelength or in any interval of wavelengths. This will be called the *focal distribution*. By *viewing conditions* we mean everything else that characterizes the overall stimulus, apart from the focal distribution. This includes the various nonspectral characteristics of the focal element, such as its size, shape, location, duration, and mode of entry through the observer's pupil, together with the spectral and nonspectral physical specifications of all the nonfocal stimulus elements.

In the interpretation of a Grassmann structure, the set A consists of all possible nonzero focal distributions. Focal distributions are functions (in fact, measure functions) and thus can be added and multiplied by scalars, to define the operations $\oplus$ and $*$. The similarity to specification of force configurations is clearest when the focal distribution is written as a density function $a(\lambda)$, defined for λ between 400 and 700 nm, the visible electromagnetic spectrum (1 nm = 10^{-9} meters). Thus λ plays the role of the angle θ in the force table. The addition and scalar multiplication are done componentwise:

$$(a \oplus b)(\lambda) = a(\lambda) + b(\lambda)$$
$$(t * a)(\lambda) = t[a(\lambda)].$$

The operation $\oplus$ has a simple physical interpretation: incoherent additive mixture of light beams. For example, suppose one region of a diffusing screen is illuminated by two projectors. If the distributions produced when they are turned on separately are a, b, then the distribution produced when they are turned on at once is $a \oplus b$. Various other optical devices exist that produce a mixture $a \oplus b$ of two given lights a, b. Likewise, multiplication by a scalar t consists of inserting or removing a neutral filter (one that attenuates all visible wavelengths by the same fraction).

For light, as for weights on the force table, there is an extensive measurement structure (for each wavelength, or each interval of wavelengths) in which radiant power is an additive representation with respect to the concatenation given by incoherent additive mixture of lights. Thus the fact that $\langle A, \oplus, * \rangle$ is a convex cone follows from the extensive measurement laws.

The final relation in the Grassmann structure is for this interpretation a perceptual one, called *metameric* matching. Let the viewing conditions be held constant so that only the focal distributions vary among stimuli:

$$a \sim b \qquad \text{iff} \qquad a, b \text{ are identical in color.}$$

As we have mentioned, identity in color means identity in hue, in saturation, and in brightness; for further discussion of color identity and of these color attributes, see Sections 15.4.4, 15.5.1, and 15.6.3.

Metameric matching is also called *symmetric* matching because the viewing conditions are identical for the two focal distributions that are matched. Asymmetric matching, which follows different laws, is discussed in Section 15.4.4.

Since viewing conditions were held constant in the above definition, one might suspect that there is a different relation $\sim$ for each possible choice of the constant viewing conditions. Surprisingly and fortunately this is not the case; the relation $\sim$ is the same for a wide variety of viewing conditions. We shall return later to a closer consideration of this point. For the moment we can be content with a satisfactory interpretation of the structure $\langle A, \oplus, *, \sim \rangle$.

Just as two force configurations that are physically distinct can be equivalent in terms of resultants (i.e., can be in equilibrium with the same force), so two focal distributions that are physically very different can yield identical color perceptions. Such a pair of distributions is called a metameric pair. For example, any broadband light distribution can be matched by a light with all radiant power concentrated at just two wavelengths. In particular, light from an ordinary tungsten-filament lamp can be matched by a mixture of wavelengths approximately at 470 and 580 nm (blue and orange-yellow under ordinary viewing conditions).

The empirical facts concerning metameric matching are largely summarized by the statement that $\langle A, \oplus, *, \sim \rangle$ is a proper trichromatic Grassmann structure. (To make it proper we exclude from consideration the null distribution and keep in mind that $*$ involves multiplication only by positive scalars, not 0. This convention is useful since, unlike the case of force, there is no nonnull distribution that matches the null one.)

As in the case of force, these empirical laws (Axioms 1–4 of Definition 2) are an idealization. For example, if $a \sim b$, then $t * a \sim t * b$ (Axiom 3) breaks down when t approaches 0, reaching radiant power levels in the neighborhood of the threshold of the retinal cones. Axiom 3 also breaks down at very bright levels due to bleaching of photopigments (Alpern, 1979; Brindley, 1953; see Section 15.5.2). The axioms hold with remarkable accuracy over a wide range if presentation of lights to the observer is controlled carefully and the color matches are made carefully. Brindley (1970) reviews evidence on this point. Some of the precautions that are required are mentioned below.

Theorems 3 and 4 now tell us that focal distributions can be represented by three-dimensional vectors, with metameric pairs mapping onto the same vector. The coordinates are sometimes called *tristimulus* coordinates. The next subsection goes into some detail concerning this representation for color stimuli.

15.4.2 Tristimulus Colorimetry

In this and the next subsection, we restate the material in the preceding section in a more concrete manner and try to anticipate some of the misunderstandings that often arise.

Color-matching experiments are often done with a visual tristimulus colorimeter. Different colorimeters operate on different optical principles, but their net effect is to produce a bipartite field of view. Often, the field is circular, small (diameter subtending $2°$), and divided along a vertical diameter. In one half of the field (say, the right half), the light is a variable linear combination of three fixed suitable lights, a_1, a_2, a_3. Theoretically, the criterion of suitability is that given by Axiom 4(ii), i.e., no nontrivial color match can be produced using just a_1, a_2, a_3; in practice, they are selected to be distinctly different. Often the lights are red, green, and blue, with most of the energy near a single wavelength; in the experiments of Wright (1928–1929), the wavelengths were 650, 530, and 460 nm. Energy controls are provided that can produce $u_1 * a_1$, $u_2 * a_2$, and $u_3 * a_3$ for a sizable range of values of u_i; and a mixing device produces, in the right-hand field, any linear combination $\sum_{i=1}^{3} u_i * a_i$. In the other half of the field, a test light appears that can also be mixed with any one of the lights a_i; that is, any combination $a \oplus (t_i * a_i)$ can be produced for various test lights a, for $i = 1, 2,$ or 3. The experimenter or the observer manipulates knobs to control energies of the lights a_i until the color appears uniform over the entire circular field, i.e., the vertical dividing line all but disappears as both sides appear to become identical. This sets up a match, for example,

$$a \oplus (t_1 * a_1) \sim (u_2 * a_2) \oplus (u_3 * a_3). \tag{3}$$

Note that this does not have quite the form given in Axiom 4(i), or in

Equation (1), which would require

$$(t * a) \oplus \sum_{i=1}^{3} (t_i * a_i) \sim (u * a) \oplus \sum_{i=1}^{3} (u_i * a_i). \tag{1'}$$

However, by Axiom 2 of Definition 2, Equation (3) is equivalent to the equation obtained by adding $a \oplus a_1 \oplus a_2 \oplus a_3$ to both sides, i.e.,

$$(2 * a) \oplus [(1 + t_1) * a_1] \oplus a_2 \oplus a_3$$
$$\sim a \oplus a_1 \oplus [(1 + u_2) * a_2] \oplus [(1 + u_3)] * a.$$

This equation does have form (1), and the calculation of the coordinates, given by Equation (2), is

$$\varphi(a) = (-t_1, u_2, u_3).$$

This is, of course, obtained directly from Equation (3).

The lights a_1, a_2, a_3 are called the physical, or instrumental, *primaries*. Depending on the test light a, all three primaries can be mixed to match the test light alone, or one of them can be added to the test light and two mixed in the other half of the field. For some selections of primaries, it would be necessary to add two primaries to the test light, but this is avoided in practice.[1] It is never necessary to add all three primaries to the test light; that is what it means to say that we have a proper Grassmann structure, not an equilibrium structure.

The usual statement of the Law of Trichromacy is the following. (i) Given any four lights, either a linear combination of three of them can be found that matches the fourth, or a linear combination of two of them matches a linear combination of the other two; (ii) there exist three lights such that no linear combination of any two of them matches the third one. It can now be seen that given a Grassmann structure (Axioms 1–3 of Definition 2), parts (i) and (ii) are logically equivalent respectively to Axioms 4(i) and 4(ii). The equivalence is proved by adding or subtracting $\Sigma_{i=0}^{3} a_i$ or $\Sigma_{i=1}^{3} a_i$ to both sides of any match.

Another way to look at the matter is to consider the two halves of the bipartite field produced by a colorimeter to be a single configuration (a, b),

[1] Avoidance is based on the use of monochromatic primaries and on the empirical fact that the spectrum locus—the projection into a plane of the coordinates of all the monochromatic colors—is convex.

where a is the light in the left half and b is in the right half. These configurations can be added, $(a, b) \oplus (a', b') = (a \oplus a', b \oplus b')$, and scalar multiplied, $t * (a, b) = (t * a, t * b)$. The set $E = \{(a, b) | a \sim b\}$ is closed under $\oplus$ and $*$; and for any (a, b), there exists another pair, namely, (b, a), such that $(a, b) \oplus (b, a)$ is in E. These definitions make $\langle A \times A, \oplus, *, E \rangle$ an equilibrium structure. In color matching, we take a configuration $(a, 0)$ [or equivalently, if we want to avoid zero, take $(t * a, u * a)$] and add a configuration of primaries $(\Sigma t_i * a_i, \Sigma u_i * a_i)$ that produces an equilibrium element. In practice, if $t_i > u_i$, we simply put $(t_i - u_i) * a$ on the left and nothing on the right, or vice versa if $u_i > t_i$. The coordinates are the numbers $(u_i - t_i)/(t - u)$, given by Equation (2); these are uniquely determined by a.

In colorimetry, the trichromacy property is tested simply by observing that a match can always be made. Additivity, Axioms 2 and 3, can easily be tested directly. However, in direct tests of Axiom 2, care must be taken to ensure that the criterion for accepting a "match" in a bipartite color field does not shift in such a way as to produce a seeming violation. A valuable control is to use isomeric or physical matches as well as metameric ones. Thus, if adding c to both sides of an isomeric match $a \sim a$, produces a slight mismatch $a \oplus c \not\sim a \oplus c$, then the problem is one of method rather than a violation of Axiom 2. See Alpern, Kitihara, and Krantz (1983a) for a discussion in the case of a dichromat's matching. The assumption that $\sim$ is an equivalence relation is an adequacy requirement that is imposed on $\sim$. Care must be taken to avoid appreciable errors that might produce intransitivities in $\sim$. Moreover, the same physical stimulus may look slightly different in the two halves of the bipartite field, but failure of symmetry and reflexivity of $\sim$ are avoided by defining $a \sim b$ to mean that test lights a and b can be matched by the same mixture of the primaries. This definition eliminates any small asymmetry in the viewing conditions due to the use of different halves of the small bipartite field.

The results of measurements of this sort have been averaged and converted into a set of internationally accepted standards (see Wright, 1964; Wyszecki and Stiles, 1982) by the Commission Internationale d'Eclairage (CIE). Because any light can be approximated by a linear combination of near-monochromatic lights, tristimulus coordinates are determined for the visible spectrum and are calculated for other lights from these measurements.

The CIE system does not use the coordinates that resulted from measurements with a particular set of instrumental primaries; rather, it employs several alternative coordinate systems. Recall the uniqueness part of Theorem 3: the homomorphism φ is unique up to nonsingular linear transformations (vector-space isomorphisms). For representations in three-dimensional vector spaces over Re, such transformations are represented by 3×3 matrices with nonzero determinants (α_{ij}), $i = 1, 2, 3$, $j = 1, 2, 3$. That is,

$$\varphi_i' = \sum_{j=1}^{3} \alpha_{ij}\varphi_j, \qquad \text{for } i = 1, 2, 3,$$

or in matrix form,

$$\begin{pmatrix} \varphi_1' \\ \varphi_2' \\ \varphi_3' \end{pmatrix} = \begin{pmatrix} \alpha_{11} & \alpha_{12} & \alpha_{13} \\ \alpha_{21} & \alpha_{22} & \alpha_{23} \\ \alpha_{31} & \alpha_{32} & \alpha_{33} \end{pmatrix} \begin{pmatrix} \varphi_1 \\ \varphi_2 \\ \varphi_3 \end{pmatrix}. \qquad (4)$$

Any coordinate system φ', definable from φ by (4) for some (α_{ij}), is as good as any other from the standpoint of representing the structure $\langle A, \oplus, *, \sim \rangle$. The choice of a particular coordinate system has been made on one of three criteria: ease of standardization, computational convenience, or representation of additional empirical relations on colors beyond $\oplus$, $*$, and $\sim$. These criteria have led to three standard coordinate systems adopted by the CIE.

One of these, the R, G, B system, is based on red, green, and blue instrumental primaries that can be easily standardized. Another, the X, Y, Z system, corresponds to no possible set of instrumental primaries; rather, the matrix (α_{ij}) is chosen such that all the coordinates φ_i' are nonnegative and such that the second (φ_2' or Y) coordinate represents luminance (approximately, another empirical relation on colors based on certain special methods of brightness matching). This system is extremely convenient computationally because of positive numbers and because the luminance of the lights is directly represented by one coordinate. The third set of coordinates is the uniform chromaticity system (UCS), which attempts to give an approximate representation of empirical color-discriminability and color-similarity relations for colors of constant luminance by means of Euclidean distances in a projective plane. (See the later discussion of *chromaticity coordinates*.)

The idea of further specializing the coordinate system to represent additional relations is the main topic of Sections 15.5.4 and 15.6. In this subsection, we deal briefly with two more matters: the use of noninstrumental primaries and the two-dimensional projections of tristimulus coordinates.

Given two sets of primaries a_1, a_2, a_3 and b_1, b_2, b_3, there are color-matching equations:

$$a_j \oplus \sum_{i=1}^{3} (t_{ij} * b_i) \sim \sum_{i=1}^{3} (u_{ij} * b_i), \qquad j = 1, 2, 3. \tag{5}$$

It can readily be shown, using Axioms 2 and 3 of Grassmann structures (see Exercise 5), that a color match between c and the a_j can be transformed into a match between c and the b_i, and that the coordinates φ_i' in terms of the b_i are related to the coordinates φ_j in terms of the a_j by Equation (4), with $\alpha_{ij} = u_{ij} - t_{ij}$, i.e., α_{ij} is the ith coordinate of a_j relative to the b_i.

It is not the case, however, that any coordinates obtained from a transformation (4) correspond to a new set of instrumental primaries. A coordinate system that does not correspond to any instrumental primaries is sometimes said to involve "imaginary" primaries. However, the present framework allows a simple interpretation of such "imaginary" primaries as *pairs* of colors. Using the equilibrium structure $\langle A \times A, \oplus, *, E \rangle$ defined above, we take as primaries any three independent *pairs*, (a_i, b_i), $i = 1, 2, 3$. The pairs must be subject to no nontrivial color match $\sum_{i=1}^{3} t_i * (a_i, b_i) \in E$, where we allow t_i to be positive, negative, or zero, defining $t * (a, b) = (|t| * b, |t| * a)$ for $t < 0$ and defining $0 * (a, b) = (a, a)$. It can be shown that any coordinate system φ is given by some such set of primary pairs, where for any light c,

$$(c, 0) \oplus \sum_{i=1}^{3} \varphi_i(c) * (b_i, a_i) \in E.$$

Finally, we should mention that tristimulus coordinates φ are often projected into a two-dimensional chromaticity diagram by defining $\bar{\varphi}_i(c) = \varphi_i(c)/\sum_{j=1}^{3}\varphi_j(c)$. Since the $\bar{\varphi}_i$ sum to unity, we obtain a representation in a plane, using, say $\bar{\varphi}_1$ and $\bar{\varphi}_2$ as orthogonal coordinates. (Readers familiar with projective geometry will recognize the $\bar{\varphi}_i$ as barycentric coordinates of points in the projective plane. As was pointed out in Chapter 12, nonsingular linear transformations of the φ_i induce projective transformations of the $\bar{\varphi}_i$; and conversely, any projective transformation corresponds to a nonsingular linear transformation that is uniquely determined up to multiplication by a nonzero constant.) Note that $\bar{\varphi}_i(t * c) = \bar{\varphi}_i(c)$. The term *chromaticity coordinates* thus arises from the fact that they are determined by the shape of the energy distribution c, independent of energy level.

Figure 2 shows the chromaticity diagram for the X, Y, Z coordinate system of the CIE. The abscissa is $\bar{\varphi}_i = x = X/(X + Y + Z)$; the ordinate

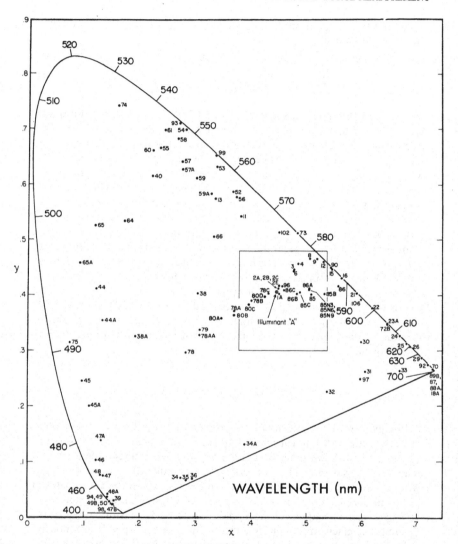

FIGURE 2. Chromaticity coordinates showing the spectrum locus and loci of lights from a tungsten lamp (CIE Standard Source "A") modified by various Kodak gelatin filters. (From *Kodak filters for scientific and technical uses*, Eastman Kodak Co. B-3, 1970.)

is $\bar{\varphi}_2 = y = Y/(X + Y + Z)$. The curved locus of monochromatic lights (spectrum locus) is shown together with the loci of a number of other lights. Point A is the locus of the light distribution from a tungsten-filament lamp (metameric to black-body radiation at 2842 °K.), and a number of other points (32, 34A, etc.) correspond to that light distribution after modification by Kodak Wratten gelatin filters with the corresponding designations. Figure 3 shows the percentage-transmission curve for Kodak Wratten filter #97. For each wavelength, the output of the tungsten lamp has been multiplied by the transmittance factor given in Figure 3, and the tristimulus coordinates for that wavelength at the resultant energy level have been taken from the CIE standards. Each tristimulus coordinate is then added over all wavelengths present to get the tristimulus coordinates for the light as modified by the filter. Those are then converted to chromaticity coordinates. (Figures 2 and 3 are adapted from *Kodak filters for scientific and technical uses*, Eastman Kodak Co., 1970.)

Chromaticity coordinates can be thought of as the intersection of the ray from the origin through the tristimulus coordinates with the plane whose equation is $\Sigma \varphi_i = 1$. Given lights a, b, the set of all positive linear combinations $(t * a) \oplus (u * b)$ corresponds to a quadrant of a plane in tristimulus space. The intersection of this quadrant with the plane $\Sigma \varphi_i = 1$ is a line segment that joins the chromaticity loci of a and b and contains the chromaticity loci on the corresponding line segment. This segment runs inside the (convex) curve formed by the spectrum locus and the line of "purples" joining its endpoints (see Figure 2). Any third light can be mixed

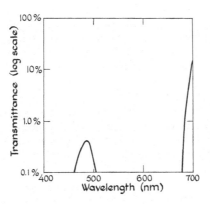

FIGURE 3. Transmittance curve for Kodak gelatin filter #97 (see Figure 2 for locus of tungsten light modified by #97).

with any mixture represented on the a-to-b segment. Thus linear combinations of primaries a_1, a_2, a_3 cover the triangle with vertices a_1, a_2, a_3. If light a lies outside this triangle but within the cone formed by the rays a_1a_2 and a_1a_3, then some mixture of a with a_1 lies on the segment a_2a_3, and so there is a color match of a and a_1 with a_2 and a_3. Similarly, the locus labeled 97 in Figure 2 is a linear combination of wavelengths in the 485 and 700 nm regions corresponding to the transmission curve of Figure 3 and taking into account the much larger output of the light source at the longer wavelengths.

15.4.3 Four Ways to Misunderstand Color Measurement

The first point to note is that the color-measurement procedures discussed above apply to lights not to objects. Ordinarily one thinks of color as a property of objects, e.g., one says "That flower is red." After donning green glasses, which modify dramatically the light arriving at the eye from the red flower, one says "Now the flower looks black," implying that it really *is* red but *looks* different. This common sense standpoint has a deep basis in human perceptual function, having to do with perceptual construction of an "objective" world outside the observer's body.[2] Nevertheless, it must be abandoned to understand color measurement. Tristimulus coordinates are determined by the light that reaches the observer's eye. One must take into account the wavelength distribution in the illuminant, its modification by the colored object, and the effects of any intervening media such as green glasses or smog. In particular, two colored objects that produce a metameric color match in one situation may no longer do so when the illuminant or the intervening media changes. Interobserver differences may also be significant: lights metameric for one observer may not be so for another. This can result from differences in transmittance of parts of the eye or differences in retinal function.

The second point is that tristimulus colorimetry predicts when two lights will look the same but not *how* they will look. If *color* means hue, saturation, and brightness, then *color measurement* and *colorimetry* are in fact misnomers. Hue, saturation, and brightness are perceptual attributes that are only crudely correlated with physical measurements such as wavelength, spectral bandwidth, and radiant power. As we shall discuss in Section 15.4.4, color appearance varies tremendously with viewing condi-

---

[2] A simple demonstration is obtained by looking at nearby and distant surfaces after seeing a bright flash. The afterimage of the flash appears to be part of whatever surface one looks at and becomes dramatically larger in apparent size when one looks at a distant surface. (The proportionality of perceived afterimage size to projected distance is called Emmert's Law.)

tions or context (see Jameson and Hurvich, 1959, 1961). The same light that looks orange in a dark surround may look greenish yellow in a red surround and brown in a white surround. If two lights are metameric, they continue to be so as context changes, but their common appearance may vary widely. Thus there are many empirical relations involving color appearance that are not predictable from the empirical relations $\oplus$, $*$, $\sim$ and that form a basis for more refined studies of color measurement. This is one of the main topics of the rest of the present chapter.

Closely connected with the preceding points is the common misunderstanding that the widely publicized two-color projection experiments of Land (1964) contradict trichromacy. Land's setup employs a variety of linear combinations of just two lights, projected in an irregular spatial array. These experiments usually employ lights that would look red-orange and white (or yellow) in isolation in a neutral context. Any single linear combination, viewed in isolation, has an appearance somewhere in the hue gamut from red through orange or pink to yellow or white. Yet when the total array is viewed at once, it contains a much wider gamut including blues and greens.

Note, however, that the wide gamut of hues produced by linear combinations of two lights does not contradict the literal statement of trichromacy because the latter is concerned with *exact* matches by mixtures of three primaries, not just with approximate hue matches. Also, metameric matches are made with both lights viewed in identical conditions, whereas in Land's setup, each linear combination is viewed in the context of all the others. Variation of appearance with context produces the wide spread of perceived hues. Detailed discussion of the Land effects along these lines can be found in Judd (1960) and Walls (1960).

Last, we say a word about the concept of *primary colors*. This phrase has at least five distinct meanings. The first two referents are sets of physical objects. A set of primaries for additive color mixture is simply a set of instrumental primaries for tristimulus colorimetry as described in the preceding subsection. Altogether different is a set of primary colors for *pigment* mixture, say, red, yellow, and blue paints. Mixtures of such paints produce a wide gamut of colors. However, such mixtures have little to do with mixtures of lights or with color measurement. The third referent is a set of perceived colors that are basic in some description of color perception. The six "unique" colors of opponent-colors theory—red, green, yellow, blue, white, and black (see Section 15.6)—are sometimes called the *psychological primaries*; the five hues on which the Munsell system is based (red, yellow, green, blue, and purple) can also be called primaries. The fourth and fifth referents are theoretical. One is simply a vector-space basis for the three-coordinate representation of additive color mixture. Instrumental primaries for additive color mixture form such a basis; but, as noted above, some coordinate systems cannot be obtained from instrumental primaries, though they can be realized from pairs (a_i, b_i) of lights. Finally, one can use *primaries* to refer to a special vector-space basis for coordinates, which describes some underlying physiological or psychological mechanism. For example, in Section 15.6 we discuss a coordinate system $(\varphi_1, \varphi_2, \varphi_3)$ in which φ_1 is a measure of redness or greenness and φ_2 is a measure of yellowness or blueness; this brings the idea of psychological primaries into the coordinate system. We also discuss coordinates (ρ_1, ρ_2, ρ_3) where ρ_i refers to a basic physiological process that is absent in the ith kind of color-defective vision. Clearly, all these meanings, except for the second (pigment mixture) are closely interrelated.

15.4.4 Asymmetric Color Matching

We have emphasized that in a metameric color match $a \sim b$, the lights a, b are viewed under identical conditions. If the viewing conditions change, the appearance of the metameric match may change drastically though it remains a match. It follows that asymmetry of the viewing conditions often produces a color mismatch between metameric or, for that matter, isomeric (identical) distributions. In fact, the perceptual effect of viewing conditions is demonstrated (as in the Land demonstrations just described and in many other experiments before and since) by showing that physically identical lights no longer match in color.

The most important fact about changes in appearance due to changes in viewing conditions is one that is so much taken for granted that it seems never to have been stated explicitly in textbooks or in the technical literature on color. It is this:

Color matches exist between focal stimuli in different viewing conditions.

In other words, it is sometimes possible to produce a color match between focal distribution b viewed under conditions β and focal distribution b' viewed under conditions β'. This is called an *asymmetric* color match, as distinct from a metameric match. For example, a light that appears white in a dark surround can appear greenish white in a red surround; that greenish white percept can be matched exactly in the dark surround by a suitably chosen light. As a contrary example, the perception of brown requires that light colors appear nearby in the field of view; hence brown cannot be matched by any light viewed under conditions of dark surround. So we assert merely that some asymmetric matches do exist.

Actually, a bit more can be asserted: not only do asymmetric matches occur, they are *nonsingular.* Roughly, this means that if a given color percept can be produced by a nonextreme focal distribution with particular viewing conditions, then any other sufficiently similar color percept can be produced by a nearby focal distribution with those same viewing conditions. (To make this statement precise we would need to specify what is meant by "nonextreme," "sufficiently similar," and "nearby.") In an asymmetric match, the same perceived focal color is obtained from two specifications of viewing conditions; nonsingularity entails that all sufficiently similar focal color percepts can also be obtained under both viewing conditions, i.e., the existence of the asymmetric match is stable for small perturbations of color appearance.

The considerable importance of asymmetric matching for theories of color vision seems to have been recognized first by Stiles (1967). The development sketched here is based on Krantz (1968b), on Alpern, Kitihara, and Krantz (1983b), and chiefly on an unpublished paper by D. H. Krantz and C. Simon.

A useful notation for asymmetric matching can be had if we let each possible specification of viewing conditions act as a function, taking as domain the set A of focal stimulus distributions and as range a subset of the set P of all possible focal colors. Thus, the general statement of an asymmetric match can be written as

$$\beta(b) = \beta'(b'),$$

where the equality is in the set P of perceived focal colors. Again, if we let β_0 denote some standard specification of viewing conditions including neutral adaptation and dark surround and let P_{brown} denote the subset of P consisting of brown colors, the nonexistence asserted above can be written as

$$\beta_0(A) \cap P_{\text{brown}} = \varnothing.$$

Note that every perceived focal color must be produced by some focal distribution b viewed under some conditions β, that is, every element of P is obtained in the form $\beta(b)$. Therefore the set P is covered by the images $\beta(A)$ for the various possible β. Since A has a three-dimensional representation via metameric matching, each of these images can be considered to be three-dimensional. If the set P itself were at least four-dimensional, then it would be possible to cover it in nonoverlapping fashion with three-dimensional slices, that is, the images $\beta(A)$ could all be pairwise disjoint: no asymmetric matches would exist. On the contrary, we assert that asymmetric matches do exist and that the regions of overlap are themselves three-dimensional. In fact, nonsingularity implies that each image $\beta(A)$ includes an open subset of P with respect to a suitable topology. (See Beals and Krantz, 1967b, for discussion of uniform topological structures based on ordering with respect to dissimilarity.) Thus the intersections for which asymmetric matches exist also include open subsets. An open subset always has the full dimensionality of P; so this dimensionality must equal 3. In short, the implication of the existence and nonsingularity of asymmetric matching is that not only color matching but also color perception is limited in dimensionality. That is, *changes in viewing conditions do not introduce new dimensions, rather, they at most create some new combinations of values* (e.g., brown) *in a fixed set of dimensions.*

A further implication is that the set P has the structure of a three-dimensional manifold, with local coordinate systems based on the three-dimensional representation of A by tristimulus matching. To establish this assertion we need to assume that each function β from A to P is continuous. This law of continuity was actually given in Grassmann's 1854 paper, though to state it rigorously requires specification of the topologies on A and on P. We have already noted above that the topology on P is based on perceptual dissimilarity. The topology on A must be taken to be the weak* topology (see, e.g., Brown and Pearcy, 1977, p. 319) since in that topology a distribution with pure wavelength components can be approximated arbitrarily well by continuous distributions.

Because they provide overlapping coordinate systems for color percepts under various viewing conditions, asymmetric matches constitute a valuable method for measuring the effects of viewing conditions on color perception. The effects of color adaptation and/or simultaneous color contrast on perception were studied in this way by Wright (1934), Hunt (1953), MacAdam (1956), Burnham, Evans, and Newhall (1957), Hurvich and Jameson (1958), and Jameson and Hurvich (1959, 1961). Alpern et al. (1983b) studied the asymmetric matches between the two eyes of an individual who had defective color vision in one eye and more or less normal vision in the other.

One can ask what sorts of axioms or general laws are satisfied by asymmetric matching. Hurvich and Jameson (1958) demonstrated directly that Grassmann's scalar multiplication law (Axiom 3 of Definition 2) can fail for asymmetric matches:

$$\beta(b) = \beta'(b') \quad \text{does not entail} \quad \beta(t*b) = \beta'(t*b').$$

This could have been inferred less directly from the studies of Hunt, MacAdam, and Burnham et al. cited above.

Similarly, Alpern *et al.* (1983b) showed that Grassmann's additivity law is grossly violated in their situation. Two asymmetric matches,

$$\beta(b) = \beta'(b') \quad \text{and} \quad \beta(c) = \beta'(c')$$

were found, the first percept being red for both the abnormal and the normal eye, the second one blue. But the color appearance of the additive mixture for the abnormal eye, $\beta(b \oplus c)$, was white whereas its counterpart for the normal eye, $\beta'(b' \oplus c')$, was of course purple (reddish blue). Alpern *et al.* also inferred, less directly, that the scalar multiplication law is violated.

As far as we know, there is no demonstration that the additivity or scalar multiplication laws are even approximately true when color matching is significantly asymmetric [i.e., where $\beta(c)$ is quite distinct from $\beta'(c)$ for some c]. Burnham *et al.* (1957) showed, however, that their adaptation effects can be well described by an affine (rather than a linear) transformation of coordinates, and Krantz (1968b) showed how such a description can be axiomatized using more complex alternatives to Grassmann's laws. This axiomatization, in fact, is similar to an axiomatization of force measurement for the three-dimensional case in which gravity is a factor and equilibria are additive only after counterweighting. However, no direct tests of Krantz's axioms have been performed.

The implications of asymmetric matching are discussed further in Section 15.5.

□15.5 THE DIMENSIONAL STRUCTURE OF COLOR AND FORCE

Trichromacy of color vision is usually interpreted in terms of the existence of three independent physiological codes for color. This interpretation is due to Thomas Young (1802). In his Bakerian Lecture, he stated that

... it becomes necessary to suppose the number [of particles contained in each sensitive point of the retina] limited, for instance, to the three principal colours... and each sensitive filament of the nerve may consist of three portions, one for each principal colour. (pp. 20–21).

This interpretation of the empirical law of trichromacy is the point of departure for almost all theories of color vision. Such theories concern the detailed nature of three independent codes and the manner in which other phenomena of color vision (e.g., color defects, simultaneous color contrast) are related to these codes.

In the next two subsections, we examine this point of departure. What, precisely, is meant by the terms *code*, *linear code*, and *independent codes*? What is the basis for assuming that there are three independent photopigments in the retinal cones? After examining these points, we proceed in Section 15.5.3 to consider the analogous theoretical reasoning with respect to force measurement. For, if the existence of three independent color codes can really be inferred from the empirical fact of trichromacy, then analogous reasoning ought to be possible in the case of a three-dimensional equilibrium structure for force configurations. Surprisingly, the clarification of this question helps also to clarify the concept of code and its application in color. On this basis, the final subsection sketches a general schema for theories of color "statics", into which most current theories can be fitted, and relates this schema to a program of developing more-refined representation and uniqueness theorems for color measurement.

15.5.1 Color Codes and Metamer Codes

We shall use the concepts of Section 15.4.4 heavily in our discussion of color codes. Throughout this section, $\langle A, \oplus, * \rangle$, Ω, P are interpreted respectively as the convex cone of focal color distributions, the set of viewing conditions under consideration, and the set of focal color percepts produced by combinations in $A \times \Omega$. Each β in Ω is a mapping of A into P. We restrict attention to a set of viewing conditions for which metameric matches are invariant; thus we assume that symmetric matches are independent of the choice of element in Ω, i.e.,

$$\beta(a) = \beta(b) \qquad \text{iff} \qquad \gamma(a) = \gamma(b)$$

for all a, b in A and β, γ in Ω. The metamerism relation is defined by $a \sim b$ if and only if $\beta(a) = \beta(b)$ for some (and hence for every) β in Ω. The resulting structure $\langle A, \oplus, *, \sim \rangle$ is assumed to be a proper trichromatic Grassmann structure. Finally, we assume that A and P have the required topologies and that the mappings in Ω have the required continuity properties so that P has the structure of a three-dimensional manifold, with local coordinates given by the inverse mappings into A and the tristimulus representation of A.

Stiles (1967) postulated that there are three "mechanisms" whose responses are equated for the two stimuli of an asymmetric match. Uttal (1967) defined a *neural code* to be any aspect of neural activity that is discriminated behaviorally. We shall use the term *code* rather than *mecha-*

nism. We formalize Stiles' and Uttal's ideas by defining a *color code* to be any real-valued function $h(b, \beta)$ such that distinct values of h yield distinct focal color percepts. That is,

$$h(b, \beta) \neq h(b', \beta') \rightarrow \beta(b) \neq \beta'(b').$$

Note that the domain of a color code is the set $A \times \Omega$ of all color stimuli. By the contrapositive of the above relation, for any asymmetric color match $\beta(b) = \beta'(b')$ the values of *every* code h must be equal for the two stimuli, as required by Stiles' postulate.

In principle, a color code might be defined in three distinct ways: $h(b, \beta)$ could be a measure of some aspect of the neural activity produced by focal distribution b viewed under conditions β; or it could be derived from perceptual data, i.e., from judgments of the focal color corresponding to distribution b viewed under conditions β; or finally, it could simply be defined mathematically as a function of other color codes. We shall refer to the first two possibilities respectively as *neural codes* and *perceptual codes*.

It is well to bear in mind that not enough is known at present about the neural basis of color vision to permit actual measurements of any neural code. Where we discuss neural codes, we are referring to the *postulated* neural computations that must be equated for the two stimuli of an asymmetric match. Perceptual codes are obtained by various color scaling techniques; Hurvich and Jameson (1955, 1956) and Jameson and Hurvich (1956, 1959, 1961) developed a model for brightness, saturation, and hue as perceptual codes.

The foregoing discussion implicitly assumes that the set Ω of viewing conditions under consideration is taken as large as possible, i.e., such that it includes all common specifications and excludes extreme specifications under which the metamerism relation changes. Thus the manifold P of color percepts includes most colors. We can, however, also include a trivial case in which Ω consists of just one element. Since in this case the only matches are symmetric, the only requirement for a code is that it distinguishes metamers, that is, it can be considered to be a function on A such that $h(a) \neq h(b)$ implies *not* $a \sim b$. We call such a function a *metamer code* as opposed to a color code. For example, any one coordinate of a tristimulus representation is a metamer code. We use this device to establish certain basic results about functions on A as corollaries of theorems on color codes.

The main goal of this section is to discuss dependencies among color codes and the cardinality of set of independent codes. A set of m color codes $h_1, \ldots, h_m$ will be regarded as a mapping from the set of color stimuli into Re^m. We shall sometimes denote such a mapping as H and refer to the code h_i as the ith component of H.

A set of color codes will be called *complete* if any two nonmatching stimuli are distinguished by at least one code in the set. A set of m color codes is said to be *independent* if no one of them is predictable, even locally, from the other $m - 1$; the simplest way to express this is to assert that $H = (h_1, \ldots, h_m)$ is *independent* if its image is an open subset of Re^m.

Suppose that $H = (h_1, \ldots, h_m)$ is a complete set of color codes and that g is any color code. Then there exists a function ψ on Re^m such that for all (b, β) in $A \times \Omega$

$$g(b, \beta) = \psi[H(b, \beta)]. \tag{6}$$

To prove this, simply use Equation (6) to define ψ. It is well defined because H is complete: if $H(b, \beta) = H(b', \beta')$, then $\beta(b) = \beta'(b')$; therefore, $g(b, \beta) = g(b', \beta')$. Conversely, whether or not H is complete, (6) defines a color code for any function ψ.

We note in addition that if $g(\ , \beta)$ and each $h_i(\ , \beta)$ are continuous and open mappings of A into Re, then the function ψ defined by (6) is a continuous and open mapping of Re^m into Re. Conversely, if the h_i and ψ are continuous and open in the above sense, then (6) defines a continuous and open code g. We shall speak of color codes as *continuous* and/or *open* in the above sense, as functions from A to Re for each fixed specification of viewing conditions. (We nowhere define or use a topological structure on the set Ω of viewing conditions.)

Next, suppose that $G = (g_1, \ldots, g_m)$ and $H = (h_1, \ldots, h_n)$ are both complete and independent sets of color codes. By the foregoing, we have an equation

$$G = \Psi(H) \tag{7}$$

where $\Psi = (\psi_1, \ldots, \psi_m)$ has each component defined by Equation (6). Moreover, since G is complete, Ψ is a one-to-one mapping from its domain in Re^n to its range in Re^m. By independence, the domain and range of Ψ are open sets. If, in addition, all the codes involved are continuous and open, then by the foregoing, Ψ is a homeomorphism and by standard results of dimension theory (Hurewicz and Wallman, 1948, pp. 91–93), $m = n$.

We have arrived at the following theorem:

THEOREM 5.   *Any two complete independent sets of continuous open color codes have the same number of codes.*

The number, obviously, is 3.

We can prove directly that any complete independent set of continuous open color codes consists of exactly 3 codes by noting that such a set of codes induces a continuous open mapping of P onto an open subset of Re^m. Under our assumptions, P is a three-dimensional manifold (hence, $m = 3$), and the mapping in question constitutes a single global coordinate system for P. This, of course, gives another proof of Theorem 5; however, we chose the proof given earlier because it bypasses the topology and the manifold structure on P. The argument can thus serve as a paradigm for consideration of neural and perceptual codes in other domains for which the coordinatization by trichromatic matching is not available. It shows in particular that the cardinality of a complete and independent set of neural codes must be the same as that of a complete and independent set of perceptual codes (assuming continuity and openness of the mappings).

The argument leading to Equation (6) gives as a corollary the fact that every metamer code can be written as a function of the tristimulus coordinates for any color-matching representation since the latter are a complete set of metamer codes.

Among metamer codes there is particular interest in those that are linear on $\langle A, \oplus, * \rangle$. By the above corollary, any linear metamer code can be written as a linear function of the coordinates of any color-matching representation. This generalizes a theorem due to Brindley (1957). In the next section we consider this result in relation to a special set of linear functions, defined by photopigment absorptions.

To summarize, because asymmetric matches exist nonsingularly, trichromatic matching gives rise to a three-dimensional coordinate system (manifold structure) for the set of all equivalence classes under asymmetric matching (set of all focal color perceptions). Invariance of topological dimension then implies that any complete and independent set of perceptual color codes and any complete and independent set of neural color codes must be 3 in number. The result comes as no surprise to perceptual theorists, who long ago constructed the "color solid" using global three-dimensional coordinate systems such as brightness, saturation, and hue.

Theorists who have emphasized underlying physiological mechanisms also anticipated this result. Yet, from the standpoint of neural computations, the consequences are much more restrictive than has generally been recognized. The restriction to three dimensions has generally been attributed to receptor mechanisms, possibly to the receptor photopigments, as will be discussed in the next section. But it is now well known that substantial effects of viewing conditions occur through alterations in the function of post-receptor mechanisms (see, e.g., Augenstein and Pugh,

1977; Cicerone and Green, 1980; and Shevell and Handte, 1983). If there were four or more distinct post-receptor neural computations, affected by viewing conditions, then asymmetric matching would require equating all these computations for the two sides of the match. This could not in general be done with mixtures of three primaries: in general, one cannot solve four or more equations in three variables. The expectation would be that asymmetric matching would be impossible or at best singular.

The conclusion seems inescapable that any neural level that is altered in function by viewing conditions must comprise only three independent computations. (A similar conclusion was reached on the basis of asymmetric matching between a normal and an abnormal eye by Alpern et al., 1983b.) This conclusion does not square well with a picture of the nervous system in which there are myriad interconnections of varying strengths, such that practically every neuron performs a computation at least somewhat distinct from every other. The visual neurons must be organized somehow so that only three distinct computations are involved in perception of color.

15.5.2 Photopigments

The effects of light on the eye are mediated by absorption by substances called photopigments. According to a generally accepted principle of photophysiology known as *univariance* (Rushton, 1972), the effect of the focal stimulus mediated by the jth photopigment depends only on the number of quanta caught by that photopigment. We write this quantum catch as a function ρ_j of the focal distribution, i.e., as $\rho_j(a)$.

These quantum-catch functions are linear on $\langle A, \oplus, * \rangle$ to a high degree of approximation. This is so because bleaching of photopigments is negligible at ordinary light levels (Rushton, 1972). If bleaching could not be neglected, then quantum catch would increase more slowly as intensity increased.

The function ρ_j is often written as an integral over the visible spectrum,

$$\rho_j(a) = \int p_j(\lambda) \, da(\lambda).$$

The function $p_j(\lambda)$ is called the quantized spectral absorption function for the jth photopigment. We shall not make use of this integral representation, only of linearity. (But the topologically minded reader should note that convergence in the weak* topology on A is equivalent to convergence of such integrals for every possible continuous photopigment absorption function.)

Suppose that n photopigments, with corresponding linear quantum-catch functions $\rho_1, \ldots, \rho_n$, mediate all phenomena of color vision. Then these linear functions are *complete* for all metameric codes. By the argument about linear metameric codes in the previous section, any color-matching representation $\varphi = (\varphi_1, \varphi_2, \varphi_3)$ can be written as linear combinations of the ρ_j,

$$\varphi = \alpha\rho, \tag{8}$$

where α is a $3 \times n$ matrix.

We next consider the inverse question. Are the photopigment functions ρ_j metameric codes? Any linear metameric code can be written as a linear combination of the φ_i; hence, if the photopigment functions are metameric codes, (8) can be inverted, and we can infer that $n = 3$. This is what Brindley (1970) called the three-pigment hypothesis.

The three-pigment hypothesis is widely believed. It may very well be true, but there is little decisive evidence that favors it. In the remainder of this section we shall examine two arguments that have been made in favor of this hypothesis. We shall see that neither is particularly strong.

First, we observe that any color code can be written as a function of tristimulus values and viewing conditions, i.e., as $h[\varphi(b), \beta]$ rather than as $h(b, \beta)$. Similarly, any color code can be written as a function of $\rho(b)$ and β, where $\rho(b)$ is the n-dimensional vector of photopigment quantum-catch values. Thus, for a complete independent set of neural color codes $H = (h_1, h_2, h_3)$, we have the relation

$$H[\varphi(b), \beta] = \Gamma[\rho(b), \beta]. \tag{9}$$

For fixed β, the function $H[\ , \beta]$ in (9) is a one-to-one function from a subset of Re^3 to itself, whereas $\Gamma[\ , \beta]$ is a function from a subset of Re^n to Re^3. Let $H^{-1}[\ , \beta]$ be the inverse of $H[\ , \beta]$, for each value of β. Then from Equations (8) and (9) we obtain

$$\alpha\rho(b) = H^{-1}\{\Gamma[\rho(b), \beta]\}. \tag{10}$$

(We have abused notation on the right side of (10) by giving the dependency on β only once, rather than in both Γ and in H^{-1}.)

Equation (10) asserts that H^{-1} undoes the effects of any nonlinearities in Γ, to yield a linear transformation α. Since we are thinking of a set H of neural codes, the function Γ expresses the neural computations linking these codes to photopigment absorptions. We might reasonably expect Γ to be highly nonlinear. For $n > 3$, except for special cases, the nonlinearities in a function from Re^n to Re^3 cannot be unraveled by a transformation in the latter space. Thus it is tempting to suppose that $n = 3$.

This argument is somewhat undercut by the possibility of linear neural color codes, i.e., Γ may in fact be linear. One specific possibility is the *three-receptor hypothesis* of Brindley (1970) and Stiles (1967). According to this, one or more of the three retinal receptor types contains several different photopigments and responds equivalently to quanta caught by any of them.

The second argument concedes the weakness of the first since it assumes that Γ is linear in (9); in addition, it assumes that the effects of viewing conditions can be described by a diagonal transformation of the photopigment quantum-catch functions. That is, a complete independent set of color codes for a reasonable family of viewing conditions is given by

$$H(b, \beta) = \Gamma \nu(\beta)\rho(b), \qquad (11)$$

where Γ is a constant $3 \times n$ matrix and $\nu(\beta)$ is an $n \times n$ diagonal matrix that varies with the viewing conditions.

Because the metamerism relation is invariant for changes in β, Equation (11) can be used to specify metameric pairs for any β. We see from (11) that $a \sim b$ if and only if $\nu(\beta)[\rho(b) - \rho(a)]$ is in the null space of Γ for every β. This means that the null space of Γ is invariant under diagonal transformations. Therefore it must contain $n - 3$ of the coordinate vectors, i.e., all but three of the photopigments actually have no effect on the values of Γ! This brings us back to the three-pigment hypothesis!

The preceding argument is mentioned in various places in the literature. A rigorous version was first given by Krantz (1968a), based on a somewhat different proof involving induced transformations in the vector space of linear transformations of $\langle A, \oplus, * \rangle$.

At one time the argument was thought strong because color adaptation was thought to be representable by a diagonal transformation on ρ. This is the so-called coefficient law of von Kries (1905). It seemed a plausible assumption because photopigments were thought to be dilute and to be bleached during light adaptation. Thus the quantum catch of the jth photopigment would simply be multiplied by a factor $\nu_i(\beta)$ representing the proportion of unbleached pigment under adaptation conditions of β. We now know, however, that photopigments are not dilute (Alpern and Wake, 1977; King-Smith, 1973; S. S. Miller, 1972) and are not appreciably bleached at ordinary levels of light adaptation (Rushton, 1972). Moreover, the failure of Grassmann's additivity and scalar multiplication laws for asymmetric matches, discussed in Section 15.4.4, provides direct evidence against (11).

This second argument, therefore, is not convincing. The three-receptor hypothesis, with more than three pigments, seems perfectly tenable.

Finally, turning away from such arguments to "direct" spectrophotometric tests, we find clear proof that there are at least two human cone pigments with absorption peaks in the middle-to-long wavelength part of the visible spectrum (Rushton, 1963, 1965). But these measurements are not sensitive enough to show that there are *only* two pigments in this spectral region; direct measurements in the short wavelength part of the spectrum are very difficult; and there is growing evidence that a single cone "type" can have a somewhat variable absorption peak (Alpern, 1976, 1979; Alpern and Pugh, 1977; Nagy, MacLeod, Heynman, and Eisner, 1981).

15.5.3 Force Measurement and Dynamical Theory

The main bridge between the empirical laws underlying color measurement and the development of color theory consists of two principles: (1) trichromacy is due to the coding of color by exactly three independent mechanisms, and (2) many or all of the major phenomena of color vision have simple descriptions in terms of these coding mechanisms. In this section we ask what are the analogous principles in mechanics. Why is the static equilibrium of forces three-dimensional? What corresponds to an interpretation in terms of three independent codes for color? The material in this section is largely based on Krantz (1973).

If lights a and b are not metameric, then the effect produced when they are seen side by side is a perception of a color difference. Such an effect is due to discrimination by at least one code. Similarly, if force configuration a is not in equilibrium, then the effect produced is acceleration in some spatial direction. The kinematic space, in which position and acceleration are measured, is three-dimensional. Therefore, the components of acceleration, in three independent directions, act like discrimination by three independent color codes. If the acceleration component vanishes in all three directions, then by a purely geometrical argument it vanishes in every direction, so the configuration of force magnitudes is in E.

In other words, we suppose the following connection between force measurement and dynamics. We have a convex cone $\langle A, \oplus, * \rangle$ of force configurations and a mapping g of A into Re^m. For any a in A, $g(a)$ represents the acceleration (observed for a particular free particle of a particular mass) that is observed when force configuration a is applied. For any real physical situation, $m = 1, 2,$ or 3; but we assume only that m is a known positive integer. We now *define* equilibrium to be absence of acceleration, i.e., $E = \{a \mid g(a) = 0\}$. This is tantamount to making Newton's first law of motion into a definition; this definition will be used to measure force via a representation for the structure $\langle A, \oplus, *, E \rangle$. After we have a representation φ onto $\langle \text{Re}^m, +, \cdot, \{0\} \rangle$, we can state Newton's second law as a postulate of dynamics, namely, $g = \varphi/M$, where M is the mass of the particle under observation.

Given that Newton's first law is a definition of E, the empirical laws that in fact must be postulated, to obtain the force-representation φ that underlies the second law, are simply the laws that make $\langle A, \oplus, *, E \rangle$ an m-dimensional equilibrium structure. The suggestion given above was that the m-dimensionality does not have to be postulated; rather, it follows somehow from the fact that g maps into Re^m. But this suggestion is too simple. The following counterexample shows that a map g into Re^m can be associated with an equilibrium structure of arbitrary, even infinite, dimensionality.

Let $\langle A, \oplus, * \rangle$ be a convex cone, let $\langle V, +, \cdot \rangle$ be an arbitrary vector space over Re, and let φ be a homomorphism of A onto V. Now let g_i be positive-definite quadratic forms on V, $i = 1, \ldots, m$. Define $g: A \to \text{Re}^m$ by

$$g(a) = [g_1\varphi(a), \ldots, g_m\varphi(a)].$$

Since each g_i is positive definite, $g_i\varphi(a) = 0$ if and only if $\varphi(a) = 0$ in V. Therefore,

$$E = \{a | g(a) = 0\} = \{a | \varphi(a) = 0\}.$$

It follows that $\langle A, \oplus, *, E \rangle$ is an equilibrium structure and that φ is a representation of this structure onto $\langle V, +, \cdot, \{0\} \rangle$. Since E was defined from a mapping G into Re^m, and V is arbitrary, the example accomplishes its purpose.[3]

---

[3] This example can be made concrete as follows. Suppose that we observe equilibria as absence of acceleration of the center ring on the force table, i.e., in two dimensions. Suppose that the weights act on the center ring through a "black box" that has three components. The first component calculates the values of n linear functionals over the configuration of force magnitudes:

$$f_j(a) = \sum_\theta \alpha_j(\theta) a(\theta), \qquad j = 1, \ldots, n,$$

where the functions $\alpha_j(\theta), \ldots, \alpha_n(\theta)$ are linearly independent weighting functions. [The vector $\varphi(a) = (f_1(a), \ldots, f_n(a))$ is the representation in V; it has arbitrary dimensionality n.] The second component calculates two positive-definite quadratic forms,

$$g_i(a) = \sum_{j,k} \gamma_{jk}^{(i)} f_j(a) f_k(a), \qquad i = 1, 2.$$

This example works because the acceleration function g is nonlinear. The question is, what sort of restriction on g or E can be formulated to avoid this kind of example? We take our clue from the above remark that acceleration vanishes if and only if it vanishes in each direction of Re^m. We assume, therefore, that *equilibrium in particular directions* is also definable by Newton's first law and also leads to (one-dimensional) equilibrium structures.

A direction in Re^m is identified as a one-dimensional subspace. Typical elements of the set of directions are denoted γ, etc., and the set of all directions is identified as the $(m-1)$-dimensional projective space P^{m-1} (Section 12.2.3). To speak of the component of the acceleration vector $g(a)$ in a particular direction, we need a metric as well as the vectorial structure in Re^m. We assume an inner product $\circ$, and for convenience, we take orthonormal coordinates relative to $\circ$, i.e., if $x = (x_1, \ldots, x_m)$ and $y = (y_1, \ldots, y_m)$ are in Re^m, then $x \circ y = \sum_{i=1}^m x_i y_i$. If R is a subspace of Re^m, we denote by $R^\perp$ the subspace of vectors orthogonal to all the vectors of R. We represent directions γ in P^{m-1} by homogeneous coordinates $(\gamma_1, \ldots, \gamma_m)$ in the orthonormal system for Re^m; thus $\gamma = (t\gamma_1, \ldots, t\gamma_m)$ for every $t \neq 0$, and $g(a)$ is in $\gamma^\perp$ if and only if $\sum_{i=1}^m \gamma_i g_i(a) = 0$.

Let E_γ be defined to be $\{a \mid g(a) \in \gamma^\perp\}$, i.e., a is in E_γ if and only if $\gamma \circ g(a) = 0$. Then E is defined to be the intersection of E_γ over all γ, i.e., a is in equilibrium if and only if it is in equilibrium for each and every direction γ. Our assumptions (testable empirical laws about static equilibria) are that $\langle A, \oplus, *, E \rangle$ is an equilibrium structure and that for each γ in P^{m-1}, $\langle A, \oplus, *, E_\gamma \rangle$ is a one-dimensional equilibrium structure. In effect, E_γ defines a linear code for $\langle A, \oplus, *, E \rangle$.

This latter assumption is violated by the counterexample presented above. There, if two configurations a, b produce no acceleration at all, then $\varphi(a \oplus b)$ also vanishes, and so $a \oplus b$ is also in E; but if a, b do produce an acceleration, with a vanishing component in a certain direction, then $a \oplus b$ can produce a nonvanishing component in that direction because of the nonlinearity of the functions g_i in that example.

Of course, the one-dimensionality of $\langle A, \oplus, *, E_\gamma \rangle$ is a strong assumption, but it is directly testable and very compelling physically. If two force configurations a, b produce opposite rectilinear accelerations along γ and cancel each other when combined, then any third configuration c, not in E_γ, can be cancelled with respect to γ by one or the other of a, b, in the

---

The third component of the black box applies $g_1(a)$, $g_2(2)$ to the center ring, as forces, in two orthogonal directions. If one observed only the external configurations a and the resulting acceleration of the center ring, statics would appear to be n-dimensional, though acceleration is two-dimensional.

sense that there exists $t > 0$ such that either $c \oplus (t * a)$ or $c \oplus (t * b)$ is in E_γ.
We have the following theorem.

THEOREM 6.   *Suppose that* $\langle A, \oplus, *, E \rangle$ *is an equilibrium structure (Definition 3) and for each* $\gamma \in P^{m-1}$, $\langle A, \oplus, *, E_\gamma \rangle$ *is a one-dimensional equilibrium structure, where* E, E_γ *are defined by a mapping g of A onto* Re^m: $a \in E_\gamma$ *iff* $\gamma \circ g(a) = 0$; $E = \cap_{\gamma \in P^{m-1}} E_\gamma$. *Then the following conclusions hold:*

(i) $\langle A, \oplus, *, E \rangle$ *is m-dimensional;*

(ii)   *there exists a homomorphism* φ *of* $\langle A, \oplus, *, E \rangle$ *onto* $\langle \mathrm{Re}^m, +, \cdot, \{0\} \rangle$ *such that for each* $\gamma \in P^{m-1}$, *the component* $\varphi_\gamma(a) = \gamma \circ \varphi(a)$ *is a homomorphism of* $\langle A, \oplus, *, E_\gamma \rangle$ *onto* $\langle \mathrm{Re}, +, \cdot, \{0\} \rangle$;

(iii)   *if* ψ *is any other function with the properties in* (ii), *then* $\psi = \alpha \varphi$, *where* $\alpha \neq 0$ *is a constant;*

(iv)   *if* φ *satisfies the properties of* (ii), *then there exists a real-valued function* α *on A such that* $\varphi(a) = \alpha(a)g(a)$ *for all* $a \in A$.

Note that this theorem asserts the existence of a force representation φ that for any configuration a represents the resultant force by a vector $\varphi(a)$ in Re^m, and represents the component force in direction γ as the component $\varphi_\gamma = \gamma \circ \varphi$. Such a representation is unique up to changes in unit and in orientation convention (multiplication by a positive constant and by ± 1). The uniqueness, of course, is relative to a fixed orthonormal coordinate system in Re^m, for position, velocity, and acceleration measurement; change in kinematic coordinates induces a corresponding change in the force measurement. Moreover, conclusion (iv) of the theorem shows that the acceleration vector $g(a)$ matches the force vector $\varphi(a)$ in direction, differing only by a multiplicative scalar factor $\alpha(a)$. The additional assumption, by which Newton's second law goes beyond statics, is that α is independent of a (and, of course, varies inversely as the mass of the particle under consideration).

Note that we must distinguish carefully between the variable θ, in the domain of force-magnitude configurations, and the variable γ, which specifies components of acceleration or of the resultant force φ. In the case of the force table (Figure 1), we take θ between 0 and 2π radians, and γ between 0 and 2π radians, and we have the following result: if a_θ denotes an element of A whose only nonzero force magnitude is in direction θ, then $\varphi_\gamma(a_\theta)$ is proportional to $\cos(\gamma - \theta)$. That is, for any fixed *direction of effect* γ, we have an input-sensitivity function that gives the sensitivity of that effect to inputs in direction θ. The form is cosinusoidal. In color vision, the role of θ is played by the wavelength variable λ. The role of γ is played by any one physiological coding mechanism or perceptual attribute of colors. Any linear metameric code has a wavelength-sensitivity function [such as $p_j(\lambda)$, Section 15.5.2] specifying the sensitivity of that code or attribute to input of wavelength λ. The structure of the input variable θ or λ plays no role, whereas the structure of the output variable γ is critical.

In the case of the force table the apparatus is specially arranged to give a simple relation between θ and γ. We could easily complicate matters by bending strings around posts, linking them with cross-ties, etc., and thus destroy this simple relation without, however, destroying the important structure of the dimensionality of the representation. (Note that even in the simple force table and its m-dimensional generalization, θ has the structure of the sphere S^{m-1}, whereas γ has the structure of the projective space P^{m-1}.)

Up to this point, the analogy between color metamers and force equilibria is nearly perfect. The measurement structures are formally almost identical, and in the theoretical account of tridimensionality, the kinematic direction variable γ plays the role of physiological codes or perceptual attributes of color. But here the analogy stops. The main concern of color theory is precisely the nature of the underlying codes, whereas in mechanics, observations of components of force and acceleration are simply taken for granted, and theoretical interest centers on dynamical questions, using Newton's second and third laws. The reason for this lies in the homogeneity of kinematic space P^{m-1} and the accessibility of E_γ to direct observation for every γ. Acceleration is always in some particular direction γ, and this is directly observable. No problem at all was associated with the introduction of the sets E_γ into the primitives, for Theorem 6. Moreover, because of this homogeneity, there are no preferred coordinate axes in the representation φ for force. Orthogonal transformations (rotations) of the kinematic space induce the same transformations of the force representation. There is really nothing more to say about the statics of a point mass.

By contrast, color codes are not very accessible to direct observation, and probably they are not homogeneous. The extreme case of inhomogeneity is postulated by the so-called Young-Helmholtz theory, which assumes that there are only three codes subject to measurement. Each code is isolated from the others, and the representation of a particular color is given by the pattern or the vector. Other color theories, especially the modern derivatives of the Hering theory, assume that besides the photopigment functions ρ_1, ρ_2, ρ_3 (Section 15.5.2), various recombinations of these are accessible to direct measurement either by neural means or by psychological experimentation. This is discussed extensively beginning with Section 15.5.4. But the number of distinctive neural recombinations and/or perceptual attributes is limited. So, unlike the case of force, the inhomogeneity and nonlinearity of color codes make color "statics" a principal topic of investigation.

15.5.4 Color Theory in a Measurement Framework

Our discussion of the breakdown of the analogy between particle statics and color statics, in the preceding subsection, suggests a general program for color theory. Just as the equilibrium structure $\langle A, \oplus, *, E \rangle$ for force measurement was enriched by adding additional observables (subsets E_γ of A) that defined linear codes $\varphi_\gamma = \gamma \circ \varphi$, so we must consider additional empirical phenomena for colors, beyond the matching relation $\sim$. The phenomena to be considered may be either physiological or perceptual. We must try to describe these phenomena in terms of codes or sets of codes, linear or nonlinear, and to discover the relations between various codes. The examples of inference about physiological codes, and especially about photopigments, from color-matching data make it clear that such a program involves a subtle interplay of reasoning from physiological to perceptual effects and (principally) vice versa.

Many perceptual phenomena of color are potentially relevant to such a program. Common defects in color vision might involve absence or weakness of just one code, others being unaffected; absence of hue (perception of black, gray, or white) might involve special "zero" values with respect to certain codes; good discriminability between two colors might involve a large difference on one code or perhaps moderate differences on two or three codes; and effects of viewing conditions might be interpreted as interactions, in which codes determining the perception of context stimuli affect those determining the perception of the focal stimulus.

This sort of program fits naturally into a measurement framework. In Theorem 6 we saw one example of the measurement approach. Adding additional empirical relations E_γ (and, for that matter, the kinematic space Re^m, its scalar product $\circ$, and the hookup to the E_γ via g) led to a refined representation theorem in which the single vector-valued function φ represented each E_γ (by $\varphi_\gamma = \gamma \circ \varphi$); requiring so much for φ made the representation unique up to multiplication by a constant (or rotations as well, if changes of coordinates in Re^m are permitted) rather than only up to arbitrary nonsingular linear transformations. Similarly, we want to have a refined theory of color measurement in which various additional empirical relations abstracted from some of the phenomena mentioned above are added to the Grassmann structure $\langle A, \oplus, *, \sim \rangle$, and a homomorphism φ on A is constructed that represents these additional empirical relations by appropriately chosen (linear or nonlinear) numerical relations. The functions of (φ, β), which enter into these additional numerical relations, will be codes. Such a representation might have a sharp uniqueness theorem for φ, say, up to diagonal transformations, in which case the coordinates of φ define three ratio-scale metameric codes in terms of which various color codes and other color phenomena have simple descriptions. We illustrate this idea by considering an example in which this program has not yet succeeded. (Section 15.6 presents more successful examples.) This is the representation of color discriminability or dissimilarity by means of so-called uniform color spaces. Given an arbitrary representation φ, there is no reason at all to suppose that the discriminability or dissimilarity of the colors corresponding to focal distributions a, b under some fixed set of viewing conditions should bear any simple relation to the Euclidean distance between the vectors $\varphi(a)$, $\varphi(b)$. In fact, the ordering of such distances is not preserved under arbitrary permissible (nonsingular linear) transformations, and so it cannot be the same as the ordering of discriminability or dissimilarity of color pairs, except for special homomorphisms φ.

Given a proximity relation $\succsim'$ on $A \times A$, where $(a, b) \succsim' (c, d)$ is interpreted to mean that, under some standard viewing conditions, the discriminability or dissimilarity of a from b is at least as great as that of c from d, one might try to impose the additional requirement on φ [besides conditions (i)–(iv) of Theorem 3]:

(v)
$$(a, b) \succsim' (c, d)$$
$$\text{iff} \quad \sum_{i=1}^{3} [\varphi_i(a) - \varphi_i(b)]^2 \geq \sum_{i=1}^{3} [\varphi_i(c) - \varphi_i(d)]^2.$$

Whether such a φ exists depends, of course, on the properties satisfied by $\succsim'$, both in isolation and in conjunction with $\oplus$, $*$. For example, it is not hard to see that a necessary property is $(a \oplus c, b \oplus c) \sim' (a, b)$, which is quite unlikely to be satisfied. Such a representation, if it existed, would be unique up to similarities (Definition 12.8). Alternatively, one could propose a power metric (Section 12.2.6), with $r \neq 2$, as the numerical representation of $\succsim'$. Since only permutations of axes and similarity transforms preserve the ordering of power-metric distances, if φ existed, the φ_i would be essentially ratio scales with a common unit. If the above representations are inappropriate, e.g., because $(a, b) \succ' (a \oplus c, b \oplus c)$, then other numerical representations must be devised that predict such an effect; any such representation would impose some restriction on the permissible transformations. Some representations that have been suggested involve nonlinear codes defined in terms of a linear representation φ. For example, Hurvich and Jameson (1955) devised a linear representation (for standard viewing conditions) in which φ_1, φ_2 are interpreted as measures of two hue components and φ_3 as a measure of whiteness (see Section 15.6.5). They then defined hue and saturation coefficients (codes) by

$$H = \frac{\varphi_1}{\varphi_1 + \varphi_2}, \qquad S = \frac{|\varphi_1| + |\varphi_2|}{|\varphi_1| + |\varphi_2| + |\varphi_3|}.$$

They described discriminability between lights of constant brightness in terms of H and S, i.e., replacing condition (v) above by

$$\begin{array}{cc} & (a, b) \succsim' (c, d) \\ \text{iff} & |H(a) - H(b)| + |S(a) - S(b)| \\ & \geq |H(c) - H(d)| + |S(c) - S(d)| \end{array}$$

A variety of other nonlinear suggestions have been offered; see Wyszecki and Stiles (1982) for a review.

The attempt to enrich color measurement by representing discriminability and dissimilarity has suffered from the absence of simple empirical laws that could function as axioms and suggest the right sort of geometrical representation. Recent studies of dissimilarity (Ismailov, 1982) suggest that a chordal metric might be appropriate in place of the metrics with additive segments that have heretofore been the focus of attention.

Color theories have differed historically mainly in which phenomena were taken as starting points for an enriched empirical structure. Helmholtz (1896) started from color discrimination (his line-element model); Hering (1878, 1920) and Hurvich and Jameson (1957) used the organization of hue sensations; König and Dieterici (1892/1903), color blindness; Stiles (1939, 1946, 1959), a combination of discrimination and adaptation; and von Kries (1905) and Wright (1946) relied on adaptation. A host of variants and additional authors are omitted from this list.

Regardless of the starting point, any color theory must eventually try to include a simple quantitative representation of all the major color phenomena. The most comprehensive attempt is that of Hurvich and Jameson (1957). Curiously, a measurement-theoretic analysis of their approach (Krantz, 1975b) shows that it is remarkably similar to the approach to force measurement in the preceding subsection. They consider special subsets of "equilibrium" or "unique" colors, which, under standard viewing conditions, yield properties of one-dimensional equilibrium structures exactly like the sets E_γ of force measurement. This aspect of their theory is presented in detail in Section 15.6, along with the König approach, which is analyzed in terms of 2-chromatic Grassmann structures. We do not, however, treat the full range of color phenomena considered by Hurvich and Jameson because the status of many phenomena is like that of the discriminability relation $\succsim'$ described above: simple laws and appropriate numerical representations have not yet been found.

15.6 THE KÖNIG AND HURVICH–JAMESON COLOR THEORIES

The theories in this section are based on a representation of a trichromatic Grassmann structure together with several lower-dimensional homomorphic images. In the case of force (Section 15.5.3), some of the information about equilibrium is obtained by observing projections in any one direction γ. The projections in three independent directions preserve all the information. Here, too, a lower-dimensional projection, which we call a *reduction structure*, preserves some information about the parent structure, and several such reduction structures recover all of the information. A reduction structure can be thought of as a multidimensional metameric code; (a one-dimensional reduction structure defines a metameric code).

In the case of force, the set of reduction structures is homogeneous in that any direction can be chosen for observation. Here, however, reduction structures are generated by special phenomena of color vision: color defects involving a missing code or attribute or direct judgments of perceptual attributes. Special coordinate systems that, it is hoped, are closely related to physiological codes are used to represent these reduction structures.

15.6.1 Representations of 2-Chromatic Reduction Structures

The König approach to color theory starts with the phenomenon of dichromatism. Dichromats are individuals whose color-matching relation can be described by a 2-chromatic Grassmann structure (Definition 2, Axioms 1–3 and Axioms 4(i) and 4(ii), with $m = 2$). A great deal is known about dichromatism and other forms of color defect (Alpern, 1974; Hurvich, 1972; Pokorny, Smith, Verriest, and Pinckers, 1979; Wright, 1946).

Dichromats generally accept normal color matches and in addition match (or "confuse") colors that normals discriminate. Classes of dichromats are distinguished by their characteristic additional confusions.

In fine detail, individuals with "normal" color vision vary in their metameric matches and so do dichromats within a single class (Alpern and Wake, 1977). For a given person with normal color vision, it is possible to find dichromats of each of the main classes that accept that individual's metameric matches. In the following discussion we shall omit this refinement and use some standard "normal" matching relation and "typical" dichromatic matching relations.

The two common types of dichromat, protanopes and deuteranopes, make metameric matches between any two wavelengths in the middle-to-long wavelength portion of the spectrum (wavelengths that, under standard viewing conditions, appear yellowish green, yellow, orange, or red to color normals). But the intensity of a long wavelength light needed to match a given middle wavelength light is much higher for a protanope than for a deuteranope because protanopes lack a photopigment with good sensitivity to long wavelengths.

Tritanopes, a less common type, can make a metameric match between certain pairs of wavelengths, one from the short and one from the middle wavelength region of the spectrum (violet and green for normals under standard viewing conditions).

The metameric matching relations of our typical protanope, deuteranope, and tritanope will be denoted $\sim_P$, $\sim_D$, and $\sim_T$, respectively. Since the typical dichromats accept the metameric matches of the typical normal, we have the set-theoretical relation,

$$\sim_P \cap \sim_D \cap \sim_T \supset \sim .$$

This inclusion relation leads to some elementary but interesting theorems.

Consider first a structure $\langle A, \oplus, *, \sim, \sim_P \rangle$ where $\langle A, \oplus, *, \sim \rangle$ is a 3-chromatic Grassmann structure, $\langle A, \oplus, *, \sim_P \rangle$ is a 2-chromatic Grassmann structure, and $\sim_P \supset \sim$. By Theorems 3 and 4, there is a homomorphism φ of $\langle A, \oplus, *, \sim \rangle$ onto $\langle C, +, \cdot, = \rangle$, in which C is a convex cone that generates a three-dimensional vector space V over the real numbers. Define a new relation, $=_P$, on C by

$$\varphi(a) =_P \varphi(b) \qquad \text{iff} \qquad a \sim_P b.$$

This is well defined since if $\varphi(a) = \varphi(a')$ and $\varphi(b) = \varphi(b')$, then $a \sim a'$ and $b \sim b'$, and hence, since $\sim_P \supset \sim$, then $a \sim_P a'$ and $b \sim_P b'$. By transitivity of $\sim_P$ it follows that $a \sim_P b$ iff $a' \sim_P b'$.

It is easy to verify that $\langle C, +, \cdot, =_P \rangle$ is a 2-chromatic Grassmann structure and that φ is a homomorphism of the reduction structure $\langle A, \oplus, *, \sim_P \rangle$ onto $\langle C, +, \cdot, =_P \rangle$. By Theorems 3 and 4 there is a homomorphism $\bar{\varphi}_P$ of $\langle C, +, \cdot, =_P \rangle$ onto $\langle C_P, +, \cdot, = \rangle$, where C_P is a convex cone that generates a two-dimensional vector space V_P over Re. Let φ_P be the composite map $\bar{\varphi}_P \varphi$ from A via C to C_P; obviously, φ_P is a two-dimensional representation of the reduction structure $\langle A, \oplus, *, \sim_P \rangle$.

We shall show how to use the mapping $\bar{\varphi}_P$ to generate a new representation $(\varphi_1, \varphi_2, \varphi_3)$ for A that combines the properties of φ and φ_P: $(\varphi_1, \varphi_2, \varphi_3)$ represents $\langle A, \oplus, *, \sim \rangle$, and (φ_1, φ_2) represents $\langle A, \oplus, *, \sim_P \rangle$. The reader should notice at this point that the critical assumption in the above argument is that $\sim_P$ includes $\sim$. This permits us to regard $\{a| \text{ for some } u \in \text{Re}^+, a \sim_P u * b\}$ as a hyperplane through $\varphi(b)$ in the representation space C, i.e., as $\{\varphi(a)|\varphi(a) =_P u \cdot \varphi(b)\}$. In the chromaticity diagram, these hyperplanes project onto lines called the dichromatic *confusion lines* for $\sim_P$. This property was paraphrased above by saying that the dichromat accepts the normal metameric matches but, in fact, the empirical point is quite delicate. Dichromats have poor color discrimination, and confusion hyperplanes or confusion lines are an idealization based on the idealized transitive matching relations $\sim_P$. Unless $\sim_P$ is defined very carefully in terms of many cross checks and averages, one obtains large confusion regions; thus "acceptance" of a normal match could be due to poor discrimination and poor technique. So the empirical property that is really invoked is that a carefully determined average normal match $a \sim b$ will also be accepted in careful determinations of the average dichromatic match.

We now derive the representation theorem for the structure $\langle A, \oplus, *, \sim, \sim_P \rangle$ considered above. Because the cone C generates V and C_P generates V_P [property (iii) of Theorem 3], the mapping $\bar{\varphi}_P$ from C to C_P can be extended to a vector-space homomorphism of V onto V_P. We simply let $\bar{\varphi}_P[\varphi(a) - \varphi(b)]$ be defined as $\bar{\varphi}_P\varphi(a) - \bar{\varphi}_P\varphi(b)$. By standard vector-space theory, there is a nonzero vector v_P in V such that v_P generates the *null space* or *kernel* or $\bar{\varphi}_P$, i.e., $\bar{\varphi}_P(v) = 0$ iff there exists $t \in \text{Re}$ with $v = t \cdot v_P$. [The null space of a homomorphism from an n-dimensional space onto an m-dimensional space is $(n - m)$-dimensional.] The vector v_P is often called the *missing primary* (*Fehlfarbe*) for $\sim_P$. It is easily seen that the confusion hyperplanes for $\sim_P$ intersect in the one-dimensional subspace $\{t \cdot v_P\}$; and so, in the chromaticity diagram, v_P projects as the intersection of the dichromatic confusion lines (Exercise 10).

Choose arbitrary vectors v_1, v_2 in V such that $\{v_1, v_2, v_P\}$ is a basis for V. Let $\varphi_1, \varphi_2, \varphi_3$ be the coordinates of $\varphi(a)$ relative to this basis, i.e., for a in A, we have the vector equation

$$\varphi(a) = \varphi_1(a) \cdot v_1 + \varphi_2(a) \cdot v_2 + \varphi_3(a) \cdot v_P.$$

The representation $(\varphi_1, \varphi_2, \varphi_3)$ of $\langle A, \oplus, *, \sim \rangle$ into $\langle \text{Re}^3, +, \cdot, = \rangle$ has the desired properties: i.e., (φ_1, φ_2) provides a representation for $\langle A, \oplus, *, \sim_P \rangle$ into $\langle \text{Re}^2, +, \cdot, = \rangle$. To prove this, suppose that $a \sim_P b$; then $\bar{\varphi}_P\varphi(a) = \bar{\varphi}_P\varphi(b)$. Thus $\varphi(a) - \varphi(b)$ is in the null space of $\bar{\varphi}_P$, i.e., $\varphi(a) - \varphi(b) = t \cdot v_P$. In terms of the φ_i we have

$$[\varphi_1(a) - \varphi_1(b)] \cdot v_1 + [\varphi_2(a) - \varphi_2(b)] \cdot v_2$$
$$+ [\varphi_3(a) - \varphi_3(b) - t] \cdot v_P = 0.$$

Since v_1, v_2, v_P are linearly independent, we have $\varphi_1(a) = \varphi_1(b)$ and $\varphi_2(a) = \varphi_2(b)$. The converse argument is similar, so $a \sim_P b$ iff $[\varphi_1(a), \varphi_2(a)] = [\varphi_1(b), \varphi_2(b)]$.

We have proved the representation part of the following theorem.

THEOREM 7. *Suppose that* $\langle A, \oplus, *, \sim \rangle$ *is a 3-chromatic Grassmann structure,* $\langle A, \oplus, *, \sim_P \rangle$ *is a 2-chromatic Grassmann structure, and* $\sim_P \supset \sim$. *Then there exists a homomorphism* $(\varphi_1, \varphi_2, \varphi_3)$ *of* $\langle A, \oplus, *, \sim \rangle$ *into* $\langle \mathrm{Re}^3, +, \cdot, = \rangle$ *such that* (φ_1, φ_2) *is a homomorphism of* $\langle A, \oplus, *, \sim_P \rangle$ *into* $\langle \mathrm{Re}^2, +, \cdot, = \rangle$. *Moreover,* $(\varphi_1, \varphi_2, \varphi_3)$ *is unique up to matrix transformations*

$$\begin{pmatrix} \varphi_1' \\ \varphi_2' \\ \varphi_3' \end{pmatrix} = \begin{pmatrix} \alpha_{11} & \alpha_{12} & 0 \\ \alpha_{21} & \alpha_{22} & 0 \\ \alpha_{31} & \alpha_{32} & \alpha_{33} \end{pmatrix} \begin{pmatrix} \varphi_1 \\ \varphi_2 \\ \varphi_3 \end{pmatrix},$$

where the matrix (α_{ij}) *has rank* 3.

The proof of uniqueness is left to the reader (Exercise 11). Some further comments about the interpretation of the representation and uniqueness parts of this theorem are presented in Section 15.6.2.

Consider next the situation in which we have the normal matching relation $\sim$ and two distinct reduction structures $\sim_P$ and $\sim_D$, each of which satisfies the hypotheses of Theorem 7. We repeat the first part of the previous argument twice, obtaining homomorphisms $\bar\varphi_P$ and $\bar\varphi_D$ of V onto 2-spaces V_P and V_D, with one-dimensional null spaces generated by Fehlfarben v_P and v_D. The vectors v_P and v_D must be linearly independent; if they were not, $\bar\varphi_P$ and $\bar\varphi_D$ would have the same null space, whence it follows easily that $\sim_P = \sim_D$. Take any third vector v_1 that is linearly independent of $\{v_P, v_D\}$, and let $(\varphi_1, \varphi_2, \varphi_3)$ be the coordinates of φ relative to the basis $\{v_1, v_D, v_P\}$. From the proof of Theorem 7, we see that (φ_1, φ_2) is a representation of $\langle A, \oplus, *, \sim_P \rangle$, and (φ_1, φ_3) is a representation of $\langle A, \oplus, *, \sim_D \rangle$. The representation $(\varphi_1, \varphi_2, \varphi_3)$ satisfying these properties is unique up to the transformations

$$\begin{pmatrix} \varphi_1' \\ \varphi_2' \\ \varphi_3' \end{pmatrix} = \begin{pmatrix} \alpha_{11} & 0 & 0 \\ \alpha_{21} & \alpha_{22} & 0 \\ \alpha_{31} & 0 & \varphi_{33} \end{pmatrix} \begin{pmatrix} \varphi_1 \\ \varphi_2 \\ \varphi_3 \end{pmatrix}.$$

We do not formulate this result as a theorem but go on to the case of three dichromatic types. Note, however, that the coordinate φ_1, determined from $\sim_P$ and $\sim_D$, which is the coordinate common to both two-dimensional representations, is determined as a ratio scale. That means, as we shall see in Section 15.6.2, that the metameric code that is missing in tritanopia can be determined from comparison of normal, protanopic, and deuteranopic color matches.

Given relations $\sim_P$, $\sim_D$, and $\sim_T$, each satisfying the hypotheses of Theorem 7, we construct homomorphisms $\bar{\varphi}_P, \bar{\varphi}_D$, and $\bar{\varphi}_T$, with null spaces generated by vectors v_P, v_D, and v_T. If these are linearly independent, then they form a basis $\{v_T, v_D, v_P\}$, relative to which the coordinates $(\varphi_1, \varphi_2, \varphi_3)$ of $\varphi(a)$ are determined essentially uniquely. We call the relations $\sim_P$, $\sim_D$, $\sim_T$ *independent* if the null spaces v_P, v_D, v_T are linearly independent. A criterion for independence is discussed below. But using this definition we have the following theorem.

THEOREM 8. *Suppose that* $\langle A, \oplus, *, \sim \rangle$ *is a 3-chromatic Grassmann structure;* $\langle A, \oplus, *, \sim_P \rangle$, $\langle A, \oplus, *, \sim_D \rangle$, *and* $\langle A, \oplus, *, \sim_T \rangle$ *are 2-chromatic Grassmann structures;* $(\sim_P \cap \sim_D \cap \sim_T) \supset \sim$; *and* $\sim_P$, $\sim_D$, $\sim_T$ *are independent. Then there exist real-valued functions* $\varphi_1, \varphi_2, \varphi_3$ *on* A *such that* $(\varphi_1, \varphi_2, \varphi_3)$ *is a homomorphism of* $\langle A, \oplus, *, \sim \rangle$ *into* $\langle \mathrm{Re}^3, +, \cdot, = \rangle$ *and* (φ_1, φ_2), (φ_1, φ_3), (φ_2, φ_3) *are respectively homomorphisms of* $\langle A, \oplus, *, \sim_P \rangle$, $\langle A, \oplus, *, \sim_D \rangle$, $\langle A, \oplus, *, \sim_T \rangle$ *into* $\langle \mathrm{Re}^2, +, \cdot, = \rangle$. *The functions are unique up to* $\varphi_i' = \alpha_i \varphi_i, \alpha_i \neq 0, i = 1, 2, 3$.

The previous discussion suffices to establish the existence of the φ_i. We sketch the proof of uniqueness, leaving details to the reader. Suppose that φ_i' are other functions with the same properties; then there is a nonsingular linear transformation ψ of V onto Re^3 such that $[\varphi_1'(a), \varphi_2'(a), \varphi_3'(a)] = \psi\varphi(a)$. Let a, b be chosen with $\varphi(a) - \varphi(b) = v_T$. Since $a \sim_T b, \varphi_i'(a) = \varphi_i'(b)$, $i = 2, 3$. Therefore, $\psi(v_T) = (\varphi_1'(a) - \varphi_1'(b), 0, 0)$. Analogous formulas hold for v_D, v_P, and the conclusion follows using linearity of ψ.

The interpretation of the ratio scales φ_i is discussed in Section 15.6.2.

The loose end we have left is the problem of stating the independence condition in terms of the primitives rather than in terms of the three vectors v_P, v_D, v_T in the representing space. We indicate now how this can be done. The discussion uses projective-geometry methods including Desargues' Theorem, and many readers may want to continue immediately with Section 15.6.2, in which we discuss the interpretation of Theorem 8 in relation to the König color theory. Note that the loose end is not really serious in terms of testability of the hypotheses of Theorem 8. In practice, one uses the fact that $\langle A, \oplus, *, \sim \rangle$ is a 3-chromatic Grassmann structure to obtain a three-dimensional representation. Then the dichromatic confusions are used to determine v_P, v_D, v_T as intersections of confusion loci. One can then test directly whether these vectors are linearly independent.

To discuss independence, it is simplest to work in terms of the chromaticity space (Section 15.3.3, Figure 2) in which proper subspaces of the representing space V are interpreted as points and lines in the projective space P^2. Independence of v_P, v_D, v_T holds if and only if their projections are not collinear in P^2. Let each of these points be the intersection of two confusion lines in P^2 for the corresponding type of dichromatism. By Desargues' theorem three such points are collinear if and only if any two triangles with corresponding sides on intersecting confusion lines are in perspective from a point.

The situation is depicted in Figure 4, where v_P, v_D, v_T are shown as collinear. We let a, b, c be a *confusion triangle*: $a \sim_P b$, $a \sim_D c$, and $(t * b) \sim_T c$. If a', b', c' is another confusion triangle, then the lines through (a, a'), (b, b'), and (c, c') are concurrent at point d. If d is a light, then in the configuration shown, a' is a linear combination of a and d, b' is a linear combination of b and d, and c' is a linear combination of c and d. Conversely, let a, b, c be lights that form a confusion triangle: $a \sim_P b$, $a \sim_D c$, and $(t * b) \sim_T c$. Let d be any light. Take a' to be any linear combination whatsoever of a with d. Let b' be a linear combination of b with d that matches a' for a protanope. By 2-chromacy, such a linear combination b' always exists, though possibly with a negative coefficient (add b and a to match d, or d and a to match b). Likewise, Let c' be a liner combination of c with d that matches a' for a deuteranope. Then independence fails if the tritanopic match can be predicted from the protanopic and deuteranopic matches, in the sense that the constructed b' and c' can be adjusted in energy so that they match for a tritanope. So

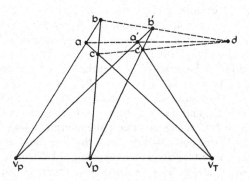

FIGURE 4. Desargues' theorem interpreted as nonindependence of Fehlfarben. The chromaticity loci v_P, v_D, and v_T are collinear if and only if the lights in a confusion triangle abc can be mixed with some light d to yield another confusion triangle $a'b'c'$.

$\sim_P$, $\sim_D$ and $\sim_T$ are independent only if, for any confusion triangle a, b, c, no linear combinations of a, b, c with a fourth light d also form a confusion triangle. This condition is also sufficient for independence.

15.6.2 The König Theory and Alternatives

In this section we shall assume that viewing conditions are specified in some standard ways. Thus we drop the dependence on β from our equations and likewise, for the time being, the distinction between metameric codes and color codes.

The Young–Helmholtz theory assumes that there are three independent linear codes (usually assumed to be identical with the photopigments functions ρ_j of Section 15.5.2) and that there are no other codes representing additional physiological mechanisms or psychological attributes of color.

Under this theory, the simplest account of dichromatism is given by the assumption that exactly one code is absent. This predicts that there can be precisely three kinds of dichromatism, one for each code that can be lacking; that each type of dichromatism gives rise to a reduction structure, as in Theorem 7; and that the three Fehlfarben are independent, as in Theorem 8. This elaboration of the Young–Helmholtz theory is usually associated with the name of König. In fact, the functions φ_1, φ_2, φ_3 of Theorem 8 form a set of three independent linear codes with exactly these properties: one code is missing in the representation of each kind of dichromatism.

Theorem 8 thus provides a measurement procedure whereby these three linear codes can be determined uniquely up to ratio-scale transformations with independent units. The procedure is based on determination of the dichromatic confusion loci and their intersections. This procedure was first carried out by König and Dieterici (1892/1903). A widely accepted modern determination is that of V. C. Smith and Pokorny (1972).

Once this is done, it remains to give a quantitative account of various other color phenomena in terms of these measured codes. A sketch of a qualitative account and of the various difficulties it encounters is given by LeGrand (1968).

One important point to emphasize in connection with both the König theory and the alternatives to be mentioned is the distinction between the Fehlfarben v_P, v_D, and v_T and the codes φ_1, φ_2, φ_3 obtained by taking the v's as a coordinate basis. (In terms of linear algebra, the distinction is that between a basis φ_1, φ_2, φ_3 for the space of all linear functionals (linear codes) on the representation space V and a dual basis for the space V itself.) To determine one of the v's, only the corresponding type of dichromatism need be considered. But, as shown in the uniqueness part of Theorem 7, this does not determine any of the φ's. Determination of a linear functional (up to multiplication by a constant) requires specification of its (two-dimensional) null space. The code φ_3, missing in protanopes but present in deuteranopes and tritanopes, vanishes on v_D and v_T; therefore, those two vectors generate its null space and determine it up to choice of unit. Thus the code φ_3 can be determined from the color matching behavior of the three classes of observers that *have* that code: normals, who specify V, and deuteranopes and tritanopes, who determine v_D, v_T. In general, under the Young–Helmholtz–König theory, the code missing in any one form of dichromatism is determined entirely from the matches of normals and of the other two types of dichromats, without reference to the behavior of the dichromat who is missing that code.

There is now fairly strong spectrophotometric evidence (Rushton, 1963, 1965) that protanopes are missing one cone pigment (we denote it ρ_3) and deuteranopes another (ρ_2). These two pigments (called erythrolabe and chlorolabe, respectively) have characteristics similar to the predictions made long ago from the König theory. Quite possibly, tritanopes are missing a third cone pigment; this is hard to determine directly by spectrophotometry, but the presumed pigment could be very accurately predicted from careful measurements of protanopic and deuteranopic confusions, by the method of Theorem 8.

If we leave the framework of the Young–Helmholtz theory, and assume that various additional codes $h_i = \Gamma_i(\rho_1, \rho_2, \rho_3)$ play significant roles in the representation of color relations, then it is not sufficient to postulate that a single code photopigment is missing in any one form of dichromatism. Rather, one must also specify how the various functions Γ_i are modified to take account of the missing photopigment.

Assuming that the φ_i, measured via Theorem 8, coincide with the photopigment codes ρ_i, a satisfactory account of color blindness must therefore specify the manner in which other codes of form $\Gamma_i(\rho_1, \rho_2, \rho_3)$ operate in dichromats. A very simple suggestion is that ρ_2 substitutes for ρ_3 in protanopes and ρ_3 substitutes for ρ_2 in deuteranopes. Thus normal code $\Gamma_i(\rho_1, \rho_2, \rho_3)$ would become $\Gamma_i(\rho_1, \rho_2, \rho_2)$ in the protanope and $\Gamma_i(\rho_1, \rho_3, \rho_3)$ in the deuteranope. The presumed mechanism of this *substitution theory* is that retinal cones of the protanope that normally contain erythrolabe are filled instead with chlorolabe but their neural connections remain unchanged. The reverse substitution takes place in the deuteranope. This suggestion predicts strong similarity between the color perceptions of protanopes and deuteranopes.

Much might be learned about normal color codes from the study of dichromatic ones, especially if there were a good basis for accepting the above substitution idea or some alternative. The development of theory along such lines has possibly been hampered by the oversimple notion that loss of a photopigment gives a full account of dichromacy. On the other hand, various theories that focused on neural codes suffer because they do not incorporate the idea of loss of photopigment.

The study of dichromatic color perception is made more difficult by uncertainty about how dichromats use language to describe colors. Important information has been obtained from a relatively small number of instances of unilateral dichromacy, in which one can assume that language is used to describe colors seen through the dichromatic eye in the same way that it is used to describe colors seen through the normal eye. In addition, one has the possibility of making asymmetric matches between eyes. Unfortunately, the study of such cases has thus far not been sufficient to make much progress. Some of the results in the literature pose severe puzzles (Alpern et al., 1983a, 1983b; MacLeod and Lennie, 1976).

15.6.3 Codes Based on Color Attributes

Just as force can be analyzed by observing components in particular directions, so color can be analyzed by judging particular attributes. In fact, "color," as used in this chapter, has been defined by certain attributes, usually hue, saturation, and brightness. That is, if two lights match in color, then they have identical hue, identical saturation, and identical brightness; and two lights that match in these three respects are said to match in color, though they may differ in other attributes, e.g., gloss or size and uniformity of the field of light being judged. Thus saturation and brightness are codes for color whereas glossiness is not. Hue can also be considered a code, though it is properly represented by a circle rather than by a linear order; one passes from red to orange, yellow, yellow-green, green, blue-green, blue, violet, purple, red-purple, and back to red.

For the analogy with force to have any bite, it is necessary that some codes, defined by judged attributes, satisfy simple laws, as do the codes defined by the sets E_γ of force-magnitude configurations. Of particular interest are judgments that lead to linear codes. The idea that brightness judgments yield a linear code has been extensively debated; the law that would have to be satisfied is called Abney's law, though it was stated by Grassmann (1854) as his fourth law (see below).

The Hering theory of opponent colors (1878, 1920) is based on the observation that certain color attributes have the property of being unitary or psychologically simple and that pairs of these unitary attributes satisfy laws of incompatibility. For example, redness is a unitary attribute, and so is yellowness; orangeness, by contrast, is a simultaneous experience of two attributes, namely, redness and yellowness. Hering noted six unitary attributes, grouped in three pairs of incompatible attributes: redness/greenness, yellowness/blueness, and whiteness/blackness. For example, brown colors are analyzed in this system as having simultaneously the attributes blackness and yellowness and, often, redness. No color is simultaneously reddish and greenish, nor yellowish and bluish, nor whitish and blackish. We emphasize that the analysis is in terms of sensation mixtures not light mixtures nor pigment mixtures. For example, yellow and blue pigments usually mix to make yellow-green; yellow and blue lights can be mixed to make white; but no *sensation* is a mixture of yellow and blue, least of all white, which has neither yellow nor blue in it.

These observations suggest that the code vector (h_1, h_2, h_3), on which color perception is based, consists of a code h_1 for redness/greenness, taking a special neutral or zero value for the so-called unique yellowish, bluish, or achromatic colors that are neither reddish nor greenish; a code h_2 for yellowness/blueness, which is zero for unique reddish, greenish, or achromatic colors that are neither yellowish nor bluish; and a code h_3 for whiteness/blackness, which is zero for mid-gray colors or for mid-lightness of colors of any hue.

The qualitative empirical facts about unitariness and incompatibility of attributes do not, by themselves, lead to any quantitative measurement of codes h_1, h_2, h_3 with the above properties. From a measurement standpoint, what is needed is some richer empirical structure, involving the above attributes, that leads to measurement procedures for h_1, h_2, and h_3. Such a structure is specified in the next two sections.

15.6.4 The Cancellation Procedure

Throughout this section we again assume a fixed standard set of viewing conditions, and so we do not include dependency on viewing conditions explicitly in codes. Below we specify more precisely what these standard conditions should be.

We denote by A_1 the set of all stimuli in A that are neutral with respect to redness/greenness, i.e., that are either yellowish, bluish, or achromatic but neither reddish nor greenish. This set is the equilibrium set for the redness/greenness code, just as, in the case of force, E_γ is the equilibrium set for components of force in direction γ. The direct measurement of a force component in direction γ can be carried out by finding the multiple $t * a$ of a fixed configuration a that brings the given configuration into equilibrium in that direction. Likewise, let a_1 be some fixed greenish light in A. For any reddish light c, form combinations $(t * a_1) \oplus c$ for different values of t. For sufficiently small t, the combination remains reddish; for sufficiently large t, it appears greenish; and for some intermediate value of t, it passes through red/green neutrality, i.e., $(t * a) \oplus c$ is in A_1. Clearly, the value of t that produces neutrality varies as a function of the reddish light c and provides some sort of index of the redness of c. This procedure of cancelling the redness by admixing varying amounts of a fixed greenish light was introduced by Jameson and Hurvich (1955). Similarly, they used varying amounts of a fixed reddish light to cancel greenness, and likewise for blueness and yellowness. In the latter cases, the equilibrium set is denoted A_2, which is the set of lights that are greenish, achromatic, or reddish, but neither yellowish nor bluish.

The quantification of opponent-colors theory by additive cancellation satisfies the requirement outlined in the preceding subsection: it combines judgment of opponent attributes (really, only judgment of neutrality with respect to an opponent pair) with the additive structure, $\oplus$, $*$ on A.

The simplest law that one could reasonably hope might be satisfied by such cancellation observations is the one satisfied by the E_γ in the case of force, namely, the law that the $\langle A, \oplus, *, A_i \rangle$ are one-dimensional equilibrium structures for $i = 1, 2$. In addition, the law corresponding to $E_\gamma \supset E$ must also hold. Here we have no overall equilibrium structure E, but a simple analogue can be stated in terms of the compatibility of A_i with the metamerism relation $\sim$. We assume that if $a \sim b$, then a is in A_i if and only if b is in A_i, $i = 1, 2$. This is surely true empirically; it is a special case of the fact that redness/greenness and yellowness/blueness are codes.

Viewing conditions not only must be specified and held fixed but also must be properly selected if there is to be any hope that $\langle A, \oplus, *, A_i \rangle$ will turn out to be an equilibrium structure. Suppose, for example, that a red surround were used. If light a is in A_1, then we know that a appears reddish in an achromatic surround. Moreover, if the energy of a is increased until it greatly exceeds that of the surround, the light again appears reddish. That is, $a \in A_1$, but $t * a \notin A_1$, violating Axiom 3 of Definition 3. The cancellation measurements of Jameson and Hurvich (1955) were, in fact, carried out with carefully selected viewing conditions: achromatic surround field and achromatic adaptation.

We note that similar considerations apply in the determination of complementary colors. The set of achromatic lights is the set $A_1 \cap A_2$: lights that appear neither reddish, greenish, yellowish, nor bluish. Again, this set varies with viewing conditions. Two lights are usually defined to be complementary if some linear combination of them is in $A_1 \cap A_2$; but under this definition, which pairs are complementary depends on viewing conditions.

If $\langle A, \oplus, *, A_i \rangle$ is a one-dimensional equilibrium structure, then by Theorems 3 and 4, it has a one-dimensional representation φ_i onto $\langle \text{Re}, +, \cdot, \{0\} \rangle$. This representation φ_i is determined by the cancellation procedure. Fix a_i, b_i, not in A_i, such that $a_i \oplus b_i$ is in A_i. Then by one-dimensionality, for any c not in A_i, there exists $t > 0$ such that one of the following holds:

$$(t * a_i) \oplus c \in A_i ; \qquad (t * b_i) \oplus c \in A_i .$$

Define $\varphi_i(c) = t$ if the first holds, $-t$ if the second holds, and 0 if c is in A_i.

It can be shown that $\langle A, \oplus, *, A_i \rangle$ is a one-dimensional equilibrium structure compatible with $\sim$ if and only if (i) there exists a_i, b_i such that every c not in A_i can be cancelled by either a_i or b_i and (ii) φ_i, determined by cancellation, is a linear code for $\langle A, \oplus, *, \sim \rangle$. Jameson and Hurvich (1955) showed, in fact, that the cancellation functions φ_i, determined as above, are well approximated by linear combinations of the CIE color-matching functions; hence, they seemed to be linear codes for $\langle A, \oplus, *, \sim \rangle$.

Another way to look at things is to note that when the conditions of an equilibrium structure are satisfied, the cancellation measure $\varphi_i(c) = t$, defined by $(t * a_i) \oplus c$ in A_i, is independent of the choice of a_i, except for change of units (Exercise 13).

The third opponent pair, whiteness/blackness, does not give rise to a cancellation method via $\oplus$, $*$. If it did, we would have three independent linear codes φ_1 φ_2, φ_3, each mapping onto all of Re. They would necessarily yield a representation of the whole Grassmann structure $\langle A, \oplus, *, \sim \rangle$ onto Re^3; thus we would have an equilibrium structure $\langle A, \oplus, *, E_- \rangle$ (see Definition 4). But the Grassmann structure $\langle A, \oplus, *, \sim \rangle$ is, in fact, proper; no element of A can be mapped onto $(0, 0, 0)$ by any representation. What this means empirically is that adding light can never add blackness, cancelling out whiteness. When lights are observed in a dark surround with the dark-adapted eye, blackness does not occur. There are no browns, deep blues, or deep purples; all colors are whitish, and the third attribute is unipolar. Only the presence of contrasting lights produces blackness.

To obtain a third linear code, corresponding to whiteness or to whiteness/blackness, one must proceed in some other way. The most satisfactory situation arises if the whiteness (or brightness) ordering is itself a linear code. That is, we have a weak ordering $\succsim_B$, on A, such that $\langle A \oplus, *, \sim_B \rangle$ is a 1-chromatic Grassmann structure. The additivity law for brightness matching, $a \sim_B b$ if and only if $a \oplus c \sim_B b \oplus c$, is known as Abney's law or Grassmann's fourth law. Much work has been devoted to analyzing differences among methods of brightness comparison and, in particular, to finding a method that satisfies this law at least approximately. We shall not enter into the details of this very complex matter here. Suffice it to say that the CIE luminance function (the Y coordinate in the X, Y, Z system; see Section 15.4.3) is a linear code (by definition) that gives quite accurate predictions of certain kinds of comparisons of brightness or lightness and that is grossly wrong for other methods of comparison. Jameson and Hurvich used this function as their whiteness code. They treated blackness in the context of a more general treatment of contrast effects.

In the next section, we state two representations, one using only the two equilibrium structures defined by cancellation, the second one using an additional relation $\sim_B$ satisfying Grassmann's fourth law.

15.6.5 Representation and Uniqueness Theorems

THEOREM 9.   Let $\langle A, \oplus, *, \sim \rangle$ be a proper trichromatic Grassmann structure (Definitions 2 and 4), and let A_1, A_2 be subsets of A. Suppose that the following axioms hold, for all $a, b \in A$, $t \in \mathrm{Re}^+$, and for $i = 1, 2$:

1. (Code axioms)   If $a \sim b$ and $a \in A_i$, then $b \in A_i$.

2. (Linearity)   If $a \in A_i$, then $t * a \in A_i$ and $b \in A_i$ iff $a \oplus b \in A_i$.

3. (Unique-hue complements)   There exist a_i, $b_i \in (A_1 \cup A_2) - A_i$ such that $a_i \oplus b_i \in A_i$.

Then the following conclusions hold:

(i) $\langle A, \oplus, *, A_i \rangle$ is a one-dimensional equilibrium structure, $i = 1, 2$.

(ii) $\langle A, \oplus, *, A_1 \cap A_2 \rangle$ is a two-dimensional equilibrium structure and $\langle A_1 \cap A_2, \oplus, *, \sim \rangle$ is a proper 1-chromatic Grassmann structure.

(iii)   There exist real-valued functions φ_1, φ_2, φ_3 on A such that $(\varphi_1, \varphi_2, \varphi_3)$ is a representation for $\langle A, \oplus, *, \sim \rangle$, (φ_1, φ_2) for $\langle A, \oplus, *, A_1 \cap A_2 \rangle$, φ_i for $\langle A, \oplus, *, A_i \rangle$ $(i = 1, 2)$, and φ_3 is a representation for $\langle A_1 \cap A_2, \oplus, *, \sim \rangle$.

Other functions φ_1', φ_2', φ_3' satisfy conclusion (iii) iff there exist $\alpha_1 \neq 0$, $\alpha_2 \neq 0$, $\beta_1, \beta_2 \in \mathrm{Re}$, and $\beta_3 \neq 0$ such that

$$\begin{pmatrix} \varphi_1' \\ \varphi_2' \\ \varphi_3' \end{pmatrix} = \begin{pmatrix} \alpha_1 & 0 & 0 \\ 0 & \alpha_2 & 0 \\ \beta_1 & \beta_2 & \beta_3 \end{pmatrix} \begin{pmatrix} \varphi_1 \\ \varphi_2 \\ \varphi_3 \end{pmatrix}. \tag{12}$$

THEOREM 10.   Let $\langle A, \oplus, *, \sim, A_1, A_2, \sim_B \rangle$ satisfy the hypotheses of Theorem 9 and $\sim_B$ be a binary relation on A such that $\sim_B \supset \sim$ and $\langle A, \oplus, *, \sim_B \rangle$ is a proper 1-chromatic Grassmann structure. Then

(i) $\sim_B = \sim$ on $A_1 \cap A_2$; and

(ii)   *the functions* φ_i *in conclusion* (iii) *of Theorem 9 can be chosen such that* φ_3 *is a representation for* $\langle A, \oplus, *, \sim_B \rangle$. *They are then unique up to* $\varphi_i' = \alpha_i \varphi_i$, *where the* α_i *are nonzero.*

We shall discuss a number of points concerning the hypotheses, conclusions, and proofs of these two theorems.

Axioms 1–3 of Theorem 9 will be referred to by name (code, linearity, and unique-hue complements). Structures satisfying the hypotheses of Theorem 9 could be called Hurvich–Jameson structures.

The code axiom and linearity axiom were discussed in the preceding section. The linearity axiom imposes a very strong constraint on the neural mechanism of the chromatic codes; it is the most interesting and vulnerable empirically. Formally, it incorporates Axioms 2 and 3 of equilibrium structures. Thus, to establish conclusion (i) of Theorem 9, it suffices to show that each element a of A can be equilibrated relative to A_i and that the equilibrium structure is one-dimensional. Let a_i', b_i' be any elements not in A_i such that $a_i' \oplus b_i'$ is in A_i. One-dimensionality means that for any a in A there is an equation

$$a \oplus (t_i * a_i') \sim_i u_i * a_i',$$

where $\sim_i$ is the Grassmann relation induced by A_i. It follows that $a \oplus [(t_i - u_i) * a_i']$ is in A_i, or a is in A_i, or $a \oplus [(u_i - t_i)*b_i']$ is in A_i, accordingly as $t_i > u_i$, $t_i = u_i$, or $t_i < u_i$. (Add $u_i * b_i'$ to both sides of the above equation.) Thus any such a_i', b_i' can be used as cancellation stimuli for all a not in A_i. We prove this for the unique-hue complements a_i, b_i, which are in $A_1 \cup A_2$ to start with (unique hues), so that $a_i \oplus b_i$ is achromatic (in $A_1 \cap A_2$). From this, conclusion (i) follows; hence, it follows that any pair a_i', b_i' with the above properties can be used as cancellation stimuli whether or not they are unique hues.

A slightly more general theorem could be proved replacing the linearity and unique-hue complements assumption by conclusion (i), made into an axiom, together with the proviso that $A_1 \neq A_2$. Conclusion (ii) could no longer be obtained, but we could still find a representation $(\varphi_1, \varphi_2, \varphi_3)$ for $\langle A, \oplus, *, \sim \rangle$ such that, for $i = 1, 2$, φ_i is a representation for $\langle A, \oplus, *, A_i \rangle$. The unique-hue complements property is necessary, however, to obtain conclusion (ii) of Theorem 9, i.e., that the set of achromatic lights defines a two-dimensional equilibrium structure. See Exercises 14 and 15 for clarification of the points raised in this paragraph.

Conclusion (iii) turns on the relation between cancellation and metameric matching. Let us choose a unique green (a_1), a unique blue (a_2) and a white $(a_3 = a_1 \oplus b_1)$ light as primaries for tristimulus colorimetry. Suppose that a is an orange light. We cancel the redness with a_1, i.e., we find t_1

such that $a \oplus (t_1 * a_1)$ is in A_1. Then we cancel the yellowness with a_2, finding t_2 such that $a \oplus (t_1 * a_1) \oplus (t_2 * a_2)$ is in A_2. Since the blue was in A_1, this is actually in $A_1 \cap A_2$, i.e., it is achromatic. We then adjust the intensity of a_3 to match, i.e., we obtain the metameric match,

$$a \oplus (t_1 * a_1) \oplus (t_2 * a_2) \sim t_3 * a_3.$$

The tristimulus coefficients are $(-t_1, -t_2, t_3)$. This shows that the cancellation coefficients, which represent the one-dimensional equilibrium structures defined by A_i, are also part of a tristimulus representation for $\sim$. (The corresponding argument for other lights a is analogous; the first two tristimulus coefficients can be positive, zero, or negative independently.) A similar argument shows, in fact, that a_1, a_2, a_3 do form a basis or a set of primaries. If there were a nontrivial color equation involving them, we could cancel out a_1, a_2 on one side by adding corresponding amounts of b_1, b_2. The resulting light would be achromatic. By the code axiom, so would the other side, with the same amounts of b_1, b_2 added. Thus the amounts of a_1, a_2 must have been the same on both sides. The match then involves a_3 alone, which is impossible (this is what it means to have a proper structure). A rigorous proof of all these statements is given in Section 15.7.2.

Note that φ_3 is not a ratio scale because any linear functions that vanish on $A_1 \cap A_2$ can be added to it, as indicated in Equation (12). Because the structure is proper, however, φ_3 can be chosen so that it does not vanish. Any such φ_3 defines an ordering of A such that $\langle A, \oplus, *, \sim_{\varphi_3} \rangle$ is a proper 1-chromatic reduction structure. Theorem 10 discusses the inverse process: given $\sim_B$ empirically, choose φ_3 to represent it. This is done by evaluating the "brightness" components of the chromatic primaries a_1, a_2. Let $a_i \sim_B \beta_i * a_3$, $i = 1, 2, 3$ ($\beta_3 = 1$). The color equation $a \sim \Sigma \varphi_i(a) * a_i$ reduces to the $\sim_B$ equation $a \sim_B [\Sigma \beta_i \varphi_i(a)] * a_3$; so $\Sigma \beta_i \varphi_i$ is the desired representation.

The basis, relative to which the coordinates are φ_1, φ_2, and $\Sigma \beta_i \varphi_i$, consists of the vectors

$$v_1 = \varphi(a_1) - \beta_i \varphi(a_3)$$
$$v_2 = \varphi(a_2) - \beta_2 \varphi(a_3)$$
$$v_3 = \varphi(a_3).$$

Intuitively, v_1 is pure green (with the whiteness removed), and v_2 is pure blue. In practice, there are no lights a_i' with $\varphi(a_i) = v_i$; chromatic components are always accompanied by whiteness.

If, in addition, we require that φ_3 preserve the brightness *ordering* $\succsim_B$, then the coefficient α_3 in Theorem 10 is constrained to be positive.

15.6.6 Tests and Extensions of Quantitative Opponent-Colors Theory

Of the axioms in the preceding section, linearity is most crucial. The code and unique-hue complement conditions are more or less obviously true, and additivity of $\sim_B$ (Theorem 10) is of less consequence. Hurvich and Jameson (1951a) found that under the conditions of neutral adaptation and surround considered here, the spectral unique hues (blue near 475 nm, green near 500 nm, and yellow near 575 nm) are invariant over a considerable range of intensities. This supported the scalar multiplication part of the axiom. Linearity was also supported roughly by the fact that their values of ϕ_1 and ϕ_2 measured by cancellation were reasonably approximated by linear combinations of the CIE Standard Observer tristimulus values that had previously been suggested by Judd (1951, p. 831) as a representation for opponent-colors theory.

After the axiomation of Theorem 9 was developed, direct tests of linearity were performed. Larimer, Krantz, and Cicerone (1974) found strong support for both the scalar multiplication and the additivity parts of the axiom, in the case of redness/greenness equilibria. This finding has since been confirmed repeatedly using many different stimulus combinations (Donnell, 1977; Ejimo and Takahashi, 1984; Shevell, 1978; Werner and Wooten, 1979). However, Larimer, Krantz, and Cicerone (1975) discovered that yellowness/blueness equilibria violate the scalar multiplication property when the color of the equilibrium light is reddish. This was also confirmed by the experiments cited above (except Shevell, who did not test yellowness/blueness cancellation).

Figures 5 and 6 show tests of linearity using data from an unpublished study by J. Larimer, L. B. Rank, M. L. Donnell and D. H. Krantz. Figure 5 shows the approximate invariance of the wavelengths of unique blue and of unique yellow for three different individuals as intensity is varied over about $4 \log_{10}$ units, a factor of 10,000. The small deviations from perfectly vertical lines are statistically reliable for each observer (except LR in the unique blue), but they are visually insignificant and inconsistent from person to person. They could be due to criterion fluctuations over the widely changing brightness levels. In fact, the data shown are averaged over four repetitions of the experiment on different days; the maximum difference between any two of the eight intensity levels is usually smaller and at most about the same as the maximum difference between replications of the experiment on different days.

It is important to test scalar multiplication not only for single wavelengths but for mixtures of the sort used in the cancellation method. Figure 6a shows the approximate invariance of intensity ratios that achieve redness/greenness equilibrium for various pairs of wavelengths, averaged over

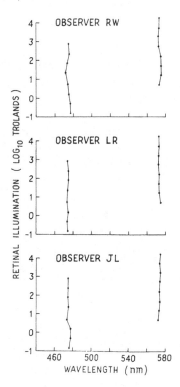

FIGURE 5. Combinations of wavelength and intensity for monochromatic lights judged to be neither reddish nor greenish by three observers. Adaptation and surround were neutral. Combinations with wavelengths near 480 nm appeared blue, those with wavelengths near 580 nm appeared yellow. Closure under scalar multiplication of redness/greenness equilibrium set A_1 requires that the curves be vertical straight lines. Deviations from this are smaller than day-to-day fluctuations in determining these combinations. (Unpublished data collected by J. Larimer, L. B. Rank, M. L. Donnell, and D. H. Krantz.)

three observers. Remarks similar to those made above are valid concerning the positive statistical significance and negative visual significance of the deviations from horizontality of the plotted lines.

When the same experiment was performed with yellowness/blueness cancellation, a very different result was obtained. Figure 6b shows the failure of invariance of intensity ratio for the mixture of 610 and 480 nm that achieves yellowness/blueness equilibrium. As intensity of the blue component (480 nm) is increased, the intensity of the yellowish component (610 nm) that cancels the blueness must be disproportionately increased.

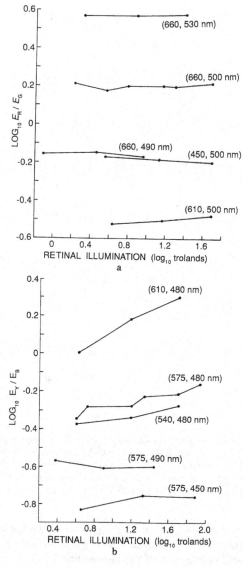

FIGURE 6. Pairs of wavelengths, and their intensities, judged to be neither reddish nor greenish (a) or judged to be neither yellowish nor bluish (b), averaged for three observers. Adaptation and surround were neutral. Closure under scalar multiplication requires horizontal straight lines, that is, the ratio of intensities for a wavelength pair should be invariant as the absolute level changes. This is approximately satisfied for redness/greenness equilibria (a), which vary in appearance from yellowish [(660,530) and (610,500)] to near-white [(660,500)] to bluish [(660,490) and (450,500)]. It is violated (b) for yellowness/blueness equilibria, especially for the wavelength pair (610,480), which appears reddish at equilibrium. (Unpublished data collected by J. Larimer, L. B. Rank, M. L. Donnell, and D. H. Krantz.)

This is the invariable finding, though there is considerable variation between individuals in its magnitude (Donnell, 1977). Such noninvariance of the intensity ratio is reduced or abolished by moderate adaptation to long-wavelength red lights (unpublished data collected by S. K. Shevell and D. H. Krantz). So it seems likely that there is a nonlinearity in the contribution of the long-wavelength sensitive cones to the yellow/blue opponent code.

From the standpoint of the measurement theorems, these results might simply mean that a more complex representation is needed for the subset A_2, lights that are in equilibrium for yellowness/blueness under the standard viewing conditions. However, there is no guarantee that the nonlinearity is one that can readily be characterized by a simple axiom system. In the absence of a good model of the nonlinearity, one cannot accurately calculate the cancellation values for complex mixtures of wavelengths from the values measured wavelength by wavelength at a fixed intensity.

The nonlinearity of yellowness/blueness cancellation is just one of many nonlinearities that arise in the phenomena of color vision. The best known is the Bezold–Bruecke phenomenon, the failure of $t * a$ to be invariant in hue as t varies. As Purdy (1931) demonstrated, there are three wavelength values that show invariance in hue, and these are roughly the same as unique blue, green, and yellow. For single wavelengths whose hue is a mixture of the two opponent attributes, yellowness/blueness has a higher threshold and once above it increases more rapidly with intensity than redness/greenness; thus, as intensity increases, orange and yellow-green become yellower and violet and blue-green become bluer.

To extend their theory to deal with this and related phenomena, Hurvich and Jameson assumed that perceived redness/greenness and yellowness/blueness are represented by codes h_1 and h_2 that vary nonlinearly with ϕ_1 and ϕ_2, respectively. It should be noted that such nonlinearities do not explain the noninvariance of yellowness/blueness equilibria; but this method of extending the theory has more severe failings. For one thing, the idea that yellowness/blueness increases faster than redness/greenness is valid for yellow-red (orange), yellow-green, blue-green, and some blue-red combinations (including those at the violet end of the spectrum), but it is false for some other blue-red combinations. This was pointed out by Purdy and confirmed in an unpublished study by C. Hill and D. H. Krantz. Second, and most critical, the postulate that the value of h_i is determined (linearly or nonlinearly) by the value of ϕ_i leads to the simple prediction that stimuli that appear equally reddish (say) will have their apparent redness cancelled by the same intensity of green. The falsity of this prediction can be shown from the data of Hurvich and Jameson (1951b), who showed that increasing intensity of desaturated broadband lights eventually produces a decrease and even a disappearance of the hue quality. But increasing intensity can only increase the cancellation coefficient. More generally, increases in the predominant attribute of a color produce decreases in other attributes. This phenomenon was identified as a problem for color theory by Krantz (1974) who used the term "veiling." Treatments of this phenomenon from various points of view can also be found in Yager and Taylor (1970), Moeller (1976), and Ingling, Russell, Rea, and Tsou (1978).

The fact that perceived redness/greenness and yellowness/blueness are not determined by the cancellation measures undercuts the original rationale upon which Jameson and Hurvich based their method, but it does not make the cancellation phenomenon less important nor does it undercut the great usefulness that derives from linearity of redness/greenness equilibria. Theorem 9 still asserts that ϕ_1 is a linear metameric code (under the correct standard viewing conditions); hence, it is a linear combination of any set of tristimulus coordinates and of the photopigment absorption functions [Equation (8)]. The fact that such a linear combination can be measured easily for an individual subject has numerous applications. This linear metameric code has been used to study effects of adaptation and backgrounds (Cicerone, Krantz, and Larimer, 1975; Shevell, 1978; Thornton, 1981) as well as normal and abnormal variations in photopigment absorptions and in linkages between photopigments and cancellation codes (Cicerone, Nagy, and Nerger, 1987; Donnell, 1977; Romeskie, 1976; Romeskie and Yager, 1978; Werner and Wooten, 1979). No doubt other applications will be found. Linearity of redness/greenness cancellation also poses a challenge and offers a potential tool for physiological investigations of neural coding.

Another set of nonlinear phenomena are found in the classical perceptual codes for color: brightness, saturation, and hue. We displayed in Section 15.5.4 the polynomial formulas proposed by Hurvich and Jameson to relate hue and saturation to the opponent attributes. The nonadditivity of brightness matching led Guth, Donley, and Marrocco (1969) to a related formula that was eventually incorporated into a proposed model for the opponent attributes themselves (Guth, Massof, and Benzschawel, 1980).

The methods of polynomial conjoint measurement (Chapter 7) might be used to explore these proposals. In addition, as was pointed out in Section 15.6.2, models for nonlinear codes in normal color vision might well apply equally to dichromatic observers, once one has taken account of appropriate losses or substitutions of photopigments or other codes.

Understanding the effects of viewing conditions is perhaps the most important undertaking of color theory. It has become clear in recent years that viewing conditions affect photopigments, receptor sensitivities, and opponent codes (Krauskopf, Williams, and Heeley, 1982; Pugh and Kirk, 1986; Shevell, 1978).

15.7 PROOFS

15.7.1 Theorem 6 (p. 266)

Let $\langle A, \oplus, * \rangle$ be a convex cone, g be a mapping of A onto Re^m, for each $\gamma \in P^{m-1}$, $E_\gamma = g^{-1}(\gamma^\perp)$ and $E = \bigcap_{\gamma \in P^{m-1}} E_\gamma$. Suppose that $\langle A, \oplus, *, E \rangle$ is an equilibrium structure, and for each $\gamma \in P^{m-1}$, $\langle A, \oplus, *, E_\gamma \rangle$ is a one-dimensional equilibrium structure. Then

(i) $\langle A, \oplus, *, E \rangle$ is m-dimensional;

(ii) there is a representation φ of $\langle A, \oplus, *, E \rangle$ such that, for $\gamma \in P^{m-1}$, $\varphi_\gamma(a) = \gamma \circ \varphi(a)$ is a representation for $\langle A, \oplus, *, E_\gamma \rangle$;

(iii) φ is unique up to multiplication by a constant; and

(iv) there is a real-valued function α on A such that $\alpha(a)g(a)$ gives the representation required in (ii).

PROOF. Choose $\gamma^{(i)} = (\gamma_1^{(i)}, \ldots, \gamma_m^{(i)}) \in P^{m-1}$ by $\gamma_j^{(i)} = 1$ if $i = j$, 0 otherwise; let $E_i = E_{\gamma^{(i)}}$. Then $g_i = g_{\gamma^{(i)}}$ is just the ith component of g.

Note that $E_\gamma \supset \bigcap_{i=1}^m E_i$ for all γ because if $a \in E_i$, then $g_i(a) = 0$, and if $g_i(a) = 0$, $i = 1, \ldots, m$, then $g_\gamma(a) = 0$. Thus $E = \bigcap_{i=1}^m E_i$.

Since g is onto, choose $a_i \in A$ with $g_j(a_i) = 1$ if $i = j$, 0 otherwise. Clearly $a_i \notin E_i$ and $a_i \in E_j$ for $j \neq i$. Choose a_i' with $a_i \oplus a_i' \in E$; then also $a_i' \notin E_i$ but $a_i' \in E_j$ for $j \neq i$.

We show that $\{a_1, \ldots, a_m\}$ is linearly independent in $\langle A, \oplus, *, E \rangle$. Suppose that

$$\sum_{i=1}^m t_i * a_i \sim_E \sum_{i=1}^m u_i * a_i.$$

Then $(\sum_{i=1}^m t_i * a_i) \oplus (\sum_{i=1}^m u_i * a_i') \in E$. From this it follows that $(t_i * a_i) \oplus (u_i * a_i') \in E_i$, and hence $t_i = u_i$, $i = 1, \ldots, m$.

Next, we show that any $m + 1$ elements of A are linearly dependent with respect to E. Let b be any element of A. By one-dimensionality of $\langle A, \oplus, *, E_j \rangle$ we have a nontrivial equation,

$$(t_j' * b) \oplus (t_j'' * a_j) \sim_{(E_j)} (u_j' * b) \oplus (u_j'' * a_j).$$

Without loss of generality, let $t_j' \geqslant u_j'$. If $t_j' = u_j'$, we obtain $(t_j'' * a_j) \oplus$

$(u''_j * a'_j) \in E_j$, whence $t''_j = u''_j$, and the equation is trivial. Thus, $t'_j > u'_j$. This yields

$$b \oplus (t_j * a_j) \oplus (u_j * a'_j) \in E_j,$$

where $t_j = t''_j/(t'_j - u'_j)$, $u_j = u''_j/(t'_j - u'_j)$. From this, in turn, follows

$$b \oplus \left(\sum_{j=1}^{m} t_j * a_j \right) \oplus \left(\sum_{j=1}^{m} u_j * a'_j \right) \in E.$$

Let $\{b_0, \ldots, b_m\}$ be any $m + 1$ elements of A, and apply the preceding reasoning to obtain equations

$$b_i \oplus \left(\sum_{j=1}^{m} t_{ji} * a_j \right) \oplus \left(\sum_{j=1}^{m} u_{ji} * a'_j \right) \in E,$$

$i = 0, \ldots, m$. Let $\alpha_{ji} = u_{ji} - t_{ji}$. The matrix (α_{ji}) has rank at most m, and so its $m + 1$ columns are linearly dependent. We can therefore choose λ_i, $\nu_i \in \text{Re}^+$, $i = 0, \ldots, m$, with $\lambda_i \neq \nu_i$ for at least one i, such that $\sum_{i=0}^{m}(\lambda_i - \nu_i)\alpha_{ji} = 0$, $j = 1, \ldots, m$, i.e.,

$$\sum_{i=0}^{m} (\lambda_i u_{ji} + \nu_i t_{ji}) = \sum_{i=0}^{m} (\nu_i u_{ji} + \lambda_i t_{ji}),$$

$j = 1, \ldots, m$. We now have

$$\sum_{i=0}^{m} \lambda_i * b_i \sim_E \left[\sum_{i,j} (\lambda_i u_{ji}) * a_j \right] \oplus \left[\sum_{i,j} (\lambda_i t_{ji}) * a'_j \right]$$

$$\sim_E \left[\sum_{i,j} (\lambda_i u_{ji} + \nu_i t_{ji}) * a_j \right] \oplus \left[\sum_{i,j} (\lambda_i t_{ji} + \nu_i t_{ji}) * a'_j \right]$$

$$= \left[\sum_{i,j} (\nu_i u_{ji} + \lambda_i t_{ji}) * a_j \right] \oplus \left[\sum_{i,j} (\lambda_i t_{ji} + \nu_i t_{ji}) * a'_j \right]$$

$$\sim_E \left[\sum_{i,j} (\nu_i u_{ji}) * a_j \right] \oplus \left[\sum_{i,j} (\nu_i t_{ji}) * a'_j \right]$$

$$\sim_E \sum_{i=0}^{m} \nu_i * b_i.$$

This is a nontrivial equation for $b_0, \ldots, b_m$. This completes the proof that $\langle A, \oplus, *, E \rangle$ is m-dimensional.

Let φ be an arbitrary representation of $\langle A, \oplus, *, E \rangle$ onto $\langle \mathrm{Re}^m, +, \cdot, \{0\} \rangle$. We want to elucidate the relationship between φ and g. We show first that φg^{-1} defines an incidence-preserving map of the subspaces of Re^m onto themselves, i.e., a projective transformation of P^{m-1} onto itself. Let R be any $(m-l)$-dimensional subspace of Re^m. We can pick l independent vectors of P^{m-1}, orthogonal to R, denoted $\bar{\gamma}^{(1)}, \ldots, \bar{\gamma}^{(l)}$. Let $F_i = E_{\bar{\gamma}^{(i)}}$, $i = 1, \ldots, l$. Then clearly $g^{-1}(R) = \cap_{i=1}^{l} F_i$. Let $\varphi[g^{-1}(R)] = S$. Obviously, if $\varphi(a), \varphi(b) \in S$, then $\varphi(a) + \varphi(b) = \varphi(a \oplus b) \in S$ (each F_i, hence their intersection, is closed under $\oplus$). Likewise, for $t > 0$, $t \cdot \varphi(a) = \varphi(t * a) \in S$. We show that S contains zero and negatives. If $a \in g^{-1}(R)$, choose b with $a \oplus b \in E$. Since $a, a \oplus b \in F_i$, $i = 1, \ldots, l$, we have $b \in F_i$, and hence $b \in g^{-1}(R)$. Thus $-\varphi(a) = \varphi(b) \in S$, and $0 \in S$. It follows that S is a subspace of Re^m. We now show that S is also $(m-l)$-dimensional. In fact, we can choose $\bar{\gamma}^{(l+1)}, \ldots, \bar{\gamma}^{(m)} \in R$ such that the entire set $\bar{\gamma}^{(1)}, \ldots, \bar{\gamma}^{(m)}$ is an orthogonal basis and can choose $\bar{a}_i \in A$ with $g(\bar{a}_i)$ orthogonal to $\bar{\gamma}^{(j)}$, $j \neq i$, but not orthogonal to $\bar{\gamma}^{(i)}$. By an argument parallel to the one above, $\bar{a}_1, \ldots, \bar{a}_m$ are linearly independent, and exactly $m - l$ of them, $\bar{a}_{l+1}, \ldots, \bar{a}_m$, are in $g^{-1}(R)$. So $\{\varphi(\bar{a}_{l+1}), \ldots, \varphi(\bar{a}_m)\}$ is a basis for S. From this it follows that the correspondence $R \to \varphi g^{-1}(R) = S$ is a projective transformation of the family of all proper subspaces of Re^m onto itself. One of the fundamental theorems of projective geometry (Schreier and Sperner, 1961, Chap. V, Theorem 6; see also Section 12.2.3) asserts that any such transformation is induced by a nonsingular linear transformation h of Re^m onto itself, i.e., $\varphi[g^{-1}(R)] = h(R)$ for every subspace R.

For any $a \in E$, φg^{-1} maps the subspace R generated by $g(a)$ into the subspace generated by $h[g(a)]$. Thus there is some nonzero constant $\alpha(a)$ such that $\varphi(a) = \alpha(a)h[g(a)]$, i.e.,

$$(h^{-1}\varphi)(a) = \alpha(a)g(a).$$

Since h is nonsingular linear, $h^{-1}\varphi$ is also a representation of $\langle A, \oplus, *, E \rangle$; and clearly, $a \in E_\gamma$ iff $\gamma \circ g(a) = 0$ iff $\gamma \circ (h^{-1}\varphi)(a) = 0$. Thus $h^{-1}\varphi$ has the required properties in (ii), and so does $\alpha(a)g(a)$, as stated in (iv).

The final point is uniqueness. Supposing that φ, ψ both satisfy (ii), then the argument given above for φ, g can be repeated to establish that $\varphi \psi^{-1}$ is a projective transformation of the subspaces of Re^m. But in this case, for any one-dimensional subspace $\gamma \in P^{m-1}$, we have $\varphi \psi^{-1}(\gamma) = \gamma$; this is true because for every $a \in A$, $\gamma \circ \varphi(a) = 0$ iff $\gamma \circ \psi(a) = 0$. This shows that the projective transformation in question is the identity, and the only linear transformations h that induce the identity in P^{m-1} are multiplications. $\diamond$

15.7.2   Theorem 9 (p. 283)

Suppose that $\langle A, \oplus, *, \sim \rangle$ is a proper 3-chromatic Grassmann structure and A_1, A_2 are subsets of A that satisfy the code axiom, linearity axiom, and existence of unique-hue complements. Then,

(i) $\langle A, \oplus, *, A_i \rangle$ *is a one-dimensional equilibrium structure;*

(ii) $\langle A, \oplus, *, A_1 \cap A_2 \rangle$ *is a two-dimensional equilibrium structure, and $\langle A_1 \cap A_2, \oplus, *, \sim \rangle$ is a proper 1-chromatic Grassmann structure;*

(iii)   *there exist ratio scales φ_1, φ_2 and a function φ_3 such that $(\varphi_1, \varphi_2, \varphi_3)$ represents $\langle A, \oplus, *, \sim \rangle$, (φ_1, φ_2) represents $\langle A, \oplus, *, A_1 \cap A_2 \rangle$, φ_i represents $\langle A, \oplus, *, A_i \rangle$ $(i = 1, 2)$, and φ_3 represents $\langle A_1 \cap A_2, \oplus, *, \sim \rangle$.*

PROOF. Let a_i, $b_i \in (A_1 \cup A_2) - A_i$ with $a_i \oplus b_i \in A_i$, $i = 1, 2$. Choose any a_3 (e.g., $a_1 \oplus b_1$) in $A_1 \cap A_2$. We show first that a_1, a_2, a_3 are a basis, i.e., satisfy the condition of Axiom 4(ii), Definition 2. If not, there is a nontrivial equation,

$$\sum_{i=1}^{3} t_i * a_i \sim \sum_{i=1}^{3} u_i * a_i. \tag{13}$$

By additivity of $\sim$ (Axiom 2, Definition 2), we can add $u_1 * b_1$ to both sides of Equation (13). By the properties of a convex cone (Definition 1), the right side of the resulting equation is equal to

$$[u_1 * (a_1 \oplus b_1)] \oplus (u_2 * a_2) \oplus (u_3 * a_3).$$

By linearity for A_1, this is in A_1; hence, by the code axiom, so is the left side of the equation. By linearity again,

$$(t_1 * a) * (u_1 * b_1) \in A_1.$$

It now follows that $t_1 = u_1$; e.g., if $t_1 > u_1$, then we could use linearity to infer $(t_1 - u_1) * a_1 \in A_1$, and hence, $a_1 \in A_1$, contrary to hypothesis. A parallel argument shows that $t_2 = u_2$. Therefore Equation (13) reduces (by additivity of $\sim$) to

$$t_3 * a_3 \sim u_3 * a_3,$$

where $t_3 \neq u_3$ by nontriviality. This is impossible in a proper Grassmann structure (Definition 4).

By the trichromacy axiom [4(i) of Definition 2], for any $a \in A$, there is a nontrivial equation involving a, a_1, a_2, a_3; since the a_i are a basis, it can be written

$$a \oplus \sum_{i=1}^{3} (t_i * a_i) \sim \sum_{i=1}^{3} u_i * a_i. \tag{14}$$

Let $\varphi_i(a)$ be the ith coordinate relative to the basis a_1, a_2, a_3, i.e., $\varphi_i(a) = u_i - t_i$.

If $a \in A_i$, then the argument used on connection with Equation (13) applies also to (14), yielding $t_i = u_i$; and conversely, $t_i = u_i$ in (14) implies $a \in A_i$. Thus $a \in A_i$ iff $\varphi_i(a) = 0$, $i = 1, 2$. If $t_i > u_i$ in (14), then adding $u_i * b_i$ to both sides now yields $a \oplus [(t_i - u_i) * a_i] \in A_i$. Similarly, if $u_i > t_i$, then $a \oplus [(u_i - t_i) * b_i] \in A_i$. This proves that every $a \notin A_i$ can be equilibrated; hence, $\langle A, \oplus, *, A_i \rangle$ is an equilibrium structure. Obviously, it is one-dimensional with representation φ_i, $i = 1, 2$.

A similar argument, adding $(u_1 * b_1) \oplus (u_2 * b_2)$ to Equation (14), shows that $\langle A, \oplus, *, A_1 \cap A_2 \rangle$ is an equilibrium structure; obviously it is two-dimensional, with representation (φ_1, φ_2).

If $a \in A_1 \cap A_2$, then Equation (14) reduces to

$$a \sim \varphi_3(a) * a_3.$$

This shows that φ_3 is positive-valued on $A_1 \cap A_2$ and is a representation of the 1-chromatic structure $\langle A_1 \cap A_2, \oplus, *, \sim \rangle$.

The uniqueness statement is immediate from the uniqueness theorems for one- and three-dimensional representations. $\diamondsuit$

15.7.3 Theorem 10 (p. 283)

Under the hypotheses of Theorem 9, if $\langle A, \oplus, *, \sim_B \rangle$ is a proper 1-chromatic reduction structure, then φ_3 can be chosen to be a ratio-scale representation for $\langle A, \oplus, *, \sim_B \rangle$.

PROOF. Let a_1, a_2, a_3 be as in the proof of Theorem 9, and let φ_1, φ_2, ψ_3 denote the coordinates relative to that basis. Since $\sim_B$ is proper, there can be no nontrivial equation $t * a_3 \sim_B u * a_3$; hence, a_3 is a basis relative to $\sim_B$. Define β_i by $a_i \sim_B \beta_i * a_3$ (thus $\beta_3 = 1$) and $\varphi_3 = \beta_1\varphi_1 + \beta_2\varphi_2 + \beta_3\psi_3$. For $a \in A$, Equation (14) yields

$$a \oplus \sum_{i=1}^{3} (t_i * a_i) \sim_B \sum_{i=1}^{3} u_i * a_i.$$

By the linearity properties of $\sim_B$, we have

$$a \oplus \sum_{i=1}^{3} (t_i \beta_i) * a_3 \sim_B \sum_{i=1}^{3} (u_i \beta_i) * a_3 .$$

Since $\sim_B$ is proper it follows that

$$\varphi_3(a) = \sum_{i=1}^{3} (u_i - t_i)\beta_i > 0$$

and $a \sim_B \varphi_3(a) * a_3$. Thus φ_3 is a representation for $\langle A, \oplus, *, \sim_B \rangle$. Uniqueness is immediate. ◇

EXERCISES

1. Let $A = (\text{Re}^+)^n$, let $\varphi_j(x) = \sum_{i=1}^{n}\alpha_{ij}x_i$, and let $x \sim y$ iff $\varphi_j(x) = \varphi_j(y)$ for $j = 1, \ldots, m$. Prove that $\langle A, \oplus, *, \sim \rangle$ is a Grassmann structure, where $\oplus$, $*$ are ordinary componentwise addition and multiplication by positive numbers. Prove that the structure is m-chromatic iff the φ_j are linearly independent; that it is proper iff there are no solutions to $\sum_{i=1}^{n}\alpha_{ij}x_i = 0$, $j = 1, \ldots, m$, with $(x_1, \ldots, x_n) \in (\text{Re}^+)^n$.   (15.2.1)

2. Formulate in terms of E the condition that the equilibrium structure $\langle A, \oplus, *, E \rangle$ is two-dimensional.   (15.2.1, 15.6.5)

3. Prove Theorem 2. What happens if one tries to extend Definitions 3 and 4 and Theorem 2 by omitting the solvability condition on the Grassmann and the equilibrium structures?   (15.2.1)

4. Let $\langle A, \oplus, * \rangle$ be a convex cone, let $\hat{E} \subset A$, and define $E = \{a \mid a \oplus b \in \hat{E}$ iff $b \in \hat{E}\}$. What conditions must $\langle A, \oplus, *, \hat{E} \rangle$ satisfy for $\langle A, \oplus, *, E \rangle$ to be an equilibrium structure? Interpret $\hat{E}$ as a set of *counterweight* configurations, i.e., configurations that eliminate the effects of constant forces such as gravity.   (15.2.1)

5. Given the color-match equations (5) linking two sets of instrumental primaries, prove that corresponding coordinate systems transform by (4), with $\alpha_{ij} = u_{ij} - t_{ij}$.   (15.4.1)

6. Formulate a representation and uniqueness theorem for chromaticity coordinates.   (15.4.1)

7. What conditions on photopigment functions $p_j(\lambda)$ are sufficient to guarantee that the spectrum locus on the chromaticity diagram is convex? (15.4.2, 15.5.2)

8. Let ρ_1, ρ_2, ρ_3 be independent linear codes for a 3-chromatic Grassmann structure $\langle A, \oplus, *, \sim \rangle$, and let f_1, f_2, f_3 be defined by

$$f_1 = \rho_1^{1/2} - \rho_2$$
$$f_2 = \rho_2 - \rho_3^{1/2}$$
$$f_3 = \rho_1 + \rho_3.$$

Suppose that $a \sim_k b$ iff $f_i(a) = f_i(b)$ for all $i \neq k$. Prove that $\langle A, \oplus, *, \sim_k \rangle$ is not a Grassmann structure, $k = 1, 2, 3$. (15.2.1, 15.5.1)

9. Why is there no direct analogue of the counterexample of Section 15.5.3 in the case of color measurement? Construct an analogue using the equilibrium structure of bipartite colorimetric fields (15.4.1) and discuss its possible empirical significance. (15.5.3)

10. Show that the confusion lines of a dichromatic reduction structure intersect in the normal chromaticity diagram at a point corresponding to the missing primary. (15.6.1)

11. Prove the uniqueness statement in Theorem 7. (15.6.1)

12. Let a dichromat have two photopigments, ρ_1', ρ_2', related to the normal ones by

$$\rho_1' = \rho_1 = \int_\lambda \rho_1(\lambda) a(\lambda) \, d\lambda$$
$$\rho_2'(a) = \int_\lambda \rho_2(\lambda - \lambda_0) a(\lambda) \, d\lambda.$$

Is it possible to discover the nature of ρ_1' and ρ_2' using the methods of 2-chromatic reduction structures? (15.6.1)

13. Without using any of the representation theorems, prove directly that if $\langle A, \oplus, *, A_i \rangle$ is a one-dimensional equilibrium structure, then changing cancellation stimuli results only in a change of units of the cancellation scale. (15.6.4)

14. Show that the unique-hue complements condition (Axiom 3 of Theorem 9) is necessary for conclusions (i) and (ii) of that theorem. (15.6.5)

15. Let $\langle A, \oplus, *, \sim, A_1, A_2 \rangle$ satisfy the conditions of Theorem 9, and let $\varphi_1, \varphi_2, \varphi_3$ be a representation satisfying conclusion (iii) of that theorem. Define $B = \{ a|\ \varphi_1(a) + \gamma\varphi_2(a) \geqslant 0 \}$, where $\gamma \in \text{Re}$, $B_i = B \cap A_i$, $i = 1, 2$, and restrict $\oplus, *, \sim$ and the φ_i to B. Show that (i) $\langle B, \oplus, *, \sim \rangle$ is a 3-chromatic Grassmann structure with representation $(\varphi_1, \varphi_2, \varphi_3)$; (ii) B_1 and B_2 satisfy the code axiom of Theorem 9; (iii) $\langle B, \oplus, *, B_i \rangle$ is a one-dimensional equilibrium structure with representation φ_i, $i = 1, 2$; and (iv) $\langle B, \oplus, *, B_1 \cap B_2 \rangle$ is not an equilibrium structure, and the unique-hue complements axiom does not hold. What would color perception be like if it were described by a structure with the properties of $\langle B, \oplus, *, \sim, B_1, B_2 \rangle$?   (15.6.5)

Chapter 16   Representations with Thresholds[1]

16.1   INTRODUCTION

Foundational theories of measurement generally assume that comparisons among objects can be idealized as a weak order; for example, $a \succsim b$, interpreted in weight measurement as "a is at least as heavy as b," or $ab \succsim cd$, interpreted in distance measurement as "the distance between a and b is at least as great as that between c and d." Although any particular method of comparison is apt to violate transitivity, this has not been of too much concern in the foundations of physical measurement, largely because it has been possible to devise increasingly more careful or refined methods of comparison for properties like weight and distance. A more refined method generally eliminates most of the intransitivities of a coarser method. Indeed, if it does not, it is assumed that some form of systematic error has intruded.

In contrast, behavioral scientists, whose data frequently violate transitivity, have been forced to confront the problem that this poses for theory construction. Some comparisons consist of choices, or comparative judgments, and the human decision maker can hardly be replaced by a more

[1] The chapter draws extensively from the excellent survey articles of Fishburn (1970b) and Roberts (1970) and from Fishburn's (1985) summary book. They also treat a number of results that are somewhat peripheral to our main focus and are not discussed here.

299

refined organism. In the absence of more refined comparison methods, theorists have had to consider what the relationship may be between the idealized weak order and the observable data.

16.1.1 Three Approaches to Nontransitive Data

At least three different lines of thought have emerged, two of which have led to the formulation of novel relational structures, some of which are covered in this and in the next chapter.

The first view, which has not led to additional measurement structures, is perhaps implicit in most of the axiomatic work on measurement, e.g., Volume I. A relational statement, such as $a \succsim b$, is not considered to be the record of a particular observation or experiment but is a theoretical assertion inferred from data, and is subject to errors of inference just like any other theoretical assertion. The problem of measurement theory is to represent such theoretical assertions by numerical ones, such as $\varphi(a) \geqslant \varphi(b)$; the problem of inference, leading from observed data to assertions of the form $a \succsim b$ is interesting and important but is not part of the foundations of measurement *per se*.

A second view is that the problems of inference and representation can usefully be combined. Relational structures are formulated that simplify the inference step or reduce it to standard forms, at least in some contexts; then, within these structures, new relations are definable that reduce the measurement problem to one that has previously been solved. Perhaps the simplest example of this line of work is as follows (a detailed treatment is in the next chapter). Suppose that the actual observations relevant to a pair (a, b) are thought of as a set of repeated trials (pair comparisons). The underlying relational structure postulates parameters $P(a, b)$, interpreted as the probability that a is judged in some fashion as being greater than b. Such parameters are estimated from observed relative frequencies; the inference problem is, thereby, reduced to a more or less standard one. The parameters $P(a, b)$ are then used to define a new relation: $a \succsim b$ if and only if $P(a, b) \geqslant 1/2$. The relation $\succsim$ is transitive just when the parameters $P(a, b)$ satisfy a condition called Weak Stochastic Transitivity (Block & Marschak, 1960). Once the relation $\succsim$ is obtained, the "nuisance" of variable data has been handled, and measurement (sometimes involving additional empirical relations such as operations or factorial structure) proceeds as usual.

Much of the work covered in the present chapter can also be thought of as representative of this view. Confronted with a symmetric but nontransitive *indifference* relation $\sim$, representing the failure of two objects a, b to be discriminated by a specific method, we use it to define a more refined, transitive indifference relation I: $a I b$ provided that for all c, $a \succsim c$ iff $b \succsim c$. That is, a, b are inferred *not* to satisfy $a I b$ if some c can be found that stands in the relation $\sim$ to one of them but not the other. Once again, the relation $\sim$ is an abstraction from the nuisance of actual data [here, insensitivity rather than inconsistency as with $P(a, b)$], and the defined relation I permits measurement to proceed more or less as usual.

Yet another example of work in this same spirit is a development by Falmagne (1976, 1985) in which a random-variable structure is introduced, and a relational structure is defined in terms of the medians of the random variables. If the random variables have appropriate properties, this results in a measurement representation of the usual sort.

The third view regards inconsistency and/or insensitivity not as a nuisance but rather as an important source of information about underlying processes. One attempts to formulate relational structures that reflect the complications of actual data and to find appropriate numerical representations for such structures. This idea traces back to Fechner's (1860/1966) notion of a just-noticeable difference (jnd) as a unit of sensation difference. In the present chapter, this Fechnerian tradition is embodied in the work on constant-threshold representations, in which the nontransitive indifference relation $\sim$ is represented by a fixed distance in the underlying scale. Other examples are found in the next chapter; they include modern versions of Fechner's idea as well as structures in which nontransitivity is represented not merely as a consequence of lack of sensitivity of a comparison method but as indicative of an underlying multidimensional or multifeature representation of the objects being compared.

16.1.2 Idea of Thresholds

Let us now turn to the types of ordering that we shall deal with in this chapter, namely, those that are suited to a representation with thresholds. The classic motivating example (Luce, 1956; Luce & Raiffa, 1957, p. 346) is preference ordering for coffee with various concentrations of sugar in it. Denote by n a standard cup of a standard brew of coffee that contains n granules of sugar. Given any two cups m and n, the subject is to express preference for one over the other or indifference between them. The subject cannot discriminate n from $n + 1$ by taste for any n and therefore expresses indifference ($n \sim n + 1$) for every n. But for some k, the subject is not indifferent between n and $n + k$. Therefore, the relation $\sim$ cannot be transitive. However, the strict preference relation, $m \succ n$, is expected to be transitive. We might hope to find a representation for this situation in terms of a function $\varphi(n)$, the "underlying hedonic value" for n, and a function $\bar{\delta}(n)$, the upper threshold at n. Thus we would have $m \succ n$

provided that the hedonic value for m is sufficiently above that for n to be noticed, i.e., $\varphi(m) > \varphi(n) + \bar{\delta}(n)$.

16.1.3 Overview

Section 16.2 is devoted to precisely the sort of ordering and threshold representation illustrated by the preceding example. As titled, it is the ordinal theory. The proofs are grouped together in Section 16.3.

In the first two subsections of Section 16.4 we consider an elaboration of the same idea that is designed to be more realistic. After all, in our coffee example, one cannot really expect the subject to express complete indifference on every test between n and $n + 1$, between n and $n + 2$, etc., and then abruptly switch to a reliable preference between n and $n + k$ when the threshold is crossed. More realistically, one can define various indifference relations in terms of a corresponding series of criteria (perhaps probabilistic ones). Thus we are led to study families of preference and indifference relations, all involving the same underlying hedonic scale φ.

In the third subsection of Section 16.4 we consider a quite different theory of intransitive indifference involving multidimensional partial orders. A partial order $\succ$ can be generated from the intersection of several simple orders $\succ_1, \ldots, \succ_t$, i.e., $b \succ c$ if and only if $b \succ_i c$ for $i = 1, \ldots, t$. An example with $t = 2$ might be a set of vacation tours of form bx, where b is the country and x is the cost. Assume that the decision maker has a strict simple order $\succ_1$ over countries; i.e., the person knows just which country is most desirable, which next, etc. Also, there is a strict simple order $\succ_2$ over cost: less cost is always preferred to more. A tour bx is preferred to cy if and only if $b \succ_1 c$ and $x \succ_2 y$. (Of course, this would depend in practice on the fact that no two costs x, y are so close together that the difference between them is neglected; likewise, we must have no competing tours bx, by to the same country, for which only cost is a factor.)

Note that in this setup, indifference can easily be intransitive. If $b \succ_1 c \succ_1 d$ and $x \succ_2 y \succ_2 z$, and if cx, dy, bz are three available packages, then $cx \succ dy$, but bz is indifferent to both cx and dy. Also note that a threshold representation may not be possible for such a setup. (See the discussion following Definition 4 and Figure 1.)

Section 16.5 is devoted to proofs of theorems in 16.4.

Section 16.6 shifts the focus from ordinal theory to additive structures with intransitive indifference relations. The theories are not very satisfactory. On the one hand, the ordinal theory leads to the definition of an underlying weak order, represented by the scale φ (hedonic strength, in our earlier example), and it is straightforward to use that weak order in conjunction with additive structure to reformulate the axiom systems studied previously, e.g., extensive and conjoint structures. However these systems are not formulated directly in terms of the observables $\succ$ and $\sim$ and the attempt to do so runs up against the lack of theories relating the magnitude of the threshold difference δ to the scale differences resulting from additive structure. Section 16.7 contains the proof of the main theorem of Section 16.6. Section 16.8 is concerned with random-variable representations including a qualitative approach that uses the classical theory of moments.

As we go along, a number of papers will be cited, but perhaps it will be helpful at the onset to list some of the historically important contributions as well as some of the more general recent treatments. Efforts at abstract formulations of thresholds are, in temporal order, Wiener (1921), Goodman (1951), Halphen (1955), Luce (1956), and Scott and Suppes (1958). The applied motivation has come mainly from psychology and economics. Among those from psychology, the initial work of Fechner (1860/1966) has been mentioned; it led in this century to further work on the problem, using functional equation techniques that were initiated by Luce and Edwards (1958) and elaborated and improved mainly by Falmagne (1971, 1974, 1977, 1985) and Krantz (1972). In economics, some of the important references are Armstrong (1939, 1948, 1950, 1951, 1957–1958), Burros (1974), Devletoglou (1965, 1968), Devletoglou and Demetriou (1967), Fishburn (1968, 1970b, 1970c, 1975), Georgescu–Roegen (1936/1966, 1958/1966), Gerlach (1957), Goodman (1951), Halphen (1955), Jamison and Lau (1973), May (1954), and Ng (1975). Some examples of threshold ideas in other domains are Roberts (1973) and Zeeman (1962). Among the most useful surveys of work in the area are Fishburn (1970b, 1985) and Roberts (1970).

16.2 ORDINAL THEORY

We start with a basic definition of (strict) *partial order* and (reflexive) *graph*.[2]

DEFINITION 1. *Suppose* $\succ$ *and* $\sim$ *are binary relations on a nonempty set* A.

(i) $\langle A, \succ \rangle$ *is a* strict partial order *iff* $\succ$ *is asymmetric and transitive.*

(ii) $\langle A, \sim \rangle$ *is a* graph *iff* $\sim$ *is reflexive and symmetric.*

(iii) $\sim$ *is the* symmetric complement *of* $\succ$ *iff* "$a \sim b$" *is equivalent to* "$not(a \succ b)$ *and* $not(b \succ a)$."

---

[2] In some treatments of graph theory, the convention is that $\sim$ is irreflexive; in others, that it is reflexive. The latter is more convenient here, and we shall omit the adjective *reflexive*.

Notational convention. In any discussion of a strict partial order $\succ$, its symmetric complement will be denoted $\sim$. If several strict partial orders are denoted with affixes to $\succ$, the corresponding affixes on $\sim$ will denote their symmetric complements. Obviously the symmetric complement of any strict partial order is a graph; thus any relation denoted by $\sim$, with or without affixes, will be a graph.

16.2.1 Upper, Lower, and Two-Sided Thresholds

We begin by considering the types of representations we are after, and only then do we turn to the question of what structures have these representations. As was noted, the main idea about how to represent nontransitive indifference dates back to the nineteenth century to Fechner's notion of a just noticeable difference or sensory threshold in psychology. [Campbell's (1920/1957) attempt to argue a somewhat similar view for physics was not widely accepted.] For certain classes of sensory comparison, such as tones ordered by loudness or lights ordered by brightness, it may be possible to represent the comparisons numerically by two functions: one describing the magnitude of the sensation and one describing the size of the region of indiscriminability. Fechner was especially concerned with representations in which the size of the region of indiscriminability is constant (Section 16.2.6, especially Theorem 11).

One important distinction we must make is between one- and two-sided representations, which involves looking either at changes in one direction (increases or decreases) that cause discrimination to occur or at both the increases and the decreases. These correspond to two structural concepts called *interval orders* and *semiorders*. As we proceed, it will be necessary to specify other more delicate distinctions having to do with what exactly goes on at the threshold boundary.

In a one-sided, upper-threshold representation there are two numerical functions $\bar{\varphi}$ and $\bar{\delta}$ such that the sum, $\bar{\varphi}(b) + \bar{\delta}(b)$, is a boundary between values of $\bar{\varphi}(a)$ such that $a \succ b$ and values such that $a \sim b$. We require that $\bar{\delta}$ be nonnegative, so this boundary is always $\geqslant \bar{\varphi}(b)$. Similarly, a lower threshold representation involves $\underline{\varphi}$ and $\underline{\delta}$, the latter $\leqslant 0$, such that $\underline{\varphi}(b) + \underline{\delta}(b)$ is a boundary between $\underline{\varphi}(a)$ values with $a \sim b$ and those with $a \prec b$.

An asymmetric relation $\succ$ on A can have both an upper-threshold representation $\langle \bar{\varphi}, \bar{\delta} \rangle$ and a lower one $\langle \underline{\varphi}, \underline{\delta} \rangle$, and it is entirely possible for these to be incompatible in the sense that $\bar{\varphi}$ and $\underline{\varphi}$ induce different orders on A. Put another way, there may be no strictly increasing function relating $\bar{\varphi}$ to $\underline{\varphi}$. For an example, see Exercise 1. If, however, there are upper and lower representations with $\bar{\varphi} = \underline{\varphi}$, then we say there is a "two-sided representation," and we use the notation $\langle \varphi, \bar{\delta}, \underline{\delta} \rangle$, where $\varphi = \bar{\varphi} = \underline{\varphi}$. In such a representation, $a \sim b$ implies that $\varphi(a)$ lies in the interval $[\varphi(\underline{b}) + \underline{\delta}(b), \varphi(b) + \bar{\delta}(b)]$.

Note that the definitions just given do not say exactly what happens at the boundary. For example, in an upper representation, if $\bar{\varphi}(a) = \bar{\varphi}(b) + \bar{\delta}(b)$, then we do not know if $a \succ b$ or $a \sim b$; either is possible. We shall arrange our definitions so that the behavior at any particular boundary depends only on $\bar{\varphi}$ values, i.e., it is consistent for a, a' with $\bar{\varphi}(a) = \bar{\varphi}(a')$. Similar remarks hold for lower and two-sided representations.

The requirement just mentioned applies to each specific boundary and does not preclude changing the rule at different boundaries. A stronger form is uniformly to require the same rule everywhere. Following Roberts (1968) we shall refer to a representation as *strong* if $\sim$ always holds at the boundary and *strong** if $\sim$ never holds at the boundary. Swistak (1980) used the term *closed* for strong. Other representations, ones that are neither strong nor strong*, are sometimes called *weak* to contrast them with the other two concepts.

All these concepts are summarized formally as follows.

DEFINITION 2. *Let $\succ$ be an asymmetric binary relation on A. A pair $\langle \bar{\varphi}, \bar{\delta} \rangle$ of real-valued functions on A is an* upper-threshold representation *iff $\bar{\delta}$ is nonnegative and for all a, b, c in A the following hold:*

(i) *If $a \succ b$, then $\bar{\varphi}(a) \geqslant \bar{\varphi}(b) + \bar{\delta}(b)$.*

(ii) *If $\bar{\varphi}(a) > \bar{\varphi}(b) + \bar{\delta}(b)$, then $a \succ b$.*

(iii) *If $\bar{\varphi}(a) = \bar{\varphi}(b)$, then $a \succ c$ iff $b \succ c$.*

A pair $\langle \underline{\varphi}, \underline{\delta} \rangle$ is a lower-threshold representation *iff $\underline{\delta}$ is nonpositive and properties (i)–(iii) hold with $\succ$, $>$, $\geqslant$ replaced by $\prec$, $<$, $\leqslant$, respectively.*

A triple $\langle \varphi, \bar{\delta}, \underline{\delta} \rangle$ of real-valued functions on A is a two-sided threshold representation *iff $\langle \varphi, \bar{\delta} \rangle$ is an upper one and $\langle \varphi, \underline{\delta} \rangle$ is a lower one.*

An upper-threshold representation $\langle \bar{\varphi}, \bar{\delta} \rangle$ is said to be strong (strong*) *iff property iv (iv*) holds:*

(iv) $a \succ b$ *iff* $\bar{\varphi}(a) > \bar{\varphi}(b) + \bar{\delta}(b)$.

(iv*) $a \succ b$ *iff* $\bar{\varphi}(a) \geqslant \bar{\varphi}(b) + \bar{\delta}(b)$.

The definitions for lower and two-sided representations are analogous.

Note that either (iv) or (iv*) implies all of (i)–(iii) (see Exercise 2).

All these representations can be generalized to the comparison of elements drawn from different sets thus necessitating different functions on each set. Such a generalization is motivated by the example of ability testing, where the set of individuals tested is scaled by ability and the set of test questions by difficulty. It is taken up briefly in Section 16.2.4.

16.2.2 Induced Quasiorders: Interval Orders and Semiorders

The principal idea in the ordinal theory of thresholds is that any asymmetric binary relation induces quasiorders, and when these quasiorders are actually weak orders (connected), the basic result of the ordinal theory of weak orders (Section 2.1) can be invoked. Using $\succ$, we define the *upper quasiorder* $\overline{Q}$ by: $a\,\overline{Q}\,b$ iff for all c in A, if $b \succ c$, then $a \succ c$. This simply says that a dominates everything that b does. Similarly, the *lower quasiorder* is defined by: $a\,\underline{Q}\,b$ iff for all c in A, $c \succ a$ implies $c \succ b$. A structure in which both of these quasiorders are connected is called an *interval order*, a concept first discussed by Fishburn (1970c) and independently by Mirkin (1970a, 1970b, 1972). For a major survey of results, see Fishburn (1985). As we shall see, this is exactly the correct concept for the existence of upper or lower thresholds. For a two-sided representation to exist, not only must the two quasiorders be connected but also they must not provide contradictory information. In that case, the structure giving rise to the condition is called a *semiorder*, a concept first formulated by Luce (1956).

Another way to look at what is going on empirically is that the information embodied in $\sim$ is being refined by examining how two indifferent elements relate to third elements. First, if $a \succ b$, then by the transitivity of $\succ$, $b \succ c$ implies $a \succ c$, so $a\,\overline{Q}\,b$. Next, if $a \sim b$ suppose we find a c such that $a \succ c$ and $b \sim c$. This we interpret as evidence that a truly lies above b; and we assert that if the opposite evidence is never found (i.e., whenever $b \succ c$, then $a \succ c$), then there is reason to treat a as above-or-equivalent to b, which is what $\overline{Q}$ means. Similarly, $\underline{Q}$ summarizes the results of testing a, b with third elements that lie above them: $a\,\underline{Q}\,b$ if there is no c that provides contrary evidence, lying above a but not above b (i.e., whenever $c \succ a$, then $c \succ b$); and the two-sided quasiorder Q involves both kinds of evidence, i.e., third elements lying above or below. This is formulated formally in Corollaries 2 and 3 following Theorem 3.

Along with $\overline{Q}$, $\underline{Q}$, and Q, it is useful to consider the corresponding equivalence relations, obtained by symmetrizing: $a\,\overline{I}\,b$ when both $a\,\overline{Q}\,b$ and $b\,\overline{Q}\,a$ (i.e., exactly the same c's satisfy $a \succ c$ as $b \succ c$), and likewise $\underline{I}$ and I.

We draw together these ideas in a formal definition.

DEFINITION 3. *Let* $\succ$ *be an asymmetric binary relation on A. Define the* upper *and* lower *quasiorders induced by* $\succ$, $\overline{Q}$ *and* $\underline{Q}$, *by:*

$$a\,\overline{Q}\,b \text{ iff for all } c \text{ in } A, \text{ if } b \succ c \text{ then } a \succ c;$$
$$a\,\underline{Q}\,b \text{ iff for all } c \text{ in } A, \text{ if } c \succ a \text{ then } c \succ b.$$

The two-sided quasiorder induced by $\succ$ *is* $Q = \overline{Q} \cap \underline{Q}$. *Define* $a\,\overline{I}\,b$ *iff* $a\,\overline{Q}\,b$ *and* $b\,\overline{Q}\,a$; *define* $\overline{P} = \overline{Q} - \overline{I}$. *Define* $\underline{I}$, I *and* $\underline{P}$, P *analogously in terms of* $\underline{Q}$, Q.

Calling these relations quasiorders is justified since it is trivial to show that they are transitive.

THEOREM 1. *Suppose* $\succ$ *is an asymmetric binary relation on A. If* $\langle A, \succ \rangle$ *has an upper (lower) (two-sided) threshold representation, then the upper (lower) (two-sided) quasiorder induced by* $\succ$ *is connected and so is a weak order.*

The proof is almost immediate. Suppose $\langle \overline{\varphi}, \overline{\delta} \rangle$ is an upper representation. If $\overline{\varphi}(a) \geq \overline{\varphi}(b)$, then $a\,\overline{Q}\,b$ since if $b \succ c$ we know $\varphi(a) \geq \varphi(b) \geq \varphi(c) + \overline{\delta}(c)$. By parts (ii) and (iii) of Definition 2, we conclude $a \succ c$. Since $\geq$ is connected, so is $\overline{Q}$.

The following example may be useful both as an exercise for Definition 3 and Theorem 1 and as an illustration for some of our subsequent definitions and comments.

EXAMPLE. Let $A = \{a, b, c, d\}$. We shall consider two different asymmetric relations on A:

$$\succ_1 = \{(a, d), (b, d)\},$$
$$\succ_2 = \{(a, d), (a, b)\}.$$

Note that $\sim_i$ ($i = 1$ or 2) holds for the remaining eight unordered pairs in each case. For $\succ_1$, the upper quasiorder $\overline{Q}_1$ does not distinguish a from b since both of these are $\succ_1 d$ and $\sim_1 c$. That is, $a\,\overline{I}_1\,b$. It also does not distinguish c from d since neither of these is $\succ_1$ anything. (This illustrates the dependence of these quasiorders on the total set of elements available for comparison: if there were a fifth element e such that $c \succ_1 e$ but $d \sim_1 e$, then c would be distinguished from d in $\overline{Q}_1$.) But a, b are ranked above c in $\overline{Q}_1$, even though $\sim_1$ holds, because of the comparisons with d. To summarize this, we assign ranks 1 to 4 from low to high, relative to $\overline{Q}_1$. These ranks could be considered values of a function $\overline{\varphi}_1$. As shown in the

TABLE 1

Ranks of Four Elements in Various Quasiorders
Based on $a \succ_1 d$, $b \succ_1 d$, $a \succ_2 d$, $a \succ_2 b$

Elements	Single quasiorders				Combined quasiorders			
	$\overline{\varphi}_1$	$\underline{\varphi}_1$	$\overline{\varphi}_2$	$\underline{\varphi}_2$	φ_1	φ_2	$\overline{\varphi}$	$\underline{\varphi}$
a	3.5	3	4	3.5	3.5	4	4	3.5
b	3.5	3	2	1.5	3.5	1.5	3	2
c	1.5	3	2	3.5	2	3	1.5	3.5
d	1.5	1	2	1.5	1	1.5	1.5	1

first column of Table 1, c and d get rank 1.5 (tied for 1, 2) and a and b get rank 3.5. In other words, $\overline{I}_1$ holds for ties in rank, and $\overline{P}_1$ holds between higher and lower ranks.

Next, consider the lower quasiorder Q_1. It is clear that a, b, c must all be tied since nothing is $\succ_1$ any of them. The ranks are shown in the $\underline{\varphi}_1$ column of the table. The lower quasiorder fails to distinguish b from c but does distinguish c from d on the basis of comparisons from above with a or with b. Finally, consider Q_1, the intersection of these two quasiorders. This is connected because the information in $\overline{Q}_1$ and $\underline{Q}_1$ is noncontradictory. The lower quasiorder breaks the c, d tie from the upper one, and the upper one breaks the b, c tie from the lower one; a, b are tied in both and, hence, in Q_1. The ranks are shown as φ_1 in the fifth column of Table 1.

The ranks $\overline{\varphi}_1$ can be used for an upper-threshold representation of $\succ_1$, if we let $\overline{\delta}(c) = 3$ and $\overline{\delta}(d) = 1$, for example. Likewise, the ranks $\underline{\varphi}_1$ can be used for a lower-threshold representation of $\succ_1$, letting $\underline{\delta}$ be -1 for a, b and -3 for c. [The reader might verify that $\overline{\varphi}_1$ cannot be used for a lower, nor $\underline{\varphi}_1$ for an upper, representation because of the clause (iii) in Definition 2.]

Finally, the ranks shown as φ_1 can be used for either an upper or a lower representation of $\succ_1$; in fact, this can be done with $\overline{\delta} = 2$ or $\underline{\delta} = -2$ for all elements of A.

In a similar fashion, the relations $\overline{Q}_2$, $\underline{Q}_2$, and Q_2 defined from $\succ_2$ are shown in columns 3, 4, and 6 of the table. It is a useful exercise to verify the information shown in these columns.

The final two columns of the table are based on combining information from the two upper quasiorders $\overline{Q}_1$ and $\overline{Q}_2$, or from the two lower ones. This goes beyond the ideas of Definitions 2 and 3 and will be discussed in detail in Section 16.4.1, so these two columns can be ignored for the present. Our next task is to characterize the connectedness of the induced quasiorders by conditions stated directly in terms of the relation $\succ$. Connectedness requires that there be no contradictory evidence as to the ordering of two elements. To show that either $a\,\overline{Q}\,b$ or $b\,\overline{Q}\,a$, we must exclude the possibility of c and d such that both $a \succ c \sim b$ and $b \succ d \sim a$. Restated, if $a \succ c$ and $b \succ d$, then either $b \sim c$ or $a \sim d$ is false, and because of the transitivity of $\succ$ this means that either $b \succ c$ or $a \succ d$. This implication we embody as a formal definition of interval orders.

DEFINITION 4.   *Suppose* $\succ$ *is an irreflexive binary relation on* A. $\langle A, \succ \rangle$ *is an* interval order *iff for all* a, b, c, d *in* A,

$$\text{if} \quad a \succ c \quad \text{and} \quad b \succ d, \quad \text{then either} \quad a \succ d \quad \text{or} \quad b \succ c. \tag{1}$$

For an alternative formulation, see Mirkin (1972) and Rabinovitch (1977).

Figure 1 illustrates the four-element subrelation that violates Equation (1). From the standpoint of c, a is above b, but from that of d, b is above a.

A concrete example that violates Equation (1) might involve four different vacation packages:

Package	Place	Cost
a	Paris	2000
b	Chicago	1000
c	San Francisco	2200
d	Moscow	1200

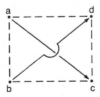

FIGURE 1. A configuration on four elements that violates the interval order property (Equation (1) in Definition 4). The two arrows represent $a \succ c$ and $b \succ d$; the dotted lines represent $\sim$.

If the relation $\succ$ represents dominance on both dimensions, then a person might have $a \succ c$ (Paris preferred to San Francisco and costs less) and $b \succ d$ (Chicago preferred to Moscow and costs less). In comparison with d, Package b looks better than a because b dominates c and a does not (higher cost). But in comparison with c, a looks better than b, for analogous reasons.

Theorem 2 shows that Equation (1) is enough to guarantee that all comparisons yield consistent information: both $\overline{Q}$ and $\underline{Q}$ are connected and, hence, weak orders.

THEOREM 2. *Suppose* $\succ$ *is a binary relation on* A. *The following statements are equivalent*:

(i) $\succ$ *is asymmetric and* $\overline{Q}$ *is connected.*

(ii) $\succ$ *is asymmetric and* $\underline{Q}$ *is connected.*

(iii) $\langle A, \succ \rangle$ *is an interval order.*

A proof is provided in Section 16.3.1.

Note that transitivity of $\succ$ was not assumed in Definitions 3 and 4. Theorem 2 shows that transitivity follows from the interval-order condition, Definition 4. If the quasiorders are defined using an asymmetric but nontransitive relation $\succ$, they are still transitive but of course need not be connected.

We turn to the condition needed to make Q connected or (which is the same thing) to make $\overline{Q}$ and $\underline{Q}$ uncontradictory. The following would be contradictory concerning the relation between b and d: from the standpoint of c, $b\overline{P}d$ (i.e., $b \succ c$ but not $d \succ c$) whereas from the standpoint of a, $d\underline{P}b$ (i.e., $a \succ b$ but not $a \succ d$). We can exclude such a contradiction by postulating that if $a \succ b \succ c$, then either $a \succ d$ or $d \succ c$. This condition, added to that for an interval order, implies that $Q = \overline{Q} \cap \underline{Q}$ is connected. For suppose that not $b Q d$: then either $d\overline{P}b$ or $d\underline{P}b$ must hold. If $d\overline{P}b$, then the above condition implies not $b\underline{P}d$; so by connectedness of $\underline{Q}$, $d\underline{Q}b$, i.e., $d Q b$. Likewise, $d\underline{P}b$ leads to $d Q b$. This suggests the following definition.

DEFINITION 5. *Suppose* $\langle A, \succ \rangle$ *is an interval order. Then it is a semiorder iff for all* a, b, c, d *in* A,

$$\text{if } a \succ b \quad \text{and} \quad b \succ c, \quad \text{then either } a \succ d \quad \text{or} \quad d \succ c. \qquad (2)$$

This form of the definition of a semiorder is due to Scott and Suppes (1958); for alternative formulations see Mirkin (1972) and Rabinovitch

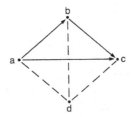

FIGURE 2. A configuration on four elements that violates the additional property (Equation (2) of Definition 5) that defines a semiorder. Again, a solid arrow represents $\succ$ and a dotted line represents $\sim$.

(1977). Chipman (1971) called Equation (2) *semitransitivity*; however, that seems a poor term since, given irreflexivity, it implies transitivity and more (Exercise 5).

Another way to state the definition is to note that an interval order is a semiorder exactly when P is the union of $\bar{P}$ and $\underline{P}$. That is, the asymmetric parts of the upper and lower weak orders combine disjunctively to yield a relation that is again asymmetric. To see that this is so, note that in an interval order, the converse complement of Q (the set $-Q^*$) is just $\bar{P} \cup \underline{P}$, whereas P = Q − I by definition. Thus, if these are equal, $-Q^* \subseteq Q$ or Q is connected.

An example of a violation of Equation (2) is shown in Figure 2. Here b lies above d (in the sense of $\bar{P}$) from the standpoint of c, whereas d lies above b (in the sense of $\underline{P}$) from the standpoint of a.

THEOREM 3. *Suppose* $\succ$ *is a binary relation on A. The following statements are equivalent*:

(i) $\succ$ *is asymmetric and* $Q = \bar{Q} \cap \underline{Q}$ *is connected.*

(ii) $\langle A, \succ \rangle$ *is an interval order and* $P = \bar{P} \cup \underline{P}$.

(iii) $\langle A, \succ \rangle$ *is a semiorder.*

Given Theorem 2 for interval orders, the proof of Theorem 3 is embodied in the remarks preceding and following Definition 5.

16.2.3 Compatible Relations

Another way of looking at this situation is to consider a threshold representation $\langle \varphi, \delta \rangle$ and to consider how the ordering induced by φ is related to $\succ$. It is not difficult to see that the following definition captures this idea.

DEFINITION 6. *Suppose $\succ$ and R are binary relations on A and $\succ$ is asymmetric. R is* upper compatible with $\succ$ *iff for every a, b, c in A, aRb and $b \succ c$ imply $a \succ c$. R is* lower compatible with $\succ$ *iff for every a, b, c in A, $a \succ b$ and bRc imply $a \succ c$. R is* fully compatible with $\succ$ *iff it is both upper and lower compatible with $\succ$.*

The definition of full compatibility in slightly different form was given by Roberts (1968, 1971a), and it played a role in his discussion of threshold representations of graphs.

There are several almost trivial observations.

1. Suppose R and S are relations on A. We denote by RS the relation defined by: $aRSb$ iff there exists c in A such that aRc and cSb. In this notion, R is upper compatible with $\succ$ means $R \succ \subseteq \succ$ and lower compatible means $\succ R \subseteq \succ$.

2. Any subset of an upper (lower) (fully) compatible relation is itself upper (lower) (fully) compatible.

3. The upper (lower) (two-sided) quasiorder $\overline{Q}(\underline{Q})(Q)$ is, by definition, upper (lower) (fully) compatible with an asymmetric $\succ$. Indeed, each is the maximal (i.e., least-refined) relation having the corresponding compatibility with $\succ$.

4. In general, Q may be a proper subset of $\overline{Q}$ and $\underline{Q}$ and so may be a more refined upper or lower compatible relation than $\overline{Q}$ and $\underline{Q}$, respectively. This is illustrated in Table 1, where column 5 refines columns 1 and 2, or 6 refines 3 and 4. In that example, it can also be checked that the ranking in column 7 is upper compatible with both $\succ_1$ and $\succ_2$ (Exercise 8), and the ranking in column 8 is lower compatible with $\succ_1$ and $\succ_2$.

The results of this section can be thought of as supplementary to the first three theorems, and so we refer to them as corollaries.

COROLLARY 1. *Suppose $\langle A, \succ \rangle$ has an upper (lower) (two-sided) representation $\langle \varphi, \delta \rangle$. Define R by: for all a, b in A, aRb iff $\varphi(a) \geqslant \varphi(b)$. Then R is upper (lower) (fully) compatible with $\succ$.*

COROLLARY 2. *$\langle A, \succ \rangle$ is an interval order iff there is a weak order that is either upper or lower compatible with $\succ$. Specifically, for $\succ$ an interval order, $\succ_U$ defined by $a \succ_U b$ iff $(\exists c)(a \sim c \ \& \ c \succ b)$ is an upper-compatible weak order, and $\succ_L$ defined by $a \succ_L b$ iff $(\exists d)(a \succ d \ \& \ d \sim b)$ is a lower-compatible one.*

COROLLARY 3. *$\langle A, \succ \rangle$ is a semiorder iff there is a weak order that is fully compatible with $\succ$. Specifically, for a semiorder, $\succ_F$ defined by $a \succ_F b$ iff either $a \succ_U b$ or $a \succ_L b$ is a fully compatible weak order.*

The easy proofs are left as Exercises 9, 10, and 11.

The major interest in compatible orderings is the possibility that they may use additional sources of information and may require a more refined representation φ. For example, even if we are interested only in the upper-threshold representation of a semiorder, the fact remains that the relation Q may be more refined and valuable than $\overline{Q}$ is alone. Another case of this would be the existence of a second asymmetric relation, such as $\succ_2$ in the example of Table 1, that provides further information leading to refinements, as in columns 7 and 8 of the table. In Section 16.4.1, we study families of interval orders and semiorders having a common, compatible weak order.

Other ways of constructing weak orders from interval orders and semiorders are discussed by Fishburn (1973b) and Fishburn and Gehrlein (1974, 1975).

16.2.4 Biorders: A Generalization of Interval Orders

As Ducamp and Falmagne (1969) first pointed out, the ideas being described here can be generalized to situations involving a relation between two distinct sets. For example, if A denotes a set of people and X a set of test items, then we may define a relation $R \subseteq A \times X$ as follows: for a in A and x in X, then aRx if and only if individual a answers test item x correctly. A natural question is whether there is some simple numerical representation of such order relations. Since two (possibly) distinct sets are involved, there will certainly have to be more than one function; and as we have seen, some relations on a single set take two functions to describe. So one can envisage the following type of representation: there exist functions φ on A and ψ on X such that for all a in A and x in X:

$$aRx \quad \text{iff} \quad \varphi(a) > \psi(x). \tag{3}$$

For the case where A and X are both finite, Ducamp and Falmagne (1969) characterized this representation as follows: Equation (3) holds iff for all a, b in A and x, y in X,

$$\text{if} \quad aRx \quad \text{and} \quad bRy, \quad \text{then either} \quad aRy \quad \text{or} \quad bRx. \tag{4}$$

The proof is left as Exercise 15. A relation satisfying Equation (4) is called a *biorder* or a *Ferrers relation* (Monjardet, 1978).

Note that Equation (4) is identical to Equation (1) with c, d replaced by x, y. So when $A = X$, a biorder is an interval order. In this case, a threshold can be defined from Equation (3) by $\overline{\delta}(x) = \psi(x) - \varphi(x)$; however, there is no assurance that $\overline{\delta}$ is nonnegative when constructed in this way.

A number of additional results about biorders can be found in Bridges (1983, 1985), Doignon (1984, 1987), Doignon, Ducamp, and Falmagne (1984, 1987), Doignon and Falmagne (1984), Ducamp (1978), and Gensemer (1989).

16.2.5 Tight Representations

What we have shown in the previous two sections, in various guises, is that a necessary condition for the construction of a one-sided, threshold representation of $\succ$ is that $\langle A, \succ \rangle$ is an interval order, and that a necessary condition for a two-sided one is that $\langle A, \succ \rangle$ is a semiorder. Another necessary condition is fairly obvious, once the theory of ordinal measurement (Section 2.1) is invoked in connection with the induced weak order: the set of equivalence classes A/I (say, for the two-sided case) must contain a finite or countable order-dense subset.

We now turn to the representation and uniqueness theorems. Can we construct a threshold representation for any interval order or semiorder whose equivalence classes have a finite or countable order-dense subset? And how unique is such a representation? It turns out that the answer to the first question is yes and that the obvious way of constructing the representation leads to a class of representations with special properties. Within this special class, we have a partial solution to the second (uniqueness) question.

Suppose that $\langle A, \succ \rangle$ is an interval order, that $\overline{Q}$ is the induced upper quasiorder, and that $A/\overline{I}$ contains a finite or countable order-dense subset. The obvious way to construct a representation starts with the construction of an order homomorphism $\overline{\varphi}$ of $\langle A, \overline{Q} \rangle$ into $\langle \mathrm{Re}, \geqslant \rangle$. This is possible by Theorem 2, Chapter 2. Moreover, we can let the range of $\overline{\varphi}$ be contained in a bounded interval of Re, and we can then proceed to construct $\overline{\delta}$ by

$$\overline{\delta}(a) = \sup\{\overline{\varphi}(a') - \overline{\varphi}(a) \mid a' \in A \text{ and } a' \sim a\}.$$

Analogous constructions are possible for lower and two-sided thresholds.

The representations constructed in this way are special in two ways: the scale φ is as coarse as it can be (i.e., uses no information except the relevant induced quasiorder), and the threshold δ is as small in absolute value as it can be. Such representations are called *tight* (Definition 7 below).

For the example of Table 1 (p. 308), we get a tight, upper representation for $\succ_1$ by using the $\overline{Q}_1$ ranks ($\overline{\varphi}_1$ in column 1 of the table) and setting $\overline{\delta} = 0$ except for element c, where $\overline{\delta}(c) = \overline{\varphi}(a) - \overline{\varphi}(c) = 2$. After reading Definition 7, the reader might try to define $\overline{\delta}$ and $\underline{\delta}$ so as to give a tight, two-sided representation of $\succ_1$ in Table 1, using the Q_1 ranks (φ_1 in column 5).

DEFINITION 7. *Suppose* $\langle \varphi, \bar{\delta}, \underline{\delta} \rangle$ *is a two-sided threshold representation of* $\langle A, \succ \rangle$. *It is said to be* tight *iff for all* a, b, c *in* A:

(i) *If* $a I b$, *then* $\varphi(a) = \varphi(b)$.

(ii) $\bar{\delta}(a) = \sup\{\varphi(a') - \varphi(a) | a' \in A \text{ and } a' \sim a\}$.

(iii) $\underline{\delta}(a) = \inf\{\varphi(a') - \varphi(a) | a' \in A \text{ and } a' \sim a\}$.

An upper (lower) *representation is* tight *if* I *is replaced by* $\bar{I}$ ($\underline{I}$) *and condition* (ii) [(iii)] *holds.*

Several minor observations should be made.

1. The converse of condition (i) is part of the definition of any representation [condition (iii) of Definition 2]. So in a tight representation, equivalence (I, $\bar{I}$, or $\underline{I}$) is represented by equality of φ values.

2. In a general representation, the order induced by φ is merely compatible with $\succ$ whereas in a tight representation it is the maximal (coarsest) compatible ordering. (See Exercise 16.)

3. Note that if $\langle \varphi, \bar{\delta}, \underline{\delta} \rangle$ is a tight representation, it is quite possible for $\langle \varphi, \bar{\delta} \rangle$ and/or $\langle \varphi, \underline{\delta} \rangle$ not to be tight. The reason (as illustrated in Table 1) is that $\bar{I}$, $\underline{I}$, and I need not agree, and so condition (i) may fail. In fact, if both one-sided representations are tight, then $\bar{Q} = \underline{Q} = Q$ (see Theorem 9).

4. Although tight representations seem desirable, they can be misleading. For example, suppose the structure under study is, in fact, a proper subset of the structure of interest. It is easy to see that, in general, a tight representation of a substructure is not the restriction of a tight representation of the larger structure.

As a preliminary to the main theorems, we consider the construction of a threshold when a homomorphism for $\langle A, \bar{Q} \rangle$ exists.

LEMMA 1. *Suppose* $\langle A, \succ \rangle$ *is an interval order and* φ *is a bounded real-valued function on* A *such that for all* a, b *in* A,

$$\text{if } \varphi(a) \geqslant \varphi(b), \quad \text{then } a\bar{Q}b.$$

Define

$$\bar{\delta}(a) = \sup\{\varphi(a') - \varphi(a) | a' \in A \text{ and } a' \sim a\}.$$

Then $\bar{\delta}$ *is well defined,* $\bar{\delta} \geqslant 0$, *and* $\langle \varphi, \bar{\delta} \rangle$ *is an upper-threshold representation for* $\langle A, \succ \rangle$. *Moreover,* $\langle \varphi, \bar{\delta} \rangle$ *is tight iff* φ *is a homomorphism of* $\langle A, \bar{Q} \rangle$ *into* $\langle \text{Re}, \geqslant \rangle$.

THEOREM 4. _Suppose $\langle A, \succ \rangle$ is an interval order for which the induced weak order $\langle A, \overline{Q} \rangle$ has a finite or countable order-dense subset. Then $\langle A, \succ \rangle$ has a tight upper-threshold representation._

The analogous result holds for lower thresholds.

As has been noted, both the interval-order condition and the order-dense subset are necessary for a one-sided representation (tightness plays no role in that proof). The proof of Theorem 4 is an immediate consequence of Theorem 2.2 and Lemma 1.

A fully satisfactory uniqueness theorem is not known. For tight representations, two partial results are given. Theorem 5 provides sufficient conditions for f to be an admissible transformation, namely, f is strictly increasing and semicontinuous from the left (right) for an upper (lower) representation. Neither property is, however, necessary for an admissible transformation; but in case the representation is strong, then monotonicity is necessary, as stated in Theorem 6. Clearly, we would like to know the necessary and sufficient conditions, but they appear to be very complex. For example, different weak representations can behave differently at the threshold boundaries. Even if one restricts oneself to strong representations, some discontinuous transformations preserve all the properties, including tightness. For an example, let $A = \mathrm{Re}^+$, $a \succ b$ iff $a > b + 1$, $\varphi(a) = a$, $\delta(a) = 1$. Define

$$f(x) = \begin{cases} x - 2, & 0 < x < 1 \\ 0, & x = 1 \\ x, & x > 1. \end{cases}$$

Both (φ, δ) and (φ', δ'), where the latter is defined by Equation (5), are strong and tight.

THEOREM 5. _Suppose $\langle A, \succ \rangle$ is an interval order with a tight upper-threshold representation $\langle \varphi, \overline{\delta} \rangle$. Suppose f is a strictly increasing, left-semicontinuous function defined over the union of the ranges of φ and $\varphi + \overline{\delta}$. Define_

$$\varphi' = f(\varphi), \qquad \overline{\delta}' = f(\varphi + \overline{\delta}) - f(\varphi). \tag{5}$$

Then $\langle \varphi', \overline{\delta}' \rangle$ is also a tight upper-threshold representation for $\langle A, \succ \rangle$. If $\langle \varphi, \overline{\delta} \rangle$ is strong (or strong*), then so is $\langle \varphi', \overline{\delta}' \rangle$.

The proof is left as Exercise 17.

THEOREM 6. *Let* $\langle \varphi, \bar{\delta} \rangle$ *and* $\langle \varphi', \bar{\delta}' \rangle$ *be strong, tight, upper-threshold representations for* $\langle A, \succ \rangle$. *Then there exists a strictly increasing function* f *such that Equation* (5) *holds.*

Lower-threshold analogues of Theorem 5 (using right semicontinuity) and 6 also hold. The next result formulates the two-sided version of Theorems 4, 5, and 6.

THEOREM 7. *Suppose* $\langle A, \succ \rangle$ *is a semiorder such that* $\langle A, Q \rangle$ *has a finite or countable order-dense subset. Then* $\langle A, \succ \rangle$ *has a tight, two-sided threshold representation* $\langle \varphi, \bar{\delta}, \underline{\delta} \rangle$. *If* f *is strictly increasing and continuous, then* $\langle \varphi', \bar{\delta}', \underline{\delta}' \rangle$, *where*

$$\varphi' = f(\varphi), \qquad \bar{\delta}' = f(\varphi + \bar{\delta}) - f(\varphi), \qquad \underline{\delta}' = f(\varphi + \underline{\delta}) - f(\varphi) \quad (6)$$

is another tight representation. Conversely, any two strong, tight representations are connected by Equation (6) *for some strictly increasing* f.

The proof is a simple consequence of the preceding ones because the two-sided Q is connected (Theorem 3). φ, $\bar{\delta}$, and $\underline{\delta}$ are constructed as in Theorem 4, with φ a homomorphism of $\langle A, Q \rangle$. The uniqueness follows as in Theorems 5 and 6.

The next two theorems concern the issue of when two-sided thresholds follow from the existence of one-sided ones.

THEOREM 8. *Suppose* $\langle A, \succ \rangle$ *has a strong* (*strong**) *upper-threshold representation* $\langle \varphi, \bar{\delta} \rangle$. *Then*

(i) $\langle \varphi + \bar{\delta}, -\bar{\delta} \rangle$ *is a strong* (*strong**) *lower representation;*

(ii) *if* $\langle \varphi, \bar{\delta} \rangle$ *is tight, then* $\varphi + \bar{\delta}$ *is a homomorphism of* $\langle A, Q \rangle$ *into* $\langle \mathrm{Re}, \geqslant \rangle$.

Note that part (i) follows directly from Definition 2 and does not rest on any later material. An analogous result holds for lower-threshold representations. The proof is left as Exercise 18.

Observe that in Theorem 8, $\varphi + \bar{\delta}$ is linked to Q when φ is linked to $\bar{Q}$. This means that the orderings Q and $\bar{Q}$ are the same provided $\varphi + \bar{\delta}$ is a monotonic function of φ, which leads to the following definition.

DEFINITION 8. *Suppose* $\langle \varphi, \delta \rangle$ *is a one-sided threshold representation. It is said to be* monotonic *iff* $\varphi + \delta$ *is a strictly increasing function of* φ.

THEOREM 9.

(i) *If* $\langle A, \succ \rangle$ *has a tight, monotonic, one-sided representation, then it is a semiorder.*

(ii) *If* $\langle A, \succ \rangle$ *has a strong (or strong*), tight, monotonic, one-sided representation, then* $\underline{Q} = \overline{Q} = Q$ *and the representation can be extended to a two-sided one.*

(iii) *If* $\langle A, \succ \rangle$ *has a strong (or strong*), tight, two-sided representation and* $\underline{Q} = \overline{Q} = Q$, *then the one-sided representations are each monotonic.*

Basically, the message of this result is that the existence of strong, tight, two-sided representations goes with a semiorder and with the monotonicity of the one-sided representations. The proof of part (i) is left as Exercise 19.

16.2.6 Constant-Threshold Representations

We turn now to a different and important class of representations in which tightness may be absent, and we ask whether φ and δ can be chosen so that δ is a constant. As was noted in the introduction to the chapter, the motives for studying the two cases are somewhat different. In the preceding section we were concerned primarily with the induced weak order Q and its representation by a real-valued function, and the threshold function is really a nuisance factor that keeps the observed order $\succ$ from being Q. Now we shall regard the constant threshold as a "natural" unit in terms of which φ can be constructed by looking at successive differences.

The result, embodied in Theorems 10 and 11, is closely related to other material: difference measurement (Chapter 4), uniform structures (Section 6.7), and simple scalability (Section 17.4.1). Because the uniqueness result (Theorem 11) is not satisfactory for a single threshold, we are led in the next two sections to consider cases in which we have two or more orderings that are presumed to arise as different thresholds applied to the same underlying scale.

A one-sided threshold representation $\langle \varphi, \delta \rangle$ is called a *constant-threshold representation* iff δ is a nonzero constant, say k. We use the notion $\langle \varphi, k \rangle$ to emphasize the constancy of the threshold. It is easy to see that $\langle \varphi, k \rangle$ is a constant upper-threshold representation iff $\langle \varphi, -k \rangle$ is a constant lower-threshold representation, and so $\langle \varphi, k, -k \rangle$ is a two-sided representation and the corresponding structure $\langle A, \succ \rangle$ is a semiorder. Thus it suffices in this section to consider only semiorders and only upper thresholds. Note, also, that all such representations are monotonic (Definition 8).

Our first result is a slightly improved version of one of the earliest representation theorems in the modern literature on measurement and is due to Scott and Suppes (1958; see also Rabinovitch, 1977; Scott, 1964). We follow Rabinovitch's (1977) proof.

THEOREM 10. *Suppose $\langle A, \succ \rangle$ is a semiorder for which A/I is finite. Then there exists a strong* constant-threshold representation $\langle \varphi, k \rangle$ such that φ is a homomorphism of $\langle A, Q \rangle$ into $\langle \text{Re}, \geqslant \rangle$.*

The proof (Section 3.6) is readily modified to show that the representation is strong as well as strong*. The example in Table 1 with φ_1 in column 5 and $k = 2$ is an illustration of this theorem. Roy (1980) reported that a necessary and sufficient condition for the existence of a strong, constant-threshold representation when X is finite is that each cycle of the relation $\succsim \; = \; \succ \; \cup \sim$ has more pairs in $\sim$ than in $\succ$.

It is important to realize that this constant representation is not necessarily tight. The following is an example of a semiorder that has no tight monotonic representations and thus no tight constant-threshold representation.

EXAMPLE. Let $A = \{1.0, 1.7, 2.2, 3.3\}$ and $a \succ b$ iff $a > b + 1$. Since $\langle \iota, 1 \rangle$, where ι is the identity function, is a constant-threshold representation, $\langle A, \succ \rangle$ must be a semiorder. Observe that according to Definition 3, 2.2 $\bar{P}$ 1.7. Thus, if $\langle \varphi, \delta \rangle$ is any tight, upper-threshold representation, it follows that $\varphi(2.2) > \varphi(1.7)$. Since $3.3 \succ 2.2$,

$$\delta(2.2) = \sup\{\varphi(a) - \varphi(2.2)|a \sim 2.2\} = 0$$

and since $3.3 \succ 1.7$ and $2.2 \sim 1.7$, $\delta(1.7) = \varphi(2.2) - \varphi(1.7)$, whence

$$\varphi(2.2) + \delta(2.2) = \varphi(2.2) = \varphi(1.7) + \delta(1.7),$$

proving that $\langle \varphi, \delta \rangle$ is not monotonic.

The conclusion of Theorem 10 may fail when A/I is countable rather than finite. The reason is that the threshold, in the countable case, can approach 0 for a subset that has a bound. An example will make the problem clear. Let A consist of the rationals in $[0, 1]$, and define $\bar{\delta}$, $\underline{\delta}$, and $\succ$ by

$$\bar{\delta}(a) = \frac{1}{2}(1 - a)$$

$$\underline{\delta}(a) = a - 1$$

$$a \succ b \quad \text{iff} \quad a \geqslant b + \bar{\delta}(b) \quad \text{iff} \quad a + \underline{\delta}(a) \geqslant b.$$

Then $\langle A, \succ \rangle$ is a semiorder with representation $\langle \iota, \bar{\delta}, \underline{\delta} \rangle$. There can be no constant-threshold representation, for if $\langle \varphi, k \rangle$ were such, let $a_n = 1 - 2^{-n}$ for $n = 0, 1, 2, \ldots$. We have $a_n + \delta(a_n) = a_{n+1}$; hence, $\varphi(a_{n+1}) - \varphi(a_n) = k$. From this, $\varphi(a_{n+1}) = \varphi(0) + nk \to +\infty$, contradicting the fact that $\{\varphi(a_{n+1})\}$ must be bounded by $\varphi(1)$.

The following lists a number of other representation theorems, including ones for the countable case. Fishburn (1970c) showed for the countable case that there is always an upper-threshold representation that is strong and strong*, but it need not be tight or constant. In (1973a) he provided necessary and sufficient conditions for a strong representation. A generalization in which ~ is not necessarily the symmetric complement of ≻ was considered by Roubens and Vincke (1983), and a further generalization to biorders was reported by Doignon (1987). Manders (1981) showed that a sufficient condition for a countable semiorder to have a constant jnd representation is that there is always a finite chain of indifferences between any two points of the structure.

The finite case considered in Theorem 10 has limited practical value. For one thing, there is no satisfactory uniqueness theorem; moreover, if the finite structure is a part (sample) of a larger structure, a representation of the finite substructure may not extend to one for the superstructure. For these reasons, it seems wise to also consider the case of infinite structures with tight representations and to ask for the conditions under which such a representation can be transformed into a tight, constant one. From the uniqueness result of Theorem 6, we know that any strong, tight, constant-threshold representation $\langle \varphi', k \rangle$ must be related to a given strong, tight representation $\langle \varphi, \delta \rangle$ by a strictly increasing function f for which

$$f(\varphi + \delta) - f(\varphi) = k. \tag{7}$$

If δ is a well-defined function of φ, i.e., $\varphi(a) = \varphi(b)$ implies $\delta(a) = \delta(b)$, then $\varphi + \delta = G(\varphi)$ for some function G, in which case Equation (7) becomes the famous functional equation of Abel. The existence and uniqueness of solutions to this equation in measurement contexts have been much studied (Bellman, 1965, pp. 494–496; Falmagne, 1971, 1974, 1977, 1985; Levine, 1970, 1972, 1975; Luce & Edwards, 1958; Section 6.8.2). These results will be used to prove the following theorem about the existence of constant-threshold representations.

THEOREM 11. *Existence: Suppose that $\langle A, \succ \rangle$ has a tight, monotonic, upper-threshold representation $\langle \varphi, \bar{\delta} \rangle$, that the range of φ is dense in some interval $[x_0, x_1]$, that $\bar{\delta}$ is a continuous function of φ, and that $\bar{\delta}$ has a positive lower bound on $[x_0, x_1]$.[3] Then there exists a continuous, strictly increasing, real-valued function f on $[x_0, x_1]$ and a constant $k > 0$ such that $\langle f(\varphi), k \rangle$ is a tight, constant-threshold representation of $\langle A, \succ \rangle$.*

---

[3] We allow $x_0 = -\infty$ and $x_1 = +\infty$ as possible values, in which case that side of the interval is treated as open rather than closed.

Uniqueness: f', k' have the same properties as f, k iff there is a continuous periodic function p with period k [i.e., $p(x + k) = p(x)$ for all real x] that satisfies the semi-Lefschetz condition:

$$p(x) - p(y) > -(k'/k)(x - y) \quad for \quad x - y > 0,$$

and is such that

$$f' = (k'/k)f + p(f).$$

Aside from assuming that φ is dense on an interval and that $\bar{\delta}(\varphi)$ is continuous and bounded away from 0 on that interval, all of which permits one to extend the whole problem to $(-\infty, \infty)$, the key fact in the theorem is this: the monotonicity assumption, which is obviously necessary for δ to be a function of φ, is also sufficient for this to be so. Once the extension is performed, then we invoke Theorem 6.8 to get the result. Note that, by Theorem 9, if the representation is strong or strong*, then the initial hypothesis of Theorem 11 can be the existence of a two-sided representation.

The uniqueness result is what always arises with Abel's equation. The verification is simple: suppose $f(\varphi + \delta) - f(\varphi) = k$, then

$$
\begin{aligned}
f'(\varphi + \delta) &- f'(\varphi) \\
&= (k'/k)f(\varphi + \delta) + p[f(\varphi + \delta)] - (k'/k)f(\varphi) - p[f(\varphi)] \\
&= (k'/k)[f(\varphi) + k] + p[f(\varphi) + k] - (k'/k)f(\varphi) - p[f(\varphi)] \\
&= k'.
\end{aligned}
$$

The converse, which is left as Exercise 21, is only slightly more difficult. The role of the semi-Lefschetz condition is simply to guarantee that f' is also strictly increasing.

The failure to achieve interval-scale uniqueness is easily understood. It arises from the fact that f can be any increasing function over the interval $[\varphi, \varphi + \delta]$ provided that it increases exactly k on the interval. Only differences of size k are determined by the two functions φ and $\varphi + \delta$. It is the same situation we faced in the problem of uniform systems (Section 6.7), and the solution is the same—we must have more than one threshold function. We take this up in the first two subsections of Section 4 which is devoted to various extensions of the ordinal theory.

The uniqueness of interval and semiorders having continuous representations is characterized in Gensemer (1987).

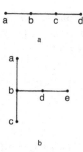

b

FIGURE 3. See text for discussion.

16.2.7 Interval and Indifference Graphs

Up to this point we have emphasized the asymmetric relation $\succ$ and its threshold representations. However, a good deal of information about $\succ$ can sometimes be recovered from its symmetric complement $\sim$. The symmetric complement of an interval order is called an *interval graph*, that of a semiorder is called an *indifference graph*. In this subsection we cover selected results about these two types of graphs. We emphasize the question that fits most closely with our previous development: to what extent is the asymmetric relation $\succ$ uniquely recoverable from $\sim$? We also mention the standard graph-theory axiomatizations of interval and indifference graphs and describe the standard representations of these structures by mapping *vertices* to *intervals*.[4]

The publication of Fishburn's authoritative and highly readable book (1985) makes it easy for us to give a brief treatment of selected points, referring to his book for other topics and for proofs.

It is clear that a symmetric graph cannot distinguish between an asymmetric relation $\succ$ and its converse $\prec$. For example, the graph of Figure 3a is compatible with oppositely directed semiorders. It is not hard to show that these two semiorders are the *only* two interval orders that give rise to that graph. For example, one cannot have an interval order with $a \succ c$ and $d \succ b$ since Equation (1) would then imply that either $a \succ b$ or $d \succ c$, both of which are excluded in Figure 3a.

Figure 3b illustrates a more complex situation. By symmetry, it is clear that even if the direction from b to e is established, two opposite relations

---

[4] Interval graphs and indifference graphs are most commonly defined in terms of their representability by relations among intervals rather than as complements of asymmetric relations; in the present context, the latter definition seems more natural, however. The term *interval graph* was coined to refer to the representation by intervals.

can hold between a and c. In fact, corresponding to the two directional possibilities each for (a, c) and (b, e), there are *exactly* four interval orders that give rise to Figure 3b as their symmetric complement. The proof that no others are possible follows from the same argument as for Figure 3(a): a choice for (b, e) determines the directions for (a, d), (a, e), (c, d), and (c, e). Finally, note that these four interval orders are *not* semiorders; in fact, no indifference graph can contain the subgraph on $\{a, b, c, d\}$ induced by Figure 3b. This subgraph, labeled $K_{1,3}$ in Figure 4, is one of the forbidden subgraphs in the standard characterization of indifference graphs (Roberts, 1969; Theorem 13 following). As an exercise, the reader should construct an upper-threshold representation for one of the interval orders compatible with Figure 3b and note that it is not monotonic.

The example of Figure 3a shows one extreme, in which essentially complete information about $\succ$ (except for an arbitrary choice of direction) is recoverable from $\sim$. At the opposite extreme is a transitive symmetric complement $\sim$, in which case the equivalence classes under $\sim$ can be ordered arbitrarily by $\succ$; essentially no order information is recoverable. Figure 3b illustrates one of the numerous possibilities falling between these two extremes. These possibilities can be analyzed in more detail in terms of maximal cliques in the symmetric graph; Fishburn's book gives a clear treatment. The most natural case from the standpoint of measurement is that in which there are enough intransitivities in $\sim$ so that it is a connected graph, i.e., any two elements are connected by a finite chain of $\sim$ relations. With connectedness, the situation in Figure 3a prevails: a semiorder is fully recoverable, except for direction, from its indifference graph. This beautiful result is due to Roberts (1971a):

THEOREM 12. *A connected indifference graph is the symmetric complement of exactly two (oppositely directed) semiorders.*

For a proof we refer to Roberts' paper or to Fishburn's book (1985, p. 53). The proof also establishes that there can be no interval order that is distinct from these two semiorders and yet is compatible with their (connected) indifference graph.

The uniqueness results for interval graphs are a bit more complex and assume finiteness (see Fishburn, 1985, pp. 54–56). It seems desirable to eliminate the latter hypothesis.

We have been considering recovery of order information from symmetric binary relations. Since the direction of the order is never determined, it is really betweenness information that is recovered. It would be natural, therefore, to introduce a ternary relation of betweenness $a|b|c$ and to break the recovery of order into two steps: (1) recovery of a ternary relation from the symmetric binary one and (2) recovery of an asymmetric binary relation from the ternary one. Results relating to both these steps are in the literature and are summarized in Chapter 4 of Fishburn's book. Some results concerning step (2) are also found in axiomatic geometry (see Section 13.2.1).

We turn now to axiomatic characterization of indifference and interval graphs. The following nice result for indifference graphs is due to Roberts (1969).

THEOREM 13. *A graph is an indifference graph iff it contains no subgraph isomorphic to any of the graphs $K_{1,3}$, G_1, G_2, or F_n ($n = 4, 5, \ldots$) of Figure 4.*

For a proof of Theorem 13 we refer once more to Fishburn's book (1985, p. 51) or to Roberts' paper.

Again, the axiomatization of interval graphs is more complicated. Graphs G_1, G_2, and the family F_n ($n \geqslant 4$) are also forbidden in interval graphs, but $K_{1,3}$ is not. However, the exclusion of G_1, G_2, and F_n ($n \geqslant 4$) is not sufficient. Lekkerkerker and Boland (1962) present an axiomatization in which there are three countably infinite families of forbidden subgraphs, F_n and two others, of which G_1 and G_2 are the smallest members, respectively. Other axiomatizations have been given by Gilmore and Hoffman (1964) and Roberts (1969). The various results are reviewed in Fishburn's book.

Finally, we turn to the representation theorems for interval and indifference graphs. To motivate them, consider an interval order $\langle A, \succ \rangle$ with an upper-threshold representation $\langle \varphi, \bar{\delta} \rangle$. If $a \sim b$, then the closed intervals

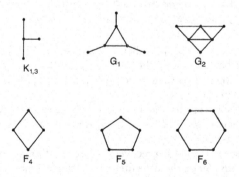

FIGURE 4. The forbidden subgraphs of an indifference graph.

$[\varphi(a), \varphi(a) + \bar{\delta}(a)]$ and $[\varphi(b), \varphi(b) + \bar{\delta}(b)]$ must intersect. Conversely, if the open intervals with those same endpoints intersect, then $a \sim b$. This suggests that a desirable representation for the graph $\langle A, \sim \rangle$ would be a function that maps each element of A to an *interval* such that $a \sim b$ if and only if the intersection of the intervals corresponding to a and to b is nonempty. In the case of a semiorder, $\varphi + \bar{\delta}$ is monotonic with φ; so for an indifference graph we expect the additional condition that the upper endpoint of the representing interval is a strictly increasing function of the lower endpoint (or vice versa).

If we permit representations by intervals in an arbitrary simple order (not necessarily one embeddable in the reals) we obtain the following theorems, which restate results due to Fishburn (1973a) and Roberts (1969), respectively.

THEOREM 14. $\langle A, \sim \rangle$ *is an interval graph iff there exists a simple order* $\langle X, \geqslant \rangle$ *and a function* (φ_L, φ_U) *from A into* $X \times X$ *such that*

(i) $\varphi_U \geqslant \varphi_L$,

(ii) $a \sim b$ *iff the closed intervals* $[\varphi_L(a), \varphi_U(a)]$ *and* $[\varphi_L(b), \varphi_U(b)]$ *have nonempty intersection in* X.

THEOREM 15. $\langle A, \sim \rangle$ *is an indifference graph iff there exists a simple order* $\langle X, \geqslant \rangle$, *a function* φ *from A into* X, *and a nondecreasing function* G *from* X *into* X *such that* $(\varphi, G\varphi)$ *satisfy conditions* (i) *and* (ii) *of Theorem 14.*

As we noted earlier, these theorems are usually taken as the definitions of interval and indifference graphs; in that case, the theorem is the assertion that $\sim$ is the symmetric complement of some interval order or semiorder. It seems slightly more natural here to take the latter conditions as definitions and to state the representations as theorems. We restate and prove (the proof is brief, using the concept of induced weak orders) the theorems in Section 16.3.7 by asserting the equivalence of the two alternative defining conditions for each type of graph.

It may be a surprise that these representations use only closed intervals because the intuitive justification, starting from a representation $\langle \varphi, \bar{\delta} \rangle$, would lead one to expect both closed and open intervals (unless the threshold representation is either strong or strong*). This comes under the heading of dirty tricks: as long as we are defining a simple order to use for the representation, we include among its elements both closed and open rays, ordered by inclusion. We do this in order to follow the mathematical literature on the subject of intersection graphs, but a more natural representation would be one that permitted mixed types of intervals.

16.3 PROOFS

16.3.1 Theorem 2 (p. 310)

For $\succ$ a binary relation on A, the following are equivalent:

(i) $\succ$ *is asymmetric and $\bar{Q}$ is connected;*
(ii) $\succ$ *is asymmetric and Q is connected; and*
(iii) $\langle A, \succ \rangle$ *is an interval order.*

PROOF. (i) implies (iii). Suppose $a \succ c$ and $b \succ d$. Since $\bar{Q}$ is connected, either $a\,\bar{Q}\,b$, in which case $a \succ d$, or $b\,\bar{Q}\,a$, in which case $b \succ c$. Since asymmetry implies irreflexivity, (iii) holds.

(ii) implies (iii). Similar.

(iii) implies (i). Suppose $\succ$ is not asymmetric, and so for some a, b, both $a \succ b$ and $b \succ a$ hold. By the interval-order condition, either $a \succ a$ or $b \succ b$, which contradicts irreflexivity. Now, suppose that for some a, b neither $a\,\bar{Q}\,b$ nor $b\,\bar{Q}\,a$. Thus c, d exist for which $b \succ d$ but not $a \succ d$, and $a \succ c$ but not $b \succ c$, which contradicts the interval-order property.

(iii) implies (ii). Similar.   ◇

16.3.2 Lemma 1 (p. 315)

Suppose $\langle A, \succ \rangle$ is an interval order and φ is a bounded mapping of $\langle A, Q \rangle$ into $\langle \text{Re}, \geqslant \rangle$ such that $\varphi(a) \geqslant \varphi(b)$ implies $a\,\bar{Q}\,b$. Let $\bar{\delta}(a) = \sup\{\varphi(a') - \varphi(a) | a' \in A, a' \sim a\}$. Then, $\bar{\delta}$ is well defined and nonnegative, and $\langle \varphi, \bar{\delta} \rangle$ is an upper-threshold representation. $\langle \varphi, \bar{\delta} \rangle$ is tight iff φ is an order homomorphism.

PROOF. Since $\varphi(a') - \varphi(a)$ is bounded and $\varphi(a) - \varphi(a) = 0$, $\bar{\delta}$ is well defined and nonnegative. Turning to Definition 2, we show the three parts:

(i) Suppose $a \succ b$ and $\varphi(b) + \bar{\delta}(b) > \varphi(a)$. By definition of $\bar{\delta}$, there exists b' in A with $b' \sim b$ and $\varphi(b') > \varphi(a)$. Thus $b'\,\bar{Q}\,a$, which with $a \succ b$ implies $b' \succ b$, contradicting $b' \sim b$. So $\varphi(a) \geqslant \varphi(b) + \bar{\delta}(b)$.

(ii) Suppose $\varphi(a) > \varphi(b) + \bar{\delta}(b)$. By definition of $\bar{\delta}$, $a \sim b$ is impossible. If $b \succ a$, then it is not the case $a\,\bar{Q}\,b$, and so $\varphi(b) > \varphi(a)$, a contradiction. Thus $a \succ b$.

(iii) Suppose $\varphi(a) = \varphi(b)$. By assumption this means $a\,\bar{I}\,b$, which in turn is equivalent by Definition 3 to $a \succ c$ iff $b \succ c$.

Concerning tightness, condition (ii) of Definition 7 is satisfied by construction; condition (i) follows by definition if φ is a homomorphism. ◇

16.3.3 Theorem 6 (p. 327)

Suppose $\langle A, \succ \rangle$ *has two strong, tight, upper-threshold representations* $\langle \varphi, \bar{\delta} \rangle$ *and* $\langle \varphi', \bar{\delta} \rangle$. *Then there exists a strictly increasing function f such that* $\varphi' = f(\varphi)$ *and* $\bar{\delta}' = f(\varphi + \bar{\delta}) - f(\varphi)$.

PROOF. By tightness, φ and φ' are both order homomorphisms of $\langle A, \bar{Q} \rangle$ into $\langle \text{Re}, \geqslant \rangle$. Therefore, there exists a strictly increasing function f, on the range of φ, such that $\varphi' = f(\varphi)$.

If $\varphi(a) + \bar{\delta}(a)$ is in the range of φ, then we show that $\bar{\delta}'(a) = f[\varphi(a) + \bar{\delta}(a)] - f[\varphi(a)]$. To this end, take b with $\varphi(b) = \varphi(a) + \bar{\delta}(a)$. By strongness, $b \sim a$, and by the upper-threshold representation property [property (ii) of Definition 2], $\varphi(b) \geqslant \varphi(c)$ [and hence, $\varphi'(b) \geqslant \varphi'(c)$] for every c such that $c \sim a$. Therefore, by tightness,

$$\begin{aligned}
\bar{\delta}'(a) &= \sup\{\varphi'(c) - \varphi'(a) | c \sim a\} \\
&= \varphi'(b) - \varphi'(a) \\
&= f[\varphi(b)] - f[\varphi(a)] \\
&= f[\varphi(a) + \bar{\delta}(a)] - f[\varphi(a)],
\end{aligned}$$

as required.

Finally, suppose that $\varphi(a) + \bar{\delta}(a)$ is not in the range of φ. Define $f[\varphi(a) + \bar{\delta}(a)] = \varphi'(a) + \bar{\delta}'(a)$. To show that this is well defined, suppose $\varphi(a) + \bar{\delta}(a) = \varphi(b) + \bar{\delta}(b)$. By strongness, $c \succ a$ iff $c \succ b$. Thus

$$\begin{aligned}
\varphi'(a) + \bar{\delta}'(a) &= \sup\{\varphi'(c) | c \sim a\} \\
&= \sup\{\varphi'(c) | c \sim b\} \\
&= \varphi'(b) + \bar{\delta}'(b).
\end{aligned}$$

To see that this extension of f remains strictly increasing, note that $\varphi(a) + \bar{\delta}(a) > \varphi(b) + \bar{\delta}(b)$ implies, by tightness, the existence of c such that $c \sim a$ and $c \succ b$; and it follows that $\varphi'(a) + \bar{\delta}'(a) \geqslant \varphi'(c) > \varphi'(b) + \bar{\delta}'(b)$. This argument also shows that $\varphi(a) + \bar{\delta}(a) \geqslant (<)\varphi(c)$ iff the corresponding inequality holds for $\langle \varphi', \bar{\delta}' \rangle$. ◇

16.3.4 Theorem 9 (p. 318)

(i) *If* $\langle A, \succ \rangle$ *has a tight, monotonic, one-sided representation, then it is a semiorder.*

(ii) *If $\langle A, \succ \rangle$ has a strong (or strong*), tight, monotonic, one-sided representation, then $Q = \overline{Q} = \underline{Q}$, and the representation can be extended to a two-sided one.*

(iii) *Conversely, if $\langle A, \succ \rangle$ has a strong (or strong*), tight, two-sided representation and $\overline{Q} = \underline{Q} = Q$, then the one-sided representations are both monotonic.*

PROOF.

(i) Exercise 19.

(ii) Suppose that $\langle \varphi, \overline{\delta} \rangle$ is strong or strong*, tight, and monotonic. Using Theorem 8(ii) gives

$$
\begin{array}{lll}
a\,\overline{Q}\,b & \text{iff} & \varphi(a) \geqslant \varphi(b) \\
& \text{iff} & \varphi(a) + \overline{\delta}(a) \geqslant \varphi(b) + \overline{\delta}(b) \\
& \text{iff} & a\,\underline{Q}\,b.
\end{array}
$$

Thus $\overline{Q} = \underline{Q} = Q$. The lower-threshold representation can be constructed as in Theorem 7. The proof is similar starting with $\langle \varphi, \underline{\delta} \rangle$.

(iii) Since $\overline{Q} = \underline{Q}$, both one-sided representations are also tight. By Theorem 8(ii), $\varphi + \overline{\delta}$ is monotonic with φ, since

$$
\varphi(a) \geqslant \varphi(b) \qquad \text{iff} \qquad a\,\underline{Q}\,b \qquad \text{iff} \qquad \varphi(a) + \overline{\delta}(a) \geqslant \varphi(b) + \overline{\delta}(b).
$$

Similarly, $\varphi + \underline{\delta}$ is monotonic.   ◇

16.3.5 Theorem 10 (p. 319)

A semiorder $\langle A, \succ \rangle$ for which A/I is finite has a strong*, constant threshold representation $\langle \varphi, k \rangle$ such that φ is a homomorphism of $\langle A, Q \rangle$ into $\langle \mathrm{Re}, \geqslant \rangle$.

PROOF. Relabel the elements of A/I by their ranks according to Q, with 1 the minimal equivalence class, 2 the next, etc. Define φ inductively as follows. $\varphi(1) = 0$. Suppose $\langle \varphi, k \rangle$ is a strong* constant representation of $\{1, 2, \ldots, n\}$ and $(n + 1) \in A/I$. Since $(n + 1)Qn$, there are three cases to consider.

(i) $(n + 1) \succ n$. Let $\varphi(n + 1) = \varphi(n) + k$.

(ii) $(n + 1) \sim 1$. Since $(n + 1)Qn$, we have $n \sim 1$. By the induction hypothesis φ is strong*, and so for some $\epsilon > 0$, $\varphi(n) = \varphi(1) + k - \epsilon$. Define $\varphi(n + 1) = \varphi(n) + \epsilon/2$.

(iii) $1 \prec (n + 1) \sim n$. Thus there exists an integer m with $1 \leqslant m < n$ such that for all integers p, $(n + 1) \succ p$ if $p \leqslant m$ and $(n + 1) \sim p$ if $p \geqslant (m + 1)$. As in (ii), there exists $\epsilon_0 > 0$ such that $\varphi(n) = \varphi(m + 1) + k - \epsilon_0$ and $\epsilon_1 > 0$ such that $\varphi(m + 1) = \varphi(m) + \epsilon_1$. Define

$$\varphi(n + 1) = \varphi(m + 1) + k - \frac{1}{2}\min(\epsilon_0, \epsilon_1).$$

First,

$$\begin{aligned}
\varphi(n + 1) - \varphi(n) &= \varphi(m + 1) + k - \frac{1}{2}\min(\epsilon_0, \epsilon_1) - \varphi(m + 1) \\
&\quad - k + \epsilon_0 \\
&\geqslant \frac{\epsilon_0}{2} \\
&> 0,
\end{aligned}$$

and so φ is strictly increasing on $\{1, \ldots, n + 1\}$. Second,

$$\begin{aligned}
\varphi(n + 1) - \varphi(m) &= \varphi(m + 1) + k - \frac{1}{2}\min(\epsilon_0, \epsilon_1) - \varphi(m + 1) + \epsilon_1 \\
&> k + \frac{\epsilon_1}{2} \\
&> k,
\end{aligned}$$

and

$$\begin{aligned}
\varphi(n + 1) - \varphi(m + 1) &= k - \frac{1}{2}\min(\epsilon_0, \epsilon_1), \\
&< k,
\end{aligned}$$

and so the representation is strong* and constant.   ◇

16.3.6 Theorem 11 (p. 320)

Suppose $\langle \varphi, \bar{\delta} \rangle$ is a tight, monotonic, upper-threshold representation of $\langle A, \succ \rangle$ for which the range of φ is dense in $[x_0, x_1]$, that $\bar{\delta}$ is a continuous function of φ, and $\bar{\delta}$ has a positive lower bound on $[x_0, x_1]$. Then there exists a continuous strictly increasing function f on $[x_0, x_1]$ and a constant $k > 0$ such that $\langle f\varphi, k \rangle$ is a tight constant-threshold representation of $\langle A, \succ \rangle$. $\langle f'\varphi, k' \rangle$ is another such representation iff there is a continuous periodic function p with period k satisfying $p(x) - p(y) > -(k'/k)(x - y)$, for all $x > y$, and such that $f' = (k'/k)f + p(f)$.

PROOF. Because $\bar{\delta}$ is a continuous function of φ and the range of φ is dense in $[x_0, x_1]$, the function $G[\varphi(a)] = \varphi(a) + \bar{\delta}(a)$ can be extended to a continuous function on all of $[x_0, x_1]$. Because $\bar{\delta}$ has a lower bound ϵ, $G(x) - x \geqslant \epsilon > 0$. By monotonicity, G is strictly increasing. Extend G to Re by:

$$G(x) = \begin{cases} x + G(x_0) - x_0 & \text{for } x < x_0 \\ x + G(x_1) - x_1 & \text{for } x > x_1. \end{cases}$$

Thus, by Theorem 6.8 the family $\{G, \iota\}$ is uniform, and so by Definition 6.9 there exists a parallelizer f which is strictly increasing and continuous on Re such that for some positive constant k,

$$G[f^{-1}(y)] = f^{-1}(y + k).$$

Setting $y = f[\varphi(a)]$,

$$\begin{aligned} \varphi(a) + \bar{\delta}(a) &= G[\varphi(a)] \\ &= G\{f^{-1}f[\varphi(a)]\} \\ &= f^{-1}\{f[\varphi(a)] + k\}, \end{aligned}$$

and so

$$f[\varphi(a) + \bar{\delta}(a)] = f[\varphi(a)] + k.$$

Since f is strictly increasing and continuous, we know by Theorem 7 that $\langle f\varphi, k \rangle$ is tight.

Uniqueness follows from that of Theorem 6.8. $\diamond$

16.3.7 Theorems 14 and 15 (p. 325)

$\langle A, \sim \rangle$ *is the symmetric graph of an interval order iff it is representable as a subgraph of the intersection graph for the closed intervals of a simple order. The interval order is a semiorder iff the upper endpoints of the representing closed intervals can be taken to be a nondecreasing function of the lower endpoints.*

(This restates both theorems.)

PROOF. Let $\langle A, \succ \rangle$ be an interval order, with $\sim$ and $\bar{Q}$ as usual. For each nonempty subset B of A, define $\bar{B}$ by:

$$\bar{B} = \{a \mid b\,\bar{Q}\,a \text{ for some } b \in B\}.$$

Let $X = \{\,\overline{B}|A \subseteq B \neq \varnothing\,\}$. It is easy to show that X is simply ordered by inclusion. We construct a representation of $\sim$ by closed intervals in $\langle X, \supseteq \rangle$.

For each $a \in A$, let $B(a) = \{b|\text{not}(b \succ a)\}$. Then $a_1 \sim a_2$ iff the closed intervals $[\overline{\{a_i\}}, \overline{B}(\bar{a}_i)]$ $(i = 1, 2)$ have a nonempty intersection in X. First suppose $a_1 \sim a_2$; then $a_i \in B(a_j)$, $i, j = 1, 2$, and the intervals intersect. Conversely, suppose $a_1 \succ a_2$; then for $b \in B(a_2)$, we have $a_1 \, \overline{P} \, b$, whence, $\overline{\{a_1\}} \supset \overline{B}(\bar{a}_2)$ and the intervals do not intersect.

Thus $\varphi_L(a) = \overline{\{a\}}$ and $\varphi_U(a) = \overline{B}(\bar{a})$ satisfy conditions (i) and (ii) of Theorem 14.

Note further that $\overline{\{a_1\}} \supseteq \overline{\{a_2\}}$ iff $a_1 \, \overline{Q} \, a_2$ but $\overline{B}(\bar{a}_1) \supseteq \overline{B}(\bar{a}_2)$ iff $a_1 \, \underline{Q} \, a_2$. Thus $\overline{B}(\bar{a})$ is a nondecreasing function of $\overline{\{a\}}$ iff $\langle A, \succ \rangle$ is a semiorder; it is strictly increasing iff $\overline{Q} = \underline{Q}$.

Finally, the converses are easy. If $\langle A, \sim \rangle$ is representable by intersections of closed intervals, then define $a \succ b$ iff $\varphi_L(a) > \varphi_U(b)$, etc. $\diamondsuit$

To shed some light on the trick in the above proof, observe the following. Let $A = \text{Re}$ and $a \succ b$ iff $a > b + 1$. The simple order $\langle X, \supseteq \rangle$ constructed above is isomorphic to $\text{Re} \times \{0, 1\}$, ordered lexicographically; it has no countable order-dense subset since $(x, 0), (x, 1)$ are the endpoints of a gap for each real x.

16.4   ORDINAL THEORY FOR FAMILIES OF ORDERS

This section is concerned with combinations of several different order relations on the same underlying set. Two different sorts of questions have been studied. First, when can several different orders be represented by means of a common weak order that is more refined than the weak order corresponding to any one of them? Second, when can a single partial order be represented as a conjunction of simple orders?

To motivate questions of the first sort, we recall that in a semiorder the induced asymmetric weak order P is a disjunctive combination $\overline{P} \cup \underline{P}$ of the one-sided asymmetric weak orders. That is, information from either $\overline{P}$ or $\underline{P}$ may be used to distinguish elements of the underlying set; each may break some equivalences ($\underline{I}$ or $\overline{I}$) that exist in the other. Analogously, if we have several interval-order relations on the same set, $\succ_1$, $\succ_2$, ..., we can seek to combine the asymmetric weak orders disjunctively, $\overline{P}_1 \cup \overline{P}_2 \cup \cdots$ or $\underline{P}_1 \cup \underline{P}_2 \cup \cdots$. In Section 16.4.1 we discuss this in detail for the case of two interval orders, with remarks on the extension to more than two.

One motive for considering disjunctive combinations of the above sort arises from the artificiality of the concept of threshold as an abrupt cutoff or transition point. It is often more useful to regard the threshold as an arbitrary cutoff imposed on an underlying continuum. Then several different arbitrary cutoffs can be tried, giving rise to several thresholds. In particular, if $P(a, b)$ denotes the probability that a exceeds b, then one may define a relation $\succ_\lambda$ by

$$a \succ_\lambda b \quad \text{iff} \quad P(a, b) > \lambda,$$

for each λ in the interval $(\frac{1}{2}, 1)$. Section 16.4.2 considers such nested families of asymmetric relations. (They are nested because if $\lambda > \mu$, then $\succ_\lambda$ is a subrelation of $\succ_\mu$.) Chapter 17 treats other kinds of models for such binary probabilities.

The second type of question has given rise to a sizable literature on the dimension of a partial order. The dimension is the least number of simple orders needed for a conjunctive representation of a given partial order. In Section 16.4.3 we briefly cite a few main results concerning the dimension of interval orders and semiorders; we give no proofs.

16.4.1 Finite Families of Interval Orders and Semiorders

We formulate the case of two relations which brings out most of the basic ideas.

DEFINITION 9. *Suppose* $\succ_1$, $\succ_2$ *are asymmetric relations on* A, *and* φ, δ_1, δ_2 *are real-valued functions on* A. $\langle \varphi, \delta_1, \delta_2 \rangle$ *is a* homogeneous, upper representation *of* $\langle A, \succ_1, \succ_2 \rangle$ *iff* $\langle \varphi, \delta_i \rangle$ *is an* upper representation *of* $\langle A, \succ_i \rangle$, $i = 1, 2$. Homogeneous, lower *and* two-sided representations *are defined analogously.*

The generalization to any number of relations is obvious. Roberts (1971b) introduced the concept and the term *homogeneous*,[5] which refers obviously to the existence of a single underlying weak order that is compatible with all of the asymmetric relations. Monjardet (1984) gives a number of properties equivalent to this definition.

The question to be confronted is the conditions under which a homogeneous representation exists. Consider elements a, b, c, d for which $a \succ_1 c$ and $b \succ_2 d$. Does this place any constraint on any other comparison? For example, is it possible for both not($a \succ_2 d$) and not($b \succ_1 c$) to hold? The answer, in the homogeneous case, is no because the hypotheses yield, under

---

[5] This term has different meanings in different contexts, and they may not be closely related conceptually. For example, its meaning in Chapter 20 has nothing to do with the present one.

an upper-threshold representation, the following set of inconsistent inequalities:

$$\varphi(a) > \varphi(c) + \delta_1(c)$$
$$\varphi(a) \leqslant \varphi(d) + \delta_2(d)$$
$$\varphi(b) > \varphi(d) + \delta_2(d)$$
$$\varphi(b) \leqslant \varphi(c) + \delta_1(c).$$

This and an analogous argument for lower representations leads to the following necessary conditions:

DEFINITION 10. *Suppose* $\succ_1$ *and* $\succ_2$ *are asymmetric relations on* A. *They satisfy* upper- (lower-) interval homogeneity *iff for all* a, b, c, d *in* A, *whenever* $a \succ_1 c$ *and* $b \succ_2 d$, *then either* $a \succ_2 d$ *or* $b \succ_1 c$ ($a \succ_1 d$ *or* $b \succ_2 c$).

The reason for the term *interval* in the definition is that when $\succ_1 = \succ_2$, the conditions reduce to the definition of an interval order. Roberts (1971b) attributed to A. A. J. Marley the condition of lower-interval homogeneity, which he called the first semiorder condition. The happy fact is that for interval orders, these conditions are sufficient as well as necessary for homogeneous representations.

THEOREM 16. *Suppose* $\langle A, \succ_i \rangle$, $i = 1, 2$, *are interval orders with upper-* (*lower-*) *threshold representations. Then* $\langle A, \succ_1, \succ_2 \rangle$ *has a homogeneous upper-* (*lower-*) *threshold representation iff it satisfies upper-* (*lower-*) *interval homogeneity.*

A number of comments on this theorem are needed.

1. It is proved by showing that the induced weak orders $\overline{Q}_1$ and $\overline{Q}_2$ are not contradictory in the sense that if $a\overline{P}_1 b$ holds, then either $a\overline{P}_2 b$ or $a\overline{I}_2 b$. Put another way, the order inferred from $\succ_1$ does not directly contradict that inferred from $\succ_2$. This means that $\overline{Q} = \overline{Q}_1 \cap \overline{Q}_2$ is connected and therefore is a weak order. Equivalently, $\overline{P}_1 \cup \overline{P}_2$ is asymmetric, and hence, $\overline{P} = \overline{P}_1 \cup \overline{P}_2$. Once we have that, the existence of the representation follows because by hypothesis $A/\overline{I}_i$ contains a countable order-dense subset B_i, and it is easy to show that $B = B_1 \cup B_2$ is countable and order dense in $A/\overline{I}$.

2. This method of combining information from two noncontradictory weak orders is analogous to the construction of a two-sided representation for a semiorder, in which case $Q = \overline{Q} \cap Q$. Consider the example of Table 1 (p. 308). The relations $\succ_1$ and $\succ_2$ satisfy both upper and lower

homogeneity, as shown by the fact that $\overline{Q}_1$ and $\overline{Q}_2$ merely break each other's ties, but do not contradict each other (columns 1, 3) and likewise for $\underline{Q}_1$ and $\underline{Q}_2$ (columns 2, 4). The combined upper ordering is shown in column 7, and the combined lower one in column 8 of the table.

3. For a single interval order $\succ$, both an upper and a lower representation exist, though they may not agree, i.e., $\overline{Q}$ and $\underline{Q}$ may be contradictory if $\succ$ is not a semiorder. For two interval orders, a homogeneous upper representation can exist without a homogeneous lower one or vice versa. Exercise 22 gives an example in which $\underline{Q} = \underline{Q}_1 \cap \underline{Q}_2$ is connected, though $\overline{Q} = \overline{Q}_1 \cap \overline{Q}_2$ is not, i.e., the $\overline{Q}_i$ are contradictory. In that example, the $\succ_i$ are not semiorders; but the same phenomena can arise in semiorders if $\overline{Q}_i \neq \underline{Q}_i$. Finally, even if $\underline{Q}_1 \cap \underline{Q}_2$ and $\overline{Q}_1 \cap \overline{Q}_2$ are both connected, with $\succ_1$ and $\succ_2$ both semiorders, there can be a contradiction between $\overline{Q}_1$ and $\overline{Q}_2$ (or between $\underline{Q}_1$ and $\underline{Q}_2$). This is illustrated in Table 1, where columns 1 and 4 are contradictory.

Thus further conditions are needed for homogeneous two-sided representations. We shall not take the trouble to spell these out for the general case of this section; the situation is simpler, as we shall see in the next section, when the relations arise from probability cutoffs.

4. Recall that in the case of a tight two-sided representation, φ is a homomorphism of $\langle A, Q \rangle$ into $\langle \mathrm{Re}, \geqslant \rangle$, but it is not in general a homomorphism of $\langle A, \overline{Q} \rangle$ or of $\langle A, \underline{Q} \rangle$ unless, of course, $\overline{Q} = \underline{Q} = Q$. That means, in general, that the one-sided representations are not tight. The present situation is similar. The homomorphism φ of $\langle A, \overline{Q}_1 \cap \overline{Q}_2 \rangle$ is not one of $\langle A, \overline{Q}_i \rangle$ unless $\overline{Q}_1$ and $\overline{Q}_2$ are the same, and so in general the representations $\langle \varphi, \delta_i \rangle$ need not be tight. Put another way, $\overline{Q}$ is upper compatible with each $\succ_i$, but it is not necessarily the coarsest ordering compatible with either $\succ_i$ alone. Of course, a tight homogeneous representation can be defined analogous to a two-sided one, i.e., $\varphi(a) > \varphi(b)$ iff $a(\overline{P}_1 \cup \overline{P}_2)b$. In that way, Theorem 16 can be extended to show the existence of a tight homogeneous representation.

5. Let us return to the problem left hanging at the end of Section 16.2.6, namely, does the existence of a homogeneous representation help to improve the uniqueness of a constant-threshold representation? If $\langle \varphi, \delta_1, \delta_2 \rangle$ is a homogeneous representation of $\langle A, \succ_1, \succ_2 \rangle$ and if there exists a transformation f such that $\langle f\varphi, k_1, k_2 \rangle$ is a homogeneous constant-threshold representation, then letting $G_i(\varphi) = \varphi + \delta_i(\varphi)$, $i = 1, 2$, we have, by Equation (7)

$$G_i(\varphi) = f^{-1}[f(\varphi) + k_i],$$

whence,

$$G_1[f^{-1}(\varphi)] = f^{-1}(\varphi + k_1) = f^{-1}(\varphi + k_1 - k_2 + k_2)$$
$$= G_2[f^{-1}(\varphi + k_1 - k_2)],$$

and

$$G_i^{-1}(x) = f^{-1}[f(x) - k_i].$$

This means that the G_i form a uniform system (Section 6.7), and so the elements of the associated group are of the form

$$f^{-1}[f(x) + mk_1 + nk_2], \qquad m, n \quad \text{integers.}$$

This group is not cyclic provided k_1/k_2 is irrational, in which case the transformation f is, by Theorem 6.9, an interval scale.

The problem is to know when this can come about, i.e., when it is possible to transform a homogeneous representation $\langle \varphi, \delta_1, \delta_2 \rangle$ into a homogeneous constant-threshold representation $\langle f\varphi, k_1, k_2 \rangle$, assuming k_1/k_2 is irrational. A solution to this problem is provided by Cozzens and Roberts (1982, Theorems 6 and 7). It is moderately complex.

6. We turn next to the generalization of Theorem 16 to more than two relations. In the case of either finitely or countably many relations the answer is simple: the necessary and sufficient condition is that of Theorem 12 for every pair of orderings. In that case the upper relation $\overline{Q} = \cap_{k=1}^{\infty} \overline{Q}_k$ is compatible with each $\succ_i$, is connected and transitive, and $A/\overline{I}$ has a countable order-dense subset relative to $\overline{Q}$.

The Monjardet (1984) result mentioned earlier is stated for any finite number of relations. Doignon (1987) has generalized the results of Cozzens and Roberts to more than two relations and has provided conditions for when they can be represented by constant thresholds.

In the uncountable case, the only problem that arises in generalizing the proof is that we have no assurance that $A/\overline{I}$ has a *countable* order-dense subset. It is unclear what the optimal theorem is. Certainly an obvious sufficient condition is to assume there is a finite or countable subfamily of orderings that includes all of the information contained in the uncountable family, i.e. $\cap_{i=1}^{\infty} Q_i = \cap Q_\lambda$, $\lambda \in \Lambda$. It seems hardly worthwhile to formulate this as a formal theorem.

The case that is most fully understood is the one arising from probability cutoffs. We treat that in the next subsection.

16.4.2 Order Relations Induced by Binary Probabilities

A real-valued function P on $A \times A$ is called a *binary probability function* if it satisfies both

$$P(a, b) \geqslant 0, \tag{8}$$
$$P(a, b) + P(b, a) = 1. \tag{9}$$

[Note that Equation (9) implies $P(a, a) = \frac{1}{2}$.] From such a function, we define a binary relation $\succ_\lambda$, for each λ in the interval $(\frac{1}{2}, 1)$, by

$$a \succ_\lambda b \quad \text{iff} \quad P(a, b) > \lambda. \tag{10}$$

This is a *nested* family of asymmetric relations; obviously, if $\lambda > \mu$, then $\succ_\lambda$ is a subrelation of $\succ_\mu$.

Several questions arise. When are all of these relations semiorders? Is the entire family compatible with a common weak order? When can a probability function be (uniquely) inferred from a given family of relations? Such questions were first raised by Roberts (1971b) and were dealt with further by Fishburn (1973c). The results presented here are substantially theirs in a slightly different guise.

Intuitively, we expect the relation $\succ_\lambda$ to be an interval order since it involves a "threshold" cutoff $\lambda > \frac{1}{2}$. Whether it is one depends on properties of the function P. We can use P to define two natural binary relations that we might expect to be weak orders:

$$a \, W \, b \quad \text{iff} \quad P(a, b) \geqslant \frac{1}{2}. \tag{11}$$
$$a \, S \, b \quad \text{iff} \quad P(a, c) \geqslant P(b, c) \quad \text{for all} \quad c. \tag{12}$$

Clearly, W is weaker than S, since $a \, S \, b$ entails that $P(a, b) \geqslant P(b, b) = \frac{1}{2}$. We think of $a \, W \, b$ as "weak" dominance of a over b and $a \, S \, b$ as "strong" dominance. These relations were first studied by Block and Marschak (1960) and Luce (1956); Luce called S the "trace of P."

Relation W includes all of the relations $\succ_\lambda$ with $\lambda > \frac{1}{2}$. It is connected but not necessarily transitive. The narrower relation S is transitive by definition and is fully compatible with every $\succ_\lambda$, but it is not necessarily connected: c, d could exist such that $P(a, c) > P(b, c)$ but $P(a, d) < P(b, d)$, so that neither $a \, S \, b$ nor $b \, S \, a$ holds. Thus, if S is connected, every $\succ_\lambda$ is fully compatible with the same weak order; and so we have a homogeneous family of semiorders. Finally, note that S is the maximal binary relation compatible with every $\succ_\lambda$. (This is proved as part of Theorem 17.) Thus it is not only sufficient but also necessary that S be connected in order for $\succ_\lambda$ to be a homogeneous family of semiorders.

Properties of the relations W and S are related to versions of *stochastic transitivity* holding for P as is shown by the following definition and theorem.

DEFINITION 11. *Let P be a binary probability function on $A \times A$. We define the following for all a, b, c, d in A:*

1. Weak stochastic transitivity: *If $P(a, b) \geqslant \frac{1}{2}$ and $P(b, c) \geqslant \frac{1}{2}$, then $P(a, c) \geqslant \frac{1}{2}$.*

2. Weak independence: *If $P(a, c) > P(b, c)$, then $P(a, d) \geqslant P(b, d)$.*

3. Strong stochastic transitivity: *If $P(a, b) \geqslant \frac{1}{2}$ and $P(b, c) \geqslant \frac{1}{2}$, then $P(a, c) \geqslant \max[P(a, b), P(b, c)]$.*

The properties 1 and 3 were studied by Block and Marschak (1960); property 2 was introduced by Fishburn (1973c). Obviously 3 implies 1. It is less obvious that 3 implies 2, but that is a corollary of Theorem 17. Further references are given in Section 17.2.1.

THEOREM 17. *Let P be a binary probability function on $A \times A$, and let $\succ_\lambda$, W, S be defined as above.*

(i) *Weak stochastic transitivity holds iff W is transitive.*

(ii) *Weak independence holds iff S is connected.*

(iii) *Strong stochastic transitivity holds iff $W = S$. (Therefore strong stochastic transitivity implies weak independence; the two are equivalent if $P(a, b) \neq \frac{1}{2}$ for $a \neq b$.)*

(iv) *S is the maximal binary relation fully compatible with all $\succ_\lambda$. Therefore, weak independence holds iff the $\succ_\lambda$ are a homogeneous family of semiorders.*

One may ask, in the spirit of the previous sections, what about interval orders? It is possible to define relations $\bar{S}$ and $\underline{S}$ whose intersection is S and that are respectively upper-compatible and lower-compatible with all the relations $\succ_\lambda$; however, these relations are somewhat odd-looking:

$$a \bar{S} b \quad \text{iff} \quad P(a, c) \geqslant P(b, c) \quad \text{whenever} \quad P(b, c) > \tfrac{1}{2};$$
$$a \underline{S} b \quad \text{iff} \quad P(a, c) \geqslant P(b, c) \quad \text{whenever} \quad \tfrac{1}{2} > P(a, c).$$

Such restrictions on binary probabilities seem too unnatural to be very interesting. In applications, $\bar{S} = \underline{S}$ would seem a natural assumption. So we shall not consider homogeneous families of interval orders arising from P. It is possible, however, to define a stochastic transitivity condition somewhat weaker than weak independence that is equivalent to the $\succ_\lambda$ all being interval orders, not necessarily homogeneous. This result is due to Fishburn (1973c).

DEFINITION 12. P *satisfies* interval stochastic transitivity *iff*

$$\max[P(a, d), P(b, c)] \geqslant \min[P(a, c), P(b, d)].$$

THEOREM 18.   *A binary probability function P on A × A satisfies interval stochastic transitivity iff each $\succ_\lambda$ is an interval order.*

The proof is left as Exercise 24.

Next, we turn to the converse question: starting from a nested family of asymmetric relations, can we define a binary probability function? The obvious procedure is to use Equation (10) in reverse as a definition of $P(a, b)$: for probabilities greater than $\frac{1}{2}$, i.e., let $P(a, b) = \sup\{\lambda | a \succ_\lambda b\}$. For this to work we need to assume that $\sim$, rather than $\succ$, holds at the supremum, as is true in Equation (10): if $P(a, b) = \gamma < 1$, then $a \sim_\gamma b$. So we have the following theorem, in which the first two requirements are obvious and the third is also necessary to guarantee that the supremum construction of $P(a, b)$ yields $\sim$ rather than $\succ$ at the boundary.

THEOREM 19.   *Suppose that A is a nonempty set that for each $\lambda \in (\frac{1}{2}, 1)$, $\succ_\lambda$ is a binary relation on A, and that the following properties hold:*

(i)   *each $\succ_\lambda$ is asymmetric;*

(ii)   *if $\lambda > \mu$, then $\succ_\mu \supseteq \succ_\lambda$ (nesting);*

(iii) *if $a \succ_\lambda b$, then for some $\mu > \lambda$, $a \succ_\mu b$.*

Then there exists a binary probability function P on A × A such that Equation (10) is valid.

The proof is left as Exercise 25.

Finally, we note some other possibilities and some problems. First, it is perfectly possible for two different binary probability functions to give rise to the same family $\succ_\lambda$. Thus the supremum construction sketched before Theorem 19 does not necessarily yield the only binary probability function satisfying Equation (10). Theorems 17 and 18, however, apply to any P related to the $\succ_\lambda$ by Equation (10); thus, if the $\succ_\lambda$ are a homogeneous family of semiorders, any such P will satisfy weak independence; or if the

$\succ_\lambda$ are interval orders, P will satisfy interval stochastic transitivity. Presumably there are further conditions that would yield uniqueness in Theorem 19, i.e., only one binary probability function P satisfying Equation (10).

Second, it is possible to generalize Theorem 19 to index sets more general than $(\frac{1}{2}, 1)$. Roberts (1971b) has done this. To obtain results comparable to the above, one must have a nesting property corresponding to a total ordering of the index set. If the index set has a countable order-dense subset, then it can be embedded in $(\frac{1}{2}, 1)$.

16.4.3 Dimension of Partial Orders

In Section 3.12 of Volume I we considered the representation of a partial order $\langle A, \succ \rangle$, suggested by thermodynamics, in which there are several real-valued functions φ and φ_λ, $\lambda \in \Lambda$, such that

$$a \succ b \quad \text{iff} \quad \varphi(a) > \varphi(b) \quad \text{and} \quad \varphi_\lambda(a) = \varphi_\lambda(b), \quad \text{for } \lambda \in \Lambda.$$

On purely formal grounds, this suggests another type of representation, namely,

$$a \succ b \quad \text{iff} \quad \varphi_\lambda(a) > \varphi_\lambda(b), \quad \text{for } \lambda \in \Lambda.$$

Since, of course, each φ_λ induces a weak order on A, we can think of $\succ$ as the intersection of this family of weak orders. For various reasons, it turns out to be better to work with simple orders. We have the following definition of Dushnik and Miller (1941):

DEFINITION 13. *A partial order $\langle A, \succ \rangle$ has dimension l iff l is the smallest cardinal such that there is a family of strict simple orders $\{ \succ_\lambda \mid \lambda \in \Lambda \}$ on A with $|\Lambda| = l$ and such that*

$$a \succ b \quad \text{iff} \quad a \succ_\lambda b \quad \text{for all } \lambda \in \Lambda.$$

Note: if A is a finite set, $|A|$ is the number of its elements.

The most general result is that due to Hiraguchi (1951) and proved in different ways by Bogart (1973), Bogart and Trotter (1973), and Rabinovitch (1978a).

THEOREM 20. *Suppose $\langle A, \succ \rangle$ is a partial order and $4 \leqslant |A| < \infty$. Then the dimension of $\langle A, \succ \rangle$ is at most $\frac{1}{2}|A|$.*

Ducamp (1967) explored the question: for a partial order of dimension l, what families of l strict simple orders generate the partial order? He called each such family a *basis* of the partial order. Trotter and Bogart (1976) have studied the case of dimensions based on interval orders rather than simple orders. Further results seem to arise only by specialization. A good deal is known about partial orders of dimension $\leqslant 2$, and a good deal is known about the dimensions of interval orders and semiorders. Partial orders of dimension $\leqslant 2$ can be characterized in a number of ways. We give two of them.

We say that a graph $\langle A, \sim \rangle$ is a *comparability graph* if and only if there is a partial order $\langle A, P \rangle$ such that, for all $a, b \in A$, $a \neq b$,

$$a \sim b \quad \text{iff} \quad a P b \quad \text{or} \quad b P a.$$

This concept is due to Gilmore and Hoffman (1964) and Ghouila-Houri (1962).

We say that a partial order $\langle A, \succ \rangle$ is an *inclusion order* if and only if there exists a mapping Δ into the closed intervals of a simply ordered set $\langle B, \succsim' \rangle$ such that, for all $a, b \in A$,

$$a \succ b \quad \text{iff} \quad \Delta(a) \supset \Delta(b).$$

Note $\Delta(a) \neq \Delta(b)$. This notion, although not the term, is due to Dushnik and Miller (1941).

THEOREM 21. *If $\langle A, \succ \rangle$ is a partial order and $\sim$ is its symmetric complement, then the following are equivalent*:

(i)   *The dimension of $\langle A, \succ \rangle$ is $\leqslant 2$.*

(ii) $\langle A, \sim \rangle$ *is a comparability graph.*

(iii) $\langle A, \succ \rangle$ *is an inclusion order.*

The equivalence of (i) and (iii) was proved by Dushnik and Miller (1941) and that of (i) and (ii) was shown by Baker, Fishburn, and Roberts (1972) to follow from another of their results. The paper by Baker *et al.* includes six other equivalent forms. Additional results can be found in Fishburn (1970a).

Not a great deal seems to be known about partial orders of dimension > 2. This may be related to the fact, shown by Yannakakis (1982), that for dimension > 2 the problem of determining the dimension is computationally complex (technically, NP-complete) whereas for dimension $\leqslant 2$ it is relatively simple (polynomially decidable). However, if attention is confined to interval orders and semiorders something is known. Rabinovitch (1978a, 1978b) gives a number of results, including some that were independently discovered by Mirkin (1972), but the most relevant one is the following.

Define a *chain* in a partial order to be any subsystem in which every pair of elements is comparable. The *height of a partial order* is one less than the number of elements in a chain of maximal length.

THEOREM 22. *If* $\langle A, \succ \rangle$ *is an interval order of height* $2^n - 1$, *then the dimension of* $\langle A, \succ \rangle$ *is* $\leqslant (n + 1)$.

Turning to even stronger structures, Rabinovitch (1978a) has shown among other things:

THEOREM 23. *If* $\langle A, \succ \rangle$ *is a semiorder, then its dimension is* $\leqslant 3$. *There are semiorders of dimension* 3. *If, in addition, the height is* 1, *then the dimension is* $\leqslant 2$.

The various notions of dimensions can be extended from order relations to biorders, and somewhat analogous results have been developed by Bouchet (1971), Cogis (1976, 1980, 1982a,b), and Doignon *et al.* (1984, 1987).

16.5 PROOFS

16.5.1 Theorem 16 (p. 333)

If two interval orders $\succ_1$ *and* $\succ_2$ *on* A *have upper-* (*lower-*) *threshold representations, then* $\langle A, \succ_1, \succ_2 \rangle$ *has a homogeneous upper-* (*lower-*) *threshold representation iff upper-* (*lower-*) *interval homogeneity holds.*

PROOF. Suppose, first, that $\langle A, \succ_1, \succ_2 \rangle$ has a homogeneous representation $\langle \varphi, \bar{\delta}_1, \bar{\delta}_2 \rangle$ and consider $a \succ_1 c$ and $b \succ_2 d$. If $\varphi(a) \geqslant \varphi(b)$, then $a \bar{Q}_2 b$ and so $a \succ_2 d$; if $\varphi(b) \geqslant \varphi(a)$, then $b \bar{Q}_1 a$ and so $b \succ_1 c$. Thus upper-interval homogeneity holds.

Conversely, assume upper-interval homogeneity. First, we show $\bar{P}_1 \cup \bar{P}_2 \subseteq \bar{Q}_1 \cap \bar{Q}_2 = \bar{Q}$. Suppose $a \bar{P}_1 b$, then by definition there is some c in A with $a \succ_1 c$ and $c \sim_1 b$. For any d in A, upper-interval homogeneity implies that if $b \succ_2 d$, then $a \succ_2 d$, whence $a \bar{Q}_2 b$. So $\bar{P}_1 \subseteq \bar{Q}_2 \cap \bar{P}_1 \subseteq \bar{Q}_2 \cap \bar{Q}_1$. Similarly, $\bar{P}_2 \subseteq \bar{Q}$. Thus $\bar{Q}$ is connected since if $(a, b) \notin \bar{Q}$, then $(b, a) \in \bar{P}_1 \cup \bar{P}_2 \subseteq \bar{Q}$; and since the intersection of transitive relations is transitive, $\bar{Q}$ is a weak order.

Next, we show that $A/\bar{I}$, where $\bar{I}$ is the symmetric part of $\bar{Q}$, has a finite or countable order-dense subset relative to the simple order induced by $\bar{Q}$.

Since $\langle A, \succ_i \rangle$ has an upper-threshold representation, $A/\bar{I}_i$ has a finite or countable order-dense subset B_i relative to $\bar{Q}_i$. $B = B_1 \cup B_2$ is finite or countable, and we show it is order dense in $A/\bar{I}$. Suppose $a\bar{P}c$, then $a\bar{P}_i c$ for either $i = 1$ or 2. Thus there is a b in B_i such that $a\bar{Q}_i b$ and $b\bar{Q}_i c$. If $a\bar{P}_i b$ and $b\bar{P}_i c$, then since $\bar{P}_i \subseteq \bar{Q}$ we have $a\bar{Q}b$ and $b\bar{Q}c$, as required. If not, then either $a\bar{I}_i b$ or $b\bar{I}_i c$, and so a or c is in B_i and order density holds.

By Theorem 2.2, there is a homomorphism φ of $\langle A, \bar{Q} \rangle$ into $\langle \mathrm{Re}, \geqslant \rangle$. Since the image of φ can be chosen to be bounded, Lemma 1 (16.3.2) implies upper thresholds $\bar{\delta}_i$ can be constructed. $\diamondsuit$

16.5.2   Theorem 17 (p. 337)

Let P be a binary probability, $a\,W\,b$ iff $P(a, b) \geqslant \frac{1}{2}$, and $a\,S\,b$ iff for all c, $P(a, c) \geqslant P(b, c)$. Then:

(i)   Weak stochastic transitivity holds iff W is transitive.

(ii)   Weak independence holds iff S is connected.

(iii)   Strong stochastic transitivity holds iff $W = S$; strong stochastic transitivity implies weak independence; and the two are equivalent if $P(a, b) \neq \frac{1}{2}$ for $a \neq b$.

(iv) S is the maximal binary relation fully compatible, for every $\lambda \in (\frac{1}{2}, 1)$, with $\succ_\lambda = \{(a, b) | P(a, b) > \lambda\}$.

Therefore, weak independence holds iff the $\succ_\lambda$ are a homogeneous family of semiorders.

PROOF.

(i)   Part (i) follows from the definition of W.

(ii)   Part (ii) follows from the definition of S.

(iii)   Suppose strong stochastic transitivity holds and $a\,W\,b$. For c such that $P(b, c) \geqslant \frac{1}{2}$, $P(a, c) \geqslant P(b, c)$ follows immediately from strong stochastic transitivity. If $P(a, c) \geqslant \frac{1}{2} \geqslant P(b, c)$, there is nothing to show. Finally, assume $P(a, c) \leqslant \frac{1}{2}$. Then $P(c, a) \geqslant \frac{1}{2}$, and strong stochastic transitivity implies $P(c, b) \geqslant P(c, a)$, and so, by Equation (9), $P(a, c) \geqslant P(b, c)$. Thus $a\,S\,b$, which yields $S \supseteq W$. The opposite inclusion is immediate.

Conversely, suppose $W = S$, $P(a, b) \geqslant \frac{1}{2}$, and $P(b, c) \geqslant \frac{1}{2}$. Then we obtain $a\,S\,b$ and $b\,S\,c$; so $P(a, c) \geqslant P(b, c)$ and $P(b, a) \geqslant P(c, a)$. It follows that $P(a, c) \geqslant \max[P(a, b), P(b, c)]$.

Since W is connected, $W = S$ implies S is connected, and therefore strong stochastic transitivity implies weak independence. If weak independence holds and $P(a, b) \neq \frac{1}{2}$ for $a \neq b$, then suppose $P(a, b) \geq \frac{1}{2}$ and $P(b, c) \geq \frac{1}{2}$. If both inequalities are strict, then $P(a, b) > P(b, b)$ yields $P(a, c) \geq P(b, c)$ and, likewise, $P(b, c) > P(c, c)$ yields $P(b, a) \geq P(c, a)$; therefore $P(a, c) \geq \max[P(a, b), P(b, c)]$. Otherwise, either $a = b$ or $b = c$, and so $P(a, c) \geq \max[P(a, b), P(b, c)]$ is immediate.

(iv) If weak independence holds, then S is connected; hence it is a weak order, fully compatible with every $\succ_\lambda$. Conversely, suppose R is any binary relation fully compatible with every $\succ_\lambda$. We show that $S \supseteq R$.

If not $a S b$, then for some c, $P(b, c) > P(a, c)$. First suppose $P(b, c) > \frac{1}{2}$, and choose λ with

$$P(b, c) > \lambda > \max[\tfrac{1}{2}, P(a, c)].$$

Then $b \succ_\lambda c$ and not $a \succ_\lambda c$, since R is upper-compatible with $\succ_\lambda$, not $a R b$. Second, if $\frac{1}{2} \geq P(b, c)$, then $P(c, a) > P(c, b) \geq \frac{1}{2}$. Choose λ between $P(c, a)$ and $P(c, b)$; again, since R is lower-compatible with $\succ_\lambda$, not $a R b$. Thus we have $S \supseteq R$.

Therefore, if the $\succ_\lambda$ are a homogeneous family of semiorders, there is a weak order R fully compatible with all $\succ_\lambda$; since $S \supseteq R$, we have that S is connected; hence weak independence is satisfied. $\diamondsuit$

16.5.3 Theorem 18 (p. 338)

A binary probability function satisfies interval stochastic transitivity iff $\succ_\lambda$ is an interval order for each $\lambda \in (\frac{1}{2}, 1)$.

PROOF. Suppose that interval stochastic transitivity holds and that $a \succ_\lambda c$ and $b \succ_\lambda d$. Then either $P(a, d) > \lambda$ or $P(b, c) > \lambda$, i.e., either $a \succ_\lambda d$ or $b \succ_\lambda c$, proving $\succ_\lambda$ is an interval order.

Conversely, suppose that

$$\min[P(a, c), P(b, d)] > \max[P(a, d), P(b, c)].$$

If the left-hand side is equal to $m > \frac{1}{2}$, then for any λ between m and $\frac{1}{2}$, $\succ_\lambda$ is not an interval order. If the left-hand side is less than or equal to $\frac{1}{2}$, Equation (9) leads to

$$\min[P(d, a), P(c, b)] > \max[P(d, b), P(c, a)]$$

with the left side $> \frac{1}{2}$; and this leads similarly to the conclusion that some $\succ_\lambda$ is not an interval order. $\Diamond$

16.5.4   Theorem 19 (p. 338)

Suppose $\{ \succ_\lambda |\frac{1}{2} < \lambda < 1 \}$ *is a nested family of asymmetric relations such that if* $a \succ_\lambda b$, *then for some* $\mu > \lambda$, $a \succ_\mu b$. *Then there exists a binary probability function P such that*

$$\succ_\lambda = \{ (a, b) | P(a, b) > \lambda \}.$$

PROOF. For any a, b, if there exists λ such that $a \succ_\lambda b$, then let $P(a, b) = \sup \{ \lambda | a \succ_\lambda b \}$. If there exists λ such that $b \succ_\lambda a$, let $P(a, b) = 1 - P(b, a)$. [By the asymmetry and nesting properties, no conflict can arise between these two provisions, i.e., for $\lambda, \mu > \frac{1}{2}$, $a \succ_\lambda b$ and $b \succ_\mu a$ cannot both hold.] Finally, if $a \sim_\lambda b$ for every λ, let $P(a, b) = \frac{1}{2}$. Obviously, P is a binary probability function on $A \times A$. If $P(a, b) > \lambda$, then for some $\mu > \lambda$, $a \succ_\mu b$; hence, $a \succ_\lambda b$ by nesting. Conversely, if $a \succ_\lambda b$, then there is some μ such that $\mu > \lambda$ and $a \succ_\mu b$; hence, $P(a, b) \geqslant \mu > \lambda$. So Equation (10) is valid. $\Diamond$

16.6   SEMIORDERED ADDITIVE STRUCTURES

Throughout both Volume I and the previous chapters of this volume, as well as in Chapters 19, 20, and 22 of Volume III, weak orders combine with other relations to form measurement structures having interval- or ratio-scale representations. In the previous sections of this chapter we have seen how weak orders can be defined from interval orders and semiorders. Thus, in one sense, it is trivial to form measurement structures that combine an interval order or semiorder with other relations; just use the defined weak order in place of the primitive weak order in any measurement structure previously developed, find the corresponding representation, and define the threshold in terms of it.

Although this is the route we shall ultimately follow, it is one that must be pursued with appropriate caution, as we explain in considerable detail for one special case, namely, semiordered probability structures. Many of the remarks apply far beyond that case (to interval orders and to nonadditive or geometric structures), but it is easier to make them clear in one specific context. In Subsection 1 we consider the various possible ways we might proceed with a semiordered probability structure, pointing out what some consider to be major difficulties, with several possible approaches that attempt to draw on the results presented in Chapter 5. Ultimately, this leads us to consider in Subsection 2 a constructive approach to the weakly ordered case, one that uses the so-called approximate standard families in lieu of exact standard sequences. One should think of this section as an addition to Chapter 5; it was not available at the time Volume I was prepared. Finally, in Subsection 3 we apply this new result to get a reasonably acceptable axiomatization of semiordered probability structures. Similar theorems should be possible for both semiordered extensive and semiordered conjoint structures, but we do not work them out in detail here.

16.6.1 Possible Approaches to Semiordered Probability Structures

Suppose $\mathscr{E}$ is an algebra of events on a set X and $\succ$ is a semiorder over $\mathscr{E}$. Then under certain added restrictions, we shall call $\langle X, \mathscr{E}, \succ \rangle$ a semiordered probability structure.[6] The interpretation is exactly as in Chapter 5, except that indifference is no longer transitive. Our goal is to axiomatize it. The most natural thing to attempt first is to write down axioms in terms of the primitives $\succ$ and X. For example, thinking back to Chapter 5, a natural monotonicity condition to consider is that, for each A, B, C in $\mathscr{E}$ with $C \cap (A \cup B) = \varnothing$,

$$A \succ B \quad \text{iff} \quad A \cup C \succ B \cup C. \tag{13}$$

Unfortunately, this property does not appear empirically reasonable when $\succ$ is a semiorder (but not a weak order). To see the potential difficulty, suppose that A and B are both unlikely events but that, relative to A, B is far more unlikely. If C is an event that is about as likely as its complement (a so-called 50-50 event), then $A \cup C$ and $B \cup C$ are also about 50-50 each, and it is not difficult to imagine that they fall within the threshold of discrimination at the 50-50 level even though A and B, which are near the 0 level, do not. In that case, they would violate Equation (13). The equation can also fail in the opposite direction. For suppose C is about 50-50, $A \cup C$ is practically certain, and $B \cup C$ is only highly likely. Events

---

[6] This notation, which follows Chapter 5, is not the usual one for a relational structure; it is analogous to the usual notation for probability spaces, $\langle X, \mathscr{E}, P \rangle$. The more precise notation would be $\langle \mathscr{E}, \succ, \cup, -, X \rangle$.

$A \cup C$ and $B \cup C$ may be discriminable by examining their complements, which consist of an almost impossible event and a highly unlikely one. Removing C, however, leaves two events A and B that are both about 50-50 and possibly are not, by themselves, discriminable.

In fact, Equation (13) is very closely related to the idea of a constant threshold throughout the domain, and in the case of bounded domains (probability is one example), constant, nonzero thresholds are not really possible. The problem is that as one reaches probabilities of 0 and 1, the size of the threshold is reduced because of the bound on the scale.

So, it is clear that something a bit more delicate than Equation (13) is needed. Several approaches have been suggested, of which the first two have the common drawback, to be discussed, that they assume some form of exact standard sequence. The third, which avoids that postulate, is very hard to realize empirically, although it may be acceptable as an idealization.

The first approach, due to Suppes (1974), begins with a weakly ordered probability structure $\langle X, \mathscr{E}, \succ \rangle$ that has a special finite subalgebra $\mathscr{S}$ of events that corresponds to an exact standard sequence in the following sense: for S, T in $\mathscr{S}$,

(i) if $S \neq \varnothing$, then $S \succ \varnothing$,

(ii) if $S \succsim T$, then there exists V in $\mathscr{S}$ such that $S \sim T \cup V$.

He then showed that there is a probability measure P that represents the subsystem $\langle \mathscr{S}, \succsim, X \rangle$ in such a way that each minimal element of $\mathscr{S}$ is assigned the same value. Suppose that for A in $\mathscr{E} - \mathscr{S}$, the open interval (S, S'), $S' \succ S$, is the minimum one in $\mathscr{S}$ that includes A. Then we assign upper and lower probabilities to A as $P_*(A) = P(S)$ and $P^*(A) = P(S')$. For A in $\mathscr{S}$, $P_*(A) = P^*(A) = P(A)$. These are shown to exhibit the usual properties of upper and lower probabilities:

1. $P^*(A) \geqslant P_*(A) \geqslant 0$,

2. $P_*(X) = P^*(X) = 1$,

3. if $A \cap B = \varnothing$, then

$$P_*(A) + P_*(B) \leqslant P_*(A \cup B)$$
$$\leqslant P_*(A) + P^*(B)$$
$$\leqslant P^*(A \cup B)$$
$$\leqslant P^*(A) + P^*(B).$$

Finally, if the relation $\succ *$ is defined on $\mathscr{E}$ by: for each A, B in $\mathscr{E}$, $A \succ * B$ iff there exists S in $\mathscr{E}$ such that $A \succ S \succ B$, then $\succ *$ can be shown to be a semiorder, and it is related to the upper and lower probabilities as:

$$\text{if} \quad A \succ * B, \quad \text{then} \quad P_*(A) \geq P^*(B),$$
$$\text{if} \quad P_*(A) \geq P^*(B), \quad \text{then} \quad A \succsim * B.$$

The second approach begins with a semiorder and states properties in terms of its induced weak order Q, which is composed of its asymmetric part P and the equivalence relation I. (In Chapter 5, these were denoted $\succsim$, $\succ$, and $\sim$, respectively, but we are continuing to use that notation for the observed relation, and so we use a different notation for the induced weak order.) The axiom systems of Chapter 5 freely used both P and I in their statement. We shall attempt to argue here that some uses of both P and I are reasonably acceptable whereas other uses of I may not be. The reason for a difference between them lies in the nature of the definitions of P and of I. Recall that for A and B in $\mathscr{E}$, $A P B$ holds if and only if either $A \succ B$ or there exists an element C such that $A \succ C$ and $C \sim B$ or there exists an element D such that $A \sim D$ and $D \succ B$. $A I B$ holds only if neither $A P B$ nor $B P A$ holds, i.e., if one cannot find elements such as C or D. The latter statement means that to infer $A I B$ in a specific case one must carry out a very fine-grained search of the domain $\mathscr{E}$. As a practical matter, one can never be sure that the search has been carried out on a sufficiently fine scale. In contrast, statements of the form $A P B$ are known to be true once a C or D is found with the requisite property; it may happen that $A P B$ is true and we fail to find evidence for it, but once found we know that the statement is correct.

Much the same point about the difficulty of verifying exact equivalences was raised in Volume I (Section 4.4.4), since the problem is not confined to weak orders arising from semiorders. It is always easier to verify an inequality, and so constructions that only use inequalities are always desirable. There we sketched a construction for algebraic difference measurement, and Exercises 9–15 of Chapter 4 developed a proof that the construction converges and yields an interval scale representation. The idea of constructions that use only inequalities originated with Adams (1965) and Krantz (1967). Krantz's paper developed a theory of extensive measurement in semiorders that used only inequalities; however, his axiom system can also be faulted. It invoked three distinct and rather complicated solvability axioms. So we turn our attention next to devising a somewhat more satisfactory system in which exact solutions are not invoked.

This asymmetry between P and I must be kept in mind. It is absolutely controlling if one intends to verify axioms empirically, but it may be far less of an issue if the axioms in which I appears are thought, in some sense, to set the stage for the empirical structure, as often is the case for both solvability and Archimedean axioms even in contexts where weak orders are assumed.

For example, Luce (1973) presented systems of axioms for semiordered additive systems, including extensive, probability, and conjoint ones[7] that included some axioms involving I. In particular, he assumed that the analog of Equation (13) held only for I, i.e., for all A, B, C in $\mathcal{E}$ with $C \cap (A \cup B) = \varnothing$,

$$A I B \quad \text{iff} \quad (A \cup C) I (B \cup C), \tag{14}$$

and he invoked the following solvability and Archimedean properties:

Solvability.   If $A P B$, then there exists an event B' in $\mathcal{E}$ such that $A \supset B'$ and $B' I B$.

Archimedean.   Let $\mathcal{B}$ be a nonempty subfamily of $\mathcal{E}$ such that each pair of distinct events B, B' in $\mathcal{B}$ satisfies $B \cap B' = \varnothing$, $B' I B$, and $B P \varnothing$. Then $\mathcal{B}$ is finite.

He argued that Equation (14) was acceptable since, in essence, it says that if two events are indistinguishable in the semiorder, then they do not become distinguishable by adding a disjoint element or by subtracting a common one. There is no doubt that refutation or verification of this axiom by accumulating judgments of $\succ$ and $\sim$ is difficult, if not impossible, for the reasons just given. He also argued that the other two axioms are idealizations that no one is likely to attempt to verify directly. Of course, there are domains for which that claim is suspect.

If one traces through his proof, it reduces the semiordered case to Theorem 5.2 of Volume I. That in turn was proved by reducing it to Theorem 3.3 on extensive measurement. And finally, that was reduced to Hölder's Theorem 2.4, the proof of which uses an explicit construction in terms of ever finer standard sequences. The key to constructing these standard sequences is, of course, the solvability axiom. In practice, we simply cannot carry out that construction with any confidence because it entails knowing when two elements are indifferent, and we can never be sure. Thus, although such results may be convincing that a representation exists, they fail to provide an effective way of finding it. For that reason, it seems advisable to devise a construction that depends only on observing P relations.

---

[7] For other examples of similar approaches, see Huber (1979), Ng (1977), and Luce (1978). For still another approach using a modified version of Scott's axioms for qualitative probability (Scott, 1964), see Domotor and Stelzer (1971).

Before turning to that, we inject an additional word of caution about not exaggerating the importance of having a constructive theory based only on inequalities. As was noted in Section 1, the usual procedure in the physical sciences is to use more refined (and expensive) methods to establish equivalence of a series of replicas. The replicas almost certainly are not perfect, but their indifference intervals are negligible compared with the errors in comparing any concatenation of them with the to-be-measured object. We do not usually worry about the inequalities of the various divisions on even a plastic ruler, much less on a carefully manufactured steel one. The main concern here is for situations in which no more refined comparison procedure is available or even imaginable, and confining ourselves to inequalities in such situations is often wise.

There is a second way in which nonconstructive results can be important when we lack constructive ones. They can tell us the class of possible representations available, and a scientist may elect to fit a body of data to the representations in that class rather than either to check axioms or to engage in a construction. Examples of this approach are familiar from various scaling procedures, including multidimensional scaling and functional measurement, and we shall see results in Chapter 19 on nonadditive concatenation structures that may be of use whether or not we have a constructive technique.

16.6.2 Probability with Approximate Standard Families

To motivate the definition of an approximate standard family, consider the usual standard sequence technique. Suppose A is an event whose probability is to be determined to within some degree of accuracy. Let E be an event of sufficiently small probability that an error of that size can be neglected. Then the question is, how many pairwise disjoint replicas of E are required to match A in the weak order as compared with the total number of possible replicas, which approximates the universal set X? The ratio of these two numbers gives us an estimate of the probability of A. Additional accuracy is gained by choosing a smaller E.

If we cannot effectively construct replicas, then approximate ones will have to do. These could be events that are "barely less than" E in the qualitative ordering or ones that are "barely greater than" E in that order. The concept *barely* needs explication since, after all, we have available only an ordering of events. Among the ways it can be made precise, we follow that of Exercise 4.10 (Volume I, p. 196). It rests on the following idea. If the number of events exceeding E that is needed to match A is almost as large as the number of events less than E that is needed to match A, then the former events must be "barely greater than" E and the latter ones must be "barely less than" E. To formulate this precisely, some notation is needed. Let A and E be given events, and let $\mathscr{B}$ be a finite family of events in $\mathscr{E}$. We say that $\mathscr{B}$ is in $\mathscr{L}(E, A)$ if each member of $\mathscr{B}$ is less than E and the union of all the members of $\mathscr{B}$ contains A. And we say that $\mathscr{B}$ is in $\mathscr{G}(E, A)$ if the elements of $\mathscr{B}$ are pairwise disjoint, each is greater than E, and the union of all the members is contained in A.

Observe that if the structure has representation φ that is order-preserving, additive over disjoint unions, and nonnegative, then the cardinality of any family $\mathscr{B}$ in $\mathscr{G}(A, E)$ is less than the cardinality of any family $\mathscr{B}'$ in $\mathscr{L}(E, A)$. This is proved simply:

$$(\text{card } \mathscr{B})\varphi(E) < \varphi(\cup\mathscr{B})$$

(pairwise disjointness, additivity of φ, and $B \succ E$ for $B \in \mathscr{B}$)

$$\leqslant \varphi(A)$$ (inclusion)

$$\leqslant \varphi(\cup\mathscr{B}')$$ (inclusion)

$$< (\text{card } \mathscr{B}')\varphi(E)$$ (subadditivity of φ, and $E \succ B'$ for $B' \in \mathscr{B}'$).

The next question to consider is: how close can these two cardinalities be? To answer this, let us begin by supposing that we can construct exact replicas, and then we modify them by removing or adding small pieces. In particular, suppose we find $B_1, \ldots, B_m$ that are pairwise disjoint, each is a subset of A, each is equivalent to E, and they are such that $B^* = A - \cup_{i=1}^{m}B_i$ is smaller than E but greater than $\varnothing$. Partition B^* into m small but positive events and adjoin them to the elements B_i. This, then, yields a family $\mathscr{B}$ of cardinality m of pairwise disjoint events, each of which is more probable than E and whose union is still contained in A. On the other hand, if we remove a small piece of each B_i and adjoin their union to B^*, then each of the modified B_i is smaller than E; and by taking the pieces sufficiently small, we maintain the modified B^* less than E. This results in a family $\mathscr{B}'$ of cardinality $m + 1$, all of whose elements are less than E and whose union is A. Thus we conclude that in a fine-grained structure, the difference in cardinality can be as small as 1.

Such a pair of families, $\mathscr{B}$ and $\mathscr{B}'$, with $\mathscr{B}$ in $\mathscr{G}(E, A)$ and $\mathscr{B}'$ in $\mathscr{L}(E, A)$ and differing only by 1 in cardinality, is called an E-approximate standard family for A. Such a family can be recognized solely by inequality observations, $B \succ E$ and $E \succ B'$ for B in $\mathscr{B}$ and B' in $\mathscr{B}'$, and by counting. The existence of such families can be made into an axiom that replaces both the old solvability and Archimedean axioms and provides a basis for constructing numerical values without using exact replicas. However, as we shall see, another solvability axiom appears to be needed. We summarize these concepts in a formal manner.

DEFINITION 14. *Let $\mathscr{E}$ be an algebra of sets and $\succ$ the asymmetric part of a weak order on $\mathscr{E}$. For A, E in $\mathscr{E}$:*

(i) *A subset $\mathscr{B}$ of $\mathscr{E}$ is said to be in $\mathscr{G}(E, A)$ iff*
  1. *the elements of $\mathscr{B}$ are pairwise disjoint,*
  2. *for each B in $\mathscr{B}$, $B \succ E$, and*
  3. *$A \supseteq \cup\mathscr{B}$.*

(ii) *A subset $\mathscr{B}'$ of $\mathscr{E}$ is said to be in $\mathscr{L}(E, A)$ iff*
  1. *for each B' in $\mathscr{B}'$, $E \succ B'$, and*
  2. *$A \subseteq \cup\mathscr{B}'$.*

(iii) *A pair $(\mathscr{B}, \mathscr{B}')$ of subsets of $\mathscr{E}$ is said to be an E-approximate standard family for A iff*
  1. *$\mathscr{B} \in \mathscr{G}(E, A)$,*
  2. *$\mathscr{B}' \in \mathscr{L}(E, A)$,*
  3. *$\mathrm{card}(\mathscr{B}') = 1 + \mathrm{card}(\mathscr{B})$.*

With these concepts available, the result we can prove is the following.

THEOREM 24. *Let X be a nonempty set, $\mathscr{E}$ an algebra of subsets of X, and $\succsim$ a binary relation on $\mathscr{E}$. Suppose that the following axioms are satisfied: For each A, A', B, B', E in $\mathscr{E}$,*

1. *$\langle \mathscr{E}, \succsim \rangle$ is a weak order.*

2. *Suppose that $A \cap A' = \varnothing$, and let $\succ$ denote the asymmetric part of $\succsim$.*
  (i) *If $A \succsim B$ and $A' \succsim B'$, then $(A \cup A') \succsim (B \cup B')$.*
  (ii) *If $A \succ B$ and $A' \succsim B'$, then $(A \cup A') \succ (B \cup B')$.*

3. *$A \succsim \varnothing$.*

4. *$X \succ \varnothing$.*

5. *If $A' \succ A$, then there exists C in $\mathscr{E}$ such that*
  (i) *$C \cap A = \varnothing$,*
  (ii) *$C \succ \varnothing$,*
  (iii) *$A' \succsim (A \cup C)$.*

6. *If $A \succ E$ and $E \succ \varnothing$, then there exists an E-approximate standard family for A.*

Then, there exists a unique real-valued function φ *on* $\mathscr{E}$ *such that*

(i) $A \succsim B$ *iff* $\varphi(A) \geq \varphi(B)$, *and*

(ii) $\langle X, \mathscr{E}, \varphi \rangle$ *is a finitely additive probability space.*

The solvability condition stated as Axiom 5 plays no role in the construction, which is completely handled by Axiom 6, and it is only used in showing that φ is order-preserving. We suspect that it may be redundant, but we have not been able to show that it is.

We outline the general nature of the proof, which is given in detail in Section 7. The first major step is to show (Lemma 4) that for each E-approximate standard family for A, $(\mathscr{B}, \mathscr{B}')$, the cardinality of $\mathscr{B}$ and of $\mathscr{B}'$ is uniquely determined by E and A. Denote the cardinality of $\mathscr{B}$ by $N(E, A)$. We then define an approximation representation $\varphi_E = N(E, A)/N(E, X)$. Such approximate representations are approximately order-preserving and approximately additive on $\mathscr{E}$. They are shown to converge to a limit φ as E is made increasingly smaller; φ is exactly order-preserving and additive on $\mathscr{E}$. The key tool in establishing convergence is an approximate multiplicative property for the function N (Lemma 8): for F smaller than E, $N(F, A)$ is approximately equal to $N(F, E)N(E, A)$. So, if one shifts from one small unit of approximation E to a still smaller one F, then the number of replicas changes by a factor of approximately $N(F, E)$. This is as one would expect.

As was noted earlier, this strategy of proof follows closely those given previously (Krantz, 1967; Volume I, Section 2.2.4 and Exercises 9–15 of Chapter 4). Two things differ slightly here: the context, which is an algebra of sets, and the formal use of approximate standard families. In the previous work they were used without being labeled explicitly and were shown to exist from approximate solvability and Archimedean axioms (Proposition 4 of Krantz, 1967, and Exercise 10, Chapter 4, of Volume I). Here, we simply assume the existence of such families and then systematically exploit their properties.

16.6.3 Axiomatization of Semiordered Probability Structures

We are now ready to state the result for semiorders.

THEOREM 25. *Let* X *be a nonempty set,* $\mathscr{B}$ *an algebra of subsets of* X, *and* $\succ$ *a binary relation on* $\mathscr{B}$. *Suppose* $\langle \mathscr{B}, \succ \rangle$ *is a semiorder and* Q *is the defined weak order. Suppose* $\langle \mathscr{B}, Q, X \rangle$ *satisfies Axioms 2–6 of Theorem 24, but with Axiom 4 still stated in terms of* $\succ$, *not* Q. *Then there exist* *unique real-valued functions* φ *and* δ *on* $\mathscr{E}$ *such that*

(i) (φ, δ) *is a tight threshold representation for* $\langle \mathscr{E}, \succ \rangle$, *and*

(ii) $\langle X, \mathscr{E}, \varphi \rangle$ *is a finitely additive probability space.*

To get the flavor of this axiomatization, one has to unpack the axioms so that they are stated in terms of the primitive $\succ$. As we remarked earlier, Axiom 2 of Theorem 24, which is universal in the primitive $\succeq$, is, in fact, an existential axiom in $\succ$. Thus it is problematic, even if highly plausible. Axiom 3 seems no problem, and Axiom 4 is stated in terms of $\succ$, not Q. The solvability embodied in Axiom 5 is moderately complex when recast in terms of $\succ$ and entails a number of existence statements. Basically, however, all it does is insure the existence of a rich family of events with very small probabilities. This is also what Axiom 6 does, and it seems surprising that both assertions are needed. The fact is that the axiomatization is not at all elegant when stated in terms of $\succ$, and we believe that an important open problem continues to be the formulation of a more satisfactory theory. This does not appear easy to do. The root difficulty is that there is no necessary interlock between the additive aspects, caught in the monotonicity property of Axiom 2, and the threshold aspects, caught in the semiorder assumption.

The proof of Theorem 25 is left as Exercise 27.

16.6.4 Weber's Law and Semiorders

As was just remarked, part of the problem in axiomatizing semiordered additive structures is in finding a plausible link between the operation and the semiorder. The simplest classical link is one in which the threshold is proportional to the scale value, in which case Weber's law is said to obtain. One way to formulate the qualitative conditions underlying Weber's law has been given by Narens (1980, 1985, p. 88), and we present a slightly modified and simpler version of his results. The basic idea is to suppose that the induced weak order forms an extensive structure, define the threshold in terms of the additive representation and the operation (rather than in terms of indifference alone), and then show that this has the appropriate properties. The proof of the result is outlined in Exercise 28–33.

THEOREM 26. *Suppose* $\langle A, \succ, \bigcirc \rangle$ *is a structure such that* $\langle A, \succ \rangle$ *is a semiorder and* $\langle A, Q, \bigcirc \rangle$, *where* Q *is the induced weak order, is a closed extensive structure with no minimal element and with an additive representation* φ. *Suppose, further, that the following two properties hold for all a, b, c in A*:

   (i)   *if aQb and $b \sim b\bigcirc c$, then $a \sim a\bigcirc c$;*

  (ii) *if $a \succ c$ and $b \sim d$, then $a\bigcirc b \succ c\bigcirc d$;*

 (iii) *if $a\bigcirc b \succ c\bigcirc d$, then either $a \succ c$ or $b \succ d$.*

Define $\delta(a) = \sup\{\varphi(b) | a \sim a\bigcirc b\}$. Then, $\langle \varphi, \delta \rangle$ is a tight threshold representation of $\langle A, \succ \rangle$, and for some $k > 0$ and each a in A, $\delta(a) = k\varphi(a)$.

This result is, of course, subject to the earlier criticism that in practice the induced weak order Q cannot be precisely estimated, and so neither the assumption that $\langle A, Q, \bigcirc \rangle$ is extensive nor the displayed assumption (i) can be really verified. Nevertheless, it does afford some insight into the qualitative properties underlying the class of threshold representations that satisfy Weber's law.

Other approaches to the problem are given by Falmagne (1985) and Falmagne and Iverson (1979).

16.7 PROOF OF THEOREM 24 (p. 351)

Suppose an algebra $\mathscr{E}$ of subsets of X ordered by $\succsim$ satisfies the following axioms:

1. $\langle \mathscr{E}, \succsim \rangle$ *is a weak order.*

2. *If $A \cap A' = \varnothing$, then*
   (i) $A \succsim B$ *and* $A' \succsim B'$ *imply* $(A \cup A') \succsim (B \cup B')$,
  (ii) $A \succ B$ *and* $A' \succsim B'$ *imply* $(A \cup A') \succ (B \cup B')$.

3. $A \succsim \varnothing$.

4. $X \succ \varnothing$.

5. *If $A' \succ A$, then there exists B in $\mathscr{E}$ such that $B \cap A = \varnothing$, $B \succ \varnothing$, and $A' \succsim (B \cup A)$.*

6. *If $A \succ E$ and $E \succ \varnothing$, then there exists an E-approximate standard family for A.*

Then there exists a unique, finitely additive, probability measure φ on $\mathscr{E}$ that preserves the order $\succsim$.

LEMMA 2. *If $A \supseteq B$, then $A \succsim B$.*

PROOF. By Axiom 1, $B \succsim B$, and by Axiom 3, $(A - B) \succsim \varnothing$. Since $B \cap (A - B) = \varnothing$, the result follows from Axiom 2(i). $\diamondsuit$

LEMMA 3. *If $\mathscr{B} \in \mathscr{G}(E, A)$ and $\mathscr{B}' \in \mathscr{L}(E, A)$, then* $\mathrm{card}(\mathscr{B}') >$ $\mathrm{card}(\mathscr{B})$.

PROOF. Suppose $\mathscr{B} = \{B_1, \ldots, B_m\}$ and $\mathscr{B}' = \{B_1', \ldots, B_n'\}$, and m $\geqslant n$. Then $B_i \succ E$ and $E \succ B_i'$ for $i = 1, \ldots, n$, and the B_i are pairwise disjoint; so Axiom 2(i) yields $(\bigcup_{i=1}^{m} B_i) \succ (\bigcup_{i=1}^{n} B_i')$. But $\bigcup_{i=1}^{n} B_i' = \bigcup \mathscr{B}' \supseteq A$ $\supseteq \bigcup_{i=1}^{m} B_i$; so Lemma 2 yields $(\bigcup_{i=1}^{n} B_i') \succsim (\bigcup_{i=1}^{m} B_i)$. Thus $m \geqslant n$ is impossible. $\diamondsuit$

LEMMA 4. *If $(\mathscr{B}_1, \mathscr{B}_1')$ and $(\mathscr{B}_2, \mathscr{B}_2')$ are E-approximate standard families for A, then* $\mathrm{card}(\mathscr{B}_1) = \mathrm{card}(\mathscr{B}_2)$.

PROOF. Using Definition 14 (iii.3) and Lemma 3, we have

$$\mathrm{card}(\mathscr{B}_1) = \mathrm{card}(\mathscr{B}_1') - 1 \geqslant \mathrm{card}(\mathscr{B}_2) = \mathrm{card}(\mathscr{B}_2') - 1 \geqslant \mathrm{card}(\mathscr{B}_1).$$

$\diamondsuit$

Thus, for E with $A \succ E$ and $E \succ \varnothing$, we define $N(E, A) = \mathrm{card}(\mathscr{B})$, where $(\mathscr{B}, \mathscr{B}')$ is any E-approximate standard family for A. The following characterization of $N(E, A)$ is also useful.

LEMMA 5. *If $A \succ E$ and $E \succ \varnothing$, then*

  (i) $N(E, A) = \max\{\mathrm{card}(\mathscr{B}) | \mathscr{B} \in \mathscr{G}(E, A)\}$
  (ii) $N(E, A) + 1 = \min\{\mathrm{card}(\mathscr{B}') | \mathscr{B}' \in \mathscr{L}(E, A)\}$.

PROOF. Immediate from Lemmas 3 and 4 and the definition of $N(E, A)$. $\diamondsuit$

The following lemma shows that $N(E, \cdot)$ preserves weak inequality for the subset $\{A \in \mathscr{E} | A \succ E\}$.

LEMMA 6. *If $A_1 \succsim A_2$, $A_2 \succ E$, and $E \succ \varnothing$, then* $N(E, A_1) \geqslant$ $N(E, A_2)$.

PROOF. By Lemma 5, there exists $\mathscr{B}_1' = \{B_1', \ldots, B_n'\}$ in $\mathscr{L}(E, A_1)$ such that $n = \mathrm{card}(\mathscr{B}_1') = N(E, A_1) + 1$, and similarly there exists $\mathscr{B}_2 = \{B_1, \ldots, B_m\}$ in $\mathscr{G}(E, A_2)$ with $m = N(E, A_2)$. If $m \geqslant n$, then by Axiom 2, $(\bigcup_{i=1}^{m} B_i) \succ (\bigcup \mathscr{B}_1')$; and this, with Lemma 2, yields $A_2 \succ A_1$, a contradiction. $\diamondsuit$

The next two lemmas exhibit the approximate additive and multiplicative properties of N.

LEMMA 7. *Suppose that $A_1 \succ E$, $A_2 \succ E$, $A_1 \cap A_2 = \varnothing$, and $E \succ \varnothing$. Then $N(E, A_1) + N(E, A_2) + 1 \geqslant N(E, A_1 \cup A_2) \geqslant N(E, A_1) + N(E, A_2)$.*

PROOF. Let $(\mathscr{B}_i, \mathscr{B}_i')$ be an E-approximate standard family for A_i, $i = 1, 2$. Since $A_1 \cap A_2 = \varnothing$, $\mathscr{B}_1 \cup \mathscr{B}_2 \in \mathscr{G}(E, A_1 \cup A_2)$ and has cardinality $\text{card}(\mathscr{B}_1) + \text{card}(\mathscr{B}_2) = N(E, A_1) + N(E, A_2)$. Thus the right-hand inequality follows from Lemma 5(i). Furthermore, $\mathscr{B}_1' \cup \mathscr{B}_2' \in \mathscr{L}(E, A_1 \cup A_2)$ and has cardinality at most $N(E, A_1) + N(E, A_2) + 2$. Thus the left-hand inequality follows from Lemma 5(ii). $\Diamond$

LEMMA 8. *Suppose that $A \succ E$, $E \succ F$, and $F \succ \varnothing$. Then*

$$[N(F, E) + 1][N(E, A) + 1] > N(F, A) \geqslant N(F, E)N(E, A).$$

PROOF. Let $(\mathscr{B}, \mathscr{B}')$ be an E-approximate standard family for A, with $\mathscr{B} = \{B_1, \ldots, B_m\}$, $m = N(E, A)$. Let $(\mathscr{C}_i, \mathscr{C}_i')$ be an F-approximate standard family for B_i, $i = 1, \ldots, m$. Since $B_i \succ E$, by Lemma 6, $N(F, B_i) = \text{card}(\mathscr{C}_i) \geqslant N(F, E)$ for each i. Since the B_i are pairwise disjoint,

$$\text{card}\left(\bigcup_{i=1}^{m} \mathscr{C}_i\right) = \sum_{i=1}^{m} \text{card}(\mathscr{C}_i) \geqslant N(F, E)N(E, A);$$

and since $\bigcup_{i=1}^{m} \mathscr{C}_i$ is in $\mathscr{G}(F, A)$, the right-hand inequality follows from Lemma 5(i).

Next, let $\mathscr{B}' = \{B_1', \ldots, B_{m+1}'\}$, and let $(\mathscr{D}_i, \mathscr{D}_i')$ be an F-approximate standard family for B_i'. Again, using Lemma 6, $\text{card}(\mathscr{D}_i') = N(F, B_i') + 1 \leqslant N(F, E) + 1$ for each $i = 1, \ldots, m + 1$. Therefore,

$$\text{card}\left(\bigcup_{i=1}^{m+1} \mathscr{D}_i'\right) \leqslant \sum_{i=1}^{m+1} \text{card}(\mathscr{D}_i') \leqslant [N(F, E) + 1][N(E, A) + 1].$$

Since $\bigcup_{i=1}^{m+1} \mathscr{D}_i' \supseteq \bigcup \mathscr{B}' \supseteq A$, we have $\bigcup_{i=1}^{m+1} \mathscr{D}_i'$ in $\mathscr{L}(F, A)$; so the left-hand inequality follows from Lemma 5(ii). $\Diamond$

Two main steps remain in the proof. First we show how Axiom 6 and Lemma 8 can be used to construct an infinite sequence $\{E_k\}$ in $\mathscr{E}$ that converges to $\varnothing$ in the sense that for every A such that $A \succ \varnothing$ and for all sufficiently large k,

$$A \succ E_k, \quad E_k \succ \varnothing, \quad \text{and} \quad \lim_{k \to \infty} N(E_k, A) = +\infty. \quad (15)$$

Second, we show that for any sequence satisfying Equation (15) and for any

$A \succ \varnothing$,

$$\varphi(A) = \lim_{k \to \infty} \frac{N(E_k, A)}{N(E_k, E)}$$

exists, is positive, and is independent of the sequence $\{E_k\}$. The function φ is the desired representation.

There is a trivial case in which every A in $\mathscr{E}$ satisfies either $A \sim X$ or $A \sim \varnothing$. That means that no A, E exist with $A \succ E$ and $E \succ \varnothing$, and so Axiom 6 never can be used. But in that case, just define $\varphi(A) = 1$ if $A \sim X$ and $\varphi(A) = 0$ if $A \sim \varnothing$.

Otherwise, let A be any element of $\mathscr{E}$ such that $X \succ A$ and $A \succ \varnothing$. Start the construction of $\{E_k\}$ by letting $E_0 = X$. By Axiom 2 we have $X \succ (-A)$ and $(-A) \succ \varnothing$. If $(-A) \succsim A$, let $E_1 = A$; otherwise, let $E_1 = -A$. We have constructed, for $k = 1$, a sequence $\{E_0, \ldots, E_k\}$ with the following two properties:

(i) $E_{i-1} \supset E_i$

(ii) $(E_{i-1} - E_i) \succsim E_i$ and $E_i \succ \varnothing$.

(That is, each E_i is nested in E_{i-1}, is above $\varnothing$, and is at most half the size of E_{i-1}.)

Suppose, inductively, that $\{E_0, \ldots, E_k\}$ satisfy (i) and (ii) for some $k \geqslant 1$. Construct E_{k+1} as follows. By Axiom 6, there exists a $\mathscr{B}'$ in $\mathscr{L}(E_k, E_{k-1})$. Note that by property (ii), $\mathscr{B}'$ must have cardinality at least 3, and so $N(E_k, E_{k-1}) \geqslant 2$. It is impossible that $\varnothing \sim (E_k \cap B)$ for every B in $\mathscr{B}'$, for if that were true, Axiom 2 together with $E_k \cap (\cup \mathscr{B}') = E_k$ would imply $\varnothing \sim E_k$. Hence, there exists a $B \in \mathscr{B}'$ such that $(E_k \cap B) \succ \varnothing$. Moreover, since $E_k \succ B$ by definition of $\mathscr{L}$, we have $(E_k - B) \succ \varnothing$. Now let $E_{k+1} = E_k \cap B$ provided that $(E_k - B) \succsim (E_k \cap B)$; otherwise, let $E_{k+1} = E_k - B$. The sequence $\{E_0, \ldots, E_{k+1}\}$ now also satisfies (i) and (ii). By induction, we have an infinite sequence $\{E_k\}$ satisfying these two properties. In addition we have established that $N(E_k, E_{k-1}) \geqslant 2$; by Lemma 8 we have more generally $N(E_l, E_k) \geqslant 2^{l-k}$.

Now consider any A in $\mathscr{E}$ such that $A \succ \varnothing$. It is impossible that $E_k \succsim A$ for every k, for if that were so, then $E_k \succ A$ for every k, and thus

$$N(A, E_k) \geqslant N(A, E_l)N(E_l, E_k) \geqslant 2^{l-k}.$$

But the left side is fixed with k and $2^{l-k} \to +\infty$ with l. Thus, for all sufficiently large k, $A \succ E_k$, and

$$N(E_l, A) \geqslant N(E_l, E_k)N(E_k, A) \geqslant 2^{l-k} \to \infty.$$

Thus, we have a sequence satisfying Equation (15).

For the second step, we again use Lemma 8, this time to obtain the inequalities

$$\frac{[N(E_l, E_k) + 1][N(E_k, A) + 1]}{N(E_l, E_k)N(E_k, X)} > \frac{N(E_l, A)}{N(E_l, X)}$$

$$> \frac{N(E_l)N(E_k)}{[N(E_l, E_k) + 1][N(E_k, X) + 1]}.$$

For any sequence $\{E_k\}$ satisfying Equation (15) and for any A with $A \succ \varnothing$, these inequalities hold for all sufficiently large k and sufficiently large $l > k$.

Letting $l \to \infty$, we have for all sufficiently large k,

$$\frac{N(E_k, A) + 1}{N(E_k, X)} \geqslant \limsup_{l \to \infty} \frac{N(E_l, A)}{N(E_l, X)} \geqslant \liminf_{l \to \infty} \frac{N(E_l, A)}{N(E_l, X)}$$

$$\geqslant \frac{N(E_k, A)}{N(E_k, X) + 1}.$$

Now, letting $k \to \infty$ shows that the lim sup and lim inf are equal and that the limit is positive.

Given two sequences $\{E_k\}$ and $\{E_l'\}$ satisfying Equation (15), the above inequalities still hold for sufficiently large k and sufficiently large l with E_l replaced throughout by E_l'. The same method of proof shows that the limit is the same for both sequences. We define this limit to be $\varphi(A)$. Clearly $\varphi(X) = 1$, and we complete the definition by letting $\varphi(A) = 0$ whenever $A \sim \varnothing$. The fact that φ is additive for disjoint union, and hence $\langle X, \mathscr{E}, \varphi \rangle$ is a finitely additive probability space, is immediate from Lemma 7.

To show that φ is order-preserving, suppose first that $A_1 \succsim A_2$. It is immediate from Lemma 6 that $\varphi(A_1) \geqslant \varphi(A_2)$. For the converse, suppose that $A_1 \succ A_2$. We show that $\varphi(A_1) > \varphi(A_2)$. By Axiom 5, there exists a B such that $B \cap A_2 = \varnothing$, $B \succ \varnothing$, and $A_1 \succsim (B \cup A_1)$. So, by Lemmas 6 and 7,

$$N(E_k, A_1) \geqslant N(E_k, B \cup A_2)$$
$$\geqslant N(E_k, B) + N(E_k, A_2).$$

So, dividing by $N(E_k, X)$ and taking limits,

$$\varphi(A_1) \geqslant \varphi(B) + \varphi(A_2).$$

Since $B \succ \varnothing$, we know from the construction that $\varphi(B) > 0$, whence $\varphi(A_1) > \varphi(A_2)$.   ◇

The reader should note the similarity of this proof to that of Theorem 2.4 in Volume I.

16.8 RANDOM-VARIABLE REPRESENTATIONS

Up to this point the analysis of thresholds arising from intrinsic variability or errors has been algebraic in character. As much scientific practice indicates, there are many reasons for having probabilistic rather than algebraic representations. In fact, the probabilistic analysis of errors goes all the way back to fundamental work of Simpson, Lagrange, and Laplace; for a good historical overview, see the classic work of Todhunter (1865/1949). The early efforts were especially directed at the problem of increasing the accuracy of the measurement of astronomical quantities.

In psychology, the analysis of thresholds in probabilistic terms has a history throughout this century. Thurstone (1927a, 1927b) emphasized the representation of threshold phenomena by probability distributions, as did other investigators. (Thurstone's ideas are discussed in some detail in Section 17.6.) The theoretical view that there is no fixed cutoff point with a deterministic numerical representation defining a sensory threshold is widespread and supported by a variety of experimental studies. Representation of sensory phenomena by random variables is a natural generalization, applicable as well to physical procedures of measurement involving error.

In practice, observations produced by a measurement procedure are almost invariably regarded, in one way or another, as realizations of random variables. Three examples illustrate some of the possibilities.

Suppose we follow the procedure for extensive measurement suggested by our proof of Theorem 2.4 in Volume I: A small element a is selected to define the gradation of a standard sequence; it is our unit of measurement. For any larger element b, we approximate its numerical value by counting the number $N(a, b)$ of copies of a that, when concatenated approximately, match b in the sense that $N(a, b)a$ does not exceed b and $[N(a, b) + 1]a$ does. This count may vary from one attempt to another. Suppose that a rule is marked in millimeter gradations but is otherwise unmarked, and consider measuring a table by counting the number of millimeters. Because of the variability one is led to regard the count not as a real number but as a random variable $N(a, b)$.

Second, let us follow the procedure suggested by the proof of Theorem 1, Chapter 6 (see Lemma 6, Volume I, p. 265), for generating a standard sequence in additive conjoint measurement. Let the structure be $(A \times P, \succsim)$. At each step, we seek the solution to an equivalence such as: given (a, p) and q from the second component, find b in the first such that $(a, p) \sim (b, q)$. In an example suggested by Falmagne (1976), the subject adjusts the voltage b into the left earphone so that the combination of sounds to the left and right ear, (b, q), matches the loudness of some other combination of sound intensities, (a, p). It is plausible that there is variability at this step and so we should regard the number (voltage) b as one realization of a random variable $B(a, p; q; k)$, where the index k distinguishes repeated presentations of a, p, and q, and B takes on values in voltages corresponding to the A-component.

Similar types of data arise whenever pointer measurement is possible (Falmagne, 1985). For example, suppose we have a spring scale whose deflections along a length scale are used to identify weights. Then, when object a is placed on the balance, a reading X_a is made. In this case, one then has the chance to study how the concatenation $X_{a \circ b}$ relates to X_a and X_b.

The third example involves the same setup as the preceding conjoint one except that equivalence is not observed directly. Rather, there is a series of binary comparisons of stimuli that can be summarized as a sequence of dichotomous random variables of the form $D(a, p; b, q; k)$, where D takes the value 1 on the kth comparison between (a, p) and (b, q) if the former is perceived as larger than the latter, and 0 if the opposite perception occurs on occasion indexed by k.

Note that in each case the randomness is introduced in a different way. In the first example, it was introduced not at the stage of constructing the standard sequence itself but in the counting process. Implicitly, we assumed that copies could be made of a with a variability that is negligible relative to that of the counting process. In a number of applications of extensive measurement, this may be a realistic model. Examples are gradation of a good steel rule and the wave crests of a laser in vacuum used for ranging a planetary satellite. For other purposes, however, it may be quite inappropriate, and it may be better to assume models either of the second type, in which variability in forming a standard sequence arises because of variability in forming individual copies, or of the third type, in which variability in forming a copy is further analyzed into variability at each binary choice.

These three examples by no means exhaust the possibilities for introducing random variables into conjoint or extensive measurement, much less the possibilities that might arise from other measurement structures, such as the geometrical ones considered in Chapters 13–15. One of the difficulties facing a theorist is the need to tailor the detailed theoretical formulation closely to specific applications.

A particularly favorable situation with respect to error is one in which the random variables have so little variability that they do not perturb the calculations seriously. In the first example, one hopes the empirical situation is such that as the gradation-defining a becomes arbitrarily small, the ratio $N(a, b)/N(a, c)$ converges in probability, for each b and c, to a number that we denote by $\varphi(b)/\varphi(c)$. This hope is not unduly optimistic, at least in the framework of classical rather than quantum physics. The underlying idea is that each refinement of measurement involves not just the choice of a finer standard sequence, but also the discovery of a method that produces random variables N whose coefficients of variation (= ratio of standard deviation to mean) are also markedly reduced. Such refinement rests on improved technique, often arising because of new discoveries. It would probably be useful to develop a representation theorem for such systems of random variables, one in which the function φ is order-preserving and additive and its ratios are the limit in probability of the random variables, as just sketched. This will entail a careful examination of exactly what is done empirically to generate the standard sequence and to provide an explicit model (like those to come) of that situation.

In favorable situations for which convergence in probability obtains, the estimates are aided by averaging. The distribution of averages,

$$\frac{1}{K} \sum_{k=1}^{K} N(a, b, k),$$

inherit the convergence properties; and, insofar as the random variables are independent, the averages converge faster than the distribution itself. Moreover, since the distribution converges, one need not be much concerned about which parameter is being estimated. An additional favorable feature is that N is approximately additive over concatenation and so the averaging is occurring on approximately the correct representation.

Our second group of examples illustrates a far more problematic situation. The fact that there is a distribution of solutions makes the empirical interpretation of the equation $(a, p) \sim (b, q)$ unclear. One possibility, suggested by Falmagne (1976), postulates that for each choice of a, p, and q, the random variables $B(a, p; q; k)$ are independently and identically distributed (in which case k can be suppressed). Since the random variables are explicitly realized in the experiment, we can in principle check this assumption; and assuming it is met, we can get some idea of the form of the distribution and how it varies with a, p, and q. Within that framework, one way to define $(a, p) \sim (b, q)$ is as the value of b that is the Pth percentile of the distribution of $\mathbf{B}(a, p; q)$. Falmagne (1976, 1985) did so, using the median $P = \frac{1}{2}$, and he proceeded in the following way.

Observe that if φ_1 is strictly increasing and if M_P denotes the Pth percentile of a distribution,

$$\varphi_1\{M_P[\mathbf{B}(a, p; q)]\} = M_P\varphi_1[\mathbf{B}(a, p; q)]. \tag{16}$$

In words, percentiles are invariant under strictly increasing transformations. If one supposes that an additive representation is approximately correct, but has an additive error, then

$$\varphi_1[\mathbf{B}(a, p; q)] = \varphi_1(a) + \varphi_2(q) - \varphi_2(p) + \epsilon(a, p; q), \tag{17}$$

where the ϵ are random variables. Suppose there is a percentile P so that independent of a, p, and q,

$$M_P[\epsilon(a, p; q)] = 0. \tag{18}$$

[Falmagne assumed a symmetric error distribution, in which case $P = \frac{1}{2}$, the median, satisfies Equation (18).] Applying Equation (16) to Equation (17) and using Equation (18) yields

$$\varphi_1(b) = \varphi_1(a) + \varphi_2(q) - \varphi_2(p), \quad \text{for} \quad b = M_p[\mathbf{B}(a, p; b)], \tag{19}$$

which is the standard additive conjoint measurement problem. Using the usual algebraic techniques, Falmagne (1976) and Gigerenzer and Strube (1983) applied the model to binaural loudness data; the additive model was rejected.

A similar development can be easily worked out for the sorts of pointer readings one might get with a concatenation structure.

A number of questions can be raised at this point. The formulation seems asymmetric for a situation that seems, itself, symmetric. It is an asymmetry often found in experiments that distinguish between the "standard" and the "variable" pairs of the matching process. To avoid the asymmetry one could, of course, introduce two families of random variables—$\mathbf{B}$ for matches in the first component and $\mathbf{Q}$ in the second component. Denote by $s = (a, p)$ the standard pair to be matched; then the complete notation for the two matches is $\mathbf{B}(s, q; k)$ and $\mathbf{Q}(s, b; k)$, where q and b are the nonvarying element of the variable pair and k indexes replications. We postulate that the distributions are both independent and identically distributed, so k can be suppressed, and that the data are symmetric in the sense that b is the Pth percentile of $\mathbf{B}(s, q)$ if and only if q is the Pth percentile of $\mathbf{Q}(s, b)$.

Should one or more of the above postulates turn out to be false, there are further questions. What if the random variables are not identically distributed? Or what if they do not satisfy the complementarity of b and q each being the Pth percentile of the other random variable? Or why do we choose to match percentiles? And which value of P should we choose? The latter would be no question if the pair (b, q) were independent of the percentile chosen but to postulate this would impose a very strong constraint on the behavior of the two families of random variables.

It is possible to consider methods that depend more on the distributions involved than do the percentiles, but then we encounter the problem that the numerical levels for the elements b, q, etc. are at best ordinal indices, and so the shapes of distributions have little significance. One can assume that the scale values that enter into a simple algebraic rule of combination also enter into a simple error theory. For example, it is sometimes postulated that the additive conjoint scale values that represent $(a, p) \gtrsim (b, a)$ as $\varphi_1(a) + \varphi_2(p) \geqslant \varphi_1(b) + \varphi_2(q)$ also represent $\varphi_1(\mathbf{B})$ and $\varphi_2(\mathbf{Q})$ as Gaussian random variables with equal variances. But there is no good reason to expect that this assumption is generally valid. (This is discussed at some length in Section 22.6.)

Let us turn next to the case where the observations are binary responses rather than continuous random variables. This situation is often modeled by supposing that the random variables,

$$\mathbf{D}(a, p; b, q; k) = \begin{cases} 1, & \text{if } ap \text{ is chosen over } bq \text{ on the } k\text{th occasion;} \\ 0, & \text{otherwise,} \end{cases}$$

are independent and identically distributed for fixed a, b, p, q, so k is suppressed from the notation, and that a probability structure P underlies the observations in the sense that

$$Pr[\mathbf{D}(a, p; b, q) = 1] = P(a, p; b, q).$$

One then attempts to arrive at theories for the structure of P. Probabilistic models of this sort are explored in some detail in Chapter 17.

To confront such a model with data, one must consider serious statistical questions. It is usual to estimate P from repeated observations, although in some domains it is difficult to achieve independent replications. Furthermore, because the accuracy is imperfect, testing the adequacy of the model poses a nontrivial statistical question. The major contributions to this topic include Falmagne (1978), Falmagne and Iverson (1979), Falmagne, Iverson, and Marcovici (1979), and Iverson and Falmagne (1985).

In the remainder of this section, we first consider weak representations, which essentially do not constrain the distributions of the random variables beyond their means or medians. We follow this with strong representations that in general fix the distributions uniquely by qualitative axioms on the moments. This approach, due to Suppes and Zanotti (1989), follows a line of development different from that of Falmagne and his collaborators by not assuming the existence of a probability measure but by deriving its existence from qualitative axioms. The proof uses various results in the classical theory of moments.

16.8.1 Weak Representations of Additive Conjoint and Extensive Structures

We begin with some of the first explicit results of importance, due to Falmagne (1976, 1985), on random conjoint measurement. The basic ideas were discussed earlier in this section from an intuitive standpoint.

First some notation. If $\mathbf{X}$ is a random variable having a unique median, then $M(\mathbf{X})$ is that median. Having in mind the comparisons (a, p) and (b, q) of additive conjoint measurement, Falmagne defines $\mathbf{U}_{pq}(a)$ as the random variable corresponding to b in the deterministic equivalence (a, p) $\sim (b, q)$, or, in terms of the representing numerical functions φ_1 and φ_2,

$$\varphi_1(a) + \varphi_2(p) = \varphi_1(b) + \varphi_2(q). \qquad (20)$$

Equation (20) is replaced by the probabilistic equation

$$\varphi_1[\mathbf{U}_{pq}(a)] = \varphi_2(p) - \varphi_2(q) + \varphi_1(a) + \epsilon_{pq}(a),$$

where $\epsilon_{pq}(a)$ is an error term assumed to have a unique median equal to zero. For simplicity of notation, following Falmagne, we define

$$m_{pq}(a) = M[\mathbf{U}_{pq}(a)]$$

and assume in the following definition that from this point on all variables are numerical.

DEFINITION 15. *Let A and P be two intervals of real numbers, and let $\mathcal{U} = \{\mathbf{U}_{pq}(a) | p, q \in P, a \in A\}$ be a collection of random variables each with a uniquely defined median. Then $\mathcal{U}$ is a structure for random additive* conjoint measurement *iff for all p, q, r in P and a in A, the medians $m_{pq}(a)$ satisfy the following axioms:* .

1. *Are continuous in all variables p, q, and a.*
2. *Are strictly increasing in a and p, and strictly decreasing in q.*
3. *Map A into A.*
4. *Satisfy the cancellation rule with respect to function composition* ∘, *i.e.,*

$$(m_{pq} \circ m_{qr})(a) = m_{pr}(a),$$

whenever both sides are defined.

Falmagne proved that these axioms imply commutativity in the sense that

$$(m_{pq} \circ m_{rs})(a) = (m_{rs} \circ m_{pq})(a),$$

when both sides are defined. Moreover, define

$$ap \geqslant bq \quad \text{iff} \quad m_{pq}(a) \geqslant b.$$

We leave it as an exercise to prove that $\langle A \times P, \geqslant \rangle$ is an additive conjoint structure in the sense of Definition 6.7. This proof is central to the following theorem of Falmagne's (1985, p. 273).

THEOREM 27. *Let $\mathcal{U}$ be a structure of random additive conjoint measurement. Then there exist real-valued continuous strictly increasing functions φ_1 and φ_2 defined on A and P, respectively, such that for any $U_{pq}(a)$ in $\mathcal{U}$,*

$$\varphi_1[U_{pq}(a)] = \varphi_2(p) + \varphi_2(q) - \varphi_1(a) + \epsilon_{pq}(a),$$

where $\epsilon_{pq}(a)$ is a random variable with a unique median equal to zero. Moreover, if φ_1' and φ_2' are two other such functions, then

$$\varphi_1'(a) = \alpha\varphi_1(a) + \beta$$

and

$$\varphi_2'(p) = \alpha\varphi_2(p) + \gamma,$$

where $\alpha > 0$.

Although this development does not provide a qualitative analysis of the conditions that are sufficient for the existence of random variables, it is a natural extension of the deterministic theory. Perhaps most important, it provides a clear theoretical framework for setting up statistical tests of the standard qualitative axioms of additive conjoint measurement. It is to be emphasized, however, as in other probabilistic theories in science, that selection of appropriate statistical tests of the theory is not in any sense automatic. Indeed, even the existence of appropriate tests needs to be proved.

Falmagne (1980) also gives several different axiom systems for probabilistic extensive measurement, which are more appropriately discussed in Section 17.2.3 where the underlying difference structure is defined.

16.8.2 Variability as Measured by Moments

The approach in this section to the distribution of the representing random variable of an object consists of developing, in the usual style of the theory of measurement, qualitative axioms concerning the *moments* of the distribution, which are represented as expectations of powers of the representing numerical random variable. The classic problem of moments in the theory of probability enters in an essential way in the developments to follow. We lay out in an explicit way the qualitative assumptions about moments that are made. The developments here follow Suppes and Zanotti (1989).

Before giving the formal developments, we address the measurement of moments from a qualitative standpoint. We outline here one approach, without any claim that it is the only way to conceive of the problem. In fact, we believe that the pluralism of approaches to measuring probability is matched by that for measuring moments, for reasons that are obvious given the close connection between the two.

The one approach we outline corresponds to the limiting relative-frequency characterization of probability, which we formulate here somewhat informally. Let s be an infinite sequence of independent trials with the outcome on each trial being heads or tails. Let $H(i)$ be the number of heads on the first i trials of s. Then, relative to s,

$$P(\text{heads}) = \lim_{i \to \infty} H(i)/i,$$

with the provision that the limit exists and that the sequence s satisfies certain conditions of randomness that need not be analyzed here. In practice, of course, only a finite initial segment of any such sequence is realized as a statistical sample; but ordinarily in the case of probability, the empirical procedure encompasses several steps. In the approach given here, the first step is to use the limiting relative-frequency characterization. The second step is to produce and analyze a finite sample with appropriate statistical methods.

Our approach to empirical measurement of qualitative moments covers the first step but not the second of giving detailed statistical methods. So, let a_0 be an object of small mass of which we have many accurate replicas; so we are assuming here that the variability in a_0 and its replicas $a_0^{(j)}$, $j = 1, 2, \ldots$, is negligible. Then we use replicas of a_0 to qualitatively weigh an object a. On each trial we force an equivalence, as is customary in classical physics. So, on each trial i, we have

$$a \sim \{ a_0^{(1)}, a_0^{(2)}, \ldots, a_0^{(m_i)} \}.$$

We symbolize the set shown on the right as $m_i a_0$. Then, as in the case of probability, we characterize a^n, the nth qualitative raw moment of a, by

$$a^n \sim \lim_{j \to \infty} \frac{1}{j} \sum_{i=1}^{j} m_i^n a_0,$$

but in practice we use a finite number of trials and use the estimate $\hat{a}^n$,

$$\hat{a}^n \sim \frac{1}{j} \sum_{i=1}^{j} m_i^n a_0,$$

and so also only estimate a finite number of moments. It is not to the point here to spell out the statistical procedures for estimating a^n. Our objective is only to outline how one can approach empirical determination of the qualitative raw moments.

There is one important observation to deal with. The observed data, summarized in the integers $m_1, m_2, \ldots, m_j$, on which the computation of the moments is based, also constitute a histogram of the distribution. Why not estimate the distribution directly? When a distribution of a particular form is postulated, there need be no conflict in the two methods, and the histogram can be of further use in testing goodness-of-fit.

The reason for working with the raw moments is theoretical rather than empirical or statistical. Various distributions can be qualitatively characterized in terms of their raw moments in a relatively simple way, as the examples in the Corollary to Theorem 28 show. Furthermore, general qualitative conditions on the moments are given in Theorem 28. Alternative qualitative approaches to characterizing distributions undoubtedly exist, and as they are developed may well supersede the one used here.

We now turn to the formal developments. In proving the representation theorem for random extensive quantities in this section, we apply a well-known theorem of Hausdorff (1923) on the one-dimensional moment problem for a finite interval.

HAUSDORFF'S THEOREM. *Let* $\mu_0, \mu_1, \mu_2, \ldots$ *be a sequence of real numbers. Then a necessary and sufficient condition that there exists a unique probability distribution F on $[0,1]$ such that μ_n is the nth raw moment of the distribution F, i.e.,*

$$\mu_n = \int_0^1 t^n \, dF, \qquad n = 0, 1, 2, \ldots, \tag{21}$$

is that $\mu_0 = 1$ and all the following inequalities hold:

$$\mu_\nu - \binom{k}{1}\mu_{\nu+1} + \binom{k}{2}\mu_{\nu+2} + \cdots + (-1)^k \mu_{\nu+k} \geq 0,$$
$$\text{for } k, \nu = 0, 1, 2, \ldots. \tag{22}$$

A standard terminology is that a sequence of numbers μ_n, $n = 0, 1, 2, \ldots$ is *completely monotonic* iff inequalities (22) are satisfied, i.e., in more compact binomial notation $\mu^\nu(1 - \mu)^k \geq 0$, for $k, \nu = 0, 1, 2, \ldots$ (for detailed analysis of many related results on the problem of moments, see Shohat and Tamarkin, 1943).

It is important to note that we do not need an additional separate specification of the domain of definition of the probability distribution in Hausdorff's theorem. The necessary and sufficient conditions expressed in the inequalities (22) guarantee that all the moments lie in the interval $[0, 1]$; so this may be taken to be the domain of the probability distribution without further assumption.

16.8.3 Qualitative Primitives for Moments

The idea, then, is to provide a qualitative axiomatization of the moments for which a qualitative analog of inequalities (22) obtains and then to show that the qualitative moments have a numerical representation that permits one to invoke Hausdorff's theorem. So the qualitative structure begins, first, with a set G of objects. These are the physical objects or entities to which we expect ultimately to associate random variables. But to get at the random variables, we must generate from G a set of entities that we can think of as corresponding to the raw moments and mixed moments of the objects in G. To do that, we must suppose that there is an operation of combining $\cdot$ so that we can generate elements $a^n = a^{n-1} \cdot a$, which from a qualitative point of view will be thought of as corresponding to the nth raw moment of a. It is appropriate to think of this operation as an operation of multiplication, but it corresponds to multiplication of random variables, not to multiplication of real numbers. We shall assume as axioms that the operation is associative and commutative but that it should not be assumed to be distributive with respect to disjoint union (which corresponds to numerical addition) can be seen from the following random-variable counterexample, given in Gruzewska (1954). Let $\mathbf{X}_1, \mathbf{X}_2, \mathbf{X}_3$ be three random variables, where

$$\mathbf{X}_1 = \mathbf{X}_2 = \begin{cases} 0 \\ 1 \end{cases} \quad \text{with} \quad P_1(\mathbf{X}_2 = 0) = P_1(\mathbf{X}_1 = 1) = \tfrac{1}{2},$$

$$\mathbf{X}_3 = 1 \quad \text{with} \quad P_3(\mathbf{X}_3 = 1) = 1.$$

Then,

$$\mathbf{X}_2 + \mathbf{X}_3 = \begin{cases} 1 & \text{with} \quad P(\mathbf{X}_2 + \mathbf{X}_3 = 1) = \tfrac{1}{2} \\ 2 & \text{with} \quad P(\mathbf{X}_2 + \mathbf{X}_3 = 2) = \tfrac{1}{2} \end{cases}$$

$$\mathbf{L}_1 = \mathbf{X}_1(\mathbf{X}_2 + \mathbf{X}_3) = \begin{cases} 0 & \text{with} \quad P(\mathbf{L}_1 = 0) = \tfrac{1}{2} \\ 1 & \text{with} \quad P(\mathbf{L}_1 = 1) = \tfrac{1}{2} \cdot \tfrac{1}{2} = \tfrac{1}{4} \\ 2 & \text{with} \quad P(\mathbf{L}_1 = 2) = \tfrac{1}{2} \cdot \tfrac{1}{2} = \tfrac{1}{4} \end{cases}$$

and

$$\mathbf{X}_1\mathbf{X}_2 = \begin{cases} 0 & \text{with} \quad P(\mathbf{X}_1\mathbf{X}_2 = 0) = \tfrac{3}{4} \\ 1 & \text{with} \quad P(\mathbf{X}_1\mathbf{X}_2 = 1) = \tfrac{1}{4} \end{cases}$$

$$\mathbf{X}_1\mathbf{X}_3 = \begin{cases} 0 & \text{with} \quad P(\mathbf{X}_1\mathbf{X}_3 = 0) = \tfrac{1}{2} \\ 1 & \text{with} \quad P(\mathbf{X}_1\mathbf{X}_3 = 1) = \tfrac{1}{2} \end{cases}$$

and

$$\mathbf{L}_2 = \mathbf{X}_1\mathbf{X}_2 + \mathbf{X}_1\mathbf{X}_3 = \begin{cases} 0 & \text{with} \quad P(\mathbf{L}_2 = 0) = \tfrac{3}{8} \\ 1 & \text{with} \quad P(\mathbf{L}_2 = 1) = \tfrac{3}{8} + \tfrac{1}{8} = \tfrac{4}{8} = \tfrac{1}{2} \\ 2 & \text{with} \quad P(\mathbf{L}_2 = 2) = \tfrac{1}{8}. \end{cases}$$

(The computations make clear the assumptions of independence that were made.) As can be seen, L_1 and L_2 have different distributions, even though

$$E\left[X_1\left(X_2 + X_3\right)\right] = E\left(X_1 X_2 + X_1 X_3\right).$$

We turn now to the explicit definition of a semigroup, which contains the associative and commutative axioms of multiplication.

DEFINITION 16. *Let A be a nonempty set, G a nonempty set, $\cdot$ a binary operation on A, and 1 an element of G. Then $\mathcal{U} = \langle A, G, \cdot, 1 \rangle$ is the* free *commutative semigroup with identity 1 generated by G iff the following axioms are satisfied for every a, b, and c in G.*

1. If $a \in G$, then $a \in A$.

2. If $s, t \in A$, then $(s \cdot t) \in A$.

3. Any member of A can be generated by a finite number of applications of Axioms 1–2 from elements of G. (A is the set of finite strings with alphabet G.)

4. $a \cdot (b \cdot c) = (a \cdot b) \cdot c$.

5. $a \cdot b = b \cdot a$.

6. $1 \cdot a = a$.

Because of the associativity axiom, we omit parentheses from here on. Intuitively, the elements of A are qualitative mixed moments. Because the product operation $\cdot$ is associative and commutative, we can always write the mixed moments in a standard form involving powers of the generators. For example, $a \cdot a \cdot a \cdot c \cdot a \cdot b \cdot c = a^4 \cdot b \cdot c^2$. This expression is interpreted as the qualitative mixed moment consisting of the fourth raw moment of a times the first one of b times the second one of c.

Our last primitive is a qualitative ordering of moments. As usual we shall denote it by $\succeq$. The first question concerns the domain of this relation. For purposes of extensive measurement, it is useful to assume that the domain is all finite subsets from the elements of the semigroup A. We can state this as a formal definition:

DEFINITION 17. *Let A be a nonempty set and $\succeq$ a binary relation on $\mathscr{F}$, the family of all finite subsets of A. Then $\mathfrak{A} = \langle A, \mathscr{F}, \succeq \rangle$ is a weak extensive structure iff the following axioms are satisfied for every B, C, and*

D in $\mathscr{F}$:

1. *The relation $\succeq$ is a weak ordering of $\mathscr{F}$.*
2. *If $B \cap D = C \cap D = \varnothing$, then $B \succeq C$ iff $B \cup D \succeq C \cup D$.*
3. *If $B \neq \varnothing$, then $B \succ \varnothing$.*

Superficially, the structure just defined looks like a structure of qualitative probability in the sense of Definition 5.4 but in fact it is not. The reason is that, because A is an infinite set, we cannot assume that $\mathscr{F}$ is closed under complementation since that would violate the assumption that the subsets in $\mathscr{F}$ are finite.

An important conceptual point is that we do require the ordering in magnitude of different raw moments. One standard empirical interpretation of what it means to say that the second raw moment a^2 is less than the first a^1 was outlined above. A formal point, appropriate to make at this stage, is to contrast the uniqueness result we anticipate for the representation theorem with the usual uniqueness up to a similarity, i.e., multiplication by a positive constant, for extensive measurement, as developed in Chapter 3. Because we have in the present setup not only the extensive operation but also the semigroup multiplication for forming moments, the uniqueness result is absolute, i.e., uniqueness in the sense of the identity function. Given this strict uniqueness, the magnitude comparison of a^m and a^n for any natural numbers m and n is not a theoretical problem. It is, of course, apparent that any procedure for measurement of moments, fundamental or derived, will need to satisfy such strict uniqueness requirements in order to apply Hausdorff's or other related theorems in the theory of moments.

Within $\mathscr{F}$, we may define what it means to have n disjoint copies of $B \in \mathscr{F}$:

$$1B = B$$
$$(n + 1)B \sim nB \cup B',$$

where $nB \cap B' = \varnothing$ and $B' \sim B$, and $\sim$ is the equivalence relation defined in terms of the basic ordering $\succeq$ on $\mathscr{F}$. Axiom 3 of Definition 18 below will simply be the assumption that such a B' always exists and so nB is defined for each n. It is essential to note that this standard extensive or additive recursive definition is quite distinct from the one for moments a^n given earlier.

16.8.4 Axiom System for Qualitative Moments

Our goal is to provide axioms on the qualitative raw moments such that we can prove that object a can be represented by a random variable $\mathbf{X}_a$, and the nth raw moment a^n is represented by the nth power of $\mathbf{X}_a$, i.e., by $\mathbf{X}_a^n$.

For convenience, we shall assume the structures we are dealing with are bounded in two senses. First, the set G of objects will have a largest element 1, which intuitively means that the expectation of the random variables associated with the elements of a will not exceed that of 1. Moreover, we shall normalize things so that the expectation associated with $\mathbf{X}_1$ is 1. This normalization shows up in the axiomatization as 1 acting as the identity element of the semigroup. Second, because of the condition arising from the Hausdorff theorem, this choice means that all of the raw moments are decreasing in powers of n, i.e., if $m \leqslant n$, then $a^n \precsim a^m$. Obviously, the theory can be developed so that the masses are greater than 1 and the moments become larger with increasing n. This is the natural theory when the probability distribution is defined on the positive real line. As might be expected, the conditions are simpler for the existence of a probability distribution on a finite interval, which is also realistic from a methodological standpoint. The exponential notation for qualitative moments, a^n, is intuitively clear, but it is desirable to have the formal recursive definition:

$$a^0 = 1,$$
$$a^n = a^{n-1} \cdot a,$$

to have a clear interpretation of a^0.

Before giving the axiom system, we must discuss more fully the issue of what constitutes a qualitative analog of Hausdorff's condition, inequality (22).

Because we have only an operation corresponding to addition and not to subtraction in the qualitative system, for k an even number, we rewrite this inequality solely in terms of addition as

$$\mu_\nu + \binom{k}{2}\mu_{\nu+2} + \cdots + \mu_{\nu+k} \geqslant \binom{k}{1}\mu_{\nu+1} + \cdots + k\mu_{\nu+(k-1)}, \quad (23)$$

and a corresponding inequality for the case in which k is odd. In the qualitative system the analog to inequality (23) must be written in terms of unions of sets, as follows for k even:

$$a^\nu \cup \binom{k}{2}a^{\nu+2} \cup \cdots \cup a^{\nu+k} \succsim \binom{k}{1}a^{\nu+1} \cup \cdots \cup ka^{\nu+(k-1)}; \quad (24)$$

and when k is odd,

$$a^\nu \cup \binom{k}{2}a^{\nu+2} \cup \cdots \cup ka^{\nu+(k-1)} \succsim \binom{k}{1}a^{\nu+1} \cup \cdots \cup a^{\nu+k}. \quad (25)$$

There are several remarks to be made about this pair of inequalities. First of all, we can infer that for $a \prec 1$, as opposed to $a \sim 1$, the moments are a strictly decreasing sequence, i.e., $a^\nu \succ a^{\nu+1}$. Second, the meaning of such terms as $\binom{k}{2}a^{\nu+2}$ was recursively defined earlier, with the recursion justified by Axiom 3 below. It is then easy to see that the unions indicated in inequalities (24) and (25) are of disjoint sets. On the basis of the earlier terminology we can then introduce the following definition. A qualitative sequence $a^0, a^1, a^2, a^3, \ldots$ is *qualitatively completely monotonic* iff inequalities (24) and (25) are satisfied.

DEFINITION 18. *A structure* $\mathfrak{A} = \langle A, \mathscr{F}, G, \succsim, \cdot, 1 \rangle$ *is a* random extensive *structure with independent objects—the elements of G—iff the following axioms are satisfied for a in G, s and t in A, k, m, m', n, and n' natural numbers, and B and C in $\mathscr{F}$:*

1. *The structure* $\langle A, \mathscr{F}, \succsim \rangle$ *is a weak extensive structure.*

2. *The structure* $\langle A, G, \cdot, 1 \rangle$ *is a commutative semigroup with identity* 1 *generated by G.*

3. *There is a C' in $\mathscr{F}$ such that $C' \sim C$ and $C' \cap B = \varnothing$.*

4. *Archimedean. If $B \succ C$, then for any D in $\mathscr{F}$ there is an n such that $nB \succsim nC \cup D$.*

5. *Independence. Let mixed moments s and t have no common objects:*
   a. *If $m1 \succsim ns$ and $m'1 \succsim n't$, then $mm'1 \succsim nn'(s \cdot t)$.*
   b. *If $m1 \precsim ns$ and $m'1 \precsim n't$, then $mm'1 \precsim nn'(s \cdot t)$.*

6. *The sequence $a^0, a^1, a^2, \ldots$ of qualitative raw moments is qualitatively completely monotonic.*

The content of Axiom 1 is familiar. What is new here is, first of all, Axiom 2, in which the free commutative semigroup, as mentioned earlier, is used to represent the mixed moments of a collection of objects. Axiom 3 is needed to make the recursive definition of $(n + 1)B$ well defined as given earlier. The special form of the Archimedean axiom is the one needed when there is no solvability axiom, as discussed in Section 3.2.1. The dual form of Axiom 5 is just what is needed to prove the independence of the moments of different objects, which means that the mixed moments factor in terms of expectation. Note that it is symmetric in $\succeq$ and $\preceq$. The notation used in Axiom 5 involves both disjoint unions, as in $m1$, and the product notation for mixed moments, as in $(s \cdot t)$. Axiom 6 formulates the qualitative analog of Hausdorff's necessary and sufficient condition as discussed above.

16.8.5 Representation Theorem and Proof

THEOREM 28.   *Let* $\mathfrak{A} = \langle A, \mathscr{F}, G, \succeq, \cdot, 1 \rangle$ *be a random extensive structure with independent objects. Then there exists a family* $\{\mathbf{X}_B, B \in \mathscr{F}\}$ *of real-valued random variables such that*

(i)   *every object* a *in* G *is represented by a random variable* $\mathbf{X}_a$ *whose distribution is on* $[0, 1]$ *and is uniquely determined by its moments;*

(ii)   *the random variables* $\{\mathbf{X}_a, a \in G\}$ *are independent;*

(iii)   *for* a *and* b *in* G, *with probability one*

$$\mathbf{X}_{a \cdot b} = \mathbf{X}_a \cdot \mathbf{X}_b;$$

(iv) $E(\mathbf{X}_B) \geqslant E(\mathbf{X}_C)$ *iff* $B \succeq C$;

(v)   *if* $B \cap C = \varnothing$, *then* $\mathbf{X}_{B \cup C} = \mathbf{X}_B + \mathbf{X}_C$;

(vi)   *if* $B \neq 0$, *then* $E(\mathbf{X}_B) > 0$.

Moreover, any function φ *from* Re *to* Re *such that* $\{\varphi(\mathbf{X}_B), B \in \mathscr{F}\}$ *satisfies* (i)–(vi) *is the identity function.*

PROOF.   First we have by familiar arguments from Axioms 1, 3, and 4 the existence of a numerical assignment φ. For any B in $\mathscr{F}$ we define the set S of numbers:

$$S_B = \left\{ \frac{m}{n} \,\middle|\, m1 \succeq nB \right\}. \tag{26}$$

It is easy to show that S is nonempty and has a greatest lower bound (g.l.b.), which we use to define φ:

$$\varphi(B) = \text{g.l.b.} \ S_B.$$

It is then straightforward to show that for B and C in $\mathscr{F}$:

$$\left.\begin{array}{ll} \varphi(B) \geqslant \varphi(C) & \text{iff} \quad B \gtrsim C, \\ \text{if} \quad B \cap C = \varnothing, & \text{then} \quad \varphi(B \cup C) = \varphi(B) + \varphi(C), \\ \text{if} \quad B \neq \varnothing, & \text{then} \quad \varphi(B) > 0. \end{array}\right\} \qquad (27)$$

Second, it follows from Axiom 2 that

$$1^n = 1,$$

whence

$$\varphi(1) = 1.$$

From Axiom 6, we infer that for any object a in G the numerical sequence

$$1, \varphi(a), \varphi(a^2), \varphi(a^3), \ldots$$

satisfies inequalities (2) and hence determines a unique probability distribution for a, which we represent by the random variable X_a. The expectation function E is defined by $E(X_a^n) = \varphi(a^n)$.

The independence of mixed moments s and t that have no common object is derived from Axiom 5 by the following argument, which uses the sets S_B defined in Equation (26) and their symmetric analog sets T_B defined later. From Axiom 5a we have at once, if

$$\frac{m}{n} \in S_s \quad \text{and} \quad \frac{m'}{n'} \in S_t,$$

then

$$\frac{mm'}{nn'} \in S_{st},$$

whence

$$\varphi(s)\varphi(t) \geqslant \varphi(st). \qquad (28)$$

Correspondingly, to use Axiom 5b we define

$$T_B = \left\{ \frac{m}{n} \,\middle|\, m1 \precsim nB \right\}.$$

Each set T_B is obviously nonempty and has a least upper bound (l.u.b.). We need to show that

$$\text{l.u.b. } T_B = \text{g.l.b. } S_B = \varphi(B).$$

Suppose by way of contradiction that

$$\text{l.u.b. } T_B < \text{g.l.b. } S_B.$$

(That the weak inequality $\leqslant$ must hold is evident.) Then there must exist integers m and n such that

$$\text{l.u.b. } T_B < \frac{m}{n} < \text{g.l.b. } S_B.$$

So we have from the left-hand inequality,

$$m1 \succ nB,$$

and from the right-hand one,

$$m1 \prec nB,$$

which together contradict the weak-order properties of $\succsim$. From the definition of T_B, we can then also infer that

$$\varphi(s)\varphi(t) \leqslant \varphi(st),$$

which together with Equation (28) establishes that

$$\varphi(s)\varphi(t) = \varphi(st).$$

The above argument establishes (ii). We next want to show that with probability one,

$$\mathbf{X}_{a \cdot b} = \mathbf{X}_a \cdot \mathbf{X}_b.$$

We do this by showing that the two random variables have identical nth moments for all n. If $a \neq b$, we have independence of $\mathbf{X}_a$ and $\mathbf{X}_b$ by the argument given earlier, i.e.,

$$E(\mathbf{X}_{a \cdot b}^n) = \varphi(a^n \cdot b^n) = \varphi(a^n)\varphi(b^n) = E(\mathbf{X}_b^n)E(\mathbf{X}_b^n).$$
$$= E(\mathbf{X}_a^n \cdot \mathbf{X}_b^n) = E[(\mathbf{X}_a \cdot \mathbf{X}_b)^n],$$

which establishes (iii) for $a \neq b$. If $a = b$, we have

$$E(\mathbf{X}_{a \cdot a}^n) = \varphi[(a \cdot a)^n] = \varphi(a^n \cdot a^n) = \varphi(a^{2n})$$
$$= E(\mathbf{X}_a^{2n}) = E(\mathbf{X}_a^n \cdot \mathbf{X}_a^n) = E[(\mathbf{X}_a \cdot \mathbf{X}_a)^n],$$

which completes the proof of (iii). For the empty set, since $\varphi(\varnothing) = 0$,

$$\mathbf{X}_\varnothing = 0,$$

and for B a nonempty set in $\mathscr{F}$, define

$$\mathbf{X}_B = \sum_{s \in B} \mathbf{X}_s,$$

Since each $s \in B$ is a multinomial moment, $\mathbf{X}_B$ is a polynomial in the random variables $\mathbf{X}_a$, with a in some string $s \in B$. Such a random variable is clearly a Borel function, and so its distribution is well defined in terms of the joint product distribution of the independent random variables $\mathbf{X}_a$. Parts (iv)–(vi) of the theorem then follow at once from Equation (27).

Finally, the uniqueness of the representation follows from the fact that $\varphi(\varnothing) = 0$ and $\varphi(1) = 1$. $\diamond$

If we specialize the axioms of Definition 18 to qualitative assertions about distributions of a particular form, we can replace Axiom 6 on the complete monotonicity of the sequence of qualitative moments of an object by much simpler conditions. In fact, we know of no simpler qualitative way of characterizing distributions of a given form than by such qualitative axioms on the moments. The following corollary concerns such a characterization of the uniform, binomial, and beta distributions of [0, 1], where the beta distribution is restricted to integer-valued parameters α and β.

COROLLARY TO THEOREM 28. *Let* $\mathfrak{A} = \langle A, \mathscr{F}, G, \succeq, \cdot, 1 \rangle$ *be a structure satisfying Axioms 1–5 of Definition 18, and for any a in G, assume* $a \lesssim 1$.

I. *If the moments of an object a for $n \geq 1$ satisfy*

$$(n + 1)a^n \sim 2a,$$

then $\mathbf{X}_a$ *is uniformly distributed on* [0, 1].

II. *If the moments of an object a for $n \geq 1$ satisfy*

$$a^n \sim a,$$

then $\mathbf{X}_a$ *has a Bernoulli distribution on* [0, 1].

III.   *If the moments of an object a for n $\geqslant$ 1 satisfy*

$$(\alpha + \beta + n)a^{n+1} \sim (\alpha + \beta)a^n,$$

where α and β are positive integers, then $\mathbf{X}_a$ *has a beta distribution on* $[0,1]$.

PROOF.   We only give the proof for the Bernoulli distribution. First, we use the hypothesis $a^n \sim a$ to verify inequalities (24) and (25). For k even,

$$\left[1 + \binom{k}{2} + \cdots + 1\right]a \succsim \left[\binom{k}{1} + \cdots + k\right]a$$

certainly holds, and similarly for k odd,

$$\left[1 + \binom{k}{2} + \cdots + k\right]a \succsim \left[\binom{k}{1} + \cdots + 1\right]a,$$

which shows that Axiom 6 of Definition 18 is satisfied, and so the unique numerical function φ of Theorem 28 exists, with

$$\varphi(a) = \varphi(a^n) = p,$$

for p some real number such that $0 < p \leqslant 1$, and the distribution is uniquely determined by the moments. The moment-generating function for the Bernoulli distribution with parameter p is, for $-\infty < t < \infty$,

$$\psi(t) = pe^t,$$

and so the nth derivative of ψ with respect to t is equal to p at $t = 0$, which completes the proof. $\diamondsuit$

Note that a Bernoulli distribution of $\mathbf{X}_a$ implies that all the probability weight is attached to the end points of the interval, so that if p is the parameter of the distribution, as in standard notation; then

$$E(\mathbf{X}_a) = (1 - p) \cdot 0 + p \cdot 1 = p.$$

We remark informally that some other standard distributions with different domains can also be characterized qualitatively in terms of moments. For example, the normal distribution on $(-\infty, \infty)$ with mean equal to zero and variance equal to one is characterized by

$$a^0 \sim 1,$$
$$a^1 \sim \varnothing,$$
$$a^2 \sim 1,$$
$$a^{2(n+1)} \sim (2n+1)a^{2n} \qquad \text{for} \quad n \geqslant 1.$$

EXERCISES

1. Define $\succ$ on $A = [0,1]$ by: $a \succ b$ iff $a > b + [1 - 2b]$. Construct upper- and lower-threshold representations $\langle \overline{\varphi}, \overline{\delta} \rangle$ and $\langle \underline{\varphi}, \underline{\delta} \rangle$. Show that $\overline{\varphi}$ cannot be a monotonic function of $\underline{\varphi}$. (16.2.1)

2. Show that either property (iv) or (iv*) of Definition 2 implies properties (i)–(iii). What properties exclude the trivial representation $\varphi = $ constant and $\delta = 0$? (16.2.1)

3. Show that the property of being strong neither implies strong* nor the converse. (16.2.1)

4. Suppose $\langle A, \succ \rangle$ is an asymmetric relation with an upper threshold representation $\langle \overline{\varphi}, \overline{\delta} \rangle$. Prove that

$$a \sim b \quad \text{implies} \quad \overline{\varphi}(a) \leqslant \overline{\varphi}(b) + \overline{\delta}(b) \quad \text{and} \quad \overline{\varphi}(b) \leqslant \overline{\varphi}(a) + \overline{\delta}(a)$$

and

$$\overline{\varphi}(a) < \overline{\varphi}(b) + \overline{\delta}(b) \quad \text{and} \quad \overline{\varphi}(b) < \overline{\varphi}(a) + \overline{\delta}(a) \quad \text{implies} \quad a \sim b.$$
$$(16.2.1)$$

Note: if R and S are relations on A, RS is the relation defined by $a\,RS\,b$ iff there exists $c \in A$ such that $a\,R\,c$ and $c\,R\,b$. When $S = R$, we denote RS by R^2.

5. Suppose an irreflexive relation $\langle A, \succ \rangle$ satisfies the defining property of a semiorder (Definition 5). Prove that $\succ$ is transitive and that $\langle A, \succ^2 \rangle$ is a semiorder. (16.2.2)

6. Prove that $\langle A, \succ \rangle$ is a semiorder iff $\langle A, \succ \rangle$ is asymmetric, $\succ \sim \succ \, \subseteq \, \succ$, and $\succ^2 \cap \sim^2 = \varnothing$. (16.2.2)

7.   Define the binary relation $\succ$ on $Re^+ \times Re^+$ by: $xy \succ x'y'$ iff $x > x' + y$. Show that this is an interval order but not a semiorder. Find upper and lower thresholds for it.   (16.2.2)

8.   Suppose $\succ$ and R are relations on A and $\succ$ is asymmetric. Show that R is fully compatible with $\succ$ iff for each $a, b \in A$, $a \succ b$ implies aRb.   (16.2.3)

9.   Prove Corollary 1 (of Theorem 1).   (16.2.3)

10.   Prove Corollary 2 (of Theorem 2).   (16.2.3)

11.   Prove Corollary 3 (of Theorem 3).   (16.2.3)

In the next three exercises, the following definition is involved. Suppose $\langle A, \sim \rangle$ is a graph and $\langle A, R \rangle$ is a weak order. $\langle A, R \rangle$ is *compatible* with $\langle A, \sim \rangle$ iff for each $a, b, c \in A$, aRb, bRc, and $a \sim c$ imply $a \sim b$ and $b \sim c$.

12.   Suppose $\langle A, \succ \rangle$ is an asymmetric relation, $\sim$ is the symmetric complement of $\succ$, and $\langle A, R \rangle$ is a weak order. Prove that $\langle A, R \rangle$ is compatible with $\langle A, \succ \rangle$ iff $\langle A, R \rangle$ is compatible with $\langle A, \sim \rangle$ and for every $a, b \in A$, $a \succ b$ implies aRb.   (16.2.3)

13.   Suppose $\langle A, \sim \rangle$ is a graph and $\langle A, R \rangle$ is a compatible weak order. Prove for each $a, b \in A$, if aRb and bRa, then aIb.   (16.2.3)

14.   Suppose $\langle A, \sim \rangle$ is a graph and $\langle A, R \rangle$ is a weak order. The *weak mapping rule* (Goodman, 1951) holds iff, for each $a, b, c, d \in A$,

$$aRb, \quad bRc, \quad cRd, \quad \text{and} \quad a \sim d \quad \text{imply} \quad b \sim c.$$

Prove that the weak mapping rule is equivalent to $\langle A, R \rangle$ being compatible with $\langle A, \sim \rangle$.   (16.2.3)

15.   Prove that Equations (3) and (4) are equivalent.   (16.2.4)

16.   Suppose $\langle A, \succ \rangle$ has a tight, upper-threshold representation $\langle \bar{\varphi}, \bar{\delta} \rangle$. Prove that the weak order induced by $\bar{\varphi}$ is the maximal weak order that is compatible with $\succ$.   (16.2.5)

17.   Prove Theorem 5.   (16.2.5)

18.   Prove Theorem 8.   (16.2.5)

19.   Prove part (i) of Theorem 9.   (16.2.5)

20. Suppose $\langle A, \succ \rangle$ is an asymmetric relation with a tight, strong, upper-threshold representation $\langle \bar{\varphi}, \bar{\delta} \rangle$. Define Δ by

$$\Delta(a) = \{x | x \in \bar{\varphi}(A) \text{ and } \bar{\varphi}(a) \leqslant x \leqslant \bar{\varphi}(a) + \bar{\delta}(a)\}.$$

Prove that:

$a \succ b$ iff $\Delta(a) > \Delta(b)$ (meaning that for every x, y, if $x \in \Delta(a)$ and $y \in \Delta(b)$, then $x > y$).

$a \sim b$ iff $\Delta(a) \cap \Delta(b) \neq \varnothing$.   (16.2.5; see also 16.2.7)

21. Prove the uniqueness part of Theorem 11.   (16.2.6)

22. For n an integer, define $\succ_n$ on $\mathrm{Re}^+ \times \mathrm{Re}^+$ by: $xy \succ_n x'y'$ iff $x > x' + ny$. Show that the structure $\langle \mathrm{Re}^+ \times \mathrm{Re}^+, \succ_1, \succ_2, \ldots \rangle$ has a homogeneous *lower*-threshold representation of the form $\langle \varphi, \underline{\delta}_1, \underline{\delta}_2, \ldots \rangle$ but that no pair of the $\succ_n$ satisfy upper interval homogeneity.   (16.4.1)

23. (i) Prove that strict stochastic transitivity implies interval stochastic transitivity, but not conversely.

(ii) *Moderate stochastic transitivity* is said to hold iff for each a, b, $c \in A$, if $P(a, b) \geqslant 1/2$ and $P(b, c) \geqslant 1/2$, then $P(a, c) \geqslant \min[P(a, b), P(b, c)]$. Show that interval stochastic transitivity implies moderate, but not conversely.   (16.4.2)

24. Prove Theorem 18.   (16.4.2)

25. Prove Theorem 19.   (16.4.2)

26. Suppose P on $A \times A$ is a binary choice probability function with the following property: for some real-valued functions φ_i, $i = 1, 2$, on A and F on $\mathrm{Re} \times \mathrm{Re}$, with F strictly increasing in its first argument and strictly decreasing in the second,

$$P(a, b) = F[\varphi_1(a), \varphi_2(b)].$$

(i) Prove that P satisfies interval stochastic transitivity.

(ii) If $\varphi_1 \equiv \varphi_2$, prove that P satisfies strong stochastic transitivity. (16.4.2)

27. Prove Theorem 25.   (16.6.3)

Exercises 28–33 provide a proof of Theorem 26; the notation and assumptions are those of that theorem.   (16.6.4)

28. Show that if $a \, Q \, b$, then $\delta(a) \geqslant \delta(b)$.

29. Show that for each integer n, $a \circ b \succ c$ iff $na \circ nb \succ nc$. Hint: use induction and the fact that $\circ$ is commutative and associative.

30. Show that $\delta(na) = n\delta(a)$. Hint: Deal with the case $\delta(a) = 0$ separately from $\delta(a) > 0$.

31. Show that $\langle \varphi, \delta \rangle$ is a threshold representation of $\langle A, \succ \rangle$.

32. For $a \, Q \, b$, define $N(b, a)$ to be the unique integer such that $[N(b, a) + 1]b \, P \, a$ and $a \, Q \, N(b, a)b$. Show that there exists a sequence of elements $\{c_i\}$ such that $c_i \succ c_{i+1}$ and $\lim_{i \to \infty} \varphi(c_i) = 0$ and that for each i,

$$\frac{N(c_i, a)\delta(c_i)}{[N(c_i, a) + 1]\delta(c_i)} \leqslant \frac{[N(c_i, a) + 1]\delta(c_i)}{N(c_i, a)\delta(c_i)}$$

33. Use Exercise 32 to show the existence of $k > 0$ such that for all $a \in A$, $\delta(a) = k\varphi(a)$.

34. Prove Part I of the Corollary of Theorem 28.   (16.8.5)

35. Prove Part III of the Corollary of Theorem 28.   (16.8.5)

Chapter 17   Representation of Choice Probabilities

17.1 INTRODUCTION

Axiomatic theories of measurement are generally based on an order relation. In the analysis of choices, the idea of an ordering is naturally linked to the concept of maximization: the decision maker selects the option that is maximal relative to the constraints. The concept of maximization, however, encounters a serious difficulty, namely, the presence of inconsistency: different objects are selected on different occasions. In some analyses, inconsistency is treated as a nuisance to be eliminated by suitable experimental controls and statistical methods prior to the application of a measurement model. Other analyses, however, treat inconsistency as an important part of the empirical relational structure and attempt to explain and predict it. This chapter reviews the major choice theories that deal with inconsistent preferences and lead to the construction of measurement scales.

The present chapter is similar in character to Chapter 8 of Volume I (Conditional Expected Utility), which was devoted to the analysis of decisions under risk or uncertainty. In both chapters, the interpretation of the structure and the theoretical development are motivated by the analysis of choice behavior. The following discussion is intended primarily for readers who are not familiar with theories of choice probabilities. For reviews of the field, see Fishburn (1977), Luce (1977), Luce and Suppes

(1965), Machina (1985), and McFadden (1976, 1981); models of pair comparisons were reviewed by Falmagne (1974, 1985), Marschak (1960), Morrison (1963), and Roberts (1979b).

Throughout this chapter we represent inconsistency by a family of finite probability distributions. We begin with a universal set A comprising all objects in the domain to be considered, certain finite subsets $B \subset A$ representing possible constraints within which a choice takes place, and probability distributions $P(\cdot, B)$ on each such finite subset. This basic structure is formalized in the following definition.

DEFINITION 1. *A triple $\langle A, M, P \rangle$ is called a* structure of choice probabilities *iff A is a set, M is a nonempty set of 2^A whose members are nonempty and finite, and P is a real-valued function with domain $\{(a, B) | a \in B \in M\}$ such that*

$$P(a, B) \geqslant 0 \quad and \quad \sum_{b \in B} P(b, B) = 1.$$

The structure $\langle A, M, P \rangle$ is called finite *iff the set A is finite. It is called* closed *iff A is finite and $M = \{B \subset A | B \neq \varnothing\}$.*

17.1.1 Empirical Interpretations

Choice probabilities most commonly arise in the context of a choice experiment in which an individual makes repeated choices under essentially identical conditions, from each subset B in the family M. The family M of offered subsets is designed in accordance with the purpose of the experiment. Often, A is finite, and M consists of all subsets of cardinality 2 (complete *pair-comparison* design). If $B = \{a, b\}$, it is common to write $P(a, b)$ for $P(a, \{a, b\})$, in which case $P(a, b)$ is the probability that object a is selected when a and b are offered, and $P(b, a) = 1 - P(a, b)$ is the probability that b is selected instead. In the usual theory for such an experiment the sequence of choices of a or b from $\{a, b\}$ is viewed as a sequence of Bernoulli trials with underlying parameter $P(a, b)$. The relative frequency with which a is chosen from $\{a, b\}$ is taken as an estimate of $P(a, b)$. The multinomial generalization to sets B with three or more elements is obvious.

The overall domain A can vary greatly, as can the criterion for choice. The set A may consist of political candidates; in one experiment, an individual may be asked to choose which candidate from subset B is most likely to win an election; in another experiment, which one the individual would vote for; and in a third, which one is most conservative. Or the set A

may consist of colors, and the individual may choose which color in subset B is most similar to a standard color, which would harmonize best with it, or simply which color in B is brightest.

In these examples, and in most others, people are often inconsistent: they make different choices from the offered set B at different times. Usually the inconsistency persists even when learning, satiation, and change in taste do not provide likely explanations. Furthermore, even in essentially unique choice situations, which cannot be replicated, people typically have less than complete confidence in their decisions and feel that, in a different state of mind, they might have made different choices. Regardless of whether one is committed to a deterministic or to a nondeterministic view of choice behavior, it seems desirable to extend the deterministic models and develop theories that account for choice probabilities.

When the numbers $P(a, B)$ are regarded as parameters of multinomial distributions, the full theory of choice contains major elements that are not captured in Definition 1, e.g., stationarity and independence for successive trials with the offered set B. These theoretical simplifications impose severe constraints. Among other problems, a decision maker may remember and be affected by earlier choices. Some experiments are carefully designed to minimize this possibility through the use of alternatives that are difficult to encode into memory (especially sensory experiments) or through a complex interweaving of trials for various subsets B. In many cases, the design achieves its purpose: stationarity and independence are at least reasonable approximations. Nonetheless, there are many instances in which it is inappropriate to regard the numbers $P(a, B)$ as multinomial parameters for a sequence of choices by an individual. Either the above precautions are impractical, or they are so difficult to achieve that a reasonable sample size for the estimation of P cannot be attained. In such cases, other empirical interpretations of $P(a, B)$ can be considered. One possibility is to regard the population as composed of individuals, each of whom will choose only once, and to think of $P(a, B)$ as describing a group rather than an individual. This interpretation has been employed in the economic literature (see, e.g., McFadden, 1981). Another possibility arises when the theories being considered do not predict the exact values of P but only their ordering. Then it may be possible to substitute other dependent variables (e.g., judgments of confidence) that are believed to be monotonically related to P.

17.1.2 Probabilistic Representations

The study of probabilistic theories of choice originated with Thurstone's (1927a, 1927b) analysis of comparative judgment. Thurstone advanced the notion that the presentation of a stimulus gives rise to an internal representation that is subject to momentary random fluctuations. He proposed, therefore, that stimuli or choice objects be represented by a collection $U = \{U_a | a \in A\}$ of jointly distributed normal random variables such that $P(a, b) = Pr(U_a \geq U_b)$ for all a, b where $Pr(\cdot)$ is the probability associated with the random variables. Thurstone's binary model can be readily generalized to $P(a, B) = Pr(U_a = \max\{U_b | b \in B\})$. The notation $Pr(\cdot)$ is used to distinguish between the (observable) choice probabilities and the (theoretical) probabilities associated with the random variables. The probability of choosing a from B, in this model, equals the probability that the random variable associated with a will take a value that is greater than the values taken by the random variables associated with the other elements of B. (For continuous distributions, the probability of the random variables being equal is zero.)

Thurstone considered several models that vary in generality; his strongest model (case V) assumes that the component random variables U_a are mutually independent with constant variance (set equal to $\frac{1}{2}$). In that case, the pairwise choice probabilities can be expressed as $\Phi[u(a) - u(b)]$, where $u(a)$ is the mean of U_a, and Φ is the standard cumulative normal distribution function. The work of Thurstone stimulated both empirical and theoretical research, though most of it has been concerned with the statistical rather than with the measurement-theoretical aspects of the model. For comprehensive treatment of this work the reader can consult the books by Torgerson (1958) and by Bock and Jones (1968).

Thurstone's work exemplifies the random utility model, which represents choice alternatives by a random vector, not necessarily normal. It is discussed in Section 17.6. Later authors, notably Block and Marschak (1960), Debreu (1958), and Luce (1959), have developed alternative representations, called constant utility models, that express choice probability as a function of some constant scale values associated with the elements of A. Unlike Thurstone, who simply (but reluctantly) postulated the existence of a normal random utility model, these authors have derived the representations from potentially observable properties of choice probability. For example, Luce (1959) presented a testable property of choice probability leading to the construction of a scale φ on A satisfying

$$P(a, B) = \frac{\varphi(a)}{\displaystyle\sum_{b \in B} \varphi(b)}, \qquad a \in B \subseteq A.$$

These models are reviewed in Section 17.4.

Many of the representations of choice probability induce an ordering on the elements of A. For example, we can define $a \succsim b$ iff $Pr(\mathbf{U}_a \geqslant \mathbf{U}_b) \geqslant \frac{1}{2}$ in Thurstone's case V, or $a \succsim b$ iff $\varphi(a) \geqslant \varphi(b)$ in Luce's model. Alternatively, we may want to define an ordering on A directly in terms of the observed choice probabilities by $a \succsim b$ iff $P(a, b) \geqslant \frac{1}{2}$ or by $a \succsim b$ iff $P(a,c) \geqslant P(b, c)$ for all $c \in A$. If we assume that $\succsim$ is an order relation, these definitions impose testable constraints on $\langle A, M, P \rangle$ called stochastic transitivity conditions. The analysis of these conditions leads to the construction of ordinal models for pair comparison that represent the ordering of binary choice probabilities but not their numerical values. These representations are discussed in Section 17.2.

An alternative way to define an ordering on A is to say that $a \succsim b$ iff $P(a, B) \geqslant P(b, B)$ for any $B \subseteq A$ that contains both a and b. This definition requires, for example, that $P(a, B) \geqslant P(b, B)$ implies $P(a, b) \geqslant \frac{1}{2}$, $a, b \in B$. However, this property, which is satisfied in Thurstone's case V and in Luce's model, is often violated when the choice set contains both similar and dissimilar elements. More specifically, suppose $P(a, b) = \frac{1}{2}$ and $B = \{a, b, b'\}$, where b and b' are practically identical. Hence, we expect $P(a, B) > P(b, B)$ because the addition of b' to the set $\{a, b\}$ is likely to "hurt" b more than a. This pattern of behavior, discussed in 17.8.2, has led to the development of a different sort of representation using a Markov process (Tversky, 1972a, 1972b). The idea is that when $B = \{a, b, b'\}$ is presented, the individual first selects either the subset $\{b, b'\}$ or the subset $\{a\}$. In the latter case, the process terminates and a is chosen; but in the former case, it goes through another transition from $\{b, b'\}$ to either $\{b\}$ or $\{b'\}$. The choice probabilities $P(a, B)$ are thus regarded as absorption probabilities for a Markov process whose initial state corresponds to the offered set B. Because each transition eliminates some alternatives, we call these Markovian elimination models. They are discussed in Section 17.8.

We are thus led to consider four classes of representations: (i) ordinal models for pair comparison; (ii) constant utility models that express choice probability as a function of the scale values associated with the alternatives; (iii) random utility models that represent the alternatives as random variables; and (iv) Markovian models that characterize choice as a sequential elimination process. It should be noted that these classes of models are not mutually exclusive. Several choice models (e.g., Thurstone's case V and Luce's model) can be expressed either as constant or as random models.

Since the key primitive concept in the chapter is a numerical function P, it could be argued that these developments do not belong to the domain of fundamental measurement. Although structures of choice probabilities have sometimes been regarded as derived measurement (Suppes and Zinnes, 1963), they are treated here as empirical structures that lead to the construction of new numerical scales via appropriate representation theorems. The emphasis of this chapter is on the development of numerical representations of choice alternatives on the basis of choice probabilities. Consequently, we do not review the extensive literature on probabilistic theories of choice. In particular, we do not discuss relations among the many assumptions about choice probabilities made in the literature; instead, we concentrate on those properties that play an important role in the analysis of numerical representations.

17.2 ORDINAL REPRESENTATIONS FOR PAIR COMPARISONS

This section develops numerical representations based on the ordering of binary choice probabilities. It differs from the rest of this chapter in several respects. First, it is concerned exclusively with pair comparisons and does not extend the representation to choice probabilities from larger sets. Second, it deals only with the ordering of choice probabilities and not with their actual numerical values. Indeed, the models of the present section are closer to the algebraic models for difference measurement (developed in Chapters 4 and 14) than to most of the probabilistic models developed later in this chapter. The basic empirical structure is defined as follows.

DEFINITION 2.   A triple $\langle A, M, P \rangle$ is a pair comparison structure iff $\langle A, M, P \rangle$ is a structure of choice probability for which M is a reflexive and symmetric binary relation on A. The structure is complete iff $M = A \times A$.

It follows that for each $(a, b) \in M$, $P(a, b) + P(b, a) = 1$.

As in Definition 1, A is interpreted as the total set of alternatives or stimuli under consideration, and $P(a, b)$ is the probability of selecting alternative a when presented with $\{a, b\}$. Since M is reflexive, $P(a, a)$ equals $\frac{1}{2}$; it is interpreted as the probability of selecting alternative a from the set that consists of two replicas of that alternative. For simplicity, we write aMb instead of $(a, b) \in M$. In a pair-comparison structure aMa; also, aMb implies bMa. We first discuss the case in which $M = A \times A$ and then turn to a more general case in which only certain pairs of alternatives are comparable.

17.2.1 Stochastic Transitivity

Some properties of binary choice probabilities that generalize the algebraic notion of transitivity are introduced in the following definition, which partly overlaps Definition 16.11 (p. 337).

DEFINITION 3. *Let* $\langle A, M, P \rangle$ *be a complete structure of pair comparison, and suppose that* $P(a, b) \geqslant \frac{1}{2}$ *and* $P(b, c) \geqslant \frac{1}{2}$.

Weak stochastic transitivity (WST) *holds iff* $P(a, c) \geqslant \frac{1}{2}$.
Moderate stochastic transitivity (MST) *holds iff*

$$P(a, c) \geqslant min[P(a, b), P(b, c)].$$

Strong stochastic transitivity (SST) *holds iff*

$$P(a, c) \geqslant max[P(a, b), P(b, c)].$$

Strict stochastic transitivity (ST) *holds iff SST holds and a strict inequality in the hypotheses implies a strict inequality in the conclusion.*

The conditions of stochastic transitivity, introduced in Definition 3, are required to define order relations on A in terms of the observed binary choice probabilities. Perhaps the simplest way to define such an order is to say that $a \succsim b$ iff $P(a, b) \geqslant \frac{1}{2}$, that is, if a is chosen at least 50% of the time from $\{a, b\}$. It is easy to see that, under this definition, WST reduces to the transitivity of $\succsim$. If it fails, no simple ordering of A can be defined. Despite its theoretical appeal and apparent plausibility, WST can be systematically violated under special circumstances involving multiattribute alternatives (see Sections 17.2.4 and 17.2.5). Almost all of the models developed in this chapter, however, satisfy WST.

As was discussed in the preceding chapter (see 16.4.2), binary choice probability can be used to define an entire family of order relations by: $a \succsim_\lambda b$ iff $P(a, b) \geqslant \lambda$, $\frac{1}{2} \leqslant \lambda \leqslant 1$. We interpret $\succsim_\lambda$ as a preference relation determined by a given level λ of decisiveness or discriminability. Clearly, $\succsim_\lambda$ includes $\succsim_\mu$ iff $1 \geqslant \mu > \lambda \geqslant \frac{1}{2}$. It is easy to verify (see Fishburn, 1973a, 1977) that MST is both necessary and sufficient for the transitivity of all $\succsim_\lambda$, $\frac{1}{2} \leqslant \lambda \leqslant 1$. WST, of course, is equivalent to the transitivity of $\succsim_{\frac{1}{2}}$. Note that $\succsim_\lambda$ is generally not connected except when $\lambda = \frac{1}{2}$. Further discussion of stochastic transitivity and other testable properties of binary choice probabilities are found in Block and Marschak (1960), Fishburn (1973a, 1977), Marschak (1960), Morrison (1963), Roberts (1971b, 1979b), and Tversky and Russo (1969).

17.2.2 Difference Structures

To extract additional qualitative information from binary choice probabilities, one can define a quaternary relation $\succsim$ on A by

$$ab \succsim cd \qquad \text{iff} \qquad P(a, b) \geqslant P(c, d)$$

The defined relation is typically interpreted as a weak order of *strength-of-preference* or of *discriminability*. Thus $P(a, b) \geqslant P(c, d)$ is taken to mean that a is preferred to b at least as much as c is preferred to d, or that c and d are closer to each other than a and b. This interpretation suggests the following representation.

DEFINITION 4. *A complete structure of pair comparison satisfies the* strong-utility model *iff there exists a real-valued function φ on A such that for all $a, b, c, d \in A$*

$$\varphi(a) - \varphi(b) \geqslant \varphi(c) - \varphi(d) \qquad \text{iff} \qquad P(a, b) \geqslant P(c, d).$$

This representation has also been called the *Fechnerian model* (see Block and Marschak, 1960; Luce and Suppes, 1965; Definition 17). An equivalent definition of the model involves a cumulative distribution function F, with $F(x) + F(-x) = 1$, and a scale φ on A such that

$$P(a, b) = F[\varphi(a) - \varphi(b)].$$

This representation originates from Fechnerian psychophysics, in which $P(a, b)$ is treated as a measure of the discriminability between stimuli a and b, and φ is interpreted as a measure of subjective sensation. For critical discussions of these concepts see Falmagne (1985) and Luce and Galanter (1963). The first axiomatization of the strong-utility model was developed by Debreu (1958).

DEFINITION 5. *A structure of pair comparison $\langle A, M, P \rangle$ is a* complete difference structure *iff $M = A \times A$ and the following axioms hold for all $a, a', b, b', c, c' \in A$.*

1. *Monotonicity. If $P(a, b) \geqslant P(a', b')$ and $P(b, c) \geqslant P(b', c')$, then $P(a, c) \geqslant P(a', c')$. Moreover, if either antecedent inequality is strict, so is the conclusion.*

2. *Solvability. For any $t \in (0, 1)$ satisfying $P(a, b) \geqslant t \geqslant P(a, d)$, there exists $c \in A$ such that $P(a, c) = t$.*

The monotonicity axiom is clearly necessary for the strong-utility model. It is essentially identical to Axiom 3 of Definition 4.3, which plays a central role in the development of the algebraic difference-model. The present numerical version of the solvability axiom was introduced by Debreu (1958). Because this chapter deals with numerical empirical structures, it is natural to use the numerical form of the solvability axiom even though it is stronger than that used elsewhere in this book; it implies the Archimedean property and leads to a representation whose range is a real interval.

THEOREM 1. *If $\langle A, M, P \rangle$ is a complete difference structure, then there exists a real-valued function φ from A onto some real interval such that for all $a, b, c, d \in A$*

$$\varphi(a) - \varphi(b) \geqslant \varphi(c) - \varphi(d) \quad iff \quad P(a, b) \geqslant P(c, d).$$

Furthermore, φ is unique up to a positive linear transformation.

The only difference between Debreu's (1958) result and Theorem 1 is his use of the quadruple condition $P(a, b) \geqslant P(c, d)$ iff $P(a, c) \geqslant P(b, d)$ instead of the monotonicity axiom. It is easily seen that in a complete structure of pair comparison the two properties are equivalent (Exercise 2). We do not present a direct proof of Theorem 1 because it follows readily from a more general result, Theorem 2, to which we turn next.

17.2.3 Local Difference Structures

The major limitation of Theorem 1 is that the pair-comparison structure is assumed to be complete. In many situations, it may be theoretically impossible or practically unfeasible to satisfy this constraint. To illustrate a major difficulty associated with the complete case, let A be a set of gambles, and suppose that the subject is not perfectly consistent in the choices between gambles that do not differ much in expected value but is consistent in his choices between gambles that differ greatly in expected value. Alternatively, let A be a set of tones varying in intensity and $P(a, b)$ be the probability that tone a is judged louder than tone b. Here, again, the observed choice probabilities may equal 0 or 1 whenever the intensity difference between the two tones is sufficiently large. In such situations, one obtains a mixture of perfect and imperfect discriminations that cannot be accommodated by the strong-utility model. By selecting a, b, and c that are sufficiently far apart, we can obtain $P(a, b) = P(a, c) = P(b, c) = 1$, contrary to the strong-utility model.

To overcome this difficulty, one may want to exclude all pairs of elements that are perfectly discriminable from each other and to scale A on the basis of imperfect discriminations only, i.e., on the basis of choice probabilities that are bounded away from 0 or 1. This requires a measurement model in which a global scale over all of A is constructed from local information only.

DEFINITION 6. *A pair comparison structure* $\langle A, M, P \rangle$ *is a local difference structure iff it satisfies the following axioms for all* $a, a', b, b', c, c' \in A$. *Define* $a \succeq b$ *iff* aMb *and* $P(a, b) \geqslant \frac{1}{2}$.

1. *Comparability.* *If* $a \succeq b$ *and* $a \succeq c$ *then* bMc. *If* $b \succeq a$ *and* $c \succeq a$ *then* bMc.

2. *Monotonicity.* *Suppose* aMb, bMc, aMc, $a'Mb'$, $b'Mc'$. *If* $P(a, b) \geqslant P(a', b')$ *and* $P(b, c) \geqslant P(b', c')$, *then* $a'Mc'$ *and* $P(a, c) \geqslant P(a', c')$. *Moreover, if either antecedent inequality is strict, so is the conclusion.*

3. *Solvability.* *If* aMb, aMd, *and* $P(a, b) \geqslant t \geqslant P(a, d)$ *for some* $t \in (0, 1)$, *then there exists* $c \in A$ *such that* aMc *and* $P(a, c) = t$.

4. *Connectedness.* *For any* $a, b \in A$ *that do not satisfy* $P(a, b) = 1/2$, *there exists a sequence* $a = a_0, a_1, \ldots, a_n = b$ *in* A *such that exactly one of the following holds:* (i) $a_i \succeq a_{i+1}$ *or* (ii) $a_{i+1} \succeq a_i$, $i = 0, \ldots, n - 1$.

The comparability axiom imposes added structure on M, beyond reflexivity and symmetry. In particular, it asserts that any two elements that are bounded (from above or below) by the same third element are comparable. Thus it ensures comparability in the small. The monotonicity and solvability axioms of Definition 6 are identical to the respective axioms of Definition 5 except for the natural restrictions on the domain of P. Finally, the connectedness axiom states that any two nonequivalent elements of A are connected either by an increasing sequence or by a decreasing sequence, and not by both.

THEOREM 2. *If* $\langle A, M, P \rangle$ *is a local difference structure, then there exists a real-valued function* φ *from* A *onto some real interval such that whenever* aMb *and* cMd, $P(a, b) \geqslant P(c, d)$ *iff* $\varphi(a) - \varphi(b) \geqslant \varphi(c) - \varphi(d)$. *Furthermore,* φ *is unique up to a positive linear transformation.*

This result was established by Doignon and Falmagne (1974) in a somewhat more general setting. Their proof is based on the reduction of a local difference structure to an Archimedean, regular, positive, ordered, local semigroup (Definition 2.2). The proof presented in this chapter reduces a local difference structure to an algebraic difference structure (Definition 4.3).

Theorem 2 can be generalized by replacing the numerical scale P with a partial order $\succeq$ on $A \times A$ satisfying $ab \succeq cd$ iff $dc \succeq ba$, provided aMb and cMd. In this case, Axiom 3 of Definition 6 should be replaced by the usual solvability and Archimedean axioms (e.g., Axioms 4 and 5 of Definition 4.3). The resulting representation theorem remains unchanged except that the range of the scale is no longer a real interval. With these modifications, Theorem 2 generalizes the algebraic difference model (Theorem 4.2) to the case in which only intervals that do not exceed a given bound can be compared with each other.

As we mentioned in Section 16.8.1, Falmagne (1980) adds a concatenation operation and additional axioms to those of Definition 6 to define probabilistic difference extensive structures. The essential axiom is that

$$P(a, b) = P(a \circ c, b \circ c) = P(c \circ a, c \circ b).$$

17.2.4 Additive Difference Structures

In this section we discuss extensions of the strong-utility model to multidimensional alternatives. Suppose the object set A has a product structure $A = A_1 \times \cdots \times A_n$, so that each object is described as an n-tuple $a = a_1 \cdots a_n$. If we assume that the attributes combine additively, we obtain two natural extensions of the Fechnerian representation: the probabilistic conjoint measurement model (Falmagne, 1979) and the additive-difference model (Tversky, 1969).

Falmagne (1979) developed a family of probabilistic conjoint measurement models in which the scale value of a is expressed as an additive function of the scale values of its components. In particular, he investigated the model

$$P(a, b) = F(\varphi[f_1(a_1) + f_2(a_2)] - \varphi[f_1(b_1) + f_2(b_2)]),$$

where φ and F are strictly increasing functions. This representation is obtained by combining the strong-utility model with an additive conjoint measurement model. More specifically, it follows from the assumptions that $\langle A, M, P \rangle$ is a local difference structure (Definition 6) and that $\langle A_1 \times A_2, \succeq \rangle$ is an additive conjoint structure (Definition 6.7), where $\succeq$ is defined by $a_1 a_2 \succeq b_1 b_2$ iff $P(a, b) \geqslant \frac{1}{2}$. Falmagne (1979) explored some special cases of this model and established necessary and/or sufficient conditions under which φ is convex (concave), exponential, or logarithmic in the continuous case. Falmagne and Iverson (1979) have also studied stronger representations obtained by treating all stimulus components as real numbers and assuming various homogeneity conditions such as $P(a_1 a_2, b_1 b_2) = P(\lambda a_1 \lambda a_2, \lambda b_1 \lambda b_2), \lambda > 0$. This principle, called the conjoint Weber's law, has been supported in a psychophysical experiment in which subjects compared the loudness of two biaural stimuli (Falmagne *et al.*, 1979). For a comprehensive review of this work, the reader is referred to Falmagne's (1985) book.

Because the probabilistic additive conjoint-measurement model is a strong-utility model, it imposes many constraints such as monotonicity (Definition 5) and strict transitivity (Definition 3) on choice probabilities. The following multidimensional generalization of the strong-utility model has been motivated by the observation that choices between multiattribute alternatives sometimes violate stochastic transitivity (see Section 17.2.5).

Let $\langle A, M, P \rangle$ be a complete structure of pair comparison, with $A = A_1 \times \cdots \times A_n$. Suppose there exist real-valued functions φ_i and F_i defined, respectively, on A_i and Re, $i = 1, \ldots, n$, such that for all $a, b, c, d \in A$

$$P(a, b) \geqslant P(c, d)$$

iff

$$\sum_{i=1}^{n} F_i[\varphi_i(a_i) - \varphi_i(b_i)] \geqslant \sum_{i=1}^{n} F_i[\varphi_i(c_i) - \varphi_i(d_i)].$$

According to this representation, the additive-difference model, the choice between a and b can be described as follows. First, the subject considers the quantities $\varphi_i(a_i) - \varphi_i(b_i)$, $i = 1, \ldots, n$, each expressing the perceived (algebraic) difference between a and b along the ith dimension. These quantities are then evaluated according to the difference functions F_i, each describing the contribution of differences along the ith dimension to the overall preference between the alternatives. The resulting terms of the form $F_i[\varphi_i(a_i) - \varphi_i(b_i)]$ are added across dimensions, and the obtained sums are assumed to generate the order of the P-values.

The above preference model is closely related to the additive-difference proximity model investigated in Section 14.5.3. Both models assume two independent processes: an intradimensional subtractive process and an interdimensional additive process. The difference between the models stems from the fact that proximity judgments are assumed to be symmetric, i.e., $(a, b) \sim (b, a)$, whereas binary choice probabilities are not symmetric since $P(a, b) = 1 - P(b, a)$. As a consequence, the proximity model is based on an absolute-difference representation whereas the preference model is based on an algebraic-difference representation. The following formulation of the additive-difference model of preference is based on the work of Tversky (1969) and Tversky and Krantz (1970).

DEFINITION 7. *A complete pair comparison structure* $\langle A, M, P \rangle$ *with* $A = A_1 \times \cdots \times A_n$, *is an* additive-difference structure *iff the following axioms hold for all* $a, a', b, b', c, c', d, d', \in A$.

1. *Independence.* *If the two elements in each of the pairs* $(a, a'), (b, b'), (c, c'), (d, d')$ *have identical components on one factor, and the two elements in each of the pairs* $(a, c), (a', c'), (b, d), (b', d')$ *have identical components on all other factors, then*

$$P(a, b) \geqslant P(a', b') \quad iff \quad P(c, d) \geqslant P(c', d').$$

2. *Monotonicity.* *Suppose* a, a', b, b', c, c' *coincide on all but one factor. If* $P(a, b) \geqslant P(a', b')$ *and* $P(b, c) \geqslant P(b', c')$, *then* $P(a, c) \geqslant P(a', c')$. *Moreover, if either antecedent inequality is strict, so is the conclusion.*

3. *Solvability.* *If* $P(a, b) \geqslant t \geqslant P(a, d)$ *for some* $t \in (0, 1)$, *then there exists* $c \in A$ *that coincides with* b *and* d *on any factor on which they coincide, and satisfies* $P(a, c) = t$.

A factor A_i is essential[1] *iff there exist* $a, b \in A$ *that coincide on all other factors and for which* $P(a, b) \neq \frac{1}{2}$.

If the number of essential factors is two, then the following form of the Thomsen condition holds for all $a_i, b_i, c_i, d_i, e_i, f_i, \in A_i$, $i = 1, 2$.

4. *The Thomsen condition.* *The equations*

$$P(a_1 e_2, b_1 f_2) = P(e_1 c_2, f_1 d_2)$$

and

$$P(e_1 a_2, f_1 b_2) = P(c_1 e_2, d_1 f_2)$$

imply

$$P(a_1 a_2, b_1 b_2) = P(c_1 c_2, d_1 d_2).$$

Axioms 2 and 3 of Definition 7 are essentially identical to Axioms 1 and 2, respectively, of Definition 5. Axioms 1 and 4 are necessary conditions for interdimensional additivity, and they are essentially identical to Definitions 14.11 and 14.12, respectively. The representation theorem for an additive-difference structure reads as follows.

THEOREM 3. *Suppose* $\langle A, M, P \rangle$ *is an additive-difference structure* (Definition 7) *with* $n \geqslant 2$. *Then there exist real-valued functions* φ_i *defined on* A_i, $i = 1, \ldots, n$, *and strictly increasing real-valued functions* F_i *in one real variable each, such that for all* $a, b, c, d \in A$,

$$P(a, b) \geqslant P(c, d)$$

---

[1] Throughout this chapter, we assume that all inessential factors have been eliminated; hence, n denotes the number of essential factors.

iff

$$\sum_{i=1}^{n} F_i[\varphi_i(a_i) - \varphi_i(b_i)] \geq \sum_{i=1}^{n} F_i[\varphi_i(c_i) - \varphi_i(d_i)],$$

where the F_i are skew-symmetric, i.e., $F_i(x) + F_i(-x) = 0$, $i = 1, \ldots n$.
Furthermore, if $\varphi_1', \ldots, \varphi_n'$ and $F_1', \ldots, F_n'$ is another collection of func-
tions satisfying the above relation then there exist constants $t_i > 0$, s_i, and
$r > 0$ such that

$$\varphi_i' = t_i\varphi_i + s_i \quad and \quad F_i' = rF_i, \quad i = 1, \ldots, n.$$

The proof of Theorem 3, given in Section 17.3.2, follows the steps leading
to the construction of an additive-difference representation for a proximity
structure (see Theorem 14.8). First, we show that there exist real-valued
functions φ_i defined on A_i, $i = 1, \ldots, n$, and a real-valued function F that
increases in each of its n real arguments such that

$$P(a, b) = F[\varphi_1(a_1) - \varphi_1(b_1), \ldots, \varphi_n(a_n) - \varphi_n(b_n)];$$

that is, intradimensional subtractivity holds. Next, we establish interdimensional additivity. That is, we show that there exist real-valued functions ψ_i
defined on A_i^2, $i = 1, \ldots, n$, and a real-valued increasing function G in one
real argument such that

$$P(a, b) = G\left[\sum_{i=1}^{n} \psi_i(a_i, b_i)\right].$$

The desired representation follows from these results.

The additive-difference model of preference assumes that all (ordered)
pairs of alternatives can be ordered via P or directly with respect to relative
strength of preference. It is not assumed, however, that the alternatives
themselves can be ordered, or that the preference relation is transitive.
Indeed, the additive-difference model has been proposed (Morrison, 1962;
Tversky, 1969) to account for intransitive preferences. This raises an
interesting question, namely, under what conditions do the preferences
generated by an additive-difference model satisfy the transitivity assumption? Put differently, what constraints are imposed on the representation
established in Theorem 3 by assuming WST (Definition 3)?

Note that if all difference functions are linear, i.e., $F_i(x) = t_i x$, $t_i > 0$, $i = 1, \ldots, n$, then Theorem 3 yields

$$P(a, b) \geqslant P(c, d)$$

iff

$$\sum_{i=1}^{n} t_i \varphi_i(a_i) - \sum_{i=1}^{n} t_i \varphi_i(b_i) \geqslant \sum_{i=1}^{n} t_i \varphi_i(c_i) - \sum_{i=1}^{n} t_i \varphi_i(d_i),$$

which is a special case of the strong-utility model, with $\varphi(a) = \sum_{i=1}^{n} t_i \varphi_i(a_i)$, $a \in A$. The latter model, however, clearly implies WST. Thus, if the difference-functions are linear, the additive-difference model satisfies WST. The following theorem, due to Tversky (1969), shows that this is essentially the *only* case in which WST is generally satisfied in an additive-difference model, provided the number of essential factors is at least three.

THEOREM 4.   *Suppose* $\langle A, M, P \rangle$ *is an additive-difference structure. For each i let* Y_i *be the domain of* F_i *(see Theorem 3), and suppose* $\sup Y_n \geqslant \sup Y_i$, $i = 1, \ldots, n - 1$.

   (i)   *For* $n = 2$, *WST holds iff* $F_1(y_1) = F_2(t y_1)$, $t > 0$, $y_1 \in Y_1$.

   (ii)   *For* $n \geqslant 3$, *WST holds iff* $F_i(y_i) = t_i y_i$, $t_i > 0$, $y_i \in Y_i$, $i = 1, \ldots, n - 1$, *and* $F_n(y_n) = t_n y_n$, $t_n > 0$ *for all* $y_n \in Y_n$ *such that* $F_n(y_n) \leqslant \sum_{i=1}^{n-1} \sup F_i(y_i)$, $y_i \in Y_i$.

Theorem 4 is proved in Section 17.3.3. It shows that WST imposes severe constraints on the form of the difference functions. If the number of essential factors is 2, the difference functions must be identical except for a change of unit of their domain. If the number of essential factors is 3 or more, all difference functions must be linear, except possibly for the far tail of F_n.

17.2.5 Intransitive Preferences

Transitivity of preference is perhaps the most fundamental principle underlying most theories of choice. It is not surprising, therefore, that this principle has been the target of several empirical investigations. Since preferences exhibit a certain degree of inconsistency, observed choices cannot be expected to satisfy algebraic transitivity. The question, then, becomes whether it is possible to extract a transitive preference relation from the observed binary choice probabilities. More specifically, suppose we define the preference relation by $a \succsim b$ iff $P(a, b) \geqslant \frac{1}{2}$. Does the defined relation satisfy transitivity? Or, equivalently, do the binary choice probabilities satisfy WST?

The test of WST poses some difficulties. Due to sampling variability, a certain number of circular triads [i.e., $P(a, b) \geqslant \frac{1}{2}$, $P(b, c) \geqslant \frac{1}{2}$, and $P(a, c) < \frac{1}{2}$] are expected even if WST holds. To reject WST, therefore, one must demonstrate that the observed degree of circularity is significantly greater than that expected under WST. However, Morrison (1963) has pointed out that as the number of alternatives becomes large, the number of circular triads produced by a diabolic or a maximally intransitive subject approaches the number of circular triads obtained under WST with $P(a, b) = \frac{1}{2}$ for all $a, b \in A$. This is because in a complete pair-comparison design, combinatorial considerations limit the maximal number of circular triads. Hence, it is exceedingly difficult to reject WST on the basis of the observed number of circular triads unless the circularities can be identified in advance. Indeed, most tests of this type were inconclusive. To obtain a reasonably powerful test of WST, therefore, one needs a model to construct the alternatives and identify the critical triads.

In the following discussion, we present one such model that is a special case of the additive-difference model and show how it generates intransitive preferences. The discussion follows the development in Tversky (1969).

Consider a set of two-dimensional alternatives $A = A_1 \times A_2$, and suppose that each $a \in A$ is characterized by its numerical values on the two dimensions. For example, A may be a set of job applicants, in which each candidate is described in terms of intelligence (measured by an IQ test) and professional experience (measured in years). Suppose the decision maker regards intelligence (A_1) to be more important than experience (A_2) yet at the same time believes that the IQ test is not very reliable. In such situations, the decision maker may adopt the following decision rule. Select some threshold value $\delta > 0$. If the difference between the alternatives on the critical dimension (A_1) exceeds δ, select the alternative that is superior on that dimension; if the difference is less than or equal to δ, treat the alternatives as equivalent with respect to A_1 and choose the alternative that is superior on the other dimension.

This decision model, which is readily generalized to more than two dimensions, is called a *lexicographic semiorder* because it combines a semiorder with a lexicographic ordering of the dimensions. Relevant axiomatizations were developed by Luce (1978) and Fishburn (1980). Note that the lexicographic semiorder is a special (limiting) case of the additive-difference model in which the difference function F_1 for the critical dimension is a step function, that is, if $|x| \leqslant \delta$, then $F_1(|x|) = 0$; and if $|x| > \delta$, then $F_1(|x|) > F_2(|y|)$ for all y.

Despite its intuitive appeal, however, the lexicographic semiorder yields intransitive preferences since it fails to satisfy the constraints of Theorem 4.

To illustrate, consider the following matrix whose entries are the numerical values of the respective alternatives along the two dimensions, expressed in δ-units.

| | | Dimensions | |
		A_1	A_2
	a	2δ	6δ
Alternatives	b	3δ	4δ
	c	4δ	2δ

Because the differences between a and b and between b and c on the first dimension do not exceed δ, these choices are made on the basis of the second dimension and hence $a \succ b$, and $b \succ c$. But since the difference between a and c on the first dimension exceeds δ, we have $c \succ a$, thereby violating transitivity. The lexicographic semiorder provides a plausible model that generates intransitivities in a predicted fashion. The following experiment attempted to realize the lexicographic semiorder in choice between gambles.

All gambles were of the form (x, p), where the decision maker receives $\$x$ if a chance event with probability p occurs and nothing when the event does not occur.

Each gamble was displayed on a card showing a circle divided into a black and a white sector. The probability of winning corresponded to the relative size of the black sector, and the payoff was printed at the top of each card. The gambles employed in this study are displayed in Table 1. Note that, unlike the payoffs, the probabilities were displayed pictorially rather than numerically. Consequently, no exact calculation of expected values was possible. The gambles were constructed so that the expected value increased with probability and decreased with payoff.

TABLE 1
Gambles Used in the Test of WST

Gamble	Probability of winning	Payoff (in $)	Expected value
a	7/24	5.00	1.46
b	8/24	4.75	1.58
c	9/24	4.50	1.69
d	10/24	4.25	1.77
e	11/24	4.00	1.83

Because the display renders the evaluation of payoff differences easier than that of probability differences, it was hypothesized that some subjects would ignore small probability differences and choose between adjacent gambles on the basis of payoffs. (Gambles are called *adjacent* if they are a step apart along the probability or the payoff dimension). Because expected value, however, is negatively correlated with payoff, it was further hypothesized that choices between gambles lying far apart in the chain would be made on the basis of expected values or probability of winning. Such a pattern of preferences must violate transitivity somewhere along the chain from *a* to *e*.

Eight subjects were selected for the experiment on the basis of their behavior in a preliminary study. Each pair of gambles was presented 20 times to each subject in a total of five experimental sessions. On each trial the subject was asked to choose which of the two gambles presented to him he would rather play.

Six out of eight subjects exhibited systematic and significant intransitivities of the predicted type. The data of one subject is displayed in Figure 1, in which arrows denote direction of preferences with the associated choice probabilities.

The figure shows that the subject chose between adjacent gambles according to payoffs and between the nonadjacent gambles according to probability of winning or expected value, thus violating WST in the predicted manner. Similar intransitivities were demonstrated in choice between college applicants. It is interesting to note that no subject realized that the preferences were intransitive, and some subjects even denied this possibility emphatically (Tversky, 1969).

For other probabilistic models especially designed for choice between gambles, see Becker, DeGroot, and Marschak (1963a, 1963b); Fishburn (1976); Luce and Suppes (1965); and Machina (1985).

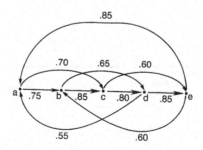

FIGURE 1. Choice probabilities for a stochastically intransitive subject.

17.3 PROOFS

17.3.1 Theorem 2 (p. 392)

If $\langle A, M, P \rangle$ is a local difference structure, then there exists a real-valued function φ from A onto some real interval such that whenever aMb and cMd,

$$P(a, b) \geqslant P(c, d) \quad \text{iff} \quad \varphi(a) - \varphi(b) \geqslant \varphi(c) - \varphi(d).$$

Furthermore, φ is an interval scale.

PROOF. The proof of Theorem 2 is divided into three parts. In part (1) we extend the partial order induced by P to a weak order of $A \times A$. For convenience we shall use the symbol $\succsim$ to denote this constructed weak order on $A \times A$ (rather than as in Definition 6 for a binary relation on A). In part (2) we embed a local difference structure in an algebraic-difference structure. In part (3) we show that the range of the resulting scale is a real interval.

Part (1): *Extension of order.* Let $A^{(0)}$ be the set of all pairs $(a, b) \in A \times A$ such that $P(a, b) = \frac{1}{2}$. Let $A^{(+1)}$ be the set of all pairs $(a, b) \in A \times A$ that do not belong to $A^{(0)}$ and for which there exists a finite sequence $(a = a_0, a_1, \ldots, a_n = b)$ such that $P(a_{i-1}, a_i) > \frac{1}{2}$, $i = 1, \ldots, n$. Let $A^{(-1)}$ be the set of all pairs $(a, b) \in A \times A$ that do not belong to $A^{(0)}$ and for which there exists a finite sequence $(a = a_0, a_1, \ldots, a_n = b)$ such that $P(a_i, a_{i-1}) > \frac{1}{2}$, $i = 1, \ldots, n$. It follows from connectedness (Axiom 4) that the sets $A^{(+1)}$, $A^{(0)}$, $A^{(-1)}$ form a partition of $A \times A$, that $(a, b) \in A^{(+1)}$ iff $(b, a) \in A^{(-1)}$, and that $(a, b) \in A^{(0)}$ iff $(b, a) \in A^{(0)}$. Let $A_n = (a_1, \ldots, a_n)$, $P(a_i, a_{i+1}) > \frac{1}{2}$, $i = 1, \ldots, n - 1$, and $B_m = (b_1, \ldots, b_m)$, $P(b_j, b_{j+1}) > \frac{1}{2}$, $j = 1, \ldots, m - 1$, be two (decreasing) sequences. The sequence A_n is said to be *nested* in the sequence B_m iff there exists a subsequence $(b_{A(1)}, \ldots, b_{A(n)})$ of B_m with $A(1) = 1$, such that $P(a_i, a_{i+1}) = P(b_{A(i)}, b_{A(i+1)})$, $i = 1, \ldots, n - 1$.

Let A_n and B_m be two decreasing sequences. We show that they can both be nested in another decreasing sequence C_k, $k \leqslant m + n - 1$ such that either $c_1 = a_1$ or $c_1 = b_1$. The proof proceeds by induction on $m + n$. If either m or n is less than 2, the result is trivial. Suppose $m = n = 2$. With no loss of generality we assume that $P(a_1, a_2) \geqslant P(b_1, b_2)$. To construct C_k, set $c_1 = a_1$, $c_3 = a_2$, and let c_2 be the solution to the equation $P(c_1, c_2) = P(b_1, b_2)$, which exists (by solvability) since $P(a_1, a_1) < P(b_1, b_2) \leqslant P(a_1, a_2)$. It is easily verified that $P(c_2, a_2) \geqslant \frac{1}{2}$ as required.

Let A_n, B_m be two decreasing sequences, in which $P(a_1, a_2) \geqslant P(b_1, b_2)$. Construct a' by solvability, so that $P(a_1, a') = P(b_1, b_2)$, and consider the sequences $A'_n = (a', a_2, \ldots, a_n)$ and $B'_m = (b_2, \ldots, b_m)$. By the induction hypothesis there exists a sequence $(c_1, \ldots, c_k)$ in which both A'_n and B'_m are nested. Select c_0 such that $P(c_0, c_1) = P(a_1, a')$, by letting $c_0 = a_1$ if $c_1 = a'$, and $c_0 = b_1$ otherwise. To show that both A_n and B_m are nested in $C_k = (c_0, c_1, \ldots, c_k)$, note that $P(c_0, c_1) = P(a_1, a')$, $P(c_1, c_j) = P(a', a_2)$ for some j and $a_1 M a_2$. Hence, by monotonicity, $P(c_0, c_j) = P(a_1, a_2)$ as required.

Using these results, we define $\succsim$ on $A \times A$ as follows. If $(a, b) \in A^{(i)}$, $(c, d) \in A^{(j)}$, and $i \neq j$, then $(a, b) \succsim (c, d)$ iff $i > j$. If $(a, b), (c, d) \in A^{(0)}$, then $(a, b) \sim (c, d)$. If $(a, b), (c, d) \in A^{(-1)}$, then $(a, b) \succsim (c, d)$ iff $(d, c) \succsim (b, a)$. To complete the definition, therefore, we only have to define $\succsim$ on $A^{(+1)}$. For any $(a, b), (c, d) \in A^{(+1)}$, construct the sequence $a = a_0, a_1, \ldots, a_n = b$, $P(a_{i-1}, a_i) > \frac{1}{2}$, $i = 1, \ldots, n$, denoted A_n; and the sequence $c = c_0, c_1, \ldots, c_m = d$, $P(c_{j-1}, c_j) > \frac{1}{2}$, $j = 1, \ldots, m$, denoted C_m. Let $E_l = (e_0, e_1, \ldots, e_l)$, $P(e_{k-1}, e_k) > \frac{1}{2}$, $k = 1, \ldots, l$, be a sequence in which both A_n and C_m are nested. Define $(a, b) \succsim (c, d)$ iff $A(n) \geqslant C(m)$. Thus, the order of the pairs is defined by the order of the indices that correspond to a_n and c_m in the nesting sequence E_l. To show that the definition is independent of the choice of a nesting sequence, suppose A_n and B_m are nested in E' as well as in E''. Since there exists a sequence in which both E' and E'' are nested, it is easy to show, by induction, that $\succsim$ is well defined.

Part (2): *Embedding of structure.* We show that, with the above definition of $\succsim$, the structure $\langle A \times A, \succsim \rangle$ is an algebraic-difference structure. We check the axioms of Definition 4.3 in turn.

1. Reflexivity and connectedness are immediate. To prove transitivity, assume $(a, b) \succsim (c, d) \succsim (e, f)$. If $(a, b) \in A^{(i)}$ and $(e, f) \in A^{(j)}$ where $i > j$, the result is immediate. If $i = j = 0$, it is also immediate. Since the cases $i = j = 1$ and $i = j = -1$ are equivalent, we shall deal with the former only. Consider the sequences:

$$a_0 = a, a_1, \ldots, a_n = b, \qquad P(a_{i-1}, a_i) > \tfrac{1}{2}, i = 1, \ldots, n.$$

$$c_0 = c, c_1, \ldots, c_m = d, \qquad P(c_{j-1}, c_j) > \tfrac{1}{2}, j = 1, \ldots, m.$$

$$e_0 = e, e_1, \ldots, e_l = f, \qquad P(e_{k-1}, e_k) > \tfrac{1}{2}, k = 1, \ldots, l.$$

To establish transitivity, construct the sequence G in which A_n, C_m, E_l are nested such that $P(a_0, a_1) = P(g_0, g_{A(1)})$, $P(c_0, c_1) = P(g_0, g_{C(1)})$, and $P(e_0, e_1) = P(g_0, g_{E(1)})$. Consequently, $A(n) \geqslant C(m)$ and $C(m) \geqslant E(l)$ imply $A(n) \geqslant E(l)$ as required.

2. If $(a, b) \succsim (c, d)$, it is immediate that $(d, c) \succsim (b, a)$.

3. Suppose $(a, b) \succsim (a', b')$ and $(b, c) \succsim (b', c')$. We want to prove that $(a, c) \succsim (a', c')$. Consider the following decreasing sequences:

$$a_0 = a, a_1, \ldots, a_n = b, \qquad a_{n+1} = b_1, \ldots, a_{n+m} = b_m = c,$$
$$a_0' = a', a_1', \ldots, a_{n'}' = b', \qquad a_{n'+1}' = b_1', \ldots, a_{n'+m'}' = b_{m'}' = c',$$

and let E_k be another decreasing sequence in which both sequences are nested. Since $(a, b) \succsim (a', b')$, $A(n) \geqslant A'(n')$. Furthermore, since $(b, c) \succsim (b'c')$, it can be shown using induction that $A(m + n) \geqslant A'(m' + n')$ as required. The proof for other cases is similar.

4. Suppose $(a, b) \succsim (c, d) \succsim (a, a)$. It follows from the results established in part (1) of the proof, that there exist d' and d'' such that

$$(a, d') \sim (c, d) \sim (d'', b).$$

5. Suppose $B = (b_0, b_1, \ldots, b_i, \ldots)$ is a decreasing bounded standard sequence. That is, $(b_i, b_{i+1}) \sim (b_0, b_1) \succ (b_0, b_0)$ and $(a, c) \succ (b_0, b_i) \succ (c, a)$, for all i. Suppose B is infinite, and let $D_n = (d_0, \ldots, d_n)$ be a decreasing connected sequence such that $d_0 = a$ and $d_n = c$. It can be shown, using part (1), that there exists a decreasing sequence $E = (e_0, e_1, \ldots, e_k, \ldots)$ such that $(b_0, b_i) \sim (e_0, e_{B(i)})$ and $(d_0, d_j) \sim (e_0, e_{D(j)})$. Thus both B and D_n are nested in E. Since B is infinite and D_n is finite, there exists an infinite sequence $B' = (b_l, b_{l+1}, \ldots, b_{l+i}, \ldots) \subseteq B$ that lies between d_j and d_{j+1} for some j in the nesting sequence E. Since d_j and d_{j+1} are connected by construction, so are all elements of B' by the comparability axiom. Let $t = \sup\{P(b_0, b_{l+i}) | i = 1, 2, \ldots\}$. Since $(d_j, d_{j+1}) \succ (b_l, b_{l+i})$ for all i, there exist f and g satisfying $P(a, f) = t$ and $P(g, f) = P(b_l, b_{l+i})$. Consequently, $P(a, f) > P(a, g)$, and by monotonicity there exists an i such that $P(b_0, b_{l+i}) > P(a, g)$ and hence, $P(b_0, b_{l+i+1}) > P(a, f) = t$, a contradiction. Therefore, any bounded standard sequence is finite.

We have shown so far that the defined order on $A \times A$ satisfies all the axioms of Definition 4.3 and hence that the desired representation and uniqueness follow at once from Theorem 4.2.

Part (3): Continuity. To complete the proof of Theorem 2 we show that the range of φ, denoted Y, is a real interval. That is, if $x, z \in Y$, and $z > y > x$, then $y \in Y$. Suppose that $x = \varphi(u)$ and $z = \varphi(v)$, then there exists an increasing sequence $e_0, \ldots, e_n$ of connected elements joining u and v. If there is no element e_j in the sequence such that $\varphi(e_j) = y$, then there must exist some i such that $\varphi(e_i) < y < \varphi(e_{i+1})$, $i = 1, \ldots, n - 1$. Let $b = e_i$ and $c = e_{i+1}$. Clearly $c \gtrsim b$. Let $A_{bc} = \{ a \in A | P(c, b) > P(c, a) > P(c, c) \}$. We show that $\langle A_{bc}, M, P \rangle$ is a complete difference structure in the sense of Definition 5. Clearly, monotonicity holds, and any two elements of A_{bc} are comparable. Thus we only need to establish the solvability axiom. Let $a_i \in A_{bc}$, $i = 1, 2, 3$, and suppose $P(a_1, a_2) > t > P(a_1, a_3)$, and $P(a_1, d) = t$. We show that $d \in A_{bc}$. If $t = \frac{1}{2}$, set $d = a_1$. If $t > \frac{1}{2}$, then $a_1 \gtrsim a_2$, $a_1 \gtrsim d$, and dMa_2. From $P(a_1, a_2) > P(a_1, d) > P(a_1, a_3)$ we obtain $\varphi(a_3) > \varphi(d) > \varphi(a_2)$, and since dMa_2, $P(d, a_2) > \frac{1}{2}$, i.e., $d \gtrsim a_2$. Furthermore, since $a_3 \gtrsim a_2$, dMa_3 and hence dMa_i, $i = 1, 2, 3$. The case for which $t < \frac{1}{2}$ is proved similarly. Since $d \gtrsim a_2$ and $c \gtrsim a_2$, dMc, and since $\varphi(a_3) > \varphi(d) > \varphi(a_2)$,

$$P(c, b) \geqslant P(c, a_2) > P(c, d) > P(c, a_3) \geqslant \tfrac{1}{2},$$

and hence $d \in A_{bc}$. Thus $\langle A_{bc}, M, P \rangle$ is a complete difference structure. The final part of the proof consists of showing that in a complete difference structure the range of φ (denoted Y) is a real interval. By the above argument Y is dense in itself. Next we show that Y has no holes. Let $a_1, a_2, \ldots$ be a sequence such that $a_i \in A$, $\varphi(a_i) > \varphi(a_{i+1}) > \lim_{i \to \infty} \varphi(a_i) = y$, and there exists $c \in A$ such that $y > \varphi(c)$. See Figure 2. We want to show that there exists $b \in A$ such that $\varphi(b) = y$. Since $\varphi(a_i) > \varphi(a_{i+1}) > \varphi(c)$, we have $P(a_i, c) > P(a_{i+1}, c) > P(c, c)$. Let $r = \inf_{i \in I^+} P(a_i, c)$. Consequently, for any $i \in I^+$, $P(a_i, c) > r \geqslant P(c, c)$, and by solvability there exists $b \in A$ such that $P(b, c) = r$. We prove that

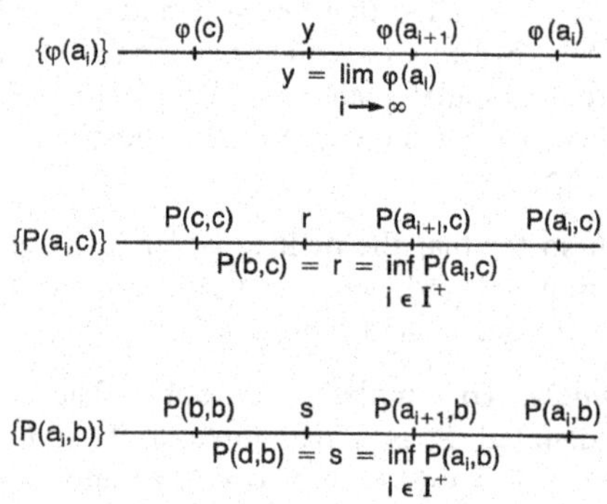

FIGURE 2. Graphic illustration of the construction employed in the proof of Theorem 2.

$\varphi(b) = y$. Because $P(a_i, c) > P(b, c)$, $\varphi(a_i) > \varphi(b)$ for all $i \in I^+$, and hence $y \geqslant \varphi(b)$ since y is the limit of the sequence $\varphi(a_i)$, $i \in I^+$. Next we show that the assumption $y > \varphi(b)$ leads to a contradiction.

Clearly, $P(a_i, b) > P(a_{i+1}, b) > P(b, b)$. Let $s = \inf_{i \in I^+} P(a_i, b)$. By strong solvability there exists $d \in A$ such that $P(d, b) = s$. Furthermore, since $\varphi(a_i) > y$ for all i, there exist $u, v \in A$ such that $y - \varphi(b) > \varphi(u) - \varphi(v) > 0$. Consequently, for all $i \in I^+$, $P(a_i, b) > P(u, v) > \frac{1}{2}$, and $s = P(d, b) > \frac{1}{2}$. But since $P(a_i, b) > P(d, b) = s$ for all $i \in I^+$, it follows that $\varphi(a_i) > \varphi(d)$ (see Figure 2). Recalling the proof of $y \geqslant \varphi(b)$, it follows similarly that $y \geqslant \varphi(d)$. Note that $P(d, c) > P(b, c) = r$. Since r is the limit of the sequence $P(a_i, c)$, $i \in I^+$, there exists some $j \in I^+$ so that $P(d, c) > P(a_j, c)$ and consequently $\varphi(d) > \varphi(a_i)$ for any $i > j$, contrary to the facts that $\varphi(a_i) > y$, $i \in I^+$, and $y \geqslant \varphi(d)$.

We have shown thus far that Y is dense in itself and closed in the interval $(\inf_{y \in Y} y, \sup_{y \in Y} y)$. By a well-known result, therefore, Y is a real interval.

$\diamondsuit$

17.3.2 Theorem 3 (p. 395)

If $\langle A, M, P \rangle$ is an additive-difference structure, with $n \geqslant 2$ essential factors, then it satisfies the additive-difference model. Furthermore, the φ_i are interval scales, and the F_i are skew-symmetric ratio scales, $i = 1, \ldots, n$.

PROOF. *Intradimensional subtractivity.* Let e be some fixed element of A, and let $E_j = \{ a \in A | a_i = e_i \text{ for all } i \neq j \}$. It follows from Axioms 2 and 3 of Definition 7 that for any $e \in A$ and any $j = 1, \ldots, n$, the structure $\langle E_j, P \rangle$ satisfies the hypotheses of Theorem 1, and hence there exists a real-valued function φ_j on A_j such that for all $a, b, c, d \in E_j$,

$$P(a, b) \geqslant P(c, d) \quad \text{iff} \quad \varphi_j(a_j) - \varphi_j(b_j) \geqslant \varphi_j(c_j) - \varphi_j(d_j).$$

Let T_j be the range of φ_j. Define a real-valued function F of n real variables on $T_1 \times \cdots \times T_n$ by

$$F[\varphi_1(a_1) - \varphi_1(b_1), \ldots, \varphi_n(a_n) - \varphi_n(b_n)] = P(a, b).$$

The proof that F is well defined is essentially the same as the analogous result for an additive-difference proximity structure (see the proofs of Theorems 14.5 and 14.6).

Interdimensional additivity. Let $A^* = A_1^2 \times \cdots \times A_n^2$. Thus, for any pair of n-tuples $a = a_1 \cdots a_n$, $b = b_1 \cdots b_n \in A$, the n-tuple of pairs $ab =$

$(a_1b_1)\cdots(a_nb_n)$ is in A^*. Define $\succsim$ on A^* by $ab \succsim cd$ iff $P(a, b) \geqslant$ $P(c, d)$. As in the proof of Theorem 14.7, $\langle A_1^2, \ldots, A_n^2 \rangle$ is an n-component structure of additive conjoint measurement in the sense of Definition 6.13. Hence, it follows from Theorem 6.13 that there exist interval scales ψ_i on A_i^2, $i = 1, \ldots, n$ such that for all $a, b, c, d \in A$,

$$P(a, b) \geqslant P(c, d) \qquad \text{iff} \qquad \sum_{i=1}^{n} \psi_i(a_ib_i) \geqslant \sum_{i=1}^{n} \psi_i(c_id_i).$$

To establish interdimensional additivity for $n = 2$, note that Axiom 4 of Definition 7 implies the Thomsen condition, whence the desired representation follows from Theorem 6.2.

Combining these results yields

$$P(a, b) = F[\varphi_1(a_1) - \varphi_1(b_1), \ldots, \varphi_n(a_n) - \varphi_n(b_n)]$$
$$= G\left[\sum_{i=1}^{n} \psi_i(a_i, b_i)\right],$$

for some strictly increasing real-valued function G. Let $\psi_i(a_i, b_i) = 0$ whenever $a_i = b_i$. Suppose a and b differ only on the jth component, whence

$$P(a, b) = F[0, \ldots, 0, \varphi_j(a_j) - \varphi_j(b_j), 0, \ldots, 0]$$
$$= G[\psi_j(a_j, b_j)].$$

Define a real-valued function F_j, $j = 1, \ldots, n$, by

$$F_j[\varphi_j(a_j) - \varphi_j(b_j)] = G^{-1}\{F[0, \ldots, 0, \varphi_j(a_j) - \varphi_j(b_j), 0, \ldots, 0]\}.$$

By the above equation,

$$F_j[\varphi_j(a_j) - \varphi_j(b_j)] = \psi_j(a_j, b_j).$$

It follows directly from the independence axiom that each F_j, $j = 1, \ldots, n$ is well defined and strictly increasing. Substituting F_j for ψ_j yields

$$P(a, b) = G\left\{\sum_{i=1}^{n} F_i[\varphi_i(a_i) - \varphi_i(b_i)]\right\}.$$

Skew-symmetry. Suppose $A = A_1 \times A_2$, and set $F_1(0) = F_2(0) = 0$. Let $x = \varphi_1(a_1) - \varphi_1(b_1)$, $y = \varphi_1(c_1) - \varphi_1(d_1)$, and $z = \varphi_2(a_2) - \varphi_2(b_2)$. It follows from the additive-difference (preference) model and solvability (Axiom 3 of Definition 7) that there exists a constant $\alpha > 0$ with the following property: for all $a_1, b_1, c_1, d_1 \in A_1$ for which $|F_1(x) - F_1(y)| \leq \alpha$, there exist $a_2, b_2 \in A_2$ such that

$$P(a_1a_2, b_1b_2) = P(c_1b_2, d_1a_2),$$

and hence,

$$P(b_1b_2, a_1a_2) = P(d_1a_2, c_1b_2).$$

Consequently,

$$F_1(x) + F_2(z) = F_1(y) + F_2(-z),$$
$$F_1(-x) + F_2(-z) = F_1(-y) + F_2(z),$$

and, by adding the equations,

$$F_1(x) + F_1(-x) = F_1(y) + F_1(-y).$$

This relation holds provided $|F_1(x) - F_1(y)| \leq \alpha$; hence the quantity $F_1(x) + F_1(-x)$ must be independent of x. By construction, therefore, $F_1(x) + F_1(-x) = 2F_1(0)$. The extensions of this result to the second factor and to the case of n factors are straightforward. By setting $F_i(0) = 0$, $i = 1, \ldots, n$, the F_i become unique up to multiplication by a common positive constant. $\diamond$

17.3.3 Theorem 4 (p. 397)

Assume an additive-difference structure. For each i, let Y_i be the domain of F_i, and suppose $\sup Y_n \geq \sup Y_i$, $i = 1, \ldots, n - 1$. For $n = 2$ WST holds iff $F_1(y_1) = F_2(ty_1)$, $t > 0$, $y_1 \in Y_1$. For $n \geq 3$, WST holds iff $F_i(y_i) = t_iy_i$, $t_i > 0$, $y_i \in Y_i$, $i = 1, \ldots, n - 1$, and $F_n(y_n) = t_ny_n$, $t_n > 0$ for all $y_n \in Y_n$ such that $F_n(y_n) \leq \sum_{i=1}^{n-1} \sup F_i(y_i)$, $y_i \in Y_i$.

PROOF. By definition $Y_i = \{ y \in Re | y = \varphi_i(a_i) - \varphi_i(b_i), a_i, b_i \in A_i \}$. It follows from Theorem 1 that each Y_i is a real interval, symmetric around 0. Let $Y = Y_1 \times \cdots \times Y_n$, and define H on Y by the equation $H = \sum_{i=1}^{n} F_i$.

It follows from solvability that for any $\alpha, \beta, \alpha + \beta \in Y$, there exist $a, b, c \in A$ such that, for $i = 1, \ldots, n$,

$$\alpha_i = \varphi_i(a_i) - \varphi_i(b_i)$$
$$\beta_i = \varphi_i(b_i) - \varphi_i(c_i)$$
$$\alpha_i + \beta_i = \varphi_i(a_i) - \varphi_i(c_i).$$

To demonstrate, consider the case $\alpha_i, \beta_i \geq 0$. Let $\alpha_i + \beta_i = \varphi_i(a_i) - \varphi_i(c_i)$, and $\alpha_i = \varphi_i(a_i') - \varphi_i(b_i')$. Clearly, $\alpha_i + \beta_i \geq \alpha_i, \beta_i$. Suppose $a_j = a_j' = c_j = c_j'$ for all $j \neq i$. Hence, $P(a, c) \geq P(a', b') \geq P(a, a) = \frac{1}{2}$, and by solvability there exists $b \in A$ such that $P(a, b) = P(a', b')$ where $b_i = a_i = c_i$ for all $i \neq j$. By subtraction, therefore, $\beta_i = \varphi_i(b_i) - \varphi_i(c_i)$. The proofs of the other cases proceed similarly.

It follows readily from WST that if $P(a, b) = \frac{1}{2}$ and $P(b, c) = \frac{1}{2}$ then $P(a, c) = \frac{1}{2}$. Hence, $H(\alpha) = 0$ and $H(\beta) = 0$ imply $H(\alpha + \beta) = 0$, provided $\alpha, \beta, \alpha + \beta \in Y$. Let $N = \{\alpha \in \text{Re}^n | H(\alpha) = 0\}$, and let V be the smallest subspace of Re^n that contains N. We show first that the dimensionality of V is less than n.

Let $K_u = \{\alpha \in Y| \, |\alpha_i| < u\}$. We want to prove that for any $u > 0$, V has a basis whose elements belong to $N \cap K_u$. Define $u(\alpha) = \sum_{i=1}^n [|\alpha_i|/u] + 1$, where $[x]$ is the integral part of x, i.e., $[x] = \sup\{i \in I | i \leq x\}$. Thus $K_u = \{\alpha \in Y | u(\alpha) = 1\}$. We show that for any $\alpha \in N$ and any $u > 0$ there exist $\beta, \gamma \in N$ such that $\alpha = \beta + \gamma$ and $u(\alpha) > u(\beta), u(\gamma)$. Suppose $\alpha \in N$ and $\alpha \notin K_u$, say $\alpha_i \geq u$ for some i. Define $\alpha' = (\alpha_1, \ldots, \alpha_{i-1}, \alpha_i/2, \alpha_{i+1}, \ldots, \alpha_n)$, $\alpha^0 = (0, \ldots, 0, \alpha_i/2, 0, \ldots, 0)$. Clearly, $H(\alpha^0) > 0 = H(\alpha) > H(\alpha')$. Hence by strong solvability there exists some $\beta \in N$ that is (component-wise) between α' and α^0. Thus $|\alpha_j| \geq |\beta_j| \geq 0$ for any $j \neq i$ and $\beta_i = \alpha_i/2$. Define $\gamma = \alpha - \beta$. Thus we obtained $\beta, \gamma \in N$ such that $\alpha = \beta + \gamma$ and $u(\alpha) > u(\beta), u(\gamma)$. By repeating this construction we can express any $\alpha \in N$ as a sum of elements in $N \cap K_u$. Suppose $V = \text{Re}^n$, i.e., suppose N contains n linearly independent vectors. By the above argument, V contains a basis $B_u \subset N \cap K_u$ for any $u > 0$. Let G_u be the group generated by the elements of B_u, and let $Y^+ = \{\alpha \in Y | \alpha_i > 0, i = 1, \ldots, n\}$. It can be shown that we can always select sufficiently small u so that $G_u \cap Y^+$ is nonempty. Thus there exists some $\alpha \in Y^+$ that is expressible as a sum of elements in B_u. By WST, $H(\alpha) = 0$, contrary to the assumptions that $F_i(0) = 0$ and F_i is strictly increasing, $i = 1, \ldots, n$. Hence, the dimensionality of V is at most $n - 1$. By a well-known theorem from linear algebra, therefore, there exist $t_1, \ldots, t_n \in \text{Re}$ that are not all 0, such that for any $\alpha \in N$, $\sum_{i=1}^n t_i \alpha_i = 0$. Suppose $t_j = 0$ and $t_k \neq 0$. Select $\alpha \in N$ such that $\alpha_j, \alpha_k \neq 0$ and $\alpha_l = 0$ for any $l \neq j, k$. Thus,

$$0 = \sum_{i=1}^{n} t_i \alpha_i = t_j \alpha_j + t_k \alpha_k = t_k \alpha_k \neq 0,$$

a contradiction. Consequently, $t_i \neq 0$, $i = 1, \ldots, n$.

Let Z_i be the range of F_i, i.e., $Z_i = F_i(Y_i)$, $i = 1, \ldots, n$. Note that each Z_i is symmetric around 0. Consider first the case of two essential dimensions ($n = 2$), and suppose $\sup Z_2 \geqslant \sup Z_1$. Select any $\alpha_1 \in Y_1$. Hence, there exists some $\beta_2 \in Y_2$ such that $|F_2(\beta_2)| \geqslant |F_1(\alpha_1)|$. By solvability, therefore, there exists $\alpha_2 \in Y_2$ such that $F_2(\alpha_2) = -F_1(\alpha_1)$, i.e., $\alpha = (\alpha_1, \alpha_2) \in N$. By the above-mentioned theorem, therefore, there exist $t_1, t_2 \neq 0$ such that $t_1 \alpha_1 + t_2 \alpha_2 = 0$. Hence,

$$\alpha_2 = -\left(\frac{t_1}{t_2}\right) \alpha_1,$$

and

$$F_1(\alpha_1) = -F_2\left(\frac{-t_1 \alpha_1}{t_2}\right) = F_2(t \alpha_1), \qquad t = -\left(\frac{t_1}{t_2}\right).$$

Next, consider the case $n \geqslant 3$. With no loss of generality we can let $\sup Z_n \geqslant \sup Z_i$, $i = 1, \ldots, n-1$, and $F_n = F$. By the above result, $F_i(\alpha_i) = F(t_i \alpha_i)$, $\alpha_i \in Y_i$ $i = 1, \ldots, n-1$. Hence,

$$\sum_{i=1}^{n} F(t_i \alpha_i) = 0 \qquad \text{implies} \qquad \sum_{i=1}^{n} t_i \alpha_i = 0.$$

Rescaling all the φ_i so that $t_i = 1$ for all i,

$$\sum_{i=1}^{n} F(\alpha_i) = 0 \qquad \text{implies} \qquad \sum_{i=1}^{n} \alpha_i = 0,$$

$$\text{implies} \qquad \alpha_n = -\sum_{i=1}^{n-1} \alpha_i$$

$$\text{and} \qquad F(\alpha_n) = F\left(-\sum_{i=1}^{n-1} \alpha_i\right).$$

On the other hand, $\sum_{i=1}^{n} F(\alpha_i) = 0$ implies $F(\alpha_n) = -\sum_{i=1}^{n-1} F(\alpha_i)$. Let $s =$

$\sum_{i=1}^{n-1} \sup Z_i$. Hence, for any real argument in $(-s, s)$, $\sum_{i=1}^{n-1} F(\alpha_i) = F(\sum_{i=1}^{n-1}\alpha_i)$. Consequently, $F(\alpha) = t\alpha$ for some $t > 0$. $\qquad\qquad\diamond$

17.4 CONSTANT REPRESENTATIONS FOR MULTIPLE CHOICE

17.4.1 Simple Scalability

This section deals with models of choice in which a scale value is assigned to each of the alternatives, and choice probability is expressed as a function of the respective scale values. This property is also shared by the Fechnerian or the strong-utility model (see Section 17.2.1); however, the models to be discussed here are not limited to the binary case. Throughout this section we assume that $\langle A, M, P \rangle$ is a closed structure of choice probabilities (Definition 1), and we use α, β, and γ, respectively, to denote the cardinality of A, B, and C.

DEFINITION 8. *A closed structure of choice probabilities* $\langle A, M, P \rangle$ *satisfies* simple scalability *iff there exists a real-valued function* φ *on* A, *and a family of real-valued functions* $\{F_\beta\}$, $2 \leqslant \beta \leqslant \alpha$, *in* β *real arguments each, such that for any* $B = \{a, b, \ldots, h\} \subseteq A$, *with* $P(a, B) \neq 0, 1$,

$$P(a, B) = F_\beta[\varphi(a), \varphi(b), \ldots, \varphi(h)],$$

where each F_β *is strictly increasing in its first argument and strictly decreasing in the other* $\beta - 1$ *arguments, in the extended sense that* $F_\beta(X_1, \ldots, X_\beta)$ $\geqslant F_\beta(Y_1, \ldots, Y_\beta)$ *if* $X_1 \geqslant Y_1$ *and* $X_i \leqslant Y_i$, $i = 2, \ldots, \beta$. *Furthermore, if any of the hypotheses is strict, so is the conclusion. If* $P(a, B)$ *is 0 or 1,* F_β *is nondecreasing in the first argument and nonincreasing in the remaining arguments.*

Simple scalability was first introduced by Krantz (1964) and further investigated by Tversky (1972a, 1972b). The present definition corrects a defect in an earlier formulation (see Colonius, 1984, Chapter 4). It postulates that the alternative-set A can be scaled (by φ) so that any choice probability $P(a, B)$ is expressible as a function (F_β) of the scale values of the elements of B. Thus the φ-values can be viewed as utilities or subjective values, and the F_β, $2 \leqslant \beta \leqslant \alpha$, can be interpreted as the composition rules that specify how the φ-values are combined to determine choice probability.

Despite the fact that the precise form of the F_β is left unspecified, simple scalability has strong testable implications. To demonstrate, suppose $B =$

$\{a, b, \ldots, h\}$ and $P(a, B) \geqslant P(b, B)$. By simple scalability, therefore,

$$F_\beta[\varphi(a), \varphi(b), \ldots, \varphi(h)] = P(a, B)$$
$$\geqslant P(b, B) = F_\beta[\varphi(b), \varphi(a), \ldots, \varphi(h)],$$

and since F increases in the first argument and decreases in the second, $\varphi(a) \geqslant \varphi(b)$. Let $C = \{a, b, \ldots, f\}$ be any subset of A that includes both a and b. Hence, by the monotonicity properties of F

$$P(a, C) = F_\gamma[\varphi(a), \varphi(b), \ldots, \varphi(f)]$$
$$\geqslant F_\gamma[\varphi(b), \varphi(a), \ldots, \varphi(f)]$$
$$= P(b, C).$$

It follows from simple scalability, therefore, that the ordering $P(a, B) \geqslant P(b, B)$ is independent of the offered set B. That is,

$$P(a, B) \geqslant P(b, B) \quad \text{iff} \quad P(a, C) \geqslant P(b, C),$$

provided none of these values is 0 or 1.

To derive another consequence of simple scalability, suppose $a, b \notin C$, $c \in C$, and $P(c, C \cup \{a\}) \geqslant P(c, C \cup \{b\})$. Hence, by simple scalability,

$$F_{\gamma+1}[\varphi(c), \ldots, \varphi(a)] = P(c, C \cup \{a\}) \geqslant P(c, C \cup \{b\})$$
$$= F_{\gamma+1}[\varphi(c), \ldots, \varphi(b)],$$

and since F decreases in its last argument, $\varphi(a) \leqslant \varphi(b)$. Hence, the ordering $P(c, C \cup \{a\}) \geqslant P(c, C \cup \{b\})$ is independent of both c and C. That is, for $a, b \notin C, D$,

$$P(c, C \cup \{a\}) \geqslant P(c, C \cup \{b\}) \quad \text{iff}$$
$$P(d, D \cup \{a\}) \geqslant P(d, D \cup \{b\})$$

provided none of these values is 0 or 1.

The following consequence of simple scalability combines the two conditions derived above.

DEFINITION 9. *A closed structure of choice probabilities satisfies* order-independence *iff for all* $a, b \in B - C$ *and* $c \in C$

$$P(a, B) \geqslant P(b, B) \text{ iff } P(c, C \cup \{a\}) \leqslant P(c, C \cup \{b\}),$$

provided the choice probabilities on the two sides of either inequality are not both 0 or 1.

Order-independence assumes that the ordering of the alternatives is independent of context. Specifically, a can be said to be at least as good as b whenever $P(a, B) \geqslant P(b, B)$ or, alternatively, whenever $P(c, C \cup \{a\})$ $\leqslant P(c, C \cup \{b\})$, $a, b \in B - C$, $c \in C$. Order-independence asserts that the two orderings coincide. This is clearly a necessary condition for simple scalability. The following theorem shows that it is also sufficient.

THEOREM 5. *A closed structure of choice probabilities satisfies order-independence (Definition 9) iff simple scalability (Definition 8) is valid. Furthermore, under order-independence, φ is an ordinal scale.*

This result was established in Tversky (1972a); it extends a previous result of Krantz (1964). The significance of Theorem 5 is that it provides a necessary and sufficient condition for simple scalability that is directly testable in terms of choice probabilities. The proof of the theorem is presented in Section 17.5.1. Some limitations of simple scalability as a model for individual choice behavior are discussed in Section 17.4.3.

In the binary case, order-independence takes a simpler form, and the representation theorem in this case can be stated as follows.

COROLLARY. *Let $\langle A, M, P \rangle$ be a complete structure of pair comparison. If for all $a, b, c \in A$,*

$$P(a, b) \geqslant \tfrac{1}{2} \quad iff \quad P(a, c) \geqslant P(b, c),$$

then and only then there exist a real-valued function φ on A and a real-valued function F in two real arguments such that for all $a, b \in A$,

$$P(a, b) = F[\varphi(a), \varphi(b)],$$

where F is strictly increasing in its first argument and strictly decreasing in the second.

Despite its generality, simple scalability does not provide an adequate basis for probabilistic theories of individual choice behavior. Theoretical considerations and empirical studies have demonstrated that order-independence (and hence simple scalability) does not always hold and that it is systematically violated in certain contexts. It is instructive to examine here some of these arguments. A more detailed review of the theoretical problems and the experimental evidence is presented in Tversky (1972b).

Problem 1. The case of the similar alternatives (Debreu, 1960). Let $A = \{a, b, c\}$, and suppose that a and b are very similar to each other whereas c is very different from both of them. For example, a and b are two recordings of the same symphony and c is a record of a very different kind. Assume that all binary choice probabilities equal one-half. It follows from simple scalability that all trinary choice probabilities equal one-third. This conclusion, however, is unacceptable. It seems unlikely that the choice between b and c is greatly affected by the addition of the very similar alternative a. Instead, one would expect $P(c, T) = \frac{1}{2}$ and $P(a, T) = P(b, T) = \frac{1}{4}$. That is, in the presence of c, a and b are treated as one alternative, contrary to order-independence. Experimental evidence for this effect was obtained by Becker *et al.* (1963b).

Problem 2. The case of the dominated alternative (Savage; see Luce and Suppes, 1965, pp. 334–335). Let a denote the option of receiving $100 or nothing depending on the toss of a fair coin; let b denote the option of receiving $99 or nothing depending on the toss of a fair coin, and let c denote the option of receiving $35 outright. Suppose one is indifferent between b and c, so that $P(b, c) = \frac{1}{2}$. Since a dominates b, $P(a, b) = 1$. By order-independence, therefore, $P(a, c) = P(a, b) = 1$. This conclusion, however, is counter-intuitive because it is unlikely that the conflict between the gamble b and the sure thing c will be completely resolved by the addition of a single dollar. Rather, one would expect that $\frac{1}{2} < P(a, c) < 1$. Empirical data (e.g., Tversky and Russo, 1969) support this hypothesis.

Perhaps the key property underlying simple scalability is the independence of the (derived) preference order. That is, if two alternatives are equivalent in one context (e.g., $P(a, b) = \frac{1}{2}$), then they are equivalent in any context (e.g., $P(a, A \cup \{a\}) = P(b, A \cup \{b\})$, $a, b \notin A$). The preceding examples show that this property does not hold in general although it may hold in many circumstances. Further theoretical developments based on simple scalability are described in the following subsection. A more general theoretical framework that does not assume order-independence is discussed in Section 17.8 where the above counterexamples are reexamined.

17.4.2 The Strict-Utility Model

Although simple scalability can be characterized purely in terms of choice probabilities, it is not fully satisfactory from a scaling standpoint because the utility scale φ is only ordinal, and the composition rule F is not specified explicitly. Indeed, most research in the field has focused on specific functional forms. Perhaps the simplest representation of choice probabilities is the strict-utility model, defined as follows.

DEFINITION 10. *A closed structure of choice probabilities satisfies the strict-utility model iff there exists a positive real-valued function φ on A such that for all $a \in B \subseteq A$,*

$$P(a, B) = \frac{\varphi(a)}{\sum_{b \in B}\varphi(b)}.$$

The strict-utility model has been studied from both psychological and statistical standpoints. The binary form of this model, i.e., $P(a, b) = \varphi(a)/[\varphi(a) + \varphi(b)]$ was employed to represent "response strength" in the early learning literature (Gulliksen, 1953; Thurstone, 1930); it was proposed as a method for scaling chess-playing ability (Zermelo, 1929); and it has been studied in the statistical literature (see, e.g., Bradley, 1954a, 1954b, 1965; Bradley and Terry, 1952; David, 1963) in connection with the analysis of pair-comparison data. A detailed theoretical analysis of the general form of the strict-utility model was developed by Luce (1959) who presented an axiomatic treatment of the model, derived many of its testable consequences, and applied it to the study of preference, learning, and psychophysics.

The strict-utility model represents the choice probability $P(a, B)$ as the ratio of the subjective value, or utility, of a to the sum of the utilities of all elements of B. One limitation of this form, however, is that it is not very well suited to handle choice probabilities that are either 0 or 1. (This difficulty cannot be properly resolved by extending the range of φ to include 0).

It follows from Definition 10 that φ is a ratio scale. To demonstrate, suppose there exists another function ψ satisfying the same relation. Hence, for all $a, b \in B \subseteq A$,

$$\frac{P(a, B)}{P(b, B)} = \frac{\varphi(a)}{\varphi(b)} = \frac{\psi(a)}{\psi(b)} \qquad \text{provided} \qquad P(b, B) \neq 0,$$

and the scale is unique up to multiplication by a positive constant. The equation also shows that, under the strict-utility model, the ratio on the left is independent of the offered set B. This property is called the constant-ratio rule (Clarke, 1957; Luce, 1959). It is formally defined as follows.

DEFINITION 11. *A closed structure of choice probabilities satisfies the* constant-ratio rule *iff for all* $a, b \in B \subseteq A$,

$$\frac{P(a, b)}{P(b, a)} = \frac{P(a, B)}{P(b, B)},$$

provided the denominators do not vanish.

The constant-ratio rule asserts, in effect, that the "strength of preference" of a over b is unaffected by the other available alternatives. Thus it can be viewed as a probabilistic analogue of the principle of independence from irrelevant alternatives (Luce and Raiffa, 1957). Note that the constant-ratio rule is considerably stronger than order-independence, according to which only the ordering of $P(a, B)$ and $P(b, B)$—not their ratio—is independent of the offered set B.

If no choice probabilities are 0 or 1, the constant-ratio rule is equivalent to the strict-utility model. To verify this assertion, define $\varphi(a) = P(a, A)$ for all $a \in A$. By the constant-ratio rule, therefore,

$$\frac{P(a, B)}{P(b, B)} = \frac{P(a, A)}{P(b, A)} = \frac{\varphi(a)}{\varphi(b)},$$

for all $a, b \in B \subseteq A$. Consequently,

$$\frac{1}{P(a, B)} = \frac{\sum_{b \in B} P(b, B)}{P(a, B)}$$

$$= \frac{\sum_{b \in B} \varphi(b)}{\varphi(a)},$$

and

$$P(a, B) = \frac{\varphi(a)}{\sum_{b \in B} \varphi(b)}$$

Luce (1959) has formulated another principle, called the *choice axiom*, that is also equivalent to the strict-utility model when all discriminations are imperfect. To introduce Luce's axiom, define, for any $C \subseteq B \subseteq A$, $P(C, B) = \sum_{c \in C} P(c, B)$. That is, $P(C, B)$ is the probability that, given an offered set B, the selected alternative will be included in $C \subseteq B$.

DEFINITION 12.   *A closed structure of choice probabilities, with $P(B, A)$*
$\neq 0$ *for all $B \subseteq A$, satisfies the* choice axiom *iff for all $C \subseteq B \subseteq A$*

$$P(C, A) = P(C, B)P(B, A).$$

Luce's choice axiom expresses in yet another form the basic principle
underlying the strict-utility model. It asserts, in effect, that the choice
process leading to the selection (of an element) of C from the total set A
can be decomposed into independent choices—the choice of C from B and
the choice of B from A. To establish the necessity of the choice axiom, note
that under the strict-utility model, with $C \subseteq B \subseteq A$,

$$P(C, A) = \frac{\Sigma_{c \in C}\varphi(c)}{\Sigma_{a \in A}\varphi(a)}$$
$$= \frac{(\Sigma_{c \in C}\varphi(c))(\Sigma_{b \in B}\varphi(b))}{(\Sigma_{b \in B}\varphi(b))(\Sigma_{a \in A}\varphi(a))}$$
$$= P(C, B)P(B, A).$$

To establish sufficiency, define $\varphi(a) = P(a, A)$; hence, by the choice
axiom, with $C = \{a\}$,

$$P(a, B) = \frac{P(a, A)}{P(B, A)}$$
$$= \frac{\varphi(a)}{\Sigma_{b \in B}\varphi(b)}.$$

It is seen that φ is unique up to multiplication by a positive constant. Thus
we have obtained the following theorem.

THEOREM 6.   *Suppose $\langle A, M, P \rangle$ is a closed structure of choice probabilities where $P(a, B) \neq 0, 1$ for all $a \in B \subseteq A$. Then the strict-utility model
(Definition 10) is satisfied iff either one of the following conditions holds:*

(i)   *The* constant-ratio rule *(Definition 11).*

(ii)   *The* choice axiom *(Definition 12).*

The restriction to nonzero choice probabilities may seem innocuous, yet
in a practical sense, it often fails to hold. In psychophysical experiments,
for example, data collection tends to concentrate on pair comparison
probabilities roughly between 0.1 and 0.9; in practice, the two stimuli from
any such pair often span only a narrow interval compared to the total range

of possible stimuli. Collecting appreciable amounts of data for stimuli much more widely separated is often inappropriate. So for choice probabilities $P(a, B)$ in the neighborhood of 0, it does not matter whether they are "really" 0 or only very small; the data are in practice apt to be sparse or missing. Thus, the construction of the strict-utility representation must be pieced together from choice probabilities that are appreciably different from 0 and 1. Luce (1959) showed how to carry out such a construction. His result is formulated in Theorem 7.

THEOREM 7.   *Suppose* $\langle A, M, P \rangle$ *is a closed structure of choice probabilities and that the following four axioms are satisfied*:

1.   *The choice axiom* (*Definition* 12) *holds in every substructure for which all discriminations are imperfect*; *that is, if* $A' \subset A$ *is such that* $P(B, A') > 0$ *for every* $B \subset A$, *then for* $C \subset B \subset A'$, $P(C, A') = P(C, B)P(B, A')$.

2.   *If* $P(c, b) = 0$, *then* c *can be dropped from choice sets containing* b, *i.e., if* $b \in B$, *then for every* $C \subset B$, $P(C, B) = P(C - \{c\}, B - \{c\})$.

3.   *Strong stochastic transitivity* (*Defintion* 3) *holds*.

4.   *Any pair of elements of* A *can be connected by a finite sequence in* A *for which each pair of successive elements is discriminated imperfectly*.

Then there exists a ratio scale φ *on* A *that gives a strict-utility representation* (*Definition* 10) *for every* B *such that* $P(b, B) > 0$ *for each* $b \in B$.

Axioms 1 and 2 in Theorem 7 are essentially the two parts of Luce's original choice axiom. Axiom 3 is needed only for the case where one or both of $P(a, b)$ and $P(b, c)$ equals 1, since SST for imperfect discriminations follows from Axiom 1. And something like Axiom 4 is obviously needed, otherwise the set A might consist of separate subsets, with perfect discrimination for pairs in different subsets, which thus would have unrelated numerical representations.

Under the assumptions of Theorem 7, all positive choice probabilities can be constructed from the positive binary choice probabilities. This fact can easily be seen in the case of a subset B to which the strict-utility representation applies: just divide numerator and denominator of the representation in Definition 10 by $\varphi(a)$, and note that, under the assumptions, $\varphi(b)/\varphi(a)$ is equal to $P(b, a)/P(a, b)$. But the fact is true more generally, as seen in the proof of Theorem 7 (sketched in Section 17.5.2).

Since all scale values can be computed from binary choices and since such data are considerably easier to collect and analyze, much of the research concerned with the strict-utility model has concentrated on the binary case, to which we now turn.

THEOREM 8. *Suppose* $\langle A, M, P \rangle$ *is a closed structure of choice probabilities where* $P(a, b) \neq 0, 1$ *for all* $a, b \in A$. *Then the following statements are equivalent:*

(i) *There exists a scale* φ *on* A *such that*

$$P(a, b) = \frac{\varphi(a)}{\varphi(a) + \varphi(b)}$$

for all $a, b \in A$.

(ii) $P(a, b)P(b, c)P(c, a) = P(a, c)P(c, b)P(b, a)$ *for all* $a, b, c \in A$.

The proof of Theorem 8 is left as an exercise. It provides a necessary and sufficient condition for the binary form of the strict-utility model. This condition, called the *product rule* (see Luce and Suppes, 1965), can be interpreted as follows. Suppose a subject is making binary choices from all pairs of elements of $A = \{a, b, c\}$, and suppose further that these choices are statistically independent. The left side of the product rule, then, is the probability of the intransitive chain $a \succ b \succ c \succ a$, and the right side of the equation is the probability of the intransitive chain $a \succ c \succ b \succ a$. The product rule asserts, in effect, that there are no reasons for an intransitive chain in one direction to be more probable than an intransitive chain in the opposite direction. This assumption contrasts sharply with the lexicographic semiorder (see Section 17.2.5), in which the probability of an intransitive chain depends critically on its direction.

Finally, note that the binary strict-utility model is a strong-utility model in the sense of Section 17.2.1. To verify, suppose all binary discriminations are imperfect; hence, for all $a, b \in A$,

$$\begin{aligned}
P(a, b) &= \frac{\varphi(a)}{\varphi(a) + \varphi(b)} \\
&= \frac{1}{1 + \varphi(b)/\varphi(a)} \\
&= \frac{1}{1 + \exp\{-[\ln \varphi(a) - \ln \varphi(b)]\}} \\
&= F[\psi(a) - \psi(b)],
\end{aligned}$$

where $F(x) = 1/(1 + e^{-x})$ is the logistic distribution function, and $\psi = \ln \varphi$.

Further developments of the strict-utility model can be found in Luce (1959). Statistical treatments of the model, including the optimal procedure for estimating scale values from binary choice probabilities, are discussed by Abelson and Bradley (1954), Bradley (1954a, 1954b, 1965), Bradley and

Terry (1952), and Ford (1957). For a comprehensive review, see Luce (1977). An extension of the model to an ordered response scale is discussed by Tutz (1986). Much of the recent work on the strict-utility model focuses on its random-utility representation that is discussed in Section 17.6.2.

17.5 PROOFS

17.5.1 Theorem 5 (p. 412).

Simple scalability (Definition 8) holds iff order-independence (Definition 9) is satisfied. Furthermore, φ is an ordinal scale.

PROOF. The necessity of order-independence is straightforward and is left as Exercise 4. To prove sufficiency, we first show that order-independence implies the following *substitutability* condition for all $a, b \notin C$.

$$P(a, b) \geq \tfrac{1}{2} \quad \text{iff} \quad P(a, C \cup \{a\}) \geq P(b, C \cup \{b\})$$

provided these probabilities are not both 0 or 1. By order-independence, with $B = \{a, b\}$, we obtain $P(a, b) \geq \tfrac{1}{2}$ iff for all $c \in C$, $P(c, C \cup \{a\}) \leq P(c, C \cup \{b\})$ iff

$$\sum_{c \in C} P(c, C \cup \{a\}) = 1 - P(a, C \cup \{a\})$$
$$\leq 1 - P(b, C \cup \{b\})$$
$$= \sum_{c \in C} P(c, C \cup \{b\})$$

iff $P(a, C \cup \{a\}) \geq P(b, C \cup \{b\})$ as required.

Define $\succsim$ on $A \times A$ by $a \succsim b$ iff $P(a, b) \geq \tfrac{1}{2}$. To establish the transitivity of $\succsim$ note that, by substitutability with $C = \{c\}$, $P(a, b) \geq \tfrac{1}{2}$ and $P(b, c) \geq \tfrac{1}{2}$ imply $P(a, c) \geq P(b, c) \geq \tfrac{1}{2}$. Define a scale φ on A by letting $\varphi(a), a \in A$, be the rank of a according to $\succsim$; thus, $\varphi(a) \geq \varphi(b)$ iff $a \succsim b$ iff $P(a, b) \geq \tfrac{1}{2}$. Next for any $\beta \in I^+$ satisfying $2 \leq \beta \leq \alpha$, define a real-valued function F_β in β real arguments by

$$F_\beta[\varphi(a), \varphi(b), \ldots, \varphi(h)] = P(a, B),$$

where $B = \{a, b, \ldots, h\}$. To show that F_β is well defined, suppose $B' = \{a', b', \ldots, h'\}$ and $\varphi(u) = \varphi(u')$, i.e., $P(u, u') = \tfrac{1}{2}$, for all $u \in B$ and $u' \in B'$. Letting $C = B - \{a\}$, it follows from substitutability that

$P(a, a') = \frac{1}{2}$ iff $P(a, B) = P(a, C \cup \{a\}) = P(a', C \cup \{a'\})$. Further-more, by order-independence, $P(b, b') = \frac{1}{2}$ iff $P(c, C \cup \{b\}) = P(c, C \cup \{b'\})$ for all $b, b' \notin C$, and $c \in C$. Consequently,

$$P(a', B') = P(a', \{a', b', \ldots, h'\})$$
$$= P(a', \{a', b', \ldots, h\}) \qquad \text{by order-independence}$$
$$\vdots$$
$$= P(a', \{a', b, \ldots, h\}) \qquad \text{by order-independence}$$
$$= P(a', C \cup \{a'\}) \qquad \text{by definition}$$
$$= P(a, B) \qquad \text{by substitutability,}$$

and the F_β, $2 \leqslant \beta \leqslant \alpha$, are well defined.

To establish the required monotonicity conditions, assume $\varphi(a) \geqslant \varphi(b)$, i.e., $P(a, b) \geqslant \frac{1}{2}$. Under order-independence, therefore, the following hold for any $C \subseteq A$ and for all $a, b \notin C$ and $c \in C$.

$$P(a, C \cup \{a\}) = F_{\gamma+1}[\varphi(a), \ldots, \varphi(c)]$$
$$\geqslant F_{\gamma+1}[\varphi(b), \ldots, \varphi(c)]$$
$$= P(b, C \cup \{b\}),$$
$$P(c, C \cup \{a\}) = F_{\gamma+1}[\varphi(c), \ldots, \varphi(a)]$$
$$\leqslant F_{\gamma+1}[\varphi(c), \ldots, \varphi(b)]$$
$$= P(c, C \cup \{b\}),$$

provided the probabilities on the two sides of either inequality are not both 0 or 1. Consequently, each F_β, $2 \leqslant \beta \leqslant \alpha$, is strictly increasing in the first argument and strictly decreasing in the other $\beta - 1$ arguments for all choice probabilities that differ from 0 or 1. It follows from order-independence that for choice probabilities that equal 0 or 1, F_β is nondecreasing in the first argument and nonincreasing in the other $\beta - 1$ arguments. It is evident from the development that φ is unique up to an order-preserving transformation, and that for any choice of φ, the F_β, $2 \leqslant \beta \leqslant \alpha$, are uniquely determined. $\diamondsuit$

17.5.2 Theorem 7 (p. 417)

Let $\langle A, M, P \rangle$ *be a closed structure of choice probabilities, If the choice axiom holds for any substructure with imperfect discrimination, each pair of elements can be linked by a finite chain of imperfectly discriminated pairs, SST holds, and* $P(c, b) = 0$ *implies that c can be dropped from choice sets containing b, then there exists a ratio scale that provides a strict utility representation for every substructure with imperfect discrimination.*

PROOF. Let a_i, $i = 0, \ldots, m$, be any sequence of elements such that successive pairs are imperfectly discriminated, i.e., $P(a_{i-1}, a_i) \neq 0, 1$. The representation requires that $P(a_i, a_{i-1})/P(a_{i-1}, a_i) = \varphi(a_i)/\varphi(a_{i-1})$. Multiplying these equations together yields

$$\frac{\varphi(a_m)}{\varphi(a_0)} = \prod_{i=1}^{m} \frac{P(a_i, a_{i-1})}{P(a_{i-1}, a_i)}.$$

This shows that ratios of φ values can be calculated uniquely from pairwise choice probabilities.

The rest of the proof turns on the fact that the product on the right of the above equation depends only on the endpoints, a_0, a_m, not on the length m or composition of the sequence. This is not hard to show (see Luce, 1959, p. 26), using Axiom 1 (but only for substructures of 3 elements), with Axioms 2 and 3 employed to assure that choice probabilities are nonzero where this is needed. ◇

17.6 RANDOM VARIABLE REPRESENTATIONS

17.6.1 The Random-Utility Model

Following Thurstone's (1927a, 1927b) notion of a discriminal process, it appears natural to represent the elements of a choice set as random variables rather than as constants (see Section 17.1.2). In this approach the subjective value of each alternative fluctuates, and the alternative with the highest momentary value is selected. Choice models of this type are called *random-utility models* although they are not restricted to preference data.

DEFINITION 13. *A structure of choice probabilities satisfies* a random-utility model *iff there exists a collection* $\mathbf{U} = \{\mathbf{U}_a | a \in A\}$ *of jointly distributed random variables, such that for* $a \in B$

$$P(a, B) = Pr(\mathbf{U}_a = \max\{\mathbf{U}_b | b \in B\}).$$

A random-utility model is independent *iff the elements of* $\mathbf{U}$ *are independent random variables. It is* continuous *iff each* $\mathbf{U}_a$, $a \in A$, *has a continuous distribution function. It is* finite *iff A is finite.*

Note that according to the preceding definition,

$$1 = P(a, b) + P(b, a) = Pr(\mathbf{U}_a \geqslant \mathbf{U}_b) + Pr(\mathbf{U}_b \geqslant \mathbf{U}_a)$$
$$= Pr(\mathbf{U}_a > \mathbf{U}_b) + Pr(\mathbf{U}_a = \mathbf{U}_b) + Pr(\mathbf{U}_b > \mathbf{U}_a) + Pr(\mathbf{U}_b = \mathbf{U}_a)$$
$$= Pr(\mathbf{U}_a \geqslant \mathbf{U}_b) + Pr(\mathbf{U}_b > \mathbf{U}_a);$$

hence, $Pr(\mathbf{U}_a = \mathbf{U}_b)$ must equal zero for $a \neq b$. A more general formulation that does not impose this constraint is discussed in Section 17.8.1.

The first characterization of the random-utility model was established by Block and Marschak (1960) using the notion of ranking. To state their result: let R be the set of all rankings of the elements of A, and let $R(a, B)$, $a \in B \subseteq A$, be the set of all rankings of A in which a is ranked above all other elements of B.

THEOREM 9. *A closed structure of choice probability satisfies a random-utility model* (Definition 13) *iff there exists a probability measure Q on R such that for all $a \in B \subseteq A$,*

$$P(a, B) = \sum_{r \in R(a, B)} Q(r).$$

A proof of Theorem 9 based on Block and Marschak (1960) is presented in Section 17.7.1. It provides a procedure for defining a random-utility model and a method for proving that a particular model is a random-utility model. However, it does not offer a testable procedure for checking whether the desired representation exists.

As noted by Block and Marschak (1960), the random-utility model implies the following observable condition. Let E_i denote the event $\mathbf{U}_a = \max\{\mathbf{U}_b | b \in B_i\}$; hence,

$$P(a, B_0 \cup B_1 \cup \cdots \cup B_n) = Pr\left(\mathbf{U}_a = \max\left\{\mathbf{U}_b | b \in \bigcup_{i=0}^{n} B_i\right\}\right)$$
$$= Pr(E_0 \cap E_1 \cap \cdots \cap E_n).$$

But since any probability measure satisfies

$$Pr(E_0) - \sum_i Pr(E_0 \cap E_i) + \sum_{i<j} Pr(E_0 \cap E_i \cap E_j)$$
$$- \cdots \pm Pr(E_0 \cap E_1 \cap \cdots \cap E_n) \geqslant 0,$$

for any collection $E_0, E_1, \ldots, E_n$ of events, we obtain Definition 14.

DEFINITION 14. *A closed structure of choice probabilities satisfies* non-negativity *iff for any* $a \in A$ *and* $B_0, B_1, \ldots, B_n \subset A$,

$$P(a, B_0) - \sum_i P(a, B_0 \cup B_i) + \sum_{i<j} P(a, B_0 \cup B_i \cup B_j)$$
$$- \cdots \pm P(a, B_0 \cup B_1 \cdots B_n) \geq 0.$$

The nonnegativity condition is due to Block and Marschak (1960) who derived it from Definition 13 and showed that it is equivalent to the random-utility model whenever A contains four or fewer elements. Note that for $n = 1$, nonnegativity reduces to $P(a, B) \geq P(a, B \cup C)$. This property, called *regularity*, asserts that choice probability cannot be increased by enlarging the offered set. The following representation theorem, due to Falmagne (1978), shows that nonnegativity is both necessary and sufficient for the representation of choice probabilities by a random-utility model.

THEOREM 10. *A closed structure of choice probabilities satisfies a random-utility model* (Definition 13) *iff it satisfies nonnegativity* (Definition 14).

It is very easy to see that the representation of a finite structure of choice probability by a random-utility model is highly nonunique. If $\{U_a | a \in A\}$ is a random-utility representation of $\langle A, M, P \rangle$, so is $\{F(U_a) | a \in A\}$, provided F is a strictly increasing function. Falmagne (1978) investigated the uniqueness of the random-utility model and showed, among other things, that it is "almost" an ordinal scale.

THEOREM 11. *If* U *and* V *are both random-utility models of the same structure then for all* $a, b, c \in A$, $Pr(U_a > U_b > U_c) = Pr(V_a > V_b > V_c)$.

For proofs of Theorems 10 and 11, the reader is referred to Falmagne (1978).

The conditions under which a pair-comparison structure can be represented by a random-utility model are not known at present. Marschak (1960) showed that the triangle condition, $P(a, b) + P(b, c) + P(c, a) \leq 2$, is necessary and sufficient for sets of five elements or less. But Cohen and Falmagne (1978; see also McFadden & Richter, 1971) showed that it is not sufficient in general and established the necessity of an additional condition. Cohen (1980) showed that Falmagne's result (Theorem 10) holds when the choice set A is countable; but when A has a higher cardinality, further assumptions are required to assure a countably additive random-utility representation. For further discussion of random-utility models, see Colonius (1984), McFadden (1981), Manski (1977), Strauss (1979), and Yellott (1977).

17.6.2 The Independent Double-Exponential Model

The generality and the lack of uniqueness associated with the random-utility model has led investigators to explore more-restrictive representations in which $U_a, U_b, U_c, \ldots$ are independent random variables. Although we do not know how to characterize this representation in terms of choice probabilities, Sattath and Tversky (1976) showed that the continuous independent random-utility model implies a testable property called the *multiplicative inequality*.

THEOREM 12.   *In a continuous independent random-utility model*

$$P(a, B \cup C) \geqslant P(a, B)P(a, C), \qquad a \in B, C \subset A.$$

This theorem, proved in Section 17.7.2, shows that the probability of choosing a from $B \cup C$ is at least as large as the probability of choosing a from both B and C on two independent choices. Combining this result with regularity, derived in the previous section, yields the following upper and lower bounds for $P(a, B \cup C)$:

$$\min[P(a, B), P(a, C)] \geqslant P(a, B \cup C) \geqslant P(a, B)P(a, C).$$

In fact, most theoretical and empirical effort has focused on two continuous independent models: the normal and the double exponential. The normality assumption, originally introduced by Thurstone, has been widely employed (see, e.g., Bock and Jones, 1968; Torgerson, 1958) because of its theoretical appeal and computational convenience. More recent work, however, associated with Luce's (1959) strict-utility model, gave rise to an alternative model based on the double-exponential distribution.

The question naturally arises whether the strict-utility model (Definition 10) is also a random-utility model. A positive answer was given by Block and Marschak (1960, p. 110) using the ranking method of Theorem 9. A more direct and informative proof, however, was obtained by Holman and Marley (see Luce and Suppes, 1965, p. 338), who established the following result.

Suppose $\langle A, M, P \rangle$ satisfies the strict-utility model, and let $\varphi(a) = P(a, A)$. Define an independent random-utility model by

$$Pr(U_a = t) = \begin{cases} \varphi(a)e^{\varphi(a)t} & \text{if } t \leqslant 0 \\ 0 & \text{otherwise} \end{cases}$$

Thus,

$$
\begin{aligned}
p(a, B) &= \frac{\varphi(a)}{\sum_{b \in B} \varphi(b)} \\
&= \int_{-\infty}^{0} \varphi(a) \exp\left[\sum_{b \in B} \varphi(b)t \right] dt \\
&= \int_{-\infty}^{0} \varphi(a) e^{\varphi(a)t} \prod_{b \in B - \{a\}} e^{\varphi(b)t} \, dt \\
&= \int_{-\infty}^{0} \varphi(a) e^{\varphi(a)t} \prod_{b \in B - \{a\}} \left[\int_{-\infty}^{t} \varphi(b) e^{\varphi(b)\tau} \, d\tau \right] dt \\
&= \int_{-\infty}^{\infty} Pr(\mathbf{U}_a = t) \prod_{b \in B - \{a\}} Pr(\mathbf{U}_b \leqslant t) \, dt \\
&= Pr(\mathbf{U}_a = \max\{\mathbf{U}_b | b \in B\}).
\end{aligned}
$$

The strict-utility model, therefore, is expressible as an independent random-utility model with a double-exponential distribution $Pr(\mathbf{U}_a \leqslant u) = \exp(-e^{u - \varphi(a)})$. (If $\mathbf{Y}$ is an exponential random variable with density e^{-y}, then the random variable $\mathbf{X} = -\log \mathbf{Y}$ has the double-exponential distribution).

We turn now to discuss the uniqueness of the double-exponential representation. Adams and Messick (1957) showed, for a pair-comparison structure, that the binary strict-utility model $P(a, b) = \varphi(a)/[\varphi(a) + \varphi(b)]$ is expressible as an independent random-utility model iff the difference variate $\mathbf{V}_{ab} = \mathbf{U}_a - \mathbf{U}_b$ has a logistic distribution $Pr(\mathbf{V}_{ab} \geqslant t) = 1/(1 + e^{-t})$. The double-exponential distribution, however, is not the only model whose difference distribution is logistic. Laha (1964) devised another distribution with the same property (see Yellott, 1977, pp. 130–131). However, Yellott showed that if M includes all pairs and triples of elements from A, then the double-exponential form is the only (generalized Thurstonian) model of the form $\mathbf{U}_a = \mathbf{U} + \mu_a$ that is equivalent to the strict-utility model. For the proof of this result, the reader is referred to Yellott's (1977) paper. A closely related result was proved independently by McFadden (1974).

Yellott (1977) has also discussed a testable property that gives rise to the double-exponential distribution. Consider a choice set $B = \{a, b, \dots\} \subset A$ consisting of n distinct alternatives, and let B^k, $k \in I^+$, denote the choice set consisting of k replicas of each of the elements of $B = B^1$. It stands to reason that the probability of choosing a replica of a from B^k equals $P(a, B)$. Yellott proved that the only generalized Thurstonian model satisfying this property (called k-invariance) is the double exponential. Hence, the conjunction of k-invariance and the nonnegativity condition (Definition

14) provides a characterization of the double-exponential model in terms of choice probabilities. Note in passing that although k-invariance is both an appealing and a plausible condition, the introduction of replicas or near replicas creates serious difficulties for an independent random-utility model. We return to this problem in Section 17.8.2. Further results concerning the double-exponential model can be found in Strauss (1979) and Dagsvik (1983).

It is instructive to compare Thurstone's normal model (Case V) to the strict-utility model in the simple case of binary comparisons among three alternatives. It follows that, in the binary strict-utility model,

$$P(a, c) = \frac{P(a, b)P(b, c)}{P(a, b)P(b, c) + P(c, b)P(b, a)},$$

whereas in Thurstone's case V,

$$P(a, c) = \Phi\{\Phi^{-1}[P(a, b)] + \Phi^{-1}[P(b, c)]\},$$

where Φ is the normal cumulative distribution function with mean 0 and variance 1. In a comparison of the two models, Burke and Zinnes (1965) showed that they coincide only for triples $[P(a, b), P(b, c), P(a, c)]$ of the form $(\frac{1}{2}, x, x)$, $(x, \frac{1}{2}, x)$, or $(x, 1 - x, \frac{1}{2})$. It is easy to show that these triples are shared by all models satisfying simple scalability. Nevertheless, the predictions of the two models are very close because of the similarity between the normal and the logistic distributions. Burke and Zinnes (1965) showed that when $P(a, b) \geqslant \frac{1}{2}$ and $P(b, c) \geqslant \frac{1}{2}$ the maximal discrepancy between the values of $P(a, c)$ predicted by the two models is 0.013. Larger discrepancies between the models exist when $P(a, b)$ is close to 0 when $P(b, c)$ is close to 1, but it is very difficult to obtain adequate estimates of choice probabilities in these regions. It appears quite difficult to choose between the models on the basis of a direct empirical test. For a discussion of some empirical data and the statistical issue of goodness-of-fit, see Burke and Zinnes (1965).

We conclude this section with an interesting development due to Thompson and Singh (1967), who showed that the normal and the logistic models are derivable from assumptions about the nature of the "sensory" process underlying the choice. Specifically, they assumed that whenever the subject encounters a stimulus, a (fixed) large number of signals are transmitted to the brain by the sense receptors. The individual signals are treated as independent random variables with a common distribution function, and are combined according to some rule to form the experienced sensation. Using the central-limit theorem, Thompson and Singh (1967) show that if the experienced sensation is the sum, the mean, or the median of the transmitted signals, then the experienced sensation has an asymptotic normal distribution, as postulated by Thurstone. If, on the other hand, the experienced sensation is the maximum of the transmitted signals, then it has an extreme-value distribution which, under reasonably general conditions on the distribution of individual signals, leads to the binary strict-utility model. For proofs and discussion of these results, the reader is referred to the original paper. Wandell and Luce (1978) have explored these ideas in detail and applied them to the analysis of psychophysical data. They have shown that the different decision rules (i.e., mean versus maximum) make distinctly different testable predictions if response times are taken into account. The results favored the normality assumption.

17.6.3 Error Tradeoff

So far, we have been concerned with the representation of probabilities obtained in choice experiments in which the subject selects on each trial a single alternative from an offered set. This section is concerned with the representation of the probabilities obtained in a series of binary identification experiments. To characterize these data, let A be a set of stimuli, e.g., tones, lights, or items stored in memory. In a binary identification experiment, the subject is presented on each trial with one stimulus from the pair (a, b), and the task is to identify the stimulus as the *high* or the *low* element of the pair with respect to the relevant dimension, e.g., sound intensity, brightness, or frequency. For notational convenience we let the ordering of the stimuli on the relevant dimension coincide with their alphabetic ordering. In the stimulus pair (a, b), for example, a denotes the low element and b the high element. The experiment is typically designed so that the subject cannot always identify correctly the *high* and *low* members of the pair. This can be accomplished by selecting highly similar stimuli or by presenting them in a manner that prevents perfect identification, e.g., for a short exposure or against a noisy background.

Let $H(b, a)$ denote the probability that a subject responds *high* when presented with stimulus b in an identification experiment confined to the pair (a, b). This probability is estimated by the proportion of trials on which the subject responded *high* when stimulus b was presented. Recall that *high* is the appropriate response to b. The results of an identification experiment are summarized by the two values $H(b, a)$ and $H(a, b)$ called the *hit rate* and the *false-alarm rate*, respectively. These terms originate from the theory of signal detection, in which b and a denote, respectively, the presence and the absence of a specified signal. Thus the hit rate $H(b, a)$ is the probability of detecting a signal when it is present, whereas the false-alarm rate $H(a, b)$ is the probability of "calling" a signal when none is present. Note that since on each trial the subject responds either *high* or *low*, the probability of *high* plus the probability of *low* add to unity for each of the two stimuli.

To complete the characterization of an identification experiment, one has to specify the *context*, i.e., the probabilities of presenting the two stimuli and the payoffs associated with the four combinations of stimuli and responses. Let J be the set of contexts under consideration, and let $H_j(a, b)$ and $H_j(b, a)$ denote the identification probabilities observed in context $j \in J$. By varying presentation probabilities and/or payoffs, one generates a series of (binary) identification experiments for the same pair of stimuli. The results of such a series are summarized by the *identification graph*

$$T_{ab} = \{(H_j(a, b), H_j(b, a)) | j \in J\}.$$

(This graph is often referred to as the ROC curve, a name from engineering that stands for *receiver-operating characteristic*.)

An identification graph is *strict* iff T_{ab} is a strictly increasing function of the unit interval $[0, 1]$ onto itself. Thus, if T_{ab} is strict, the hit rate $H(b, a)$ can be expressed as a continuously increasing function of the false-alarm rate $H(a, b)$. Formally, we write $(x, y) \in T_{ab}$ or $T_{ab}(x) = y$ iff there exists $j \in J$ such that $H_j(a, b) = x$ and $H_j(b, a) = y$. This assumption has two parts. First it asserts that the set of contexts J is sufficiently rich so that $\{H_j(a, b) | j \in J\}$ and $\{H_j(b, a) | j \in J\}$ are isomorphic to the unit interval. Since payoffs and presentation probabilities can be varied so that the identification probabilities range from 0 to 1 in an essentially continuous fashion, the above requirement appears to be an acceptable idealization of identification data.

Second, strictness asserts that the hit rate is a monotonically increasing function of the false-alarm rate. That is, for all $j, k \in J$,

$$H_j(a, b) \geqslant H_k(a, b) \quad \text{iff} \quad H_j(b, a) \geqslant H_k(b, a).$$

Note that a change of context, e.g., from j to k, does not affect the discriminability between the stimuli. By changing payoffs and/or presentation probabilities one manipulates, in effect, the subject's criterion for responding *high* rather than *low*. Thus any change of criterion that makes the subject more likely to respond *high* increases both $H(a, b)$ and $H(b, a)$. The above monoticity assumption underlies essentially all theories of signal detection, and it has received considerable empirical support (see, e.g., Green and Swets, 1966).

The following discussion, which is based primarily on the work of Marley (1971), concerns the representation of strict identification graphs by random variables.

DEFINITION 15. *A family of identification graphs* $\{T_{ab} | a, b \in A\}$ *is representable iff there exist random variables* U_a, $a \in A$, *and mappings* Γ: $A \times A \times J \to Re$, *with* $\Gamma_{ab} = \Gamma_{ba}$, *such that*

$$H_j(b, a) = Pr[U_b > \Gamma_{ab}(j)] \quad and \quad H_j(a, b) = Pr[U_a > \Gamma_{ab}(j)].$$

In a representable family of indentification graphs, each stimulus is represented by a random variable. In each identification task, the subject is assumed to select a criterion value $\Gamma_{ab}(j)$ that depends on the stimuli and the context and to respond *high* whenever the perceived value of the presented stimulus exceeds the selected criterion value.

It is seen that any single identification graph is representable. Definition 15, however, constrains the identification graphs whenever A has three or more elements. To demonstrate, consider a representable family of identification graphs, where the densities associated with the random variables U_a, U_b, U_c are displayed below. Let t be some fixed number, and let $x = Pr(U_a > t)$, $y = Pr(U_b > t)$, and $z = Pr(U_c > t)$. It is evident from the representation (see Figure 3) that if x and y are, respectively, the false-alarm rate and the hit rate for the pair (a, b), and y and z are, respectively, the false-alarm rate and the hit rate for the pair (b, c), then x and z must be the false-alarm rate and the hit rate, respectively, for the pair (a, c). This consequence of representability is formalized as follows.

DEFINITION 16. *A family of identification graphs satisfies* graph composition *iff* $(x, y) \in T_{ab}$ *and* $(y, z) \in T_{bc}$ *imply* $(x, z) \in T_{ac}$, *i.e.,* $T_{bc} \circ T_{ab} = T_{ac}$.

Graph composition asserts, in effect, that the identification graphs T_{ac} is determined by the graphs T_{ab} and T_{bc}; specifically, $T_{cb}[T_{ba}(x)] = T_{ca}(x)$.

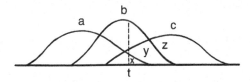

FIGURE 3. Graphic illustration showing that representability (Definition 14) implies graph composition (Definition 15).

The following theorem shows that graph composition is not only necessary but also sufficient for the representability of a family of strict identification graphs.

THEOREM 13. *A family of strict identification graphs is representable iff it satisfies graph composition. The representation is unique up to an order-preserving transformation.*

The proof of Theorem 13, based on Marley (1971), is presented in Section 17.7.3. Since he did not assume strictness, his results are somewhat more complicated and general than the present theorem. The reader is referred to his original paper for the discussion of several related results.

In light of Theorem 13 it seems that an experimental test of graph composition could be of considerable interest despite the difficulties in obtaining adequate empirical estimates of identification graphs. Note that Definition 15 does not impose any constraint on the distributions of the random variables that are associated with the stimuli. The following definition introduces a much stronger representation where these distributions are identical except for their location, i.e., where each $\mathbf{U}_a$ is expressable as $\mathbf{U} + \mu_a$.

DEFINITION 17. *A family of identification graphs* $\{T_{ab}|a, b \in A\}$ *is uniform iff there exist constants* μ_a, $a \in A$, *a random variable* $\mathbf{U}$, *and mappings* $\Gamma_{ab} : J \to \mathrm{Re}$, *with* $\Gamma_{ab} = \Gamma_{ba}$, *such that*

$$H_j(b, a) = Pr[\mathbf{U} + \mu_b > \Gamma_{ab}(j)] \qquad and$$
$$H_j(a, b) = Pr[\mathbf{U} + \mu_a > \Gamma_{ab}(j)].$$

Note that a uniform family of identification graphs is a special case of a uniform family of scales (Definition 6.9), which was investigated by Levine (1970, 1972) and is discussed in Section 6.7.

Next, we show that in a uniform family, all identification graphs are consistently ordered, i.e., the graphs do not intersect. Suppose $T_{ab}(x) \geq T_{cd}(x)$, i.e., $H_j(a, b) = H_k(c, d) = x$ and $H_j(b, a) \geq H_k(d, c)$ for some j, $k \in J$. Consequently,

$$Pr[\mathbf{U} > \Gamma_{ab}(j) - \mu_a] = Pr[\mathbf{U} > \Gamma_{cd}(k) - \mu_c],$$

and

$$\Gamma_{ab}(j) - \Gamma_{cd}(k) = \mu_a - \mu_c.$$

Similarly,

$$Pr[\mathbf{U} > \Gamma_{ab}(j) - \mu_b] \geqslant Pr[\mathbf{U} > \Gamma_{cd}(k) - \mu_d],$$

and

$$\Gamma_{ab}(j) - \Gamma_{cd}(k) \leqslant \mu_b - \mu_d.$$

Now suppose $H_i(a, b) = H_l(c, d) = y$ for some $i, l \in J$. Hence,

$$\Gamma_{ab}(i) - \Gamma_{cd}(l) = \mu_a - \mu_c \geqslant \mu_b - \mu_d,$$
$$Pr[\mathbf{U} > \Gamma_{ab}(i) - \mu_b] \leqslant Pr[\mathbf{U} > \Gamma_{cd}(l) - \mu_d],$$

and

$$H_i(b, a) = T_{ab}(y) \geqslant T_{cd}(y) = H_l(c, d).$$

Thus we derived the following condition.

DEFINITION 18.   *A family of indentification graphs is* uncrossed *iff for all* $a, b, c, d \in A$, *and* $x, y \in (0, 1)$,

$$T_{ab}(x) \geqslant T_{cd}(x) \qquad iff \qquad T_{ab}(y) \geqslant T_{cd}(y).$$

THEOREM 14.   *Let* $\{T_{ab} | a, b \in A\}$ *be a family of strict identification graphs such that for any* (x, y), $0 \leqslant x \leqslant y \leqslant 1$, *there exist* $a, b \in A$ *such that* $(x, y) \in T_{ab}$. *Then* $\{T_{ab} | a, b \in A\}$ *is uniform iff it is uncrossed. Furthermore, the representation is unique up to a positive linear transformation.*

Thus every uncrossed family of strict identification curves, which includes any point in the upper triangle $(x \leqslant y)$ of the unit square, is representable by random variables whose distributions differ by location only. Theorem 14 reduces (Exercise 10) to Theorem 6.7, which is based on Levine's (1970) work on uniform systems. Furthermore, the conditions for uniformity with a finite number of identification graphs follows readily from Theorem 6.9. Later work by Levine (1972) established a more general representation of a family of (identification) graphs, in which the stimuli are represented by random variables whose distributions are identical except for their means and variances.

17.7 PROOFS

17.7.1 Theorem 9 (p. 422)

A random-utility model is satisfied iff there exists a probability distribution Q, defined on the set R of all rankings of the elements of A, such that for all $a \in B \subseteq A$,

$$P(a, B) = \sum_{r \in R(a, B)} Q(r)$$

where $R(a, B)$ is the set of all rankings of A in which a is ranked above all other elements of $B \subseteq A$.

PROOF. Suppose, first, that the choice probabilities satisfy a random-utility model. Let $r = (a \succ b \succ \cdots \succ k)$ denote a typical element of R, and define a probability distribution on R by $Q(r) = Pr(U_a > U_b > \cdots > U_k)$. Hence,

$$P(a, B) = Pr(U_a > U_b \text{ for all } b \in B - \{a\})$$
$$= \sum_{r \in R(a, B)} Q(r).$$

Conversely, suppose this equation holds for all $a \in B \subseteq A$. First, define a one-to-one mapping H from A to the positive integers $\{1, 2, \ldots, \alpha\}$. Second, for each ranking r of A, let $r^* = (r_1^*, \ldots, r_\alpha^*)$ be the permutation of the integers $\{1, \ldots, \alpha\}$ in which r_β^*, $1 \leqslant \beta \leqslant \alpha$, is the rank of the alternative whose image under H is β according to the ranking r. Third, define a random vector U on A as follows. For any α-tuple $u = (u_1, \ldots, u_\alpha)$ of real numbers, let

$$Pr(U = u) = \begin{cases} Q(r) & \text{if } u = r^*, \\ 0 & \text{otherwise.} \end{cases}$$

Hence, for all $a \in B \subseteq A$,

$$P(a, B) = \sum_{r \in R(a, B)} Q(r)$$
$$= \sum_{\{r^* \mid r \in R(a, B)\}} Pr(U = r^*)$$
$$= Pr(U_a > U_b \text{ for all } b \in B - \{a\}). \qquad \Diamond$$

17.7.2 Theorem 12 (p. 424)

In an independent continuous random-utility model

$$P(a, B \cup C) \geqslant P(a, B)P(a, C), \quad \text{for all} \quad a \in B, C \subseteq A.$$

PROOF. Let f_a and F_a denote, respectively, the density and the distribution functions of U_a, $a \in A$. In an independent random-utility model, therefore,

$$P(a, B) = Pr(U_a > U_b \text{ for all } b \in B - \{a\})$$

$$= \int_{-\infty}^{\infty} f_a(t) \left[\prod_{b \in B - \{a\}} F_b(t) \right] dt.$$

Let $B(t) = \prod_{b \in B - \{a\}} F_b(t)$, and $a(t) = f_a(t)$. Thus,

$$P(a, B) = \int_{-\infty}^{\infty} a(t) B(t) \, dt.$$

Similarly,

$$P(a, C) = \int_{-\infty}^{\infty} a(t) C(t) \, dt.$$

On the other hand,

$$P(a, B \cup C) = Pr(U_a > U_d \text{ for all } d \in (B \cup C) - \{a\})$$

$$= \int_{-\infty}^{\infty} f_a(t) \left[\prod_{d \in (B \cup C) - \{a\}} F_d(t) \right] dt$$

$$\geqslant \int_{-\infty}^{\infty} f_a(t) \left[\prod_{b \in B - \{a\}} F_b(t) \right] \left[\prod_{c \in C - \{a\}} F_c(t) \right] dt$$

$$= \int_{-\infty}^{\infty} a(t) B(t) C(t) \, dt.$$

Thus it suffices to show that

$$\int_{-\infty}^{\infty} a(t) B(t) C(t) \, dt \geqslant \left[\int_{-\infty}^{\infty} a(t) B(t) \, dt \right] \left[\int_{-\infty}^{\infty} a(t) C(t) \, dt \right].$$

Let $\mu = \int_{-\infty}^{\infty} a(t)C(t)\, dt$. The above inequality can be expressed as

$$\int_{-\infty}^{\infty} a(t)B(t)[C(t) - \mu]\, dt \geq 0.$$

Since $C(t)$ is a product of cumulative density functions, it is continuous and nondecreasing. Let t_0 be a point for which $C(t_0) = \mu$. Thus we have to show that

$$\int_{-\infty}^{t_0} a(t)B(t)[C(t) - \mu]\, dt + \int_{t_0}^{\infty} a(t)B(t)[C(t) - \mu]\, dt \geq 0,$$

or, equivalently, that

$$\int_{t_0}^{\infty} a(t)B(t)[C(t) - \mu]\, dt \geq \int_{-\infty}^{t_0} a(t)B(t)[\mu - C(t)]\, dt.$$

But since

$$0 = \int_{-\infty}^{\infty} a(t)[C(t) - \mu]\, dt$$

$$= \int_{-\infty}^{t_0} a(t)[C(t) - \mu]\, dt + \int_{t_0}^{\infty} a(t)[C(t) - \mu]\, dt,$$

we have

$$\int_{t_0}^{\infty} a(t)[C(t) - \mu]\, dt = \int_{-\infty}^{t_0} a(t)[\mu - C(t)]\, dt.$$

Consequently,

$$\int_{t_0}^{\infty} a(t)B(t)[C(t) - \mu]\, dt \geq \int_{t_0}^{\infty} a(t)B(t_0)[C(t) - \mu]\, dt$$

$$= B(t_0)\int_{t_0}^{\infty} a(t)[C(t) - \mu]\, dt$$

$$= B(t_0)\int_{-\infty}^{t_0} a(t)[\mu - C(t)]\, dt$$

$$\geq \int_{-\infty}^{t_0} a(t)B(t)[\mu - C(t)]\, dt. \qquad \diamondsuit$$

17.7.3 Theorem 13 (p. 430)

A family of strict identification graphs is representable iff it satisfies graph composition. The representation is unique up to an order-preserving transformation.

PROOF. The necessity of graph composition is obvious. To establish sufficiency, select some fixed $e \in A$, and let U_a be a random variable with cumulative distribution function $F_a(x) = 1 - T_{ea}(1 - x)$, $0 \leqslant x \leqslant 1$. For any $a, b \in A$, $j \in J$, choose $x \in [0, 1]$ such that $H_j(a, b) = Pr(U_a > x) = 1 - F_a(x)$. This is possible because, by regularity, T_{ea} and hence F_a are onto $[0, 1]$. Consequently,

$$
\begin{aligned}
H_j(b, a) &= T_{ab}[H_j(a, b)] \\
&= T_{ab}[1 - F_a(x)] && \text{by construction} \\
&= T_{ab} \circ T_{ea}(1 - x) \\
&= T_{eb}(1 - x) && \text{by graph composition} \\
&= 1 - F_b(x) \\
&= Pr(U_b > x) && \text{by construction.}
\end{aligned}
$$

Setting $\Gamma_{ab}(j) = x$ completes the derivation.

Next, we show that the distribution functions of the representing random variables are unique up to a common monotone transformation. Suppose there exists another representation consisting of random variables V_a, $a \in A$, with cumulative distribution functions G_a and mappings Δ_{ab}: $J \to \text{Re}$, satisfying $H_j(b, a) = Pr[V_b > \Delta_{ab}(j)]$ and $H_j(a, b) = Pr[V_a > \Delta_{ab}(j)]$. We want to show that there exists a strictly increasing function φ such that $G_a = F_a \circ \varphi^{-1}$ and $\Delta = \varphi \circ \Gamma$. Let $\varphi = G_e^{-1}$. Hence,

$$
\begin{aligned}
1 - G_a[\Delta_{ea}(j)] &= H_j(a, e) \\
&= T_{ea}[H_j(e, a)] \\
&= 1 - F_a[1 - H_j(e, a)] \\
&= 1 - F_a\{G_e[\Delta_{ea}(j)]\} \\
&= 1 - F_a\{\varphi^{-1}[\Delta_{ea}(j)]\},
\end{aligned}
$$

and $G_a(x) = F_a[\varphi^{-1}(x)]$. Furthermore, since

$$
1 - F_a[\Gamma_{ab}(j)] = H_j(a, b) = 1 - G_a[\Delta_{ab}(j)],
$$
$$
F_a(\Gamma) = G_a(\Delta) = F_a(\varphi^{-1} \circ \Delta),
$$

and hence, $\Gamma = \varphi^{-1} \circ \Delta$, and $\Delta = \varphi \circ \Gamma$.   ◇

17.8 MARKOVIAN ELIMINATION PROCESSES

17.8.1 The General Model

The choice of a single alternative from an offered set can be viewed as a process of sequential elimination. According to this view the process of selection proceeds as follows. Given some offered set B, one selects a (nonempty) subset of B, say C, with probability $Q_B(C)$. Now the choice is restricted to C, from which one selects another (nonempty) subset D with probability $Q_C(D)$, and so on until the selected subset consists of a single alternative. The following definition (based on Tversky, 1972a) formalizes this process of sequential elimination.

DEFINITION 19.   *An elimination structure is a quadruple* $\langle A, M, P, Q \rangle$ *where* $\langle A, M, P \rangle$ *is a closed structure of choice probabilities, and* $Q = \{ Q_B | B \subseteq A \}$ *is the corresponding family of transition probability functions. That is, each* $Q_B \in Q$ *is a mapping from* 2^B *into* $[0, 1]$ *satisfying*

(i) $Q_B(B) = 1$ *iff* $B = \{ b \}$,

(ii) $\Sigma_{C_i \subseteq B} Q_B(C_i) = 1$, $C_i \neq \varnothing$, *and*

(iii) $P(a, B) = \Sigma_{C_i \subseteq B} Q_B(C_i) P(a, C_i)$.

Condition (i) is necessary to guarantee that the elimination process ends with the choice of a single alternative. Condition (ii) ensures that each Q_B is a proper (transition) probability distribution. Thus $Q_B(C)$ is interpreted as the probability of eliminating from B all alternatives that are not included in $C \subseteq B$. Condition (iii) states, in effect, that choice probabilities are the absorbing probabilities of the Markov chain.

Note that Definition 19 does not constrain choice probabilities since we can set

$$Q_B(C) = \begin{cases} P(a, B) & \text{if} \quad C = \{ a \} \\ 0 & \text{otherwise.} \end{cases}$$

From the standpoint of the elimination process, therefore, models of choice probabilities are characterized in terms of the constraints they impose on the transition probabilities. One example is the *discard model*, which was proposed by Luce (1960) and further investigated by Marley (1965). The discard model is characterized by the assumption that

$$Q_B(C) > 0 \quad \text{only if} \quad C = B - \{ b \}, b \in B.$$

Hence,

$$P(a, B) = \sum_{b \in B} Q_B(B - \{b\}) P(a, B - \{b\}).$$

Thus, in the discard model, exactly one alternative is eliminated at every stage. Although this model has some testable consequences, it imposes very little constraint on choice probabilities. In particular, it follows from regularity (i.e., $P(a, B \cup C) \leqslant P(a, B)$, see Marley, 1965) and it does not restrict the binary choice probabilities in any way.

Another approach to the study of elimination structures, advanced by Corbin and Marley (1974), assumes that there exists a random vector on A that generates the transition probabilities in the same way that it is assumed to generate the choice probabilities in the random-utility model.

DEFINITION 20. *An elimination structure $\langle A, M, P, Q \rangle$ satisfies a random-elimination model iff there exists a random vector $\mathbf{U}$ defined on A which satisfies*

   (i) $Pr(\mathbf{U}_a = \mathbf{U}_b) \neq 1$ *for all $a, b \in A$,*

   (ii) $Q_B(C) = Pr(\mathbf{U}_c = \mathbf{U}_d > \mathbf{U}_b$ *for all $c, d \in C$, $b \in B - C$).*

A random elimination model is Boolean *iff the components of $\mathbf{U}$ are all 0 or 1.*

Note that, unlike the random-utility model of Definition 13, in which $Pr(\mathbf{U}_a = \mathbf{U}_b) = 0$ for all $a, b \in A$, the random-elimination model allows for the possibility that $Pr(\mathbf{U}_a = \mathbf{U}_b) \in (0, 1)$. Nevertheless, Corbin and Marley (1974) have shown that the choice probabilities in a random-elimination model are compatible with a random-utility model. Their proof is based on the construction of a probability distribution over the set of rankings of A that satisfies the conditions of Theorem 9.

17.8.2 Elimination by Aspects

In this section we develop a stronger elimination model that can be interpreted as a process of elimination-by-aspects (EBA) (Tversky, 1972a, 1972b). In terms of the Markov process of Definition 19, this model imposes the following constraint on the transition probabilities.

DEFINITION 21. *A closed structure of choice probabilities satisfies* proportionality *iff there exists a family $Q = \{ Q_B | B \subseteq A \}$ of functions such that*

(i) $\langle A, M, P, Q \rangle$ is an elimination structure; and

(ii)   for all $D, C \subseteq B \subseteq A$,

$$\frac{Q_B(C)}{Q_B(D)} = \frac{\Sigma_{C_j \cap B = C} Q_A(C_j)}{\Sigma_{D_i \cap B = D} Q_A(D_i)},$$

provided that the denominators are both positive. Furthermore, if one denominator vanishes, so does the other.

Definition 20 severely restricts the transition probabilities: each transition probability of the form $Q_B(C)$, $B \subset A$, is computable, via proportionality, from the transition probabilities of the form $Q_A(C_j)$. Specifically, the probability of eliminating from B all the alternatives that do not belong to C is proportional to the sum of the probabilities of eliminating from the total set A all the alternatives that do not belong to C_j, for any C_j whose intersection with B equals C. To illustrate, let $A = \{a, b, c\}$ and $B = \{a, b\}$. Hence, proportionality asserts that

$$\frac{Q_B(a)}{Q_B(b)} = \frac{Q_A(a) + Q_A(a, c)}{Q_A(b) + Q_A(b, c)}.$$

That is, the odds of "passing" from $\{a, b\}$ to $\{a\}$ rather than to $\{b\}$ equal the odds of "passing" from $\{a, b, c\}$ to a set containing a but not b rather than to a set containing b but not a. (For simplicity we omit brackets inside the parentheses in the formula). Note that the transition probabilities coincide with the choice probabilities in the binary case, in which $Q_{\{a, b\}}(a) = P(a, b)$, but not in general.

Although the transition probabilities were treated as theoretical constructs, they can also be viewed as observable quantities on par with choice probabilities. The subject can be instructed, for example, to partition any offered set into acceptable and unacceptable subsets. The relative frequency of selecting C from B as the set of acceptable alternatives can be taken as an estimate of the transition probability $Q_B(C)$. If such a method for estimating the transition probabilities can be devised, then the proportionality assumption can be tested directly. Even when the proportionality assumption cannot be tested in terms of transition probabilities, it has many testable consequences formulated in terms of choice probabilities. These consequences are derived from the following representation, which rests on the proportionality assumption.

THEOREM 15. *A closed structure of choice probabilities satisfies proportionality iff there exists a nonnegative real-valued function F, defined on 2^A, such that for all $a \in B \subseteq A$,*

$$P(a, B) = \frac{\Sigma_{C_j \in B*} F(C_j) P(a, B \cap C_j)}{\Sigma_{B_k \in B*} F(B_k)},$$

where $B* = \{ B_k \subseteq A | B_k \cap B \neq \varnothing, B \}$.

The proof of Theorem 15 is presented in Section 17.9.1. Note that this equation, like the Markovian equation of Definition 18, is a recursive formula that expresses the probability of choosing a from B as a weighted sum of the probabilities of selecting a from the various subsets of B. The difference between the equations lies in the structure of the weights, i.e., of the transition probabilities. In the general Markov model, the transition probabilities associated with the subsets [e.g., $Q_B(C_i)$] are not constrained in any way. In the above representation, on the other hand, the various transition probabilities [e.g., $F(C_j)/\Sigma_{B_k \in B*} F(B_k)$] are all interlocked through the proportionality assumption. As we now show, this representation is interpretable as a process of elimination by aspects.

Suppose that each alternative consists of a collection of aspects and that there is a utility scale defined over all the aspects. Suppose further that at each stage in the process one selects an aspect (from those included in the available alternatives) with a probability that is proportional to its utility. The selection of an aspect eliminates all the alternatives that do not include this aspect, and the process continues until only a single alternative remains.

To illustrate the process, consider the choice of a restaurant for dinner. The first aspect selected may be seafood; this eliminates all restaurants that do not serve seafood. Given the remaining alternatives, another aspect—say, a price level—is selected, and all alternatives that do not meet this criterion are eliminated. The process continues until only one restaurant, which includes all the selected aspects, remains. Note that the aspects could represent values along some dimensions, or they could be nominal values indicating the presence or absence of a given property.

To characterize formally the process of elimination by aspects, consider a mapping that associates with each $a \in A$ a nonempty set $a' = \{ \alpha, \beta, \ldots \}$ of elements interpreted as the aspects of a. For any $B \subseteq A$, let $B' = \{ \alpha | \alpha \in a'$ for some $a \in B \}$, and $B^0 = \{ \alpha | \alpha \in a'$ for all $a \in B \}$. Thus B' is the set of aspects that belong to at least one alternative in B, and B^0 is the set of aspects that belong to all the alternatives in B. In particular, A' is the set of all aspects under consideration, and A^0 is the set of aspects shared by all the alternatives under study. Given some $\alpha \in A'$, let B_α denote the set of alternatives of B that includes α, that is, $B_\alpha = \{a \in B | \alpha \in a'\}$.

DEFINITION 22. *A closed structure of choice probabilities satisfies the model of* elimination-by-aspects, *or the EBA model,* iff *there exists a positive-valued function* f, *defined on* $A' - A^0$, *such that for all* $a \in B \subseteq A$,

$$P(a, B) = \frac{\sum_{\alpha \in a' - B^0} f(\alpha) P(a, B_\alpha)}{\sum_{\beta \in B' - B^0} f(\beta)}.$$

The next theorem summarizes the relations between Definitions 19, 20, and 21.

THEOREM 16. *The EBA model* (Definition 22), *the Boolean random-elimination model* (Definition 20), *and the proportionality condition* (Definition 21) *are equivalent.*

The proof of Theorem 16, presented in Section 17.9.2, is divided into two parts. One part establishes the equivalence between the EBA model and the Boolean random-elimination model, following Corbin and Marley (1974). Since they also showed that any random-elimination model is a random-utility model, it follows that the EBA model is a random-utility model. For a direct proof of this result see Tversky (1972a). The other part of the proof of Theorem 16 demonstrates the equivalence of the EBA model and the representation established in Theorem 15. Note that the main difference between the two representations is in the domains of the scales. In Theorem 15, the scale F is defined on the subsets of A whereas in the EBA model the scale f is defined on the set of relevant aspects. To establish the equivalence between the forms, we interpret $F(B)$ as the sum of the utilities of the aspects that are included in all the alternatives of B and are not included in any of the alternatives that do not belong to B. That is, $F(B) = \sum_{\alpha \in \bar{B}} f(\alpha)$, where $\bar{B} = \{\alpha | \alpha \in b'$ for all $b \in B$ and $\alpha \notin a'$ for any $a \notin B\}$. Thus $F(B)$ is viewed as the measure of the unique advantage of the alternatives of B. The significance of Theorem 16 lies in showing (i) that both the Boolean random-elimination model and the proportionality assumption lead to a process that is interpretable as elimination by aspects, and (ii) that the EBA model can be formulated (and tested) in terms of the subsets of A and without any reference to specific aspects. An alternative derivation of the EBA model was presented by Marley (1981), who showed that the transition probabilities of the model can be obtained as the limit of a multivariate stochastic process generated from a set of underlying independent Poisson processes.

The remainder of this section discusses several consequences of the EBA model and reviews some related developments.

Consider first some special cases of the EBA model. Suppose that all pairs of alternatives are aspect-wise disjoint, i.e., $a' \cap b' = \varnothing$ for all $a, b \in A$. In this case,

$$P(a, B) = \frac{\sum_{\alpha \in a'} f(\alpha)}{\sum_{\beta \in B'} f(\beta)}$$

since $\alpha \in a'$ implies $B_\alpha = \{a\}$, and $P(a, \{a\}) = 1$. Letting $F(a) = \sum_{\alpha \in a'} f(\alpha)$ yields

$$P(a, B) = \frac{F(a)}{\sum_{b \in B} F(b)},$$

which is the strict-utility model discussed in Section 17.4.2.

In the binary case, the EBA model reduces to

$$P(a, b) = \frac{\sum_{\alpha \in a' - b'} f(\alpha)}{\sum_{\alpha \in a' - b'} f(\alpha) + \sum_{\beta \in b' - a'} f(\beta)},$$

where $a' - b' = \{\alpha | \alpha \in a'$ and $\alpha \notin b'\}$ etc. Letting $F(a' - b') = \sum_{\alpha \in a' - b'} f(\alpha)$, and $F(b' - a') = \sum_{\beta \in b' - a'} f(\beta)$ yields

$$P(a, b) = \frac{F(a' - b')}{F(a' - b') + F(b' - a')},$$

which is the binary model advanced by Restle (1961), who first proposed the representation of choice alternatives as measurable collections of aspects. Hence, the EBA model generalizes the choice models of Luce and Restle. For a critical discussion of these models, see Indow (1975b).

It is not difficult to show that the EBA model implies MST but not SST (Tversky, 1972a). Furthermore, EBA is a random-utility model (although not an independent one), and hence, it satisfies the nonnegativity condition (Definition 14). Finally, Sattath and Tversky (1976) proved that the EBA model implies the multiplicative inequality $P(a, B \cup C) \geq P(a, B)P(a, C)$, $a \in B$, $C \subset A$. Thus the EBA model imposes considerable constraints on the observed choice probabilities despite the large number of parameters.

It is instructive to examine how the EBA model resolves the difficulties associated with simple scalability (see the discussion in Section 17.4.1 following Theorem 5). Consider first the problem of the similar alternatives. There we let $A = \{a, b, c\}$, where a and b were very similar (e.g., two different recordings of the same symphony) whereas c was dissimilar to either of them. Assume, for simplicity, that c is aspect-wise disjoint from a and b. By assumption, all binary choice probabilities are equal. This is the case in the EBA model if and only if

$$F(a) = F(b) = x, \qquad F(a, b) = y, \qquad \text{and} \qquad F(c) = x + y,$$

for some positive x, y. A graphical illustration of the relations among the alternatives is displayed in Figure 4, in which it is easy to see that all binary choice probabilities equal $\frac{1}{2}$.

According to the EBA model, however, the trinary choice probabilities in this case are not equal because

$$P(c, A) = \frac{x + y}{3x + 2y} > \frac{x + \frac{y}{2}}{3x + 2y} = P(a, A) = P(b, A).$$

As $\frac{x}{y}$ approaches 0, the left-hand side approaches $\frac{1}{2}$ and the right-hand side approaches $\frac{1}{4}$, in accord with our intuitive analysis and contrary to simple scalability.

Next, consider the problem of the dominated alternative in which a and b denote, respectively, a fifty-fifty chance to win \$100 and \$99 whereas c is the option of receiving \$35 with certainty. Since a dominates b whereas b and c are assumed indifferent, there exist some positive x, y such that $F(a) = x$ and $F(a, b) = F(c) = y$ (see Figure 5). It is easy to verify now

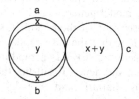

FIGURE 4. Graphic illustration of the relations among the alternatives in the case of the similar alternatives.

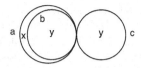

FIGURE 5. Graphic illustration of the relations among the alternatives in the case of the dominated alternative.

that, according to the EBA model,

$$P(a, b) = \tfrac{x}{x} = 1, \quad \text{and} \quad P(b, c) = \tfrac{y}{2y} = \tfrac{1}{2},$$

as required. Contrary to order-independence, however,

$$P(a, c) = \frac{x + y}{x + 2y},$$

which lies strictly between $\tfrac{1}{2}$ and 1. Thus, if the utility associated with the added dollar is relatively small, $P(a, c)$ will be close to $\tfrac{1}{2}$, in agreement with our intuitions.

The EBA model accounts for the departures from simple scalability by introducing an underlying structure of aspects (or subsets) and by adopting a sequential elimination strategy. However, the same phenomena of choice can also be explained by replacing the Markovian model by a random-utility model that treats the aspects as (positive-valued) random variables. Tversky (1972a) introduced the additive-random-aspect (ARA) model in which $Pr(U_a > U_b)$ equals the probability that the sum of the random variables associated with the aspects that belong to a but not to b exceeds the sum of the random variables associated with the aspects that belong to b but not to a. Note that the random variables that represent the aspects are independent, but those representing the objects (i.e., U_a and U_b) are not. Although EBA and ARA employ different decision rules, they both account for the dependence among alternatives while preserving independence among aspects.

The most natural distributional assumptions for a random-aspect model, of course, are the normal and the double-exponential (also known as the extreme value distribution). Indeed, Edgell and Geisler (1980) proposed an ARA model in which the aspect random variables are normally distributed and concluded that its predictions are similar to those of EBA, and

McFadden (1981) developed a multivariate random-utility model based on the generalized-extreme-value (GEV) distribution. He interpreted this model as an elimination model with nonstationary transition probabilities and suggested that it is roughly comparable to EBA in flexibility and complexity. McFadden (1981) also developed a more general model, called *elimination-by-strategy*, that generalizes EBA and is equivalent to the general random-utility model.

17.8.3 Preference Trees

The EBA model, discussed in the preceding section, resolves the difficulties due to the failure of simple scalability (Definition 8) by introducing an underlying structure of aspects. This structure gives rise to a large number of parameters (equal to the number of nonempty proper subsets of A) that complicates the estimation problem. In particular, the representation cannot be constructed from pair-comparison data alone since the number of parameters exceeds the number of data points in this case. The question arises, then, whether the EBA model can be significantly simplified by imposing some structure on the aspect space. This section describes a model, developed by Tversky and Sattath (1979), that represents the alternatives and their aspects in a treelike graph. This representation, called preference tree (Pretree), is considerably more parsimonious than EBA, yet much less restrictive than the strict-utility model.

A graph is a collection of points, called *nodes*, some of which are linked by lines called *links* or *edges*. A sequence of adjacent links without cycles is called a *path*. A (rooted) tree is a graph (containing one distinguished node called the *root*) in which any two nodes are joined by a unique path. For ease of reference, we place the root at the top of the tree and the terminal nodes at the bottom, as in Figure 6. To interpret a rooted tree as a family of aspect sets, we associate each terminal node of the tree with a single element of A and each link of the tree with the set of aspects that are shared by all the alternatives that include (or follow from) that link and

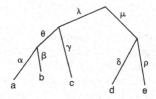

FIGURE 6. Tree representation of a choice set.

that are not shared by any of the alternatives that do not include that link. The length of each link represents the measure of the respective set of aspects. An illustrative example of a tree representation of a choice set with five elements is displayed in Figure 6, where the link labeled λ represents the aspects shared by a, b, and c, and the link labeled δ represents the unique aspects of d, etc.

Next, we associate each link λ of the tree with the set $A_\lambda = \{a \in A | \lambda \in a'\}$ of elements that include (or follow from) that link. In Figure 6, for example, $A_\theta = \{a, b\}$ and $A_\delta = \{d\}$. It is easy to verify that for any two links α, β in a tree, either $A_\alpha \supset A_\beta$ or $A_\beta \supset A_\alpha$ or $A_\alpha \cap A_\beta$ is empty. Hence, the tree induces a hierarchical clustering on A. It can also be shown that the family $A^* = \{a' | a \in A\}$ of aspect sets associated with a given set of alternatives is representable as a tree iff the following inclusion rule holds for all a, b, c in A: either $a' \cap b' \supset a' \cap c'$ or $a' \cap c' \supset a' \cap b'$. Stated differently, any subset of A with three elements contains one alternative, say c, such that $a' \cap c' = b' \cap c'$, which in turn is included in $a' \cap b'$. The inclusion rule imposes considerable constraints on the structure of aspects and reduces the number of parameters from $2^n - 2$ (the number of proper nonempty subsets of A) to $2n - 2$ (those that correspond to the maximal number of links in a tree with n terminal nodes).

If A^* is representable as a tree, the process of elimination by aspects reduces to elimination by tree (EBT). That is, one selects a link from the tree, with a probability that is proportional to its length, and eliminates all the alternatives that do not include the selected link. The same process is applied to the selected branch until one alternative remains. Note that the strict-utility model corresponds to a degenerate tree with no internal nodes except the root.

The representation of choice alternatives as a tree suggests an alternative model that interprets the tree as a hierarchy of choice points. According to this theory, called the *hierarchical elimination model* (HEM), one begins at the top of the tree and selects first among the links that are connected directly to the root. One then proceeds to the next choice point at the bottom of the selected link and repeats the process until the chosen branch contains a single alternative. The probability of choosing alternative a from an offered set B is the product of the probabilities of selecting the branch containing a at each stage of the process; and the probability of selecting a branch is proportional to its weight. Thus each internal node in the tree is interpreted as a choice point, and one proceeds in order from the top to the bottom of the hierarchy.

To define HEM in a more rigorous manner, let $A_\alpha = \{a \in A | \alpha \in a'\}$, and define $\alpha | \beta$ if β follows directly from α, that is, $A_\alpha \supset A_\beta$ and $A_\gamma \supset A_\beta$ imply $A_\gamma \supset A_\alpha$. Let $f(\alpha)$ denote the length of α, and let $m(\alpha)$ be the measure or total length of all links that follow from α, including α. In Figure 6, for example, $\theta|\alpha$, $\theta|\beta$, and $m(\theta) = f(\alpha) + f(\beta) + f(\theta)$. If A^* is a tree and $B \subset A$, then $B^* = \{a'|a \in B\}$ is also a tree, called a subtree of A. Naturally, the relation | and the measure f on A induce a corresponding relation and measure on B. Finally, let $P(C, B) = \Sigma_{a \in C}P(a, B)$, $a \in C \subset B \subset A$.

DEFINITION 23.   *A closed structure of choice probability satisfies* HEM *iff there exists a tree A^* with a measure f such that*:

(1)   *If $\alpha|\beta$ and $\beta|\gamma$, then $P(A_\alpha, A_\gamma) = P(A_\alpha, A_\beta)P(A_\beta, A_\gamma)$.*

(2)   *If $\gamma|\beta$ and $\gamma|\alpha$, then $P(A_\alpha, A_\gamma)/P(A_\beta, A_\gamma) = m(\alpha)/m(\beta)$, provided $P(A_\beta, A_\gamma) \neq 0$.*

(3)   *Conditions (1) and (2) hold for any subtree B^* of A^*, with the induced structure on B^*.*

The first part of Definition 23 implies that $P(a, B)$ equals the product of the probabilities of selecting the branches that contain a at each junction. The second part states that the probability of selecting one branch rather than another at any given junction is proportional to the weight of that branch, defined as the sum of the length of all its links. The last part of the definition ensures that (1) and (2) apply to any subtree of A.

The two models EBT and HEM represent different conceptions of the choice process that assume a tree structure. EBT describes $P(a, B)$ as a weighted sum of the probabilities $P(a, B_\alpha)$ of choosing a from the various subsets of B. HEM, on the other hand, expresses $P(a, B)$ as a product of the probabilities $P(A_\alpha, A_\beta)$, $\beta|\alpha$, of selecting a branch containing a at each level in the hierarchy. The difference in form reflects a difference in processing strategy. EBT assumes free access; that is, any aspect can be selected at any stage. HEM assumes sequential access; that is, aspects are considered in a fixed hierarchical fashion. Thus EBT permits a choice on the basis of a single (distinctive) link whereas HEM requires a systematic scanning of all links included in the selected alternatives.

Despite the difference in mathematical form and psychological interpretation, Tversky and Sattath (1979) have proved that EBT and HEM are equivalent. That is, a given set of choice probabilities satisfies EBT iff it satisfies HEM, and the two models yield the same tree structure. Furthermore, any alternating strategy that consists of a mixture of EBT and HEM is also equivalent to both of them. For example, a person may select a restaurant according to EBT and then choose an entree according to HEM or vice versa. The equivalence of the two models implies that the two strategies cannot be distinguished on the basis of choice probabilities, though they might be distinguishable on the basis of other data, e.g., reaction time or verbal protocols. It also suggests that the tree model provides a versatile representation of choice that is compatible with both free-access and sequential-access strategies. We use the term preference tree or *Pretree* to denote the choice probabilities generated by EBT or by HEM, irrespective of the particular strategy.

We next derive some properties of binary choice probabilities that characterize the tree model. To simplify the exposition, we let $R(a, b) = P(a, b)/P(b, a)$ and restrict the discussion to the case in which $P(a, b) \neq 0$, for all $a, b \in A$, so that $R(a, b)$ is always well defined. The results can be extended to deal with binary choice probabilities that equal 0 or 1. Consider first the three-alternative case, and note that any tree of three elements has the form portrayed in Figure 7a, except for the labeling of the alternatives and the possibility of vanishing links. We use parentheses to describe the structure of the tree; thus the tree in Figure 7a is described by $(ab)c$, and the tree in Figure 6 by $((ab)c)(de)$.

Using the notation in Figure 7a, it follows that $R(a, b) = \alpha/\beta$ is more extreme (i.e., further from 1) than $R(a, c)/R(b, c) = (\alpha + \theta)/(\beta + \theta)$. Any triple of alternatives, therefore, that forms a subtree $(ab)c$ satisfies the

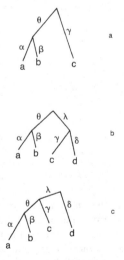

FIGURE 7. Preference trees for 3 and 4 alternatives: (a) $(ab)c$, (b) $(ab)(cd)$, and (c) $((ab)c)d$.

following *trinary condition*:

$$\text{If} \quad R(a, b) \geqslant 1, \quad \text{then} \quad R(a, b) \geqslant \frac{R(a, c)}{R(b, c)} \geqslant 1,$$

where a strict inequality in the hypothesis implies a strict inequality in the conclusion. $R(a, b)$ can be viewed as a direct measure of the strength of preference for a over b whereas $R(a, c)/R(b, c)$ can be interpreted as an indirect measure (via c) of the same quantity. The trinary condition asserts that the former is more extreme than the latter whenever $(ab)c$ holds. In this case, of course, a and b have more in common with each other than with the odd element c. Hence, commonality between alternatives (i.e., the measure of their common aspects) enhances their discriminability and produces choice probabilities that are closer to 0 or 1. Translated into binary choice probabilities, the trinary condition implies the inequality

$$P(a, b)P(b, c)P(c, a) \geqslant P(a, c)P(c, b)P(b, a),$$

whenever $(ab)c$ holds and $P(a, b) \geqslant \frac{1}{2}$. Recall that under the strict-utility model the two sides of this equation must be equal.

Let us consider the case of four alternatives. In Figures 7b and 7c, there are two distinct subtrees of four alternatives, up to permutation of elements and vanishing links. It is easy to verify that the subtree $(ab)(cd)$ satisfies

$$\frac{R(a, c)}{R(b, c)} = \frac{(\alpha + \theta)/(\gamma + \lambda)}{(\beta + \theta)/(\gamma + \lambda)} = \frac{(\alpha + \theta)/(\delta + \lambda)}{(\beta + \theta)/(\delta + \lambda)} = \frac{R(a, d)}{R(b, d)},$$

whereas the subtree $((ab)c)d$ satisfies

$$\frac{R(a, c) - R(b, c)}{R(a, d) - R(b, d)} = \frac{(\alpha - \beta)/\gamma}{(\alpha - \beta)/\delta} = \frac{(\alpha + \theta - \gamma)/\gamma}{(\alpha + \theta - \gamma)/\delta}$$
$$= \frac{R(a, c) - R(c, c)}{R(a, d) - R(c, d)}.$$

These equations are called the *quaternary conditions*. Note that in the strict-utility model the quaternary conditions hold for any four elements and not merely for these quadruples that form the appropriate subtrees. Although subtrees of five or more elements entail additional testable properties, these properties are implied by the previous conditions. Moreover, the following theorem shows that the trinary and the quaternary conditions are not only necessary but also sufficient to ensure the representation of binary choice probabilities as a preference tree.

THEOREM 17. *A complete finite pair comparison structure, with* $P(a, b) \neq 0, 1$, *satisfies Pretree with a given tree structure iff the trinary and the quaternary conditions are satisfied relative to that structure.*

This theorem of Tversky and Sattath (1979) is proved in Section 17.9.3. It shows, in effect, how to construct a scale f such that $P(a, b) = f(a' - b')/[f(a' - b') + f(b' - a')]$. Recall that $f(a' - b')$ is the measure of the aspects of a that are not included in b, which corresponds to the length of the path from the terminal node associated with a to the meeting point of a and b. Although the preceding theorem is formulated in terms of a given tree, the trinary and the quaternary conditions can be used to diagnose which tree structure, if any, is satisfied by the data. More specifically, the trinary condition must hold for every triple of alternatives in one or more permutations, and at least one permutation of every quadruple should satisfy one of the quaternary conditions. Hence, by permuting the alternatives so as to satisfy both the trinary and the quaternary conditions, any tree structure that is compatible with the data will emerge.

It is readily seen that, given a tree structure, the scale f is unique up to multiplication by a positive constant provided that $P(a, b) \neq \frac{1}{2}$. The tree structure, however, is not always unique. Tversky and Sattath (1979) showed that a pair-comparison structure can be compatible with more than one tree, but the tree structure is uniquely determined by both binary and trinary choice probabilities. Furthermore, the binary and the trinary data must be related as follows. Suppose $(ab)c$ holds (see Figure 7a), then

$$\frac{P(a, c)}{P(c, a)} = \frac{\alpha + \theta}{\gamma} \geqslant \frac{\alpha + \theta\alpha/(\alpha + \beta)}{\gamma} = \frac{P(a, \{a, b, c\})}{P(c, \{a, b, c\})},$$

and

$$\frac{P(a, b)}{P(b, a)} = \frac{\alpha}{\beta} = \frac{\alpha + \theta\alpha/(\alpha + \beta)}{\beta + \theta\beta/(\alpha + \beta)} = \frac{P(a, \{a, b, c\})}{P(b, \{a, b, c\})}.$$

In the subtree $(ab)c$, therefore, the constant-ratio rule (Definition 11) holds for the adjacent pair (a, b) but not for the split pairs (a, c) and (b, c). Since b is closer to a than to c (in the sense that $a' \cap b' \supset b' \cap c'$), the addition of b to the set $\{a, c\}$ reduces the probability of choosing a proportionally more than the probability of choosing c. But since c is equally distant from a and from b (in the sense that $a' \cap c' = b' \cap c'$), the addition of c to the set $\{a, b\}$ reduces the probabilities of choosing a and b by the same factor. The reader is invited to verify that the counterexamples to simple scalability (discussed in Section 17.8.2 in terms of the EBA model) are compatible with Pretree.

Theoretical and empirical applications of the tree model are discussed in Tversky and Sattath (1979), including an interpretation of Pretree (and EBA) as an aggregate model, a method for constructing preference trees from both similarity and preference data, and experimental tests of Pretree against the strict-utility model. McFadden (1981) has developed statistical procedures for estimating preference trees and developed a tree model based on the extreme value (i.e., double-exponential) distribution. He also explored the economic significance of the model and applied it to commuters' choices among four travel modes (auto, bus, rapid transit, car pool).

17.9 PROOFS

17.9.1 Theorem 15 (p. 439)

Proportionality holds iff there exists a nonnegative real-valued function F, defined on 2^A, such that for all $a \in B \subseteq A$,

$$P(a, B) = \frac{\sum_{C_j \in B^*} F(C_j) P(a, B \cap C_j)}{\sum_{B_k \in B^*} F(B_k)},$$

where $B^* = \{B_k | B_k \cap B \neq \theta, B\}$.

PROOF. For any $B \subseteq A$, let $F(B) = Q_A(B)$. Recall that B, C_j, etc. denote nonempty subsets of A. Thus, for any $C \subseteq B \subseteq A$,

$$Q_B(C) = \frac{Q_B(C)}{\sum_{B_i \subseteq B} Q_B(B_i)} \qquad \text{(since the denominator equals 1)}$$

$$= \frac{\sum_{C_j \cap B = C} F(C_j)}{\sum_{B_i \subseteq B} \sum_{D_i \cap B = B_i} F(D_i)} \qquad \text{(by proportionality)}$$

$$= \frac{\sum_{C_j \cap B = C} F(C_j)}{\sum_{B_k \cap B \neq \emptyset} F(B_k)},$$

since $\{D_i | D_i \cap B = B_i \text{ for some } B_i \subseteq B\} = \{B_k | B_k \cap B \neq \emptyset\}$. Substituting this expression for $Q_B(C_i)$ in the Markovian equation yields

$$P(a, B) = \sum_{C_i \subseteq B} Q_B(C_i) P(a, C_i)$$

$$= \frac{\sum_{C_i \subseteq B} \sum_{C_j \cap B = C_i} F(C_j) P(a, C_i)}{\sum_{B_k \cap B \neq \varnothing} F(B_k)}.$$

Note that $\{C_i | C_i = B \cap C_j \text{ for some } C_j\} = \{C_i | C_i \subseteq B\}$. Hence,

$$\sum_{C_i \subseteq B} \sum_{C_j \cap B = C_i} F(C_j) P(a, C_i) = \sum_{C_j \cap B \neq \varnothing} F(C_j) P(a, B \cap C_j)$$

$$= \sum_{C_j} F(C_j) P(a, B \cap C_j),$$

since $P(a, B \cap C_j) = 0$ whenever $C_j \cap B = \varnothing$, and by substitution,

$$P(a, B) = \frac{\sum_{C_j} F(C_j) P(a, B \cap C_j)}{\sum_{B_k \cap B \neq \varnothing} F(B_k)}$$

$$= \frac{\sum_{C_j \supseteq B} F(C_j) P(a, B \cap C_j) + \sum_{C_j \supseteq B} F(C_j) P(a, B)}{\sum_{B_k \cap B \neq \varnothing, B} F(B_k) + \sum_{B_k \supseteq B} F(B_k)}.$$

Letting $x = \sum_{C_j \not\supseteq B} F(C_j) P(a, B \cap C_j)$, $y = \sum_{B_k \cap B \neq \varnothing, B} F(B_k)$, and $z = \sum_{C_j \supseteq B} F(C_j)$ yields

$$P(a, B) = \frac{x + zP(a, B)}{y + z} = \frac{x}{y}, \qquad \text{provided } y \neq 0.$$

Note that for any $C_i \subset B$,

$$y = \sum_{B_k \cap B \neq \varnothing, B} F(B_k) \geqslant \sum_{B_k \cap B = C_i} F(B_k).$$

The expression on the right, however, is proportional by Definition 20 to $Q_B(C_i)$, which cannot vanish for all $C_i \subset B$ except when B contains a single element. The proof in the latter case is immediate.

Consequently,

$$P(a, B) = \frac{\sum_{C_j \in B^*} F(C_j) P(a, B \cap C_j)}{\sum_{B_k \in B^*} F(B_k)}$$

as required.

To prove the converse, suppose there exists a nonnegative function F on 2^A satisfying the above equation for all $a \in B \subseteq A$. For any $C \subseteq B \subseteq A$, let

$$Q_B(C) = \frac{\Sigma_{C_j \cap B = C} F(C_j)}{\Sigma_{B_k \cap B \neq \varnothing} F(B_k)}.$$

It is readily verified that each Q_B, $B \subset A$, is a transition probability function in the sense of Definition 18, and that $\{Q_B | B \subseteq A\}$ satisfies the proportionality condition. $\diamondsuit$

17.9.2   Theorem 16 (p. 440)

(i)   *The EBA model is equivalent to the proportionality condition.*

(ii)   *The EBA model is equivalent to the Boolean random elimination model.*

PROOF.   To prove part (i) we show that the formula of the EBA model (Definition 22) is satisfied iff the representation derived from the proportionality condition (Theorem 15) is valid. First, suppose the EBA model holds. For each $a \in A$, let a' denote the set of aspects associated with a. For any $B \subseteq A$, define

$$B' = \{\alpha | \alpha \in b' \text{ for some } b \in B\},$$
$$B^0 = \{\alpha | \alpha \in b' \text{ for all } b \in B\},$$
$$\overline{B} = \{\alpha | \alpha \in b' \text{ for all } b \in B \text{ and } \alpha \notin a' \text{ for any } a \notin B\},$$
$$B_\alpha = \{b \in B | \alpha \in b'\},$$
$$B^* = \{B_i | B_i \cap B \neq \varnothing, B\}.$$

Note that $\{\overline{B}_i | B_i \subset A\}$ forms a partition of $A' - A^0$. Let $F(B) = \Sigma_{\alpha \in \overline{B}} f(\alpha)$. Clearly, F is nonnegative and $F(B) = 0$ iff $\overline{B} = \varnothing$. If $\alpha, \beta \in C$, then for all $B \subset A$, $B_\alpha = B_\beta = B \cap C$. Furthermore, the numerator of the EBA formula for $P(a, B)$ can be expressed as

$$\sum_{\alpha \in a' - B^0} P(a, B_\alpha) f(\alpha) = \sum_{a \in C_i} \sum_{\alpha \in \overline{C}_i} P(a, B_\alpha) f(\alpha)$$
$$= \sum_{C_i \in B^*} P(a, B \cap C_i) \sum_{\alpha \in \overline{C}_i} f(\alpha)$$
$$= \sum_{C_i \in B^*} P(a, B \cap C_i) F(C_i).$$

Similarly, since $\{\beta | \beta \in B' - B^0\} = \{\beta | \beta \in \overline{B}_j$ for some $B_j \in B^*\}$, the denominator of the EBA formula for $P(a, B)$ can be expressed as $\Sigma_{\beta \in B' - B^0} f(\beta) = \Sigma_{B_j \in B^*} F(B_j)$. Hence, it follows from the EBA model that for $a \in B \subseteq A$,

$$P(a, B) = \frac{\Sigma_{C_i \in B^*} F(C_i) P(a, B \cap C_i)}{\Sigma_{B_j \in B^*} F(B_j)}.$$

Conversely, suppose proportionality is satisfied. By Theorem 15, therefore, the preceding equation holds. For any $a \in A$, let $a' = \{C_i \subseteq A | a \in C_i\}$, $B' = \{C_i \subseteq A | b \in C_i$ for some $b \in B\}$, $B^0 = \{D \subset A | D \subset B\}$. It is readily verified that the above equation reduces to

$$\frac{\Sigma_{C_i \in a' - B^0} F(C_i) P(a, B \cap C_i)}{\Sigma_{B_j \in B' - B^0} F(B_j)},$$

which is an equivalent form of the EBA model.

To prove part (ii) assume first that the Boolean random-elimination model is satisfied. Hence, for any $C_j \subset A$,

$$Q_A(C_j) = Pr(\mathbf{U}_c = 1 \text{ for all } c \in C_j \text{ and } \mathbf{U}_a = 0 \text{ for all } a \in A - C_j).$$

Thus, for any $C \subset B \subset A$,

$$\sum_{C_j \cap B = C} Q_A(C_j) = \sum_{C_j \cap B = C} Pr(\mathbf{U}_c = 1 \text{ for all } c \in C_j \text{ and }$$
$$\mathbf{U}_a = 0 \text{ for all } a \in A - C_j)$$
$$= Pr(\mathbf{U}_c = 1 \text{ for all } c \in C \text{ and }$$
$$\mathbf{U}_b = 0 \text{ for all } b \in B - C)$$
$$= Q_B(C).$$

Furthermore, it follows that

$$Q_B(B) = Pr(\mathbf{U}_b = 1 \text{ for all } b \in B \text{ or } \mathbf{U}_b = 0 \text{ for all } b \in B)$$
$$= \sum_{B_j \cap B = B, \varnothing} Q_A(B_j).$$

By Definition 18, $P(a, B) = \sum_{C \subset B} Q_B(C) P(a, C)$. Hence,

$$P(a, B)[1 - Q_B(B)] = \sum_{C \subset B} \sum_{C_j \cap B = C} Q_A(C_j) P(a, C)$$

$$= \sum_{C_j \cap B \neq \varnothing, B} Q_A(C_j) P(a, B \cap C_j),$$

and

$$P(a, B) = \frac{\sum_{C_j \in B^*} Q(C_j) P(a, B \cap C_j)}{1 - Q_B(B)}$$

$$= \frac{\sum_{C_j \in B^*} Q(C_j) P(a, B \cap C_j)}{\sum_{B_k \in B^*} Q(B_k)},$$

which is equivalent to the EBA model by part (i) of this theorem.

To prove the converse, suppose the EBA model is satisfied. Consider the representation derived from proportionality (Theorem 15), and normalize the function F so that $\sum_{B_j \subseteq A} F(B_j) = 1$. Construct a Boolean random vector $\mathbf{U}$ on A as follows. For any $B \subseteq A$, let

$$\Gamma_B(a) = \begin{cases} 1 & \text{if } a \in B \\ 0 & \text{otherwise.} \end{cases}$$

Let $A = \{a_1, \ldots, a_n\}$, and let $V = (V_1, \ldots, V_n)$ be a Boolean vector. Define the random vector $\mathbf{U}$ by

$$Pr(\mathbf{U} = V) = \begin{cases} F(B) & \text{if } \Gamma_B(a_i) = V_i, i = 1, \ldots, n \\ 0 & \text{otherwise.} \end{cases}$$

Let $R_B(C) = Pr(\mathbf{U}_d = \mathbf{U}_c > \mathbf{U}_b$ for all $d, c \in C, b \in B - C)$. It follows that

$$R_B(C) = \begin{cases} \displaystyle\sum_{C_j \cap B = C} F(C_j) & \text{if } C \neq B \\ \displaystyle\sum_{C_j \cap B = \varnothing, B} F(C_j) & \text{if } C = B, \end{cases}$$

and that $\langle A, M, P, R \rangle$ is a Boolean random-elimination model in the sense of Definition 20. ◇

17.9.3   Theorem 17 (p. 449)

A complete pair comparison structure, with $P(a, b) \neq 0, 1$ satisfies Pretree with a given tree structure iff the trinary and quaternary conditions are satisfied relative to that structure.

The proof is divided into a series of lemmas., Let $P_A = \{P(a, b)|a, b \in A\}$.

LEMMA 1.   *If $A = \{a, b, c\}$, then P_A satisfies Pretree with $(ab)c$ iff the trinary condition is satisfied in this form.*

PROOF.   Necessity is obvious. To prove sufficiency, we use the notation of Figure 7a in which $R(a, b) \geqslant 1$. Set $\alpha = 1$, $\beta = R(b, a)$, and select $\theta \geqslant 0$ so that $[R(a, c) - R(b, c)]\theta = R(b, c) - R(b, a)R(a, c)$, and let $\gamma = R(c, a)(1 + \theta)$. (Note that when $R(a, b) > 1$, θ is uniquely defined and positive and, when $R(a, b) = 1$, θ, can be chosen arbitrarily.)

Let $\overline{P}_A$ be the set of binary probabilities obtained by using the above expressions for $\alpha, \beta, \gamma, \theta$ in the defining equations of the model. It can be verified, after some algebra, that $\overline{P}_A = P_A$, as required.

Before we go further, note that if P_A satisfies Pretree with $(ab)c$ and $R(a, b) > 1$, then $\beta/\alpha = R(b, a)$. Furthermore,

$$\frac{\theta/\alpha + 1}{\theta/\alpha + \beta/\alpha} = \frac{\theta + \alpha}{\theta + \beta} = \frac{R(a, c)}{R(b, c)} \quad \text{implies}$$

$$\frac{\theta}{\alpha} = \frac{R(b, c) - R(b, a)R(a, c)}{R(a, c) - R(b, c)}$$

and

$$\frac{1 + \theta/\alpha}{\gamma/\alpha} = \frac{\alpha + \theta}{\gamma} = R(a, c) \quad \text{implies}$$

$$\frac{\gamma}{\alpha} = R(c, a)\left(1 + \frac{R(b, c) - R(b, a)R(a, c)}{R(a, c) - R(b, c)}\right) = \frac{1 - R(b, a)}{R(a, c) - R(b, c)}.$$

Hence, the lengths of all the links are determined up to multiplication by a positive constant. Furthermore, the present model readily entails the following property.

LEMMA 2.   *Suppose C and $B = \{a, b, c\}$ are sets of objects such that $b, c \in C$ and $a \notin C$, and suppose that both P_C and P_B satisfy Pretree. (It is assumed that $P(c, b)$ is the same in both structures.) Then the measures on C'*

and B' can be selected so that $f(c' - b')$—as well as $f(b' - c')$—are the same in both measures.

LEMMA 3. *Suppose $C = \{a, b, c\}$ and $B = \{b, c, d\}$ satisfy Pretree, with representing measures f_C and f_B in the forms $(ab)c$ and $(bc)d$, respectively. If $D = C \cup B = \{a, b, c, d\}$ satisfies the appropriate quaternary condition with $(ab)(cd)$ or with $((ab)c)d$, then there exists a representing measure f on D' that extends both f_C and f_B. Naturally, we assume that f_C and f_B were selected according to Lemma 2.*

PROOF. Consider the form $(ab)(cd)$ (see Figure 7b). By Lemma 2, $f_C(\beta + \theta) = f_B(\beta + \theta)$ and $f_C(\lambda + \gamma) = f_B(\lambda + \gamma)$. Hence, f_C and f_B can be used to define a measure f on D'. To show that f is a representing measure on D' we have to show that $R(a, d) = f(\theta + \alpha)/f(\lambda + \delta)$. Since D satisfies Pretree, it follows from the first quaternary condition that

$$
\begin{aligned}
R(a, d) &= R(b, d)R(a, c)R(c, d) \\
&= \frac{f(\beta + \theta)}{f(\lambda + \delta)} \frac{f(\alpha + \theta)}{f(\lambda + \gamma)} \frac{f(\lambda + \gamma)}{f(\beta + \theta)} \\
&= \frac{f(\theta + \alpha)}{f(\lambda + \delta)}.
\end{aligned}
$$

Next, consider the form $((ab)c)d$ (see Figure 7c). Here, we have to show that $R(a, d) = f(\alpha + \theta + \lambda)/f(\delta)$. It follows from the second quaternary condition that

$$
\begin{aligned}
R(a, d) &= \frac{[1 - R(a, c)]R(b, d) + R(c, d)[R(a, c) - R(b, c)]}{1 - R(b, c)} \\
&= \frac{\begin{array}{c}[1 - f(\alpha + \theta)/f(\gamma)][f(\beta + \theta + \lambda)/f(\delta)] \\ + [f(\gamma + \lambda)/f(\delta)][f(\alpha + \theta)/f(\gamma) - f(\beta + \theta)/f(\gamma)]\end{array}}{1 - f(\beta + \theta)/f(\gamma)} \\
&= \frac{f(\alpha + \theta + \lambda)}{f(\delta)},
\end{aligned}
$$

as required.   ◇

LEMMA 4. *P_A satisfies Pretree with a specified structure iff for every $E \subset A$, with four elements or less, P_E satisfies Pretree relative to the same structure.*

PROOF. Necessity is immediate. Sufficiency is proved by induction on the cardinality of A, denoted n. Suppose $n > 4$, and assume that the lemma holds for any cardinality smaller than n.

Suppose $(ab)c$ holds for any c in A. Let $C = A - \{a\}$, and $B = \{a, b, c\}$. By the induction hypothesis, both P_C and P_B satisfy Pretree with the appropriate structure. By Lemma 2 we can assume, with no loss of generality, that the measures of b and c in C' coincide with their measures in B'. Since any aspect in A' appears either in C' or in B', and since the aspects that appear in both trees have the same measure, we can define the measure of any aspect in A' by its measure in C' or in B'. Letting $\bar{P}$ denote the calculated binary probability function, we show that $\bar{P}_A = P_A$.

Since $\bar{P}_C = P_C$ and $\bar{P}_B = P_B$, it remains to be shown that $\bar{P}(a, d) = P(a, d)$ for any $d \in A - B$.

Let $D = \{a, b, c, d\}$, which satisfies Pretree, by assumption, with either $(ab)(cd)$ or $((ab)c)d$. Since $D = B \cup \{b, c, d\}$, Lemma 3 implies that the representing measure on D' coincides with the restriction to D' of the defined measure on A'. Hence, $\bar{P}(a, d) = P(a, d)$ as required.

In conclusion, Lemma 3 together with Lemma 1 show that the trinary and the quaternary conditions are necessary and sufficient for the representation of quadruples. Lemma 4 shows that if Pretree is satisfied by all quadruples, then it is satisfied by the entire object set. This completes the proof of the representation theorem.

EXERCISES

1. Show that the monotonicity axiom (Definition 5) implies SST (Definition 3).   (17.2.2 and 17.2.1)

2. Show that the monotonicity axiom is equivalent to the quadruple condition in a complete pair-comparison structure.   (17.2.2)

3. Show that MST (Definition 3) holds iff the relations $\geqslant_\lambda$ (defined by $a \geqslant_\lambda b$ iff $P(a, b) \geqslant \lambda$, $1/2 \leqslant \lambda \leqslant 1$) are transitive.   (17.2.1)

4. Show that simple scalability (Definition 8) implies order-independence (Definition 9).   (17.4.1 and 17.5.1)

5. Prove the Corollary to Theorem 5.   (17.4.1)

6. Prove Theorem 8.   (17.4.2)

7. Show that regularity does not imply the random-utility model. (17.6.1)

8. Find necessary and sufficient conditions under which $P(a, B \cup C) = P(a, B)P(a, C)$ in a continuous independent random-utility model. (17.6.2 and 17.7.2)

9. Show that Thurstone's Case III (independent normal variables with different means and variances) implies MST but not SST. (17.6.1 and Definition 3)

10. Prove Theorem 14. (17.6.3)

11. Show that the EBA model implies MST. (17.8.2)

12. Prove directly that the strict-utility model (Definition 10) satisfies proportionality (Definition 21). (17.8.2)

13. Show that the EBA model implies the multiplicative inequality $P(a, B \cup C) \geqslant P(a, B)P(a, C)$. (17.8.2)

14. Show that the EBA model is not an independent random-utility model. (17.8.2)

15. Demonstrate the equivalence of HEM (Definition 23) and EBT. (17.8.3).

References

Abelson, R. M., & Bradley, R. A. (1954). A two-by-two factorial with paired comparisons. *Biometrics*, *10*, 487–502.

Aczél, J. (1966). *Lectures on functional equations and their applications*. New York: Academic Press.

Adams, E. W. (1965). Elements of a theory of inexact measurement. *Philosophy of Science*, *32*, 205–228.

Adams, E., & Messick, S. (1957). *An axiomatization of Thurstone's successive intervals and paired comparisons scaling models* (Tech. Rep. No. 12). Stanford, CA: Stanford University, Behavioral Sciences Division, Applied Mathematics and Statistics Laboratory.

Alexandrov, A. D. (1950). On Lorentz transformations. *Uspekhi Matematicheskikh Nauk*, *5*, 3 (37), 187 (in Russian).

Alexandrov, A. D. (1975). Mappings of spaces with families of cones and space-time transformations. *Annali di Matematica Pura ed Applicata*, *103*, 229–257.

Alpern, M. (1974). What is it that confines in a world without color? *Investigative Ophthalmology 13*, 647–674.

Alpern, M. (1976). Tritanopia. *American Journal of Optometry and Physiological Optics*, *53*, 340–349.

Alpern, M. (1979). Lack of uniformity in colour matching. *Journal of Physiology (London)*, *288*, 85–105.

Alpern, M., Kitihara, K., & Krantz, D. H. (1983a). Classical tritanopia. *Journal of Physiology (London)*, *335*, 665–681.

Alpern, M., Kitihara, K., & Krantz, D. H. (1983b). Perception of color in unilateral tritanopia. *Journal of Physiology (London)*, *335*, 683–697.

Alpern, M., & Pugh, E. N. (1977). Variation in the action spectrum of erythrolabe among deuteranopes. *Journal of Physiology (London)*, *266*, 613–646.

Alpern, M., & Wake, T. (1977). Cone pigments in human deutan colour vision defects. *Journal of Physiology (London)*, *266*, 595–612.

Ames, A. (1946). Binocular vision as affected by relations between uniocular stimulus patterns in common place environments. *American Journal of Psychology*, *59*, 333.

Ames, A., Ogle, K., & Gliddon, G. H. (1932). Corresponding retinal points, the horopter and size and shape of ocular images. *Journal of the Optical Society of America*, *22*, 538–631.

Angell, R. B. (1974). The geometry of visibles. *Nôus*, *8*, 87–117.

Archimedes. (1897). *Measurement of a circle* (T. L. Heath, Trans.) (pp. 91–98). London & New York: Cambridge University Press.

Armstrong, W. E. (1939). The determinateness of the utility function. *Economics Journal*, *49*, 453–467.

Armstrong, W. E. (1948). Uncertainty and the utility function. *Economics Journal*, *58*, 1–10.

Armstrong, W. E. (1950). A note on the theory of consumer's behavior. *Oxford Economic Papers*, *2*, 119–122.

Armstrong, W. E. (1951). Utility and the theory of welfare. *Oxford Economic Papers*, *3*, 259–271.

Armstrong, W. E. (1957–1958). Utility and the "ordinalist fallacy." *Review of Economic Studies*, *25*, 172–181.

Attneave, F. (1950). Dimensions of similarity. *American Journal of Psychology*, *63*, 516–556.

Augenstein, E. J., & Pugh, E. N., Jr. (1977). The dynamics of the II_1 colour mechanism: further evidence for two sites of adaptation. *Journal of Physiology*, *272*, 247–281.

Bachmann, F. (1959). *Aufbau der Geometrie aus dem Spiegelungsbegriff*. Berlin: Springer-Verlag.

Baker, K. A., Fishburn, P. C., & Roberts, F. S. (1972). Partial orders of dimension. 2. *Networks*, *2*, 11–28.

Battro, A. M., Netto, S. P., & Rozestraten, R. J. A. (1976). Riemannian geometries of variable curvature in visual space: Visual alleys, horopters, and triangles in big open fields. *Perception*, *5*, 9–23.

Beals, R., & Krantz, D. H. (1967). Metrics and geodesics induced by order relations. *Mathematische Zeitschrift*, *101*, 285–298.

Beals, R., Krantz, D. H., & Tversky, A. (1968). Foundations of multidimensional scaling. *Psychological Review*, *75*, 127–142.

Becker, G. M., DeGroot, M. H., & Marschak, J. (1963a). Stochastic models of choice behavior. *Behavioral Science*, *8*, 41–55.

Becker, G. M., DeGroot, M. H., & Marschak, J. (1963b). An experimental study of some stochastic models for wagers. *Behavioral Science*, *8*, 199–202.

Bellman, R. (1965). Functional equations. In R. D. Luce, R. R. Bush, & E. Galanter (Eds.), *Handbook of mathematical psychology* (Vol. 3, pp. 487–513). New York: Wiley.

Berkeley, G. (1901). An essay towards a new theory of vision. In A. C. Fraser (Ed.), *Berkeley's complete works* (Vol. 1). London & New York: Oxford University Press. (Original work published 1709.)

Besl, P. J., & Jain, R. C. (1985). Three-dimensional object recognition. *ACM Computing Surveys*, *17*, 75–145.

Beth, E. W., & Tarski, A. (1956). Equilaterality as the only primitive notion of Euclidean geometry. *Indagationes Mathematicae*, *18*, 462–467.

Blank, A. A. (1953). The Luneburg theory of binocular visual space. *Journal of the Optical Society of America*, *43*, 717–727.

Blank, A. A. (1957). The geometry of vision. *British Journal of Physiological Optics*, *14*, 154–169, 213.

Blank, A. A. (1958). Analysis of experiments in binocular space perception. *Journal of the Optical Society of America*, *48*, 911–925.

Blank, A. A. (1961). Curvature of binocular visual space: An experiment. *Journal of the Optical Society of America, 51*, 335–339.

Blank, A. A. (1978). Metric geometry in human binocular perception: Theory and fact. In E. L. J. Leeuwenberg & H. F. J. M. Buffart (Eds.), *Formal theories of visual perception* (pp. 83–102). New York: Wiley.

Block, H. D., & Marschak, J. (1960). Random orderings and stochastic theories of responses. In I. Olkin, S. Ghurye, W. Hoeffding, W. Madow, & H. Mann (Eds.), *Contributions to probability and statistics* (pp. 97–132). Stanford, CA: Stanford University Press.

Blumenfeld, W. (1913). Untersuchungen über die scheinbare Grösse in Sehraume. *Zeitschrift für Psychologie und Physiologie der Sinnesorgane, 65*, 241–404.

Blumenthal, L. M. (1961). *A modern view of geometry.* San Francisco, CA: Freeman.

Bock, R. D., & Jones, L. V. (1968). *The measurement and prediction of judgment and choice.* San Francisco, CA: Holden-Day.

Bogart, K. P. (1973). Maximal dimensional partially ordered sets: T. Hiraguchi's theorem. *Discrete Mathematics, 5*, 21–31.

Bogart, K. P., & Trotter, W. T. (1973). Maximal dimensional partially ordered sets. II. *Discrete Mathematics, 5*, 33–43.

Borsuk, K., & Szmielew, W. (1960). *Foundations of geometry.* Amsterdam: North-Holland.

Bouchet, A. (1971). *Etude combinatoire des ordonnés finis. Applications.* Thèse, Université Scientifique et Médicale de Grenoble.

Bradley, R. A. (1954a). Incomplete block rank analysis: On the appropriateness of the model for a method of paired comparisons. *Biometrics, 10*, 375–390.

Bradley, R. A. (1954b). Rank analysis of incomplete block designs. II. Additional tables for the method of paired comparisons. *Biometrika, 41*, 502–537.

Bradley, R. A. (1965). Another interpretation of a model for paired comparisons. *Psychometrika, 30*, 315–318.

Bradley, R. A., & Terry, M. E. (1952). Rank analysis of incomplete block designs. I. The method of paired comparisons. *Biometrika, 39*, 324–345.

Bridges, D. S. (1983). A numerical representation of preferences with intransitive indifference. *Journal of Mathematical Economics, 11*, 25–42.

Bridges, D. S. (1985). Representing interval orders by a single real-valued function. *Journal of Economic Theory, 36*, 149–155.

Brindley, G. S. (1953). The effects on colour vision of adaptation to very bright lights. *Journal of Physiology (London), 122*, 332–350.

Brindley, G. S. (1957). Two theorems in colour vision. *Quarterly Journal of Experimental Psychology, 9*, 101–104.

Brindley, G. S. (1970). *Physiology of the retina and the visual pathway* (2nd ed.). London: Edward Arnold.

Brown, A., & Pearcy, C. (1977). *Introduction to operator theory: Vol. 1. Elements of functional analysis.* New York: Springer-Verlag.

Buneman, P. (1971). The recovery of trees from measures of dissimilarity. In F. R. Hodson, D. G. Kendall, & P. Tautu (Eds.), *Mathematics in the archaeological and historical sciences.* Edinburgh: Edinburgh University Press.

Buneman, P. (1974). A note on the metric properties of trees. *Journal of Combinatorial Theory, 17(B)*, 48–50.

Burke, C. J., & Zinnes, J. L. (1965). A paired comparison of paired comparisons. *Journal of Mathematical Psychology, 2*, 53–76.

Burnham, R. W., Evans, R. M., & Newhall, S. M. (1957). Prediction of color appearance with different adaptation illuminations. *Journal of the Optical Society of America, 47*, 35–42.

Burns, B., Shepp, B. E., McDonough, D., & Weiner-Ehrlich, W. K. (1978). The relation between stimulus analyzability and perceived dimensional structure. In G. H. Bower (Ed.), *The psychology of learning and motivation* (Vol. 12, pp. 77–115). New York: Academic Press.

Burros, R. H. (1974). Axiomatic analysis of non-transitivity of preference and of indifference. *Theory and Decision, 5*, 185–204.

Busemann, H. (1955). *The geometry of geodesics.* New York: Academic Press.

Busemann, H., & Kelly, P. J. (1953). *Projective geometry and projective metrics.* New York: Academic Press.

Campbell, N. R. (1920). *Physics: The Elements.* London & New York: Cambridge University Press. (Reprinted as *Foundations of science: The philosophy of theory and experiment.* New York: Dover, 1957.)

Carroll, J. D. (1976). Spatial, nonspatial and hybrid models of scaling. *Psychometrika, 41*, 439–463.

Carroll, J. D., & Arabie, P. (1980). Multidimensional scaling. *Annual Review of Psychology, 31*, 607–649.

Carroll, J. D., & Chang, J. (1970). Analysis of individual differences in multidimensional scaling via an N-way generalization of "Eckart-Young" decomposition. *Psychometrika, 35*, 283–319.

Chipman, J. S. (1971). Consumption theory without transitive indifference. In J. S. Chipman, L. Hurwicz, M. K. Richter, & H. Sonnenschein (Eds.), *Studies in the mathematical foundations of utility and demand theory: A symposium at the University of Michigan, 1968* (pp. 224–253). New York: Harcourt, Brace & World.

Cicerone, C. M., & Green, D. G. (1980). Light adaptation within the receptive field centre of the rat retinal ganglion cells. *Journal of Physiology 301*, 517–534.

Cicerone, C. M., Krantz, D. H., & Larimer, J. (1975). Opponent process additivity. III. Effect of moderate chromatic adaptation. *Vision Research, 15*, 1125–1135.

Cicerone, C. M., Nagy, A. L., & Nerger, J. L. (1987). Equilibrium hue judgments of dichromats. *Vision Research, 27*, 983–991.

Clarke, F. R. (1957). Constant-ratio rule for confusion matrices in speech communication. *Journal of the Acoustical Society of America, 29*, 715–720.

Cogis, O. (1976). Détermination d'un préordre total contenant un préordre et contenu dans une relation de Ferrers lorsque leur domaine commun est fini. *Comptes Rendus de l'Académie des Sciences, Serie A, 283*, 1007–1009.

Cogis, O. (1980). *La dimension Ferrers des graphes orientés.* Thèse, Université Pierre et Marie Curie, Paris.

Cogis, O. (1982a). Ferrers digraphs and threshold graphs. *Discrete Mathematics, 38*, 33–46.

Cogis, O. (1982b). On the Ferrers dimension of a digraph. *Discrete Mathematics, 38*, 47–52.

Cohen, M. (1980). Random utility systems—the infinite case. *Journal of Mathematical Psychology, 22*, 1–23.

Cohen, M., & Falmagne, J.-C. (1978). Random scale representation of binary choice probabilities: A counterexample to a conjecture of Marschak. *New York University, Mathematical Studies in Perception & Cognition, 3.*

Colonius, H. (1984). *Stochastische Theorien Individuellen Wahlverhaltens.* Berlin: Springer-Verlag.

Coombs, C. H. (1964). *A theory of data.* New York: Wiley.

Corbin, R., & Marley, A. A. J. (1974). Random utility models with equality: An apparent, but not actual generalization of random utility models. *Journal of Mathematical Psychology, 11*, 274–293.

Corter, J., & Tversky, A. (1986). Extended similarity trees. *Psychometrika, 51*, 429–451.

Coxeter, H. S. M. (1961). *Introduction to geometry.* New York: Wiley.

Coxon, A. P. M. (1982). *The user's guide to multidimensional scaling*. Exeter, NH: Heinemann Educational Books.

Cozzens, M. B., & Roberts, F. S. (1982). Double semiorders and double indifference graphs. *SIAM Journal on Algebraic and Discrete Methods*, *3*, 566–583.

Crampe, S. (1958). Angeordnete projektive Ebenen. *Mathematische Zeitschrift*, *69*, 435–462.

Cutting, J. E. (1986). *Perception with an eye for motion*. Cambridge, MA: MIT Press.

Dagsvik, J. K. (1983). Discrete dynamic choice: An extension of the choice models of Thurstone and Luce. *Journal of Mathematical Psychology*, *27*, 1–43.

Daniels, N. (1972). Thomas Reid's discovery of a non-Euclidean geometry. *Philosophy of Science*, *39*, 219–234.

David, H. A. (1963). *The method of paired comparisons*. London: Griffin.

Davies, P. M., & Coxon, A. P. M. (Eds.). (1982). *Key texts in multidimensional scaling*. Exeter, NH: Heinemann Educational Books.

Davison, M. L. (1983). *Multidimensional scaling*. New York: Wiley.

Debreu, G. (1958). Stochastic choice and cardinal utility. *Econometrica*, *26*, 440–444.

Debreu, G. (1960). Review of R. D. Luce, Individual choice behavior: A theoretical analysis. *American Economic Review*, *50*, 186–188.

Delessert, A. (1964). *Une construction de la géométrie élémentaire fondée sur la notion de réflexion*. Genève: L'Enseignement Mathématique, Imprimerie Kundig.

Dembowski, P. (1968). *Finite geometries*. New York: Springer-Verlag.

Devletoglou, N. E. (1965). A dissenting view of duopoly and spatial competition. *Economica*, *32*, 140–160.

Devletoglou, N. E. (1968). Threshold and rationality. *Kylos*, *21*, 623–636.

Devletoglou, N. E., & Demetriou, P. A. (1967). Choice and threshold: A further experiment in spatial duopoly. *Economica*, *34*, 351–371.

Dobson, J. (1974). Unrooted trees for numerical taxonomy. *Journal of Applied Probability*, *11*, 32–42.

Doignon, J.-P. (1984). Generalizations of interval orders. In E. Degreef & J. Van Buggenhaut (Eds.), *Trends in mathematical psychology* (pp. 209–217). Amsterdam: North-Holland.

Doignon, J.-P. (1987). Threshold representations of multiple semiorders. *SIAM Journal on Algebraic and Discrete Methods*, *8*, 77–84.

Doignon, J.-P., Ducamp, A., & Falmagne, J.-C. (1984). On realizable biorders and the biorder dimension of a relation. *Journal of Mathematical Psychology*, *28*, 73–109.

Doignon, J.-P., Ducamp, A., & Falmagne, J.-C. (1987). On the separation of two relations by a biorder or a semiorder. *Mathematical Social Science*, *13*, 1–18.

Doignon, J. P., & Falmagne, J.-C. (1974). Difference measurement and simple scalability with restricted solvability. *Journal of Mathematical Psychology*, *11*, 493–499.

Doignon, J.-P., & Falmagne, J.-C. (1984). Matching relations and the dimensional structure of social choices. *Mathematical Social Sciences*, *7*, 211–229.

Domotor, Z. (1972). Causal models and space-time geometries. *Synthese*, *24*, 5–57.

Domotor, Z., & Stelzer, J. H. (1971). Representation of finitely additive semiordered qualitative probability structures. *Journal of Mathematical Psychology*, *8*, 145–158.

Donnell, M. L. (1977). *Individual red/green and yellow/blue opponent-isocancellation functions: Their measurement and prediction*. Doctoral dissertation, University of Michigan, Ann Arbor (University Microfilms).

Dorling, J. Special relativity out of Euclidean geometry. *In press*.

Dubrovin, B. A., Fomenko, A. T., & Novikov, S. P. (1984). *Modern geometry—methods and applications. Part I. The geometry of surfaces, transformation groups, and fields*. New York: Springer-Verlag.

Dubrovin, B. A., Fomenko, A. T., & Novikov, S. P. (1985). *Modern geometry—methods and applications. Part II. The geometry and topology of manifolds.* Berlin & New York: Springer-Verlag.

Ducamp, A. (1967). Sur la dimension d'un ordre partiel. In *Theory of graphs* (pp. 103–112). New York: Gordon & Breach.

Ducamp, A. (1978). A note on an alternative proof of the representation theorem for bi-semiorder. *Journal of Mathematical Psychology, 18*, 100–104.

Ducamp, A., & Falmagne, J.-C. (1969). Composite measurement. *Journal of Mathematical Psychology, 6*, 359–390.

Dushnik, B., & Miller, E. W. (1941). Partially ordered sets. *American Journal of Mathematics, 63*, 600–610.

Edgell, S. E., & Geisler, W. S. (1980). A set-theoretic random utility model of choice behavior. *Journal of Mathematical Psychology, 21*, 265–278.

Eisler, H., & Knöppel, J. (1970). Relative attention in judgments of heterogeneous similarity. *Perception & Psychophysics, 8*, 420–426.

Ejimo, Y., & Takahashi, S. (1984). Bezold-Bruecke hue shift and non-linearity in opponent-color process. *Vision Research, 24*, 1897–1904.

Eschenburg, J.-H. (1980). Is binocular visual space constantly curved? *Journal of Mathematical Biology, 9*, 3–22.

Euclid. (1945). Optics (H. E. Burton, Trans.). *Journal of the Optical Society of America, 35*, 357–372.

Euler, L. (1736). *Mechanica sive motus scientia analytice exposita.* Petrogad: Imperial Academy of Science. In L. Euleri (1912), *Opera omnia,* 2nd series, *Opera mechanica et astronomica* (Vol. 1). Leipzig: Teubner.

Falmagne, J.-C. (1971). The generalized Fechner problem and discrimination. *Journal of Mathematical Psychology, 8*, 22–43.

Falmagne, J.-C. (1974). Foundations of Fechnerian psychophysics. In D. H. Krantz, R. C. Atkinson, R. D. Luce, & P. Suppes (Eds.), *Contemporary developments in mathematical psychology* (Vol. 2, pp. 127–159). San Francisco, CA: Freeman.

Falmagne, J.-C. (1976). Random conjoint measurement and loudness summation. *Psychological Review, 83*, 65–79.

Falmagne, J.-C. (1977). Weber's inequality and Fechner's problem. *Journal of Mathematical Psychology, 16*, 267–271.

Falmagne, J.-C. (1978). A representation theorem for finite random scale systems. *Journal of Mathematical Psychology, 18*, 52–72.

Falmagne, J.-C. (1979). On a class of probabilistic conjoint measurement models: Some diagnostics properties. *Journal of Mathematical Psychology, 19*, 73–88.

Falmagne, J.-C. (1980). A probabilistic theory of extensive measurement. *Philosophy of Science 47*, 277–296.

Falmagne, J.-C. (1985). *Elements of psychophysical theory.* London & New York: Oxford University Press.

Falmagne, J.-C., & Iverson, G. (1979). Conjoint Weber laws and additivity. *Journal of Mathematical Psychology, 20*, 164–183.

Falmagne, J.-C., Iverson, G., & Marcovici, S. (1979). Binaural "loudness" summation: Probabilistic theory and data. *Psychological Review, 86*, 25–43.

Fechner, G. T. (1860). *Elemente der Psychophysik.* Leipzig: Druck und Verlag von Breitkopfs Härtel [*Elements of Psychophysics* (Vol. 1). New York: Holt, Rinehart, & Winston, 1966].

Fillenbaum, S., & Rapoport, A. (1971). *Structures in the subjective lexicon.* New York: Academic Press.

Fishburn, P. C. (1968). Semiorders and risky choices. *Journal of Mathematical Psychology, 5*, 358–361.

Fishburn, P. C. (1970a). An interval graph is not a comparability graph. *Journal of Combinatorial Theory, 8*, 442–443.

Fishburn, P. C. (1970b). Intransitive indifference in preference theory: A survey. *Operations Research, 18*, 207–228.

Fishburn, P. C. (1970c). Intransitive indifference with unequal indifference intervals. *Journal of Mathematical Psychology, 7*, 144–149.

Fishburn, P. C. (1973a). Interval representations for interval orders and semiorders. *Journal of Mathematical Psychology, 10*, 91–105.

Fishburn, P. C. (1973b). On the construction of weak orders from fragmentary information. *Psychometrika, 38*, 459–472.

Fishburn, P. C. (1973c). Binary choice probabilities: On the varieties of stochastic transitivity. *Journal of Mathematical Psychology, 10*, 327–352.

Fishburn, P. C. (1975). Semiorders and choice functions. *Econometrica, 43*, 975–977.

Fishburn, P. C. (1976). Binary choice probabilities between gambles: Interlocking expected utility models. *Journal of Mathematical Psychology, 14*, 99–122.

Fishburn, P. C. (1977). Models of individual preference and choice. *Synthese, 36*, 287–314.

Fishburn, P. C. (1980). Lexicographic additive differences. *Journal of Mathematical Psychology, 21*, 191–198.

Fishburn, P. C. (1985). *Interval orders and interval graphs*. New York: Wiley.

Fishburn, P. C., & Gehrlein, W. V. (1974). Alternative methods of constructing strict weak orders from interval orders. *Psychometrika, 39*, 501–516.

Fishburn, P. C., & Gehrlein, W. V. (1975). A comparative analysis of methods for constructing weak orders from partial orders. *Journal of Mathematical Sociology, 4*, 93–102.

Foley, J. M. (1964a). Desarguesian property in visual space. *Journal of the Optical Society of America, 54*, 684–692.

Foley, J. M. (1964b). Visual space: A test of the constant curvature hypothesis. *Psychonomic Science, 1*, 9–10.

Foley, J. M. (1972). The size-distance relation and intrinsic geometry of visual space: Implications for processing. *Vision Research, 12*, 323–332.

Foley, J. M. (1978). Primary distance perception. In R. Held, H. Liebowitz, & R. Teuber (Eds.), *Handbook of sensory physiology: Vol. 8. Perception* (pp. 181–213). Berlin & New York: Springer-Verlag.

Foley, J. M. (1980). Binocular distance perception. *Physological Review, 87*, 411–434.

Ford, L. R., Jr., (1957). Solution of a ranking problem from binary comparisons. *American Mathematical Monthly, Herbert Ellsworth Slaught Memorial Papers, 64*, 28–33.

Freudenthal, H. (1965). Lie groups in the foundations of geometry. *Advances in Mathematics, 1*, 145–190.

Fu, K. S. (1974). *Syntatic methods in pattern recognition*. New York: Academic Press.

Garner, W. R. (1974). *The processing of information and structure*. Potomac, MD: Erlbaum.

Gati, I., & Tversky, A. (1982). Representation of qualitative and quantitative dimensions. *Journal of Experimental Psychology: Human Perception and Performance, 8*, 325–340.

Gati, I., & Tversky, A. (1984). Weighing common and distinctive features in perceptual and conceptual judgments. *Cognitive Psychology, 16*, 341–370.

Gensemer, S. H. (1987). On relationships between numerical representations of interval orders and semiorders. *Journal of Economic Theory, 43*, 157–169.

Gensemer, S. H. Continuous semiorder representations. *Journal of Mathematical Economics. In press.*

Georgescu-Roegen, N. (1936). The pure theory of consumer's behavior. *Quarterly Journal of Economics*, *50*, 545–593. (Reprinted in *Analytical economics: Issues and problems* (pp. 133–170). Cambridge, MA: Harvard University Press, 1966.)

Geogescu-Roegen, N. (1958). Threshold in choice and the theory of demand. *Econometrica*, *26*, 157–168. (Reprinted in *Analytical economics: Issues and problems* (pp. 228–240). Cambridge, MA: Harvard University Press, 1966.)

Gerlach, M. W. (1957). *Interval measurement of subjective magnitudes with subliminal differences* (Tech. Rep. No. 7). Stanford, CA: Stanford University, Behavioral Sciences Division, Applied Mathematics and Statistics Laboratory.

Ghouila-Houri, A. (1962). Caractérisation des graphes nonorientés dont on peut orienter les arêtes de manière à obtenir le graphe d'une relation d'ordre. *Comptes Rendus Hebdomadaires des Séances de l'Académie des Sciences*, *254*, 1370–1371.

Gigerenzer, G., & Strube, G. (1983). Are there limits to binaural additivity of loudness? *Journal of Experimental Psychology: Human Perception and Performance*, *9*, 126–136.

Gilmore, P. C., & Hoffman, A. J. (1964). A characterization of comparability graphs and of interval graphs. *Canadian Journal of Mathematics*, *16*, 539–548.

Goldblatt, R. (1987). *Orthogonality and spacetime geometry*. New York: Springer-Verlag.

Goodman, N. (1951). *Structure of appearance*. Cambridge, MA: Harvard University Press.

Grassmann, H. (1854). On the theory of compound colours. *Philosophical Magazine, Series 4*, *7*, 254–264.

Green, D. M., & Swets, J. A. (1966). *Signal detection theory and psychophysics*. New York: Wiley.

Gruzewska, H. M. (1954). L'arithmétique des variables aléatoires. *Cahiers Rhodaniens*, *6*, 9–56.

Gulliksen, H. (1953). A generalization of Thurstone's learning function. *Psychometrika*, *18*, 297–307.

Guth, S. L., Donley, J., & Marrocco, R. T. (1969). On luminance additivity and related topics. *Vision Research*, *9*, 537–575.

Guth, S. L., Massof, R. W., & Benzschawel, T. (1980). Vector model for normal and dichromatic color vision. *Journal of the Optical Society of America*, *70*, 197–212.

Guttman, L. (1968). A general nonmetric technique for finding the smallest coordinate space for a configuration of points. *Psychometrika*, *33*, 469–506.

Hakimi, S. L., & Yau, S. S. (1964). Distance matrix of a graph and its realizability. *Quarterly of Applied Mathematics*, *22*, 305–317.

Halphen, E. (1955). La notion de vraisemblance. *Publication de l'Institut de Statistique de l'Université de Paris*, *4*, 41–92.

Hardy, G. H., Littlewood, J. E., & Polya, G. (1952). *Inequalities*. London & New York: Cambridge University Press.

Hardy, L. H., Rand, G., Rittler, M. C., Blank, A. A., & Boeder, P. (1953). *The geometry of binocular space perception*. New York: Knapp Memorial Laboratories, Institute of Ophthalmology, Columbia University College of Physicians and Surgeons.

Hartigan, J. A. (1967). Representation of similarity matrices by trees. *Journal of the American Statistical Association*, *62*, 1140–1158.

Hartigan, J. A. (1975). *Clustering algorithms*. New York: Wiley.

Hausdorff, F. (1923). Momentprobleme für ein endliches Intervall. *Mathematische Zeitschrift*, *16*, 220–248.

Helmholtz, H. von (1867). *Handbuch der Physiologischen Optik* (1st ed.). Leipzig: Voss.

Helmholtz, H. von (1868). Ueber die Thatsachen, die der Geometrie zu Grunde liegen. *Göttingen Nachrichten*, *9*, 193–221.

Helmholtz, H. von (1896). *Handbuch der Physiologischen Optik* (2nd ed.). Leipzig: Voss.

Henley, N. M. (1969). A psychological study of the semantics of animal terms. *Journal of Verbal Learning and Verbal Behavior*, *8*, 176–184.

Hering, E. (1878). *Zur Lehre vom Lichtsinne*. Vienna: C. Gerold's Sohn.

Hering, E. (1920). *Grundzüge der Lehre vom Lichtsinn*. Berlin: Springer-Verlag. L. M. Hurvich & D. Jameson (Trans.), *Outlines of a theory of the light sense*. Cambridge, MA: Harvard University Press, 1964.

Heyting, A. (1963). *Axiomatic projective geometry*. New York: Wiley.

Hilbert, D. (1899). *Grundlagen der Geometrie* (8th ed., with revisions and supplements by P. Bernays, 1956). Stuttgart: Teubner.

Hillebrand, F. (1902). Theorie der scheinbaren Grösse bei binocularem Sehen. *Denkschriften der Wiener Akademie der Wissenschaften, Mathematische-Naturwissenschaftliche Classe*, *7*, 255–307.

Hiraguchi, T. (1951). On the dimension of partially ordered sets. *Science Reports of Kanazawa University*, *1*, 77–94.

Hochberg, J. (1988). Visual perception. In R. C. Atkinson, R. J. Herrnstein, G. Lindzey, & R. D. Luce (Eds.), *Handbook of Experimental Psychology* (pp. 195–276). New York: Wiley.

Hölder, O. (1901). Die Axiome der Quantität und die Lehre vom Mass. *Berichte über die Verhandlungen der Königlich Sächsischen Gesellschaft der Wissenschaften zu Leipzig, Mathematische-Physische Klasse*, *53*, 1–64.

Holman, E. W. (1972). The relation between hierarchical and Euclidean models for psychological distances. *Phychometrika*, 417–423.

Holman, E. W. (1979). Monotonic models for asymmetric proximities. *Journal of Mathematical Psychology*, *20*, 1–15.

Huber, O. (1979). Nontransitive multidimensional preferences: Theoretical analysis of a model. *Theory and Decision*, *10*.

Hudgin, R. H. (1973). Coordinate-free relativity. In P. Suppes (Ed.), *Space, time and geometry* (pp. 366–382). Dordrecht, Netherlands: Reidel.

Hunt, R. W. G. (1953). The perception of color in 1° fields for different states of adaptation. *Journal of the Optical Society of America*, *43*, 479–484.

Hurewicz, H., & Wallmann, H. (1948). *Dimension theory*. Princeton, NJ: Princeton University Press.

Hurvich, L. M. (1972). Color vision deficiencies. In D. Jameson & L. M. Hurvich (Eds.), *Handbook of sensory physiology: Vol. VII/4. Visual psychophysics* (pp. 582–624). Berlin & New York: Springer-Verlag.

Hurvich, L. M., & Jameson, D. (1951a). The binocular fusion of yellow in relation to color theories. *Science*, *114*, 199–202.

Hurvich, L. M., & Jameson, D. (1951b). A psychophysical study of white. I. Neutral adaptation. *Journal of the Optical Society of America*, *41*, 521–527.

Hurvich, L. M., & Jameson, D. (1955). Some quantitative aspects of opponent-colors theory. II. Brightness, saturation, and hue in normal and dichromatic vision. *Journal of the Optical Society of America*, *45*, 602–616.

Hurvich, L. M., & Jameson, D. (1956). Some quantitative aspects of an opponent-colors theory. IV. A psychological color specification system. *Journal of the Optical Society of America*, *46*, 416–421.

Hurvich, L. M., & Jameson, D. (1957). An opponent-process theory of color vision. *Psychological Reviews*, *64*, 384–404.

Hurvich, L. M., & Jameson, D. (1958). Further development of a quantified opponent-colour theory. In *Visual problems of colour* (Vol. 2). London: H.M. Stationery Office.

Indow, T. (1967). Two interpretations of binocular visual space: Hyperbolic and Euclidean. *Annals of the Japan Association for Philosophy of Science*, *3*, 51–64.

Indow, T. (1968). Multidimensional mapping of visual space with real and simulated stars. *Perception & Psychophysics*, *3*, 45–53.

Indow, T. (1974). Applications of multidimensional scaling in perception. In E. C. Carterette & P. Friedman (Eds.), *Handbook of perception: Vol. 2. Psychophysical judgment and measurement* (pp. 493–531). New York: Academic Press.

Indow, T. (1975a). An application of MDS to study of binocular visual space. *U.S.-Japan Seminar: Theory, methods and applications of multidimensional scaling and related techniques*. University of California, August 20–24, San Diego.

Indow, T. (1975b). On choice probability. *Behaviometrika*, *2*, 13–31.

Indow, T. (1979). Alleys in visual space. *Journal of Mathematical Psychology*, *19*, 221–258.

Indow, T. (1982). An approach to geometry of visual space with no a priori mapping functions: Multidimensional mapping according to Riemannian metrics. *Journal of Mathematical Psychology*, *26*, 204–236.

Indow, T., Inoue, E., & Matsushima, K. (1962a). An experimental study of the Luneburg theory of binocular space perception (1). The 3- and 4-point experiments. *Japanese Psychological Research*, *4*, 6–16.

Indow, T., Inoue, E., & Matsushima, K. (1962b). An experimental study of the Luneburg theory of binocular space perception (2). The alley experiments. *Japanese Psychological Research*, *4*, 17–24.

Indow, T., Inoue, E., & Matsushima, K. (1963). An experimental study of the Luneburg theory of binocular space perception (3). The experiments in a spacious field. *Japanese Psychological Research*, *5*, 10–27.

Indow, T., & Watanabe, T. (1984a). Parallel—and distance—alleys with moving points in the horizontal plane. *Perception & Psychophysics*, *35*, 144–154.

Indow, T., & Watanabe, T. (1984b). Parallel alleys and distance-alleys on horopter plane in the dark. *Perception*, *13*, 165–182.

Ingling, C. R., Jr., Russell, P. W., Rea, M. S., & Tsou, B. H. (1978). Red-green opponent spectral sensitivity: Disparity between cancellation and direct matching methods. *Science*, *201*, 1221–1223.

Ismailov, Ch. A. (1982). Uniform color space and multidimensional scaling (MDS). In H.-G. Geissler & P. Petzold (Eds.), *Psychophysical judgment and the process of perception* (pp. 52–62). Amsterdam: North-Holland.

Iverson, G., & Falmagne, J.-C. (1985). Statistical issues in measurement. *Mathematical Social Sciences*, *10*, 131–153.

Jacobson, N. (1951). *Lectures in abstract algebra: Vol. 1. Basic concepts*. Princeton, NJ: Van Nostrand.

Jacobson, N. (1953). *Lectures in abstract algebra: Vol. 2. Linear algebra*. Princeton, NJ: Van Nostrand.

Jameson, D., & Hurvich, L. M. (1955). Some quantitative aspects of an opponent-colors theory. I. Chromatic responses and spectral saturation. *Journal of the Optical Society of America*, *45*, 546–552.

Jameson, D., & Hurvich, L. M. (1956). Some quantitative aspects of an opponent-colors theory. III. Changes in brightness, saturation, and hue with chromatic adaptation. *Journal of the Optical Society of America*, *46*, 405–415.

Jameson, D., & Hurvich, L. M. (1959). Perceived color and its dependence on focal, surrounding, and preceding stimulus variables. *Journal of the Optical Society of America*, *49*, 890–898.

Jameson, D., & Hurvich, L. M. (1961). Opponent chromatic induction: Experimental evaluation and theoretical account. *Journal of the Optical Society of America*, *51*, 46–53.

Jamison, D. T., & Lau, L. J. (1973). Semiorders and the theory of choice. *Econometrica, 41,* 901–902.

Jammer, M. (1957). *Concepts of force.* Cambridge, MA: Harvard University Press.

Jardine, N., & Sibson, R. (1971). *Mathematical taxonomy.* New York: Wiley.

Johnson, E., & Tversky, A. (1984). Representations of perceptions of risk. *Journal of Experimental Psychology: General, 113,* 55–70.

Johnson, S. C. (1967). Hierarchical clustering schemes. *Psychometrika, 32,* 241–254.

Jónsson, B., & Tarski, A. (1952). Boolean algebras with operators. Part II. *American Journal of Mathematics, 74,* 127–162.

Judd, D. B. (1951). Basic correlates of the visual stimulus. In S. S. Stevens (Ed.), *Handbook of experimental psychology* (pp. 811–867). New York: Wiley.

Judd, D. B. (1960). Appraisal of Land's work on two-primary color projections. *Journal of the Optical Society of America, 50,* 254–268.

King-Smith, P. E. (1973). The optical density of erythrolabe determined by retinal densitometry using the self-screening method. *Journal of Physiology (London), 230,* 551–560.

Kirchhoff, G. (1876). *Vorlesungen ueber mathematische Physik: Mechanik.* Leipzig: Teubner.

Klein, F. (1893). A comparative review of recent researches in geometry. *Bulletin of the New York Mathematical Society, 2,* 215–249.

Kline, J. R. (1916). Double elliptic geometry in terms of point and order alone. *Annals of Mathematics, 18,* 31–44.

Klingenberg, W. (1982). *Riemannian geometry.* Berlin: de Gruyter.

König, A., & Dieterici, C. (1892). Die Grundempfindungen in normalen und anormalen Farbensystemen und ihre Intensitätvertheilung im Spectrum. *Zeitschrift für Psychologie und Physiologie der Sinnesorgane, 4,* 241–347. (Reprinted in A. König, *Gesammelte Abhandlungen zur physiologischen Optik* (pp. 214–321). Leipzig: Barth, 1903).

Kordos, M. (1969). On the syntactic form of dimension axiom for affine geometry. *Bulletin de l'Académie Polonaise des Sciences, Série des Sciences Mathématiques, Astronomiques et Physiques, 17,* 833–837.

Kordos, M. (1973). Elliptic geometry as a theory of one binary relation. *Bulletin de l'Académie Polonaise des Sciences, Série des Sciences Mathématiques, Astronomiques et Physiques, 21,* 609–614.

Krantz, D. H. (1964). *The scaling of small and large color differences.* Unpublished doctoral dissertation, University of Pennsylvania, Philadelphia. University Microfilms No. 65-5777.

Krantz, D. H. (1967). Extensive measurement in semiorders. *Philosophy of Science, 34,* 348–362.

Krantz, D. H. (1968a). A survey of measurement theory. In G. B. Dantzig & A. F. Veinott, Jr. (Eds.), *Mathematics of the decision sciences: Part 2. Lectures in applied mathematics (Vol. 12).* Providence, RI: American Mathematical Society.

Krantz, D. H. (1968b). A theory of context effects based on cross-context matching. *Journal of Mathematical Psychology, 5,* 1–48.

Krantz, D. H. (1972). Integration of just-noticeable differences. *Journal of Mathematical Psychology, 8,* 591–599.

Krantz, D. H. (1973). Fundamental measurement of force and Newton's first and second laws of motion. *Philosophy of Science, 40,* 481–495.

Krantz, D. H. (1974). Measurement theory and qualitative laws in psychophysics. In D. H. Krantz, R. C. Atkinson, R. D. Luce, & P. Suppes (Eds.), *Contemporary developments in mathematical psychology: Vol. 2. Measurement, psychophysics, and neural information processing* (pp. 160–199). San Francisco, CA: Freeman.

Krantz, D. H. (1975a). Color measurement and color theory. I. Representation theorem for Grassmann structures. *Journal of Mathematical Psychology, 12,* 283–303.

Krantz, D. H. (1975b). Color measurement and color theory. II. Opponent-colors theory. *Journal of Mathematical Psychology, 12,* 304–327.

Krantz, D. H., & Tversky, A. (1975). Similarity of rectangles: An analysis of subjective dimensions. *Journal of Mathematical Psychology, 12,* 4–34.

Krauskopf, J., Williams, D. R., & Heeley, D. W. (1982). Cardinal directions of color space. *Vision Research, 22,* 1123–1131.

Krein, M., & Milman, D. (1940). On the extreme points of regularly convex sets. *Studia Mathematica, 9,* 133–138.

Kruskal, J. B. (1964). Multidimensional scaling by optimizing goodness-of-fit to a nonmetric hypothesis. *Psychometrika, 29,* 1–28, 115–129.

Kruskal, J. B., & Wish, M. (1978). *Multidimensional scaling.* Beverly Hills, CA: Sage.

La Gournerie, J. (1859). *Traité de perspective linéaire contenant les tracés pour les tableaux plans et courbes, les bas-reliefs et les décorations théatrales, avec une théorie des effets de perspective* (1 vol. and 1 atlas of plates). Paris: Dalmont et Dunod; Mallet-Bachelier.

Laha, R. G. (1964). On a problem connected with beta and gamma distributions. *Transactions of the American Mathematical Society, 113,* 287–298.

Land, E. H. (1964). The retinex. *American Scientist, 52,* 247–264.

Larimer, J., Krantz, D. H., & Cicerone, C. M. (1974). Opponent process additivity. I. Red-green equilibria. *Vision Research, 14,* 1127–1140.

Larimer, J., Krantz, D. H., & Cicerone, C. M. (1975). Opponent-process additivity. II. Yellow-blue equilibria and nonlinear models. *Vision Research, 15,* 723–731.

Latzer, R. W. (1972). Nondirected light signals and the structure of time. *Synthese, 24,* 236–280.

LeGrand, Y. (1968). *Light, colour and vision.* London: Chapman & Hall.

Lekkerkerker, C. G., & Boland, J. C. (1962). Representation of a finite graph by a set of intervals on the real line. *Fundamental Mathematics, 51,* 45–64.

Levine, M. V. (1970). Transformations that render curves parallel. *Journal of Mathematical Psychology, 7,* 410–443.

Levine, M. V. (1972). Transforming curves into curves with the same shape. *Journal of Mathematical Psychology, 9,* 1–16.

Levine, M. V. (1975). Additive measurement with short segments of curves. *Journal of Mathematical Psychology, 12,* 212–224.

Lie, S. (1886). Bemerkungen zu Helmholtz' Arbeit über die Thatsachen, die der Geometrie zu Grunde liegen. *Berichte über die Verhandlungen der Königlich Sächsischen Gesellschaft der Wissenschaften zu Leipzig, Mathematisch-Physische Klasse, 38,* 337–342.

Lindberg, D. C. (1976). *Theories of vision from Al-Kindi to Kepler.* Chicago, IL: University of Chicago Press.

Lingenberg, R., & Baur, A. (1960). *Der synthetische und der analytische Standpunkt in der Geometrie, Grundzüge der Mathematik: Band II. Geometrie.* Göttingen: Vandenhoeck & Ruprecht.

Luce, R. D. (1956). Semiorders and a theory of utility discrimination. *Econometrica, 24,* 178–191.

Luce, R. D. (1959). *Individual choice behavior: A theoretical analysis.* New York: Wiley.

Luce, R. D. (1960). Response latencies and probabilities. In K. J. Arrow, S. Karlin, & P. Suppes (Eds.), *Mathematical methods in the social sciences, 1959* (pp. 298–311). Stanford, CA: Stanford University Press.

Luce, R. D. (1963). Detection and recognition. In R. D. Luce, R. R. Bush, & E. Galanter (Eds.), *Handbook of mathematical psychology* (Vol. 1), pp. 103–189. New York: Wiley.

Luce, R. D. (1973). Three axiom systems for additive semiordered structures. *SIAM Journal of Applied Mathematics, 25,* 41–53.

Luce, R. D. (1977). The choice axiom after twenty years. *Journal of Mathematical Psychology*, *15*, 215–233.

Luce, R. D. (1978). Lexicographic tradeoff structures. *Theory and Decision*, *9*, 187–193.

Luce, R. D., & Edwards, W. (1958). The derivation of subjective scales from just noticeable differences. *Psychological Review*, *65*, 222–237.

Luce, R. D., & Galanter, E. (1963). Discrimination. In R. D. Luce, R. R. Bush, & E. Galanter (Eds.), *Handbook of mathematical psychology* (Vol. 1, pp. 191–244). New York: Wiley.

Luce, R. D., & Raiffa, H. (1957). *Games and decisions: Introduction and critical survey*. New York: Wiley.

Luce, R. D., & Suppes, P. (1965). Preference, utility, and subjective probability. In R. D. Luce, R. R. Bush, & E. Galanter (Eds.), *Handbook of mathematical psychology* (Vol. 3, pp. 249–410). New York: Wiley.

Luneburg, R. K. (1947). *Mathematical analysis of binocular vision*. Princeton, NJ: Princeton University Press.

Luneburg, R. K. (1948). Metric methods in binocular visual perception. In *Studies and essays* (pp. 215–240). New York: Wiley (Interscience).

Luneburg, R. K. (1950). The metric of binocular visual space. *Journal of the Optical Society of America*, *40*, 627–642.

MacAdam, D. L. (1956). Chromatic adaptation. *Journal of the Optical Society of America*, *46*, 500–513.

McFadden, D. (1974). Conditional logit analyses of qualitative choice behavior. In P. Zarembka (Ed.), *Frontiers in econometrics*. New York: Academic Press.

McFadden, D. (1976). Quantal choice analysis: A survey. *Annals of Economic and Social Measurement*, *5*, 363–390.

McFadden, D. (1981). Econometric models of probabilistic choice. In C. F. Manski & D. McFadden (Eds.), *Structural analysis of discrete data*. Cambridge, MA: MIT Press.

McFadden, D., & Richter, M. K. (1971). *On the extension of a set function to a probability on the Boolean generated by a family of event, with applications*. Unpublished manuscript, University of California, Berkeley.

Machina, M. J. (1985). Stochastic choice functions generated from deterministic preferences over lotteries. *Economic Journal*, *95*, 575–594.

McKinsey, J. C. C., Sugar, A. C., & Suppes, P. (1953). Axiomatic foundations of classical particle mechanics. *Journal of Rational Mechanics & Analysis*, *2*, 253–272.

MacLeod, D. I. A., & Lennie, P. (1976). Red-green blindness confined to one eye. *Vision Research*, *16*, 691–702.

Manders, K. L. (1981). On JND representations of semiorders. *Journal of Mathematical Psychology*, *24*, 224–248.

Manski, C. F. (1977). The structure of random utility models. *Theory and Decision*, *8*, 229–254.

Marley, A. A. J. (1965). The relation between the discard and regularity conditions for choice probabilities. *Journal of Mathematical Psychology*, *2*, 242–253.

Marley, A. A. J. (1971). Conditions for the representation of absolute judgment and pair comparison isosensitivity curves by cumulative distributions. *Journal of Mathematical Psychology*, *8*, 554–590.

Marley, A. A. J. (1981). Multivariate stochastic processes compatible with "aspect" models of similarity and choice. *Psychometrika*, *46*, 421–428.

Marschak, J. (1960). Binary-choice constraints and random utility indicatos. In K. J. Arrow, S. Karlin, & P. Suppes (Eds.), *Mathematical methods in the social sciences, 1959* (pp. 312–329). Stanford, CA: Stanford University Press.

472

Matsushima, K., & Noguchi, H. (1967). Multidimensional representation of binocular visual space. *Japanese Psychological Research*, *9*, 83–94.

May, K. O. (1954). Intransitivity, utility, and the aggregation of preference patterns. *Econometrica*, *22*, 1–13.

Mehlberg, H. (1935). Essai sur la théorie causale du temps. I. *Studia Philosophica*, *1*, 119–260.

Mehlberg, H. (1937). Essai sur la théorie causale du temps. II. *Studia Philosophica*, *2*, 111–231.

Miller, G. A., & Nicely, P. E. (1955). An analysis of perceptual confusions among some English consonants. *Journal of the Acoustical Society of America*, *27*, 338–352.

Miller, S. S. (1972). Psychophysical estimates of visual pigment densities in red-green dichromats. *Journal of Physiology (London)*, *223*, 89–107.

Milne, E. A. (1948). *Kinematic relativity*. London & New York: Oxford University Press (Clarendon).

Mirkin, B. G. (1970a). Ob odnom klasse ot noshenij predpochtenija (About a class of preference relations). In K. Bagrinovkij & E. Berlantz (Eds.), *Matematitchieskija woprosy formirovnija economitcheskij modelei (Mathematical problems of forming of economical models)*, Novosibirsk, USSR.

Mirkin, B. G. (1970b). Ob odnoi axiome matematitcheskoi teorii poleznosti (About an axiom of mathematical utility theory). *Kirbernetika*, *6*, 60–62.

Mirkin, B. G. (1972). Description of some relations on the set of real-line intervals. *Journal of Mathematical Psychology*, *9*, 243–252.

Miyamoto, J. (1988). Generic utility theory: Measurement foundations and applications in multiattribute utility theory. *Journal of Mathematical Psychology*, *32*, 357–404.

Moeller, J. R. (1976). *Measuring the red/green quality of lights: A study relating the Jameson and Hurvich red/green cancellation valence to direct magnitude estimation of greenness*. Doctoral dissertation, University of Michigan, Ann Arbor (University Microfilms).

Monjardet, B. (1978). Axiomatiques et propriétés des quasi-ordres. *Mathématiques et Sciences Humaines*, *63*, 51–82.

Monjardet, B. (1984). Probabilistic consistency, homogeneous families of relations and linear Λ-relations. In E. Degreef & J. Van Buggenhaut (Eds.), *Trends in mathematical psychology* (pp. 271–281). Amsterdam: North-Holland.

Morrison, H. W. (1962). *Intransitivity of paired comparison choice*. Unpublished doctoral dissertation, University of Michigan, Ann Arbor.

Morrison, H. W. (1963). Testable conditions for triads of paired comparison choices. *Psychometrika*, *28*, 369–390.

Mulholland, H. P. (1947). On generalizations of Minkowski's inequality in the form of a triangle inequality. *Proceedings of the London Mathematical Society, Series 2*, *51*, 294–307.

Mundy, B. (1986a). Optical axiomatization of Minkowski space-time geometry. *Philosophy of Science*, *37*, 1–30.

Mundy, B. (1986b). The physical content of Minkowski geometry. *British Journal for the Philosophy of Science*, *37*, 25–54.

Nagy, A. L., MacLeod, D. I. A., Heynman, N. E. & Eisner, A. (1981). Four cone pigments in women heterozygous for color deficiency. *Journal of the Optical Society of America*, *71*, 719–722.

Narens, L. (1980). A note on Weber's law for conjoint structures. *Journal of Mathematical Psychology*, *21*, 88–92.

Narens, L. (1985). *Abstract measurement theory*. Cambridge, MA: MIT Press.

Newton, I. (1704/1730). *Opticks*. London (reprinted in London by G. Bell & Sons, 4th ed., 1931).

Ng, Y.-K. (1975). Bentham or Bergson? Finite sensibility, utility functions and social welfare functions. *Review of Economic Studies*, *42*, 543–569.

Ng, Y.-K. (1977). Sub-semiorder: A model of multidimensional choice with preference intransitivity. *Journal of Mathematical Psychology*, *16*, 51–59.

Nishikawa, Y. (1967). Euclidean interpretation of binocular visual space. *Japanese Psychological Research*, *9*, 191–198.

Osherson, D. (1987). New axioms for the contrast model of similarity. *Journal of Mathematical Psychology*, *31*, 93–103.

Patrinos, A. N., & Hakimi, S. L. (1972). The distance matrix of a graph and its tree realization. *Quarterly of Applied Mathematics*, *30*, 255–269.

Pecham, J. (1970). *Perspectiva communis* (D. C. Lindberg, Trans.). Madison: University of Wisconsin Press.

Pickert, G. (1955). *Projektive Ebenen*. Berlin: Springer-Verlag.

Pieri, M. (1908). La geometria elementare istituita sulle nozioni di 'punto' e 'sfera'. *Memorie di Matematica e de Fisica della Società Italiana delle Scienze, Ser. 3*, *15*, 345–350.

Pirenne, M. M. (1975). Vision and art. In E. C. Carterette & M. P. Friedman (Eds.) *Handbook of perception: Seeing Vol. 5*. (pp. 434–490). New York: Academic Press.

Playfair, J. (1795). *Elements of geometry*. Edinburgh.

Podehl, E., & Reidemeister, K. (1934). Eine Begründung der ebenen elliptischen Geometrie. *Abhandlungen aus dem Mathematischen Seminar der Hamburgischen Universität*, *10*, 231–255.

Pokorny, J., Smith, V. C., Verriest, G., & Pinckers, A. J. L. G. (1979). *Congenital and acquired color vision defects*. New York: Grune & Stratton.

Postnikov, M. (1982). *Lectures in geometry: Semester 1. Analytical geometry*. Moscow: Mir Publishers, (English translation by V. Shokurov, 1979 Russian edition).

Pruzansky, S., Tversky, A., & Carroll, J. D. (1982). Spatial versus tree representation of proximity data. *Psychometrika*, *47*, 3–24.

Pugh, E. N., & Kirk, D. B. (1986). The Π mechanisms of W. S. Stiles: An historical review. *Perception*, *15*, 705–728.

Purdy, D. M. (1931). Spectral hue as a function of intensity. *American Journal of Psychology*, *43*, 541–559.

Rabinovitch, I. (1977). The Scott-Suppes theorem on semiorders. *Journal of Mathematical Psychology*, *15*, 209–212.

Rabinovitch, I. (1978a). The dimensions of semiorders. *Journal of Combinatorial Theory*, *25*, 50–61.

Rabinovitch, I. (1978b). An upper bound on the "dimension of interval orders." *Journal of Combinatorial Theory*, *25*, 68–71.

Reichenbach, H. (1924). *Axiomatik der relativistischen Raum-Zeit-Lehre*. Braunschweig: Vieweg & Sons.

Reid, T. (1967). Inquiry into the human mind. In G. Olms (Ed.), *Philosophical works of Thomas Reid* (Vol. 1). Hildesheim, Germany: Verlag's Buchhandlung. (Original work published 1764.)

Restle, F. (1959). A metric and an ordering on sets. *Psychometrika*, *24*, 207–220.

Restle, F. (1961). *Psychology of judgment and choice*. New York: Wiley.

Riemann, B. (1866–1867). Ueber die Hypothesen, welche der Geometrie zu Grunde liegen. *Gesellschaft der Wissenschaften zu Göttingen: Abhandlungen*, *13*, 133–152.

Robb, A. A. (1911). *Optical geometry of motion: A new view of the theory of relativity*. Cambridge, England: Heffer. London & New York.

Robb, A. A. (1914). *A theory of time and space*. London & New York: Cambridge University Press.

Robb, A. A. (1921). *The absolute relations of time and space*. London & New York: Cambridge University Press.

Robb, A. A. (1928). On the connexion of a certain identity with the extension of conical order to *n* dimensions. *Proceedings of the Cambridge Philosophical Society, 24,* 357–374.

Robb, A. A. (1930). On a symmetrical analysis of conical order and its relation to time-space. *Proceedings of the Royal Society of London, Series A, 129,* 549–579.

Robb, A. A. (1936). *Geometry of time and space.* London & New York: Cambridge University Press.

Roberts, F. S. (1968). *Representations of indifference relations.* Unpublished doctoral dissertation, Stanford University, Stanford, CA.

Roberts, F. S. (1969). Indifference graphs. In F. Harary (Ed.), *Proof techniques in graph theory* (pp. 139–146). New York: Academic Press.

Roberts, F. S. (1970). On nontransitive indifference. *Journal of Mathematical Psychology, 7,* 243–258.

Roberts, F. S. (1971a). On the compatibility between a graph and a simple order. *Journal of Combinatorial Theory, 11,* 28–38.

Roberts, F. S. (1971b). Homogeneous families of semiorders and the theory of probabilistic consistency. *Journal of Mathematical Psychology, 8,* 248–263.

Roberts, F. S. (1973). Tolerance geometry. *Notre Dame Journal of Formal Logic, 14,* 68–76.

Roberts, F. S. (1979). *Measurement theory with applications to decision making, utility, and the social sciences.* Reading, MA: Addison-Wesley.

Robinson, R. M. (1959). Binary relations as primitive notions in elementary geometry. In L. Henkin, P. Suppes, & A. Tarski (Eds.), *The axiomatic method with special reference to geometry and physics* (pp. 68–85). Amsterdam: North-Holland.

Romeskie, M. (1976). *Chromatic response functions of anomalous trichromats.* Ph.D. dissertation, Brown University, Providence, RI (University Microfilms).

Romeskie, M., & Yager, D. (1978). Psychophysical measure and theoretical analysis of dichromatic opponent-response functions. *Modern Problems of Opthamology, 19,* 212–217.

Romney, A. K., Shepard, R. N., & Nerlove, S. B. (Eds.). (1972). *Multidimensional scaling: Theory and applications in the behavioral sciences: Vol. 2. Applications.* New York: Seminar Press.

Rosch, E. (1975). Cognitive reference points. *Cognitive Psychology, 7,* 532–547.

Rosch, E., & Mervis, C. B. (1975). Family resemblance: Studies in the internal structure of categories. *Cognitive Psychology, 7,* 573–603.

Roubens, M., & Vincke, P. (1983). Linear fuzzy graphs. *Fuzzy Sets and Systems, 10,* 79–86.

Roy, B. (1980). Préférence, indifférence, incomparabilitie. *Documents du LAMSADE Université Paris-Dauphine, 9.*

Royden, H. L. (1959). Remarks on primitive notions for elementary Euclidean and non-Euclidean geometry. In L. Henkin, P. Suppes, & A. Tarski (Eds.), *The axiomatic method with special reference to geometry and physics* (pp. 68–85). Amsterdam: North-Holland.

Rubin, H., & Suppes, P. (1954). Transformations of systems of relativistic particle mechanics. *Pacific Journal of Mathematics, 4,* 563–601.

Rushton, W. A. H. (1963). A cone pigment in the protanope. *Journal of Physiology (London), 168,* 345–359.

Rushton, W. A. H. (1965). A foveal pigment in the deuteranope. *Journal of Physiology (London), 176,* 24–37.

Rushton, W. A. H. (1972). Visual pigments in man. In H. J. A. Dartnall (Ed.), *Handbook of sensory physiology: Photochemistry of vision* Vol. VII/1. (pp. 364–394). Berlin & New York: Springer-Verlag.

Russell, B. (1936). On order in time. *Proceedings of the Cambridge Philosophical Society, 32,* 216–228.

Sattath, S., & Tversky, A. (1976). Unite and conquer: A multiplicative inequality for choice probabilities. *Econometrica*, *44*, 79–89.

Sattath, S., & Tversky, A. (1977). Additive similarity trees. *Psychometrika*, *42*, 319–345.

Sattath, S., & Tversky, A. (1987). On the relation between common and distinctive feature models. *Psychological Review*, *94*, 16–22.

Schiffman, S. S., Reynolds, M. L., & Young, F. W. (Eds.) (1981). *Introduction to multidimensional scaling*. New York: Academic Press.

Schnell, K. (1937). *Eine Topologie der Zeit in logistischer Darstellung*. Dissertation, Münster.

Schreier, O., & Sperner, E. (1961). *Projective geometry of n dimensions*. New York: Chelsea.

Schutz, J. W. (1973). *Foundations of special relativity: Kinematic axioms for Minkowski space-time*. Berlin & New York: Springer-Verlag.

Schutz, J. W. (1979, April). *An axiomatic system for Minkowski space time* (MPI-PAE/Astro 181). Munaich: Max-Planck-Institut für Physik und Astrophysik.

Schwabhäuser, W., & Szczerba, L. W. (1975). Relations on lines as primitive notions for Euclidean geometry. *Fundamenta Mathematicae*, *82*, 347–355.

Scott, D. (1956). A symmetric primitive notion for Euclidean geometry. *Indagationes Mathematicae*, *18*, 457–461.

Scott, D. (1964). Measurement structures and linear inequalities. *Journal of Mathematical Psychology*, *1*, 233–247.

Scott, D., & Suppes, P. (1958). Foundational aspects of theories of measurement. *Journal of Symbolic Logic*, *23*, 113–128.

Shepard, R. N. (1962a). The analysis of proximities: Multidimensional scaling with an unknown distance function. Part I. *Psychometrika*, *27*, 125–140.

Shepard, R. N. (1962b). The analysis of proximities: Multidimensional scaling with an unknown distance function. Part II. *Psychometrika*, *27*, 219–246.

Shepard, R. N. (1964). Attention and the metric structure of the stimulus space. *Journal of Mathematical Psychology*, *1*, 54–87.

Shepard, R. N. (1974). Representation of structure in similarity data: Problems and prospects. *Psychometrika*, *39*, 373–421.

Shepard, R. N. (1980). Multidimensional scaling, tree-fitting, and clustering. *Science*, *210*, 390–398.

Shepard, R. N. (1984). Ecological constraints on internal representation: Resonant kinematics of perceiving, imagining, thinking and dreaming. *Psychological Review*, *91*, 417–447.

Shepard, R. N. (1987). Toward a universal law of generalization for psychological science. *Science*, *237*, 1317–1323.

Shepard, R. N., & Arabie, P. (1979). Additive clustering: Representation of similarities as combinations of discrete overlapping properties. *Psychological Review*, *86*, 87–123.

Shepard, R. N., Romney, A. K., & Nerlove, S. B. (Eds.). (1972). *Multidimensional scaling: Theory and applications in the behavioral sciences: Vol. 1. Theory*. New York: Seminar Press.

Shevell, S. K. (1978). The dual role of chromatic backgrounds in color perception. *Vision Research*, *18*, 1649–1661.

Shevell, S. K., & Handte, J. P. (1983). Postreceptoral adaptation in suprathreshold color perception. In J. D. Mollon & L. T. Sharpe (Eds.), *Colour vision: Physiology and psychophysics* (pp. 399–407). London: Academic Press.

Shohat, J. A., & Tamarkin, J. D. (1943). *The problem of moments*. New York: American Mathematical Society.

Silberstein, L. (1946). The complete three-dimensional color domain and its metrics. *Philosophical Magazine, Series VII*, *37*, 126–144.

Smith, E. E., & Medin, D. L. (1981). *Categories and concepts*. Cambridge, MA: Harvard University Press.

Smith, V. C., & Pokorny, J. (1972). Spectral sensitivity of colorblind observers and the cone pigments. *Vision Research, 12,* 2059–2071.

Sokal, R. R., & Sneath, P. H. (1963). *Numerical taxonomy.* San Francisco, CA: Freeman.

Stevens, S. S. (1957). On the psychophysical law. *Psychological Review, 64,* 153–181.

Stevenson, F. W. (1972). *Projective planes.* San Francisco, CA: Freeman.

Stiles, W. S. (1939). The directional sensitivity of the retina and the spectral sensitivities of the rods and cones. *Proceedings of the Royal Society of London, Series B, 127,* 64–105.

Stiles, W. S. (1946). A modified Helmholtz line-element in brightness-colour space. *Proceedings of the Physical Society, London, 58,* 41–65.

Stiles, W. S. (1959). Colour vision: The approach through increment-threshold sensitivity. *Proceedings of the National Academy of Sciences of the U.S.A., 45,* 100–114.

Stiles, W. S. (1967). Mechanism concepts in colour theory. *Journal of the Colour Group, 11,* 105–123.

Strauss, D. (1979). Some results on random utility models. *Journal of Mathematical Psychology, 20,* 35–52.

Suppes, P. (1959). Axioms for relativistic kinematics with or without parity. In L. Henkin, P. Suppes, & A. Tarski (Eds.), *The axiomatic method with special reference to geometry and physics* (pp. 291–307). Amsterdam: North-Holland.

Suppes, P. (1972). Some open problems in the philosophy of space and time. *Synthese, 24,* 298–316. (Reprinted in P. Suppes (Ed.), *Space, time and geometry* (pp. 383–401). Dordrecht, Netherlands: Reidel, 1973.)

Suppes, P. (1974). The measurement of belief. *Journal of the Royal Statistical Society, Series B, 36,* 160–191.

Suppes, P. (1977). Is visual space Euclidean? *Synthese, 35,* 397–421.

Suppes, P. (1989). Qualitative axioms for classical and Minkowski space-time. *In press.*

Suppes, P. (1989). Philosophy and the sciences. In W. Sieg (Ed.), *Acting and reflecting: The interdisciplinary turn in philosophy.* Dordrecht: Kluwer.

Suppes, P., & Rottmayer, W. (1974). Automata. In E. C. Carterette & M. P. Friedman (Eds.), *Handbook of perception: Vol. 1. Historical and philosophical roots of perception* (pp. 335–362). New York: Academic Press.

Suppes, P., & Zanotti, M. (1989). Qualitative axioms for random-variable representation of extensive quantities. In C. W. Savage & P. Ehrlich (Eds.), *Philosophical and foundational issues in measurement theory.* Hillsdale, N.J.: Lawrence Erlbaum Associates.

Suppes, P., & Zinnes, J. L. (1963). Basic measurement theory. In R. D. Luce, R. R. Bush, & E. Galanter (Eds.), *Handbook of mathematical psychology* (Vol. 1, pp. 1–76). New York: Wiley.

Swistak, P. (1980). Some representation problems for semiorders. *Journal of Mathematical Psychology, 21,* 124–135.

Szczerba, L. W. (1984). Imbedding of finite planes. *Potsdamer Forschungen, Reihe B 41,* 99–102.

Szczerba, L. W., & Tarski, A. (1965). Metamathematical properties of some affine geometries. In Y. Bar-Hillel (Ed.), *Proceedings of the 1964 International Congress on Logic, Methodology and Philosophy of Science* (pp. 166–178). Amsterdam: North-Holland.

Szczerba, L. W., & Tarski, A. (1979). Metamathematical discussion of some affine geometries. *Fundamenta Mathematicae 104,* 115–192.

Szekeres, G. (1968). Kinematic geometry: An axiomatic system for Minkowski space-time. *Journal of the Australian Mathematical Society, 8,* 134–160.

Szmielew, W. (1959). Some metamathematical problems concerning elementary hyperbolic geometry. In L. Henkin, P. Suppes, & A. Tarski (Eds.), *The axiomatic method with special reference to geometry and physics.* Amsterdam: North-Holland.

Szmielew, W. (1983). *From affine to Euclidean geometry: An axiomatic approach*. Boston, MA: Reidel. (Translated by Mr. Mozzynska from the Polish edition. Waiszawa: PWN-Polish Scientific Publishers, 1981.)

Tarski, A. (1929). Les fondements de la géométrie des corps. *Ksiega pamiatkowa Pierwszego Polskiego Zjasdu Matematyków Kraków, 29-33*.

Tarski, A. (1956). A general theorem concerning primitive notions of Euclidean geometry. *Indagationes Mathematicae, 18*, 468-474.

Thompson, W. A., Jr., & Singh, J. (1967). The use of limit theorems in paired comparison model building. *Psychometrika, 32*, 255-264.

Thornton, J. E. (1981). Relating chromatic antagonism in Π_5 to red/green hue cancellation. Doctoral dissertation, University of Michigan, Ann Arbor (University Microfilms).

Thurstone, L. L. (1927a). A law of comparative judgment. *Psychological Review, 34*, 273-286.

Thurstone, L. L. (1927b). Psychophysical analysis. *American Journal of Psychology, 38*, 368-389.

Thurstone, L. L. (1930). The learning function. *Journal of General Psychology, 3*, 469-493.

Todhunter, I. (1949). *A history of the mathematical theory of probability*. New York: Chelsea. (Original work published 1865.)

Torgerson, W. S. (1952). Multidimensional scaling. I. Theory and method. *Psychometrika, 17*, 401-419.

Torgerson, W. S. (1958). *Theory and methods of scaling*. New York: Wiley.

Trotter, W. T., & Bogart, K. P. (1976). On the complexity of posets. *Discrete Mathematics, 16*, 71-82.

Tutz, G. (1986). Bradley-Terry-Luce models with an ordered response. *Journal of Mathematical Psychology, 30*, 306-316.

Tversky, A. (1969). Intransitivity of preferences. *Psychological Review, 76*, 31-48.

Tversky, A. (1972a). Choice by elimination. *Journal of Mathematical Psychology, 9*, 341-367.

Tversky, A. (1972b). Elimination by aspects: A theory of choice. *Psychological Review, 79*, 281-299.

Tversky, A. (1977). Features of similarity. *Psychological Review, 84*, 327-352.

Tversky, A., & Gati, I. (1978). Studies of similarity. In E. Rosch & B. Lloyd (Eds.), *Cognition and categorization*. Hillsdale, NJ: Erlbaum.

Tversky, A., & Gati, I. (1982). Similarity, separability, and the triangle inequality. *Psychological Review, 89*, 123-154.

Tversky, A., & Hutchinson, J. W. (1986). Nearest neighbor analysis of psychological spaces. *Psychological Review, 93*, 3-22.

Tversky, A., & Krantz, D. H. (1969). Similarity of schematic faces: A test of interdimensional additivity. *Perception & Psychophysics, 5*, 124-128.

Tversky, A., & Krantz, D. H. (1970). The dimensional representation and the metric structure of similarity data. *Journal of Mathematical Psychology, 7*, 572-596.

Tversky, A., Rinott, Y., & Newman, C. M. (1983). Nearest neighbor analysis of point processes: Applications to multidimensional scaling. *Journal of Mathematical Psychology, 27*, 235-250.

Tversky, A., & Russo, J. E. (1969). Substitutability and similarity in binary choices. *Journal of Mathematical Psychology, 6*, 1-12.

Tversky, A., & Sattath, S. (1979). Preference trees. *Psychological Review, 86*, 542-573.

Ullman, S. (1979). *The interpretation of visual motion*. Cambridge, MA: MIT Press.

Uttal, W. R. (1967). Evoked brain potentials: Signs or codes? *Perspectives in Biology and Medicine, 10*, 627-639.

Veblen, O. (1904). A system of axioms for geometry. *Transactions of the American Mathematical Society, 5*, 343-384.

Veblen, O., & Young, J. W. (1910). *Projective geometry* (Vol. 1). Boston, MA: Ginn. (Reprinted by Blaisdell, New York, 1938.)

Veblen, O., & Young, J. W. (1918). *Projective geometry* (Vol. 2). Boston, MA: Ginn. (Reprinted by Blaisdell, New York, 1938.)

von Kries, J. (1905). Die Gesichtsempfindungen. In W. Nagel, *Handbuch der Physiologie des Menschen* (Vol. 3). Braunschweig: Viewig.

Wagner, M. (1985). The metric of visual space. *Perception & Psychophysics, 38*, 483–495.

Walker, A. G. (1948). Foundations of relativity. Parts I and II. *Proceedings of the Royal Society of Edinburgh, Section A, Mathematical and Physical Sciences, 62* (Pt. III, No. 34), 319–335.

Walker, A. G. (1959). Axioms for cosmology. In L. Henkin, P. Suppes, & A. Tarski (Eds.), *The axiomatic method with special reference to geometry and phsyics* (pp. 308–321). Amsterdam: North-Holland.

Walls, G. L. (1960). Land! Land! *Psychological Bulletin, 57*, 29–48.

Wandell, B., & Luce, R. D. (1978). Pooling peripheral information: Average vs. extreme values. *Journal of Mathematical Psychology, 17*, 220–235.

Wang, H. (1951). Two theorems on metric spaces. *Pacific Journal of Mathematics, 1*, 473–480.

Wender, K. (1971). A test of independence of dimensions in multidimensional scaling. *Perception & Psychophysics, 10*, 30–32.

Werner, J. S., & Wooton, B. R. (1979). Opponent chromatic mechanisms: Relation to photopigments and hue naming. *Journal of the Optical Society of America, 69*, 422–434.

Weyl, H. (1923). *Mathematische Analyse des Raumproblems*. Berlin: Springer.

Wiener, N. (1914). A contribution to the theory of relative position. *Proceedings of the Cambridge Philosophical Society, 17*, 441–449.

Wiener, N. (1921). A new theory of measurement: A study in the logic of mathematics. *Proceedings of the London Mathematical Society, 19*, 181–205.

Wiener-Ehrlich, W. K. (1978). Dimensional and metric structures in multidimensional scaling. *Perception & Psychophysics, 24*, 399–414.

Wilson, E. B., & Lewis, G. N. (1912). The space-time manifold of relativity. The non-Euclidean geometry of mechanics and electromagnetics. *Proceedings of the American Academy of Arts and Sciences, 48*, 389–507.

Winnie, J. A. (1977). The causal theory of space-time. In J. Earman, C. Glymour, & J. Stachel (Eds.), *Foundations of space-time theories* (pp. 134–205). Minneapolis: University of Minnesota Press.

Wright, W. D. (1928–1929). A re-determination of the trichromatic coefficients of the spectral colours. *Transactions of the Optical Society of London, 30*, 141–164.

Wright, W. D. (1934). The measurement and analysis of colour-adaptation phenomena. *Proceedings of the Royal Society of London, Series B, 115*, 49–87.

Wright, W. D. (1946). *Researches on normal and defective colour vision*. London: Henry Kimpton.

Wright, W. D. (1964). *The measurement of colour*. London: Hilger & Watts.

Wyszecki, G., & Stiles, W. S. (1982). *Color science: Concepts and methods, quantitative data and formulae* (2nd ed.). New York: Wiley.

Yager, D., & Taylor, E. (1970). Experimental measures and theoretical account of hue scaling as a function of luminance. *Perception & Psychophysics, 7*, 360–364.

Yannakakis, M. (1982). The complexity of the partial order dimension problem. *SIAM Journal on Algebraic and Discrete Methods, 3*, 351–358.

Yellott, J. I., Jr. (1977). The relationship between Luce's choice axiom, Thurstone's theory of comparative judgment and the double exponential distribution. *Journal of Mathematical Psychology, 15*, 109–144.

Young, F. W. (1984). Scaling. *Annual Review of Psychology, 35*, 55–82.

Young, G., & Householder, A. S. (1938). Discussion of a set of points in terms of their mutual distances. *Psychometrika*, *3*, 19–22.

Young, T. (1802). On the theory of light and colours. *Philosophical Transactions of the Royal Society of London*, *92*, 12–48.

Zajaczkowska, A. (1956a). Experimental determination of Luneburg's constants σ and K. *Quarterly Journal of Experimental Psychology*, *8*, 66–78.

Zajaczkowska, A. (1956b). Experimental test of Luneberg's theory. Horopter and alley experiments. *Journal of the Optical Society of America*, *46*, 514–527.

Zeeman, E. C. (1962). The topology of the brain and visual perception. In M. K. Fort (Ed.), *The topology of 3-manifolds* (pp. 240–256). Englewood Cliffs, NJ: Prentice-Hall.

Zeeman, E. C. (1964). Causality implies the Lorentz group. *Journal of Mathematical Physics*, *5*, 490–493.

Zeeman, E. C. (1967). The topology of Minkowski space. *Topology*, *6*, 161–170.

Zermelo, E. (1929). Die Berechnung der Turnierergebnisse als ein Maximumproblem der Wahrscheinlichkeitsrechnung. *Mathematische Zeitung*, *29*, 436–460.

Author Index

Numbers in italics refer to the pages on which the complete references are listed. The letter n following a page number indicates that the entry is cited in a footnote to that page.

Subject Index